Meeting Challenges in the Diagnosis and Treatment of Glaucoma

Meeting Challenges in the Diagnosis and Treatment of Glaucoma

Guest Editors

Karanjit S. Kooner
Osamah J. Saeedi

Basel • Beijing • Wuhan • Barcelona • Belgrade • Novi Sad • Cluj • Manchester

Guest Editors

Karanjit S. Kooner
Department of Ophthalmology
University of Texas
Southwestern Medical Center
Dallas
United States

Osamah J. Saeedi
Department of Ophthalmology
and Visual Sciences
University of Maryland
Baltimore
United States

Editorial Office
MDPI AG
Grosspeteranlage 5
4052 Basel, Switzerland

This is a reprint of the Special Issue, published open access by the journal *Bioengineering* (ISSN 2306-5354), freely accessible at: www.mdpi.com/journal/bioengineering/special_issues/D4LY85E517.

For citation purposes, cite each article independently as indicated on the article page online and using the guide below:

Lastname, A.A.; Lastname, B.B. Article Title. *Journal Name* **Year**, *Volume Number*, Page Range.

ISBN 978-3-7258-3082-4 (Hbk)
ISBN 978-3-7258-3081-7 (PDF)
https://doi.org/10.3390/books978-3-7258-3081-7

© 2025 by the authors. Articles in this book are Open Access and distributed under the Creative Commons Attribution (CC BY) license. The book as a whole is distributed by MDPI under the terms and conditions of the Creative Commons Attribution-NonCommercial-NoDerivs (CC BY-NC-ND) license (https://creativecommons.org/licenses/by-nc-nd/4.0/).

Contents

About the Editors vii

Karanjit S. Kooner, Dominic M. Choo and Priya Mekala
Meeting Challenges in the Diagnosis and Treatment of Glaucoma
Reprinted from: *Bioengineering* **2024**, *12*, 6, https://doi.org/10.3390/bioengineering12010006 . . 1

Leo Yan Li-Han, Moshe Eizenman, Runjie Bill Shi, Yvonne M. Buys, Graham E. Trope and Willy Wong
Using Fused Data from Perimetry and Optical Coherence Tomography to Improve the Detection of Visual Field Progression in Glaucoma
Reprinted from: *Bioengineering* **2024**, *11*, 250, https://doi.org/10.3390/bioengineering11030250 . 6

Mark Christopher, Ruben Gonzalez, Justin Huynh, Evan Walker, Bharanidharan Radha Saseendrakumar and Christopher Bowd et al.
Proactive Decision Support for Glaucoma Treatment: Predicting Surgical Interventions with Clinically Available Data
Reprinted from: *Bioengineering* **2024**, *11*, 140, https://doi.org/10.3390/bioengineering11020140 . 22

Elizabeth E. Hwang, Dake Chen, Ying Han, Lin Jia and Jing Shan
Multi-Dataset Comparison of Vision Transformers and Convolutional Neural Networks for Detecting Glaucomatous Optic Neuropathy from Fundus Photographs
Reprinted from: *Bioengineering* **2023**, *10*, 1266, https://doi.org/10.3390/bioengineering10111266 35

Ryan Shean, Ning Yu, Sourish Guntipally, Van Nguyen, Ximin He and Sidi Duan et al.
Advances and Challenges in Wearable Glaucoma Diagnostics and Therapeutics
Reprinted from: *Bioengineering* **2024**, *11*, 138, https://doi.org/10.3390/bioengineering11020138 . 48

Bryan Chin Hou Ang, Sheng Yang Lim, Bjorn Kaijun Betzler, Hon Jen Wong, Michael W. Stewart and Syril Dorairaj
Recent Advancements in Glaucoma Surgery—A Review
Reprinted from: *Bioengineering* **2023**, *10*, 1096, https://doi.org/10.3390/bioengineering10091096 66

Jeremy C.K. Tan, Hussameddin Muntasser, Anshoo Choudhary, Mark Batterbury and Neeru A. Vallabh
Swept-Source Anterior Segment Optical Coherence Tomography Imaging and Quantification of Bleb Parameters in Glaucoma Filtration Surgery
Reprinted from: *Bioengineering* **2023**, *10*, 1186, https://doi.org/10.3390/bioengineering10101186 93

Ramona Ileana Barac, Vasile Harghel, Nicoleta Anton, George Baltă, Ioana Teodora Tofolean and Christiana Dragosloveanu et al.
Initial Clinical Experience with Ahmed Valve in Romania: Five-Year Patient Follow-Up and Outcomes
Reprinted from: *Bioengineering* **2024**, *11*, 820, https://doi.org/10.3390/bioengineering11080820 . 106

Bhoomi Dave, Monica Patel, Sruthi Suresh, Mahija Ginjupalli, Arvind Surya and Mohannad Albdour et al.
Wound Modulations in Glaucoma Surgery: A Systematic Review
Reprinted from: *Bioengineering* **2024**, *11*, 446, https://doi.org/10.3390/bioengineering11050446 . 120

Özlem Evren Kemer, Priya Mekala, Bhoomi Dave and Karanjit Singh Kooner
Managing Ocular Surface Disease in Glaucoma Treatment: A Systematic Review
Reprinted from: *Bioengineering* **2024**, *11*, 1010, https://doi.org/10.3390/bioengineering11101010 **146**

Flaviu Bodea, Simona Gabriela Bungau, Andrei Paul Negru, Ada Radu, Alexandra Georgiana Tarce and Delia Mirela Tit et al.
Exploring New Therapeutic Avenues for Ophthalmic Disorders: Glaucoma-Related Molecular Docking Evaluation and Bibliometric Analysis for Improved Management of Ocular Diseases
Reprinted from: *Bioengineering* **2023**, *10*, 983, https://doi.org/10.3390/bioengineering10080983 . **172**

Matthew Fung, James J. Armstrong, Richard Zhang, Anastasiya Vinokurtseva, Hong Liu and Cindy Hutnik
Development and Verification of a Novel Three-Dimensional Aqueous Outflow Model for High-Throughput Drug Screening
Reprinted from: *Bioengineering* **2024**, *11*, 142, https://doi.org/10.3390/bioengineering11020142 . **199**

Kateki Vinod and Sarwat Salim
Addressing Glaucoma in Myopic Eyes: Diagnostic and Surgical Challenges
Reprinted from: *Bioengineering* **2023**, *10*, 1260, https://doi.org/10.3390/bioengineering10111260 **214**

Abdelrahman M. Elhusseiny, Giuliano Scarcelli and Osamah J. Saeedi
Corneal Biomechanical Measures for Glaucoma: A Clinical Approach
Reprinted from: *Bioengineering* **2023**, *10*, 1108, https://doi.org/10.3390/bioengineering10101108 **226**

Ashley Shuen Ying Hong, Bryan Chin Hou Ang, Emily Dorairaj and Syril Dorairaj
Premium Intraocular Lenses in Glaucoma—A Systematic Review
Reprinted from: *Bioengineering* **2023**, *10*, 993, https://doi.org/10.3390/bioengineering10090993 . **252**

About the Editors

Karanjit S. Kooner

Karanjit S. Kooner, MD, PhD, MBA is a full-time glaucoma specialist in the Department of Ophthalmology at the University of Texas Southwestern Medical Center, Dallas, Texas, and an Adjunct Associate Professor in the Department of Bioengineering at the University of Texas at Dallas. Dr. Kooner's research interests include different aspects of glaucoma, such as early diagnosis of both open-angle and narrow-angle glaucoma using optical coherence tomography angiography and ultrasound biomicroscopy. In pursuing these interests, he has delivered scores of presentations, contributed to books, and published numerous academic articles. He also plays an active role in the teaching of residents and fellows, including international fellows. His humanitarian efforts have taken him to more than a dozen countries in Asia, South America, and Africa.

Osamah J. Saeedi

Osamah J. Saeedi, MD, MS, is a Professor of Ophthalmology at the University of Maryland School of Medicine in Baltimore, MD, and an Adjunct Associate Professor of Bioengineering at the University of Maryland, College Park. He completed medical school and his ophthalmology residency at the University of Texas Southwestern Medical Center and a glaucoma fellowship at the Wilmer Eye Institute, Johns Hopkins University. He completed a Master of Science in Epidemiology and Clinical Research from the University of Maryland, Baltimore. He is the recipient of numerous awards, including the NIH Director's Award and the Heidelberg Xtreme Research Award, and was recently recognized with a distinguished alumni award from the University of Maryland Graduate School. He has received funding from the National Institutes of Health as well as numerous other organizations. His research focuses on finding novel imaging biomarkers for glaucoma, specifically looking at novel techniques for assessing ocular blood flow.

Editorial

Meeting Challenges in the Diagnosis and Treatment of Glaucoma

Karanjit S. Kooner [1,2,*], Dominic M. Choo [1] and Priya Mekala [1]

1 Department of Ophthalmology, University of Texas Southwestern Medical Center, Dallas, TX 75390, USA; dominic.choo@utsouthwestern.edu (D.M.C.); priya.mekala@utsouthwestern.edu (P.M.)
2 Department of Ophthalmology, Veteran Affairs North Texas Health Care Medical Center, Dallas, TX 75216, USA
* Correspondence: karanjit.kooner@utsouthwestern.edu

Received: 16 December 2024
Accepted: 23 December 2024
Published: 25 December 2024

Citation: Kooner, K.S.; Choo, D.M.; Mekala, P. Meeting Challenges in the Diagnosis and Treatment of Glaucoma. *Bioengineering* **2025**, *12*, 6. https://doi.org/10.3390/bioengineering12010006

Copyright: © 2024 by the authors. Licensee MDPI, Basel, Switzerland. This article is an open access article distributed under the terms and conditions of the Creative Commons Attribution (CC BY) license (https://creativecommons.org/licenses/by/4.0/).

Glaucoma, a progressive and multifactorial optic neurodegenerative disease, still poses significant challenges in both diagnosis and management and remains a perpetual enigma. The disease's insidious onset, combined with our reliance on subjective diagnostic tools, hinders early detection and timely intervention. Current treatments, while beneficial, often fail to provide consistent intraocular pressure (IOP) control, and management is further complicated by variability in disease progression, non-adherence to therapy, socioeconomic challenges, complications from surgical interventions, and new risk factors. For example, chronic stress has recently been implicated as a major risk factor in the pathogenesis of glaucoma [1,2]. The current Special Issue focusing on glaucoma has 14 excellent papers that discuss notable advancements aimed at addressing several challenges regarding this disease.

Several papers highlight the potential for the use of artificial intelligence (AI) in glaucoma care. Li-Han et al. developed a model that combines perimetry and optic coherence tomography (OCT) data to detect disease progression and achieved an F1 score of 0.60 within two years [3]. The model outperformed traditional methods like Bayesian regression (0.48), showcasing its potential for precise, early detection. Another novel model developed by Christoper et al. achieved an accuracy of 85% in predicting the need for glaucoma surgical interventions up to three years in advance. It utilized data such as patient demographics, medical history, clinical measurements, OCT, and perimetry [4]. Notably, the superiority of Vision Transformer (ViT) models, compared to convolutional neural networks (CNNs), in identifying glaucomatous optic neuropathy from fundus photographs was demonstrated by Hwang et al. [5].

The use of Ahmed valve shunts in Romania, as described by Barac et al., led to a reported 60% success rate in maintaining target IOP over five years, with a mean reduction to 17 mmHg [6]. Other innovations in filtering surgery are discussed by Ang et al. [7]. For example, limited deep sclerectomy-augmented trabeculectomy was shown to outperform conventional trabeculectomy, contributing to a mean IOP reduction from 29 ± 4.6 mmHg to 12.54 ± 1.67 mmHg at 12 months. A modified scleral tunnel technique dramatically reduced glaucoma drainage device (GDD) tube exposure rates to zero, as shown by follow-up examinations after 20 months. Similarly, for microinvasive glaucoma surgeries (MIGS), the introduction of the ab-externo approach for XEN stents improved IOP control and reduced postoperative interventions. Surprisingly, in the domain of cataract surgery, premium IOLs showed promising outcomes regarding spectacle independence and user satisfaction, specifically in patients with well-controlled, early-stage glaucoma or ocular hypertension [8]. However, their usage in advanced glaucoma must be considered with caution [8]. It was stressed that IOL selection should be individualized based on disease severity, visual expectations, and co-existing ocular conditions such as pseudo-exfoliation and ocular surface disease (OSD).

The importance of newer imaging technologies in predicting surgical outcomes was showcased by Tan et al. Using anterior segment OCT (AS-OCT), they found that successful blebs after deep sclerectomy had a significantly greater height (1.48 mm vs. 1.10 mm; $p < 0.0001$) and trabecular-Descemet window length (613.7 µm vs. 450.8 µm; $p = 0.004$) compared to failures [9]. Further contributing to the advancement of surgical outcomes, Fung et al. introduced a new high-throughput system that functions as a pre-animal testing platform for anti-fibrotic compounds [10]. Using this platform, they demonstrated that Verteporfin improves surgical outcomes by modulating the TGFβ-SMAD (Small Mothers against Decapentaplegic) pathway. In the same vein, Dave et al. reviewed emerging antifibrotic therapies, highlighting integrin inhibitors, and anti-LOXL2 antibodies as safer, more effective alternatives to traditional agents like mitomycin C [11].

Novel Rho-associated protein kinase (ROCK) inhibitors, particularly ZINC000000022706 and ZINC000034800307, with high binding affinities (-10.7 kcal/mol and -9.1 kcal/mol, respectively) were identified by Bodea et al. using bibliometric analysis and molecular docking [12]. A review of OSD by Kemer et al. demonstrated that 22–78% of patients on topical glaucoma medications experienced significant clinical side effects, and they encouraged the consideration of preservative-free formulations and adjunctive therapies like cyclosporine A [13]. Other novel emerging therapies, including extraocular and intraocular sustained-drug delivery systems, photobiomodulation, gene therapy, and stem cell applications, were reviewed as well.

Advancements in wearable technologies are also highlighted by Shean et al., with a focus on the Sensimed Triggerfish (24 h IOP measuring device) and drug-eluting contact lenses [14]. Notably, continuous monitoring using Triggerfish allowed for the identification of nocturnal IOP peaks as a critical risk factor for glaucoma, and drug-eluting contact lenses provided sustained IOP reductions of up to 30% over several weeks. Another review by Elhusseiny et al. emphasizes the predictive value of corneal hysteresis (CH), showing that a lower CH is associated with faster visual field deterioration (every 1 mmHg decrease in CH was associated with an additional 0.25% decline per year in visual field index) [15]. The potential to use Brillouin microscopy as a non-contact, three-dimensional assessment of corneal elasticity was also highlighted, as it offers a significant advancement compared to traditional methods that are solely reliant on IOP.

The interesting relationship between myopia and glaucoma is discussed by Vinod et al. [16]. The presence of myopia was shown to increase glaucoma risk due to structural vulnerabilities (thinning of the sclera and lamina cribrosa), with highly myopic eyes being 7.3 times more likely to develop glaucoma than emmetropic eyes. Overlapping features (optic nerve head changes and retinal nerve fiber layer thinning) may be overcome by utilizing the ever-expanding AI and advanced imaging technologies such as OCT-Angiography (OCTA) and Swept Source-OCTA.

Future advancements in glaucoma must continue to focus on early diagnosis and accurate prognostic evaluations, medical and surgical therapeutics, patient-centered technologies, and wearable theranostic devices (e.g., smart contact lenses) [14,17]. Emerging polymer-based long-term drug delivery systems have the potential for sustained IOP control and improved adherence. However, they still require continued optimization for biocompatibility and scalability to ensure their widespread application is possible [18,19]. Exciting improvements in surgical technique are on their way, such as the combination of traditional and minimally invasive approaches achieved by new advancements in GDDs and MIGS device designs [17,20,21]. Regarding AI tools, we have just begun to scratch the surface. These tools are being validated in diverse populations and their diagnostic and risk stratification capabilities continue to improve [22–25].

At the molecular level, neuroprotective strategies targeting oxidative stress-related mitochondrial dysfunction may offer promising pathways for retinal ganglion cell preservation [18,26–30]. Similarly, the impact of oxidative stress on trabecular meshwork cells and aqueous outflow resistance could also factor into targeted therapeutic approaches [31–36]. Epigenetic biomarkers (DNA methylation and histone modifications) and emerging gene therapies have the potential to revolutionize early diagnosis and therapeutic targeting [37–42]. Understanding the role of glial cell activation and its modulation of neuroinflammatory pathways may lead to improvements in the preservation of visual function and mitigation of neurodegeneration [19,43–46]. There is increasing focus on the relationship between gut microbiota and glaucoma, with emerging theories suggesting systemic roles for microbiome dysbiosis, as well as roles in neuroinflammation, including in autoimmune mechanisms [47–55]. Future research should prioritize understanding the mechanisms through which gut microbiota influence the development and progression of glaucoma.

Reviewing the submissions to this Special Issue on glaucoma, we have achieved our underlying aim of stimulating research and dialog among the global glaucoma research community. It is our firm belief that the future for glaucoma research looks exceptionally promising.

Acknowledgments: We would like to thank all the authors for their hard work, dedication, and relentless research on glaucoma.

Conflicts of Interest: The authors declare no conflicts of interest.

References

1. Yoo, K.; Lee, C.; Baxter, S.L.; Xu, B.Y. Relationship Between Glaucoma and Chronic Stress Quantified by Allostatic Load Score in the All of Us Research Program. *Am. J. Ophthalmol.* **2024**, *269*, 419–428. [CrossRef] [PubMed]
2. McDermott, C.E.; Salowe, R.J.; Di Rosa, I.; O'Brien, J.M. Stress, Allostatic Load, and Neuroinflammation: Implications for Racial and Socioeconomic Health Disparities in Glaucoma. *Int. J. Mol. Sci.* **2024**, *25*, 1653. [CrossRef] [PubMed]
3. Li-Han, L.Y.; Eizenman, M.; Shi, R.B.; Buys, Y.M.; Trope, G.E.; Wong, W. Using Fused Data from Perimetry and Optical Coherence Tomography to Improve the Detection of Visual Field Progression in Glaucoma. *Bioengineering* **2024**, *11*, 250. [CrossRef] [PubMed]
4. Christopher, M.; Gonzalez, R.; Huynh, J.; Walker, E.; Saseendrakumar, B.R.; Bowd, C.; Belghith, A.; Goldbaum, M.H.; Fazio, M.A.; Girkin, C.A.; et al. Proactive Decision Support for Glaucoma Treatment: Predicting Surgical Interventions with Clinically Available Data. *Bioengineering* **2024**, *11*, 140. [CrossRef] [PubMed]
5. Hwang, E.E.; Chen, D.; Han, Y.; Jia, L.; Shan, J. Multi-Dataset Comparison of Vision Transformers and Convolutional Neural Networks for Detecting Glaucomatous Optic Neuropathy from Fundus Photographs. *Bioengineering* **2023**, *10*, 1266. [CrossRef]
6. Barac, R.I.; Harghel, V.; Anton, N.; Baltă, G.; Tofolean, I.T.; Dragosloveanu, C.; Leuștean, L.F.; Deleanu, D.G.; Barac, D.A. Initial Clinical Experience with Ahmed Valve in Romania: Five-Year Patient Follow-Up and Outcomes. *Bioengineering* **2024**, *11*, 820. [CrossRef]
7. Ang, B.C.H.; Lim, S.Y.; Betzler, B.K.; Wong, H.J.; Stewart, M.W.; Dorairaj, S. Recent Advancements in Glaucoma Surgery—A Review. *Bioengineering* **2023**, *10*, 1096. [CrossRef]
8. Hong, A.S.Y.; Ang, B.C.H.; Dorairaj, E.; Dorairaj, S. Premium Intraocular Lenses in Glaucoma-A Systematic Review. *Bioengineering* **2023**, *10*, 993. [CrossRef]
9. Tan, J.C.K.; Muntasser, H.; Choudhary, A.; Batterbury, M.; Vallabh, N.A. Swept-Source Anterior Segment Optical Coherence Tomography Imaging and Quantification of Bleb Parameters in Glaucoma Filtration Surgery. *Bioengineering* **2023**, *10*, 1186. [CrossRef]
10. Fung, M.; Armstrong, J.J.; Zhang, R.; Vinokurtseva, A.; Liu, H.; Hutnik, C. Development and Verification of a Novel Three-Dimensional Aqueous Outflow Model for High-Throughput Drug Screening. *Bioengineering* **2024**, *11*, 142. [CrossRef]
11. Dave, B.; Patel, M.; Suresh, S.; Ginjupalli, M.; Surya, A.; Albdour, M.; Kooner, K.S. Wound Modulations in Glaucoma Surgery: A Systematic Review. *Bioengineering* **2024**, *11*, 446. [CrossRef] [PubMed]

12. Bodea, F.; Bungau, S.G.; Negru, A.P.; Radu, A.; Tarce, A.G.; Tit, D.M.; Bungau, A.F.; Bustea, C.; Behl, T.; Radu, A.-F. Exploring New Therapeutic Avenues for Ophthalmic Disorders: Glaucoma-Related Molecular Docking Evaluation and Bibliometric Analysis for Improved Management of Ocular Diseases. *Bioengineering* **2023**, *10*, 983. [CrossRef] [PubMed]
13. Kemer, Ö.E.; Mekala, P.; Dave, B.; Kooner, K.S. Managing Ocular Surface Disease in Glaucoma Treatment: A Systematic Review. *Bioengineering* **2024**, *11*, 1010. [CrossRef] [PubMed]
14. Shean, R.; Yu, N.; Guntipally, S.; Nguyen, V.; He, X.; Duan, S.; Gokoffski, K.; Zhu, Y.; Xu, B. Advances and Challenges in Wearable Glaucoma Diagnostics and Therapeutics. *Bioengineering* **2024**, *11*, 138. [CrossRef]
15. Elhusseiny, A.M.; Scarcelli, G.; Saeedi, O.J. Corneal Biomechanical Measures for Glaucoma: A Clinical Approach. *Bioengineering* **2023**, *10*, 1108. [CrossRef]
16. Vinod, K.; Salim, S. Addressing Glaucoma in Myopic Eyes: Diagnostic and Surgical Challenges. *Bioengineering* **2023**, *10*, 1260. [CrossRef]
17. Kim, T.Y.; Mok, J.W.; Hong, S.H.; Jeong, S.H.; Choi, H.; Shin, S.; Joo, C.-K.; Hahn, S.K. Wireless theranostic smart contact lens for monitoring and control of intraocular pressure in glaucoma. *Nat. Commun.* **2022**, *13*, 6801. [CrossRef]
18. Wang, L.H.; Huang, C.H.; Lin, I.C. Advances in Neuroprotection in Glaucoma: Pharmacological Strategies and Emerging Technologies. *Pharmaceuticals* **2024**, *17*, 1261. [CrossRef]
19. Fernández-Albarral, J.A.; Ramírez, A.I.; de Hoz, R.; Matamoros, J.A.; Salobrar-García, E.; Elvira-Hurtado, L.; López-Cuenca, I.; Sánchez-Puebla, L.; Salazar, J.J.; Ramírez, J.M. Glaucoma: From Pathogenic Mechanisms to Retinal Glial Cell Response to Damage. *Front. Cell Neurosci.* **2024**, *18*, 1354569. [CrossRef]
20. Dhawale, K.K.; Tidake, P. A Comprehensive Review of Recent Advances in Minimally Invasive Glaucoma Surgery: Current Trends and Future Directions. *Cureus* **2024**, *16*, e65236. [CrossRef]
21. Spratt, A.; Lee, R.K. A review of the surgical approaches to glaucoma treatment. *Vision. Pan Am.* **2013**, *12*, 41–44.
22. Zeppieri, M.; Gardini, L.; Culiersi, C.; Fontana, L.; Musa, M.; D'esposito, F.; Surico, P.L.; Gagliano, C.; Sorrentino, F.S. Novel Approaches for the Early Detection of Glaucoma Using Artificial Intelligence. *Life* **2024**, *14*, 1386. [CrossRef] [PubMed]
23. Zhang, X.; Lai, F.; Chen, W.; Yu, C. An Automatic Glaucoma Grading Method Based on Attention Mechanism and EfficientNet-B3 Network. *PLoS ONE* **2024**, *19*, e0296229. [CrossRef] [PubMed]
24. Tonti, E.; Tonti, S.; Mancini, F.; Bonini, C.; Spadea, L.; D'esposito, F.; Gagliano, C.; Musa, M.; Zeppieri, M. Artificial Intelligence and Advanced Technology in Glaucoma: A Review. *J. Pers. Med.* **2024**, *14*, 1062. [CrossRef]
25. Zhu, Y.; Salowe, R.; Chow, C.; Li, S.; Bastani, O.; O'brien, J.M. Advancing Glaucoma Care: Integrating Artificial Intelligence in Diagnosis, Management, and Progression Detection. *Bioengineering* **2024**, *11*, 122. [CrossRef]
26. Chrysostomou, V.; Rezania, F.; Trounce, I.A.; Crowston, J.G. Oxidative stress and mitochondrial dysfunction in glaucoma. *Curr. Opin. Pharmacol.* **2013**, *13*, 12–15. [CrossRef]
27. Jassim, A.H.; Fan, Y.; Pappenhagen, N.; Nsiah, N.Y.; Inman, D.M. Oxidative Stress and Hypoxia Modify Mitochondrial Homeostasis During Glaucoma. *Antioxid. Redox Signal* **2021**, *35*, 1341–1357. [CrossRef]
28. Pinazo-Durán, M.D.; Zanón-Moreno, V.; Gallego-Pinazo, R.; García-Medina, J.J. Oxidative stress and mitochondrial failure in the pathogenesis of glaucoma neurodegeneration. *Prog. Brain Res.* **2015**, *220*, 127–153. [CrossRef]
29. Zhang, Z.Q.; Xie, Z.; Chen, S.Y.; Zhang, X. Mitochondrial dysfunction in glaucomatous degeneration. *Int. J. Ophthalmol.* **2023**, *16*, 811–823. [CrossRef]
30. McElnea, E.; Quill, B.; Docherty, N.; Irnaten, M.; Siah, W.; Clark, A.; O'brien, C.; Wallace, D. Oxidative stress, mitochondrial dysfunction and calcium overload in human lamina cribrosa cells from glaucoma donors. *Mol. Vis.* **2011**, *17*, 1182–1191. [PubMed] [PubMed Central]
31. Zhao, J.; Wang, S.; Zhong, W.; Yang, B.; Sun, L.; Zheng, Y. Oxidative stress in the trabecular meshwork (Review). *Int. J. Mol. Med.* **2016**, *38*, 995–1002. [CrossRef]
32. Saccà, S.C.; Pascotto, A.; Camicione, P.; Capris, P.; Izzotti, A. Oxidative DNA damage in the human trabecular meshwork: Clinical correlation in patients with primary open-angle glaucoma. *Arch. Ophthalmol.* **2005**, *123*, 458–463. [CrossRef] [PubMed]
33. Wang, M.; Zheng, Y. Oxidative stress and antioxidants in the trabecular meshwork. *PeerJ* **2019**, *7*, e8121. [CrossRef] [PubMed] [PubMed Central]
34. Saccà, S.C.; Izzotti, A.; Rossi, P.; Traverso, C. Glaucomatous outflow pathway and oxidative stress. *Exp. Eye Res.* **2007**, *84*, 389–399. [CrossRef] [PubMed]
35. Izzotti, A.; Bagnis, A.; Saccà, S.C. The role of oxidative stress in glaucoma. *Mutat. Res.* **2006**, *612*, 105–114. [CrossRef]
36. Yu, A.L.; Fuchshofer, R.; Kampik, A.; Welge-Lüssen, U. Effects of oxidative stress in trabecular meshwork cells are reduced by prostaglandin analogues. *Invest. Ophthalmol. Vis. Sci.* **2008**, *49*, 4872–4880. [CrossRef]
37. Tonti, E.; Dell'omo, R.; Filippelli, M.; Spadea, L.; Salati, C.; Gagliano, C.; Musa, M.; Zeppieri, M. Exploring Epigenetic Modifications as Potential Biomarkers and Therapeutic Targets in Glaucoma. *Int. J. Mol. Sci.* **2024**, *25*, 2822. [CrossRef]
38. Gauthier, A.C.; Liu, J. Epigenetics and Signaling Pathways in Glaucoma. *Biomed. Res. Int.* **2017**, *2017*, 5712341. [CrossRef] [PubMed] [PubMed Central]

39. Wiggs, J.L. The cell and molecular biology of complex forms of glaucoma: Updates on genetic, environmental, and epigenetic risk factors. *Investig. Ophthalmol. Vis. Sci.* **2012**, *53*, 2467–2469. [CrossRef]
40. Sulak, R.; Liu, X.; Smedowski, A. The concept of gene therapy for glaucoma: The dream that has not come true yet. *Neural Regen. Res.* **2024**, *19*, 92–99. [CrossRef]
41. Liu, X.; Rasmussen, C.A.; Gabelt, B.T.; Brandt, C.R.; Kaufman, P.L. Gene therapy targeting glaucoma: Where are we? *Surv. Ophthalmol.* **2009**, *54*, 472–486. [CrossRef] [PubMed] [PubMed Central]
42. Hakim, A.; Guido, B.; Narsineni, L.; Chen, D.W.; Foldvari, M. Gene therapy strategies for glaucoma from IOP reduction to retinal neuroprotection: Progress towards non-viral systems. *Adv. Drug Deliv. Rev.* **2023**, *196*, 114781. [CrossRef]
43. García-Bermúdez, M.Y.; Freude, K.K.; Mouhammad, Z.A.; van Wijngaarden, P.; Martin, K.K.; Kolko, M. Glial Cells in Glaucoma: Friends, Foes, and Potential Therapeutic Targets. *Front. Neurol.* **2021**, *12*, 624983. [CrossRef] [PubMed]
44. Salkar, A.; Wall, R.V.; Basavarajappa, D.; Chitranshi, N.; Parilla, G.E.; Mirzaei, M.; Yan, P.; Graham, S.; You, Y. Glial Cell Activation and Immune Responses in Glaucoma: A Systematic Review of Human Postmortem Studies of the Retina and Optic Nerve. *Aging Dis.* **2024**, *15*, 2069–2083. [CrossRef]
45. Miao, Y.; Zhao, G.L.; Cheng, S.; Wang, Z.; Yang, X.L. Activation of retinal glial cells contributes to the degeneration of ganglion cells in experimental glaucoma. *Prog. Retin. Eye Res.* **2023**, *93*, 101169. [CrossRef]
46. Chong, R.S.; Martin, K.R. Glial cell interactions and glaucoma. *Curr. Opin. Ophthalmol.* **2015**, *26*, 73–77. [CrossRef]
47. Ullah, Z.; Tao, Y.; Mehmood, A.; Huang, J. The Role of Gut Microbiota in the Pathogenesis of Glaucoma: Evidence from Bibliometric Analysis and Comprehensive Review. *Bioengineering* **2024**, *11*, 1063. [CrossRef]
48. Wang, L.; Cioffi, G.A.; Cull, G.; Dong, J.; Fortune, B. Immunohistologic evidence for retinal glial cell changes in human glaucoma. *Invest. Ophthalmol. Vis. Sci.* **2002**, *43*, 1088–1094.
49. Chen, J.; Chen, D.F.; Cho, K.S. The Role of Gut Microbiota in Glaucoma Progression and Other Retinal Diseases. *Am. J. Pathol.* **2023**, *193*, 1662–1668. [CrossRef]
50. Wu, Y.; Shi, R.; Chen, H.; Zhang, Z.; Bao, S.; Qu, J.; Zhou, M. Effect of the gut microbiome in glaucoma risk from the causal perspective. *BMJ Open Ophthalmol.* **2024**, *9*, e001547. [CrossRef]
51. Zhang, Y.; Zhou, X.; Lu, Y. Gut microbiota and derived metabolomic profiling in glaucoma with progressive neurodegeneration. *Front. Cell Infect. Microbiol.* **2022**, *12*, 968992. [CrossRef] [PubMed]
52. Li, C.; Lu, P. Association of Gut Microbiota with Age-Related Macular Degeneration and Glaucoma: A Bidirectional Mendelian Randomization Study. *Nutrients* **2023**, *15*, 4646. [CrossRef] [PubMed]
53. Huang, L.; Hong, Y.; Fu, X.; Tan, H.; Chen, Y.; Wang, Y.; Chen, D. The role of microbiota in glaucoma. *Mol. Aspects Med.* **2023**, *94*, 101221. [CrossRef] [PubMed]
54. Gong, H.; Zhang, S.; Li, Q.; Zuo, C.; Gao, X.; Zheng, B.; Lin, M. Gut microbiota compositional profile and serum metabolic phenotype in patients with primary open-angle glaucoma. *Exp. Eye Res.* **2020**, *191*, 107921. [CrossRef]
55. Pezzino, S.; Sofia, M.; Greco, L.P.; Litrico, G.; Filippello, G.; Sarvà, I.; La Greca, G.; Latteri, S. Microbiome Dysbiosis: A Pathological Mechanism at the Intersection of Obesity and Glaucoma. *Int. J. Mol. Sci.* **2023**, *24*, 1166. [CrossRef]

Disclaimer/Publisher's Note: The statements, opinions and data contained in all publications are solely those of the individual author(s) and contributor(s) and not of MDPI and/or the editor(s). MDPI and/or the editor(s) disclaim responsibility for any injury to people or property resulting from any ideas, methods, instructions or products referred to in the content.

Article

Using Fused Data from Perimetry and Optical Coherence Tomography to Improve the Detection of Visual Field Progression in Glaucoma

Leo Yan Li-Han [1,*], Moshe Eizenman [2,3], Runjie Bill Shi [3,4], Yvonne M. Buys [2], Graham E. Trope [2] and Willy Wong [1,4]

1. The Edward S. Rogers Sr. Department of Electrical & Computer Engineering, University of Toronto, Toronto, ON M5S 3G4, Canada
2. Department of Ophthalmology & Vision Sciences, University of Toronto, Toronto, ON M5T 3A9, Canada
3. Temerty Faculty of Medicine, University of Toronto, Toronto, ON M5S 1A8, Canada
4. Institute of Biomedical Engineering, University of Toronto, Toronto, ON M5S 3E2, Canada
* Correspondence: yyaann.li@mail.utoronto.ca

Abstract: Perimetry and optical coherence tomography (OCT) are both used to monitor glaucoma progression. However, combining these modalities can be a challenge due to differences in data types. To overcome this, we have developed an autoencoder data fusion (AEDF) model to learn compact encoding (AE-fused data) from both perimetry and OCT. The AEDF model, optimized specifically for visual field (VF) progression detection, incorporates an encoding loss to ensure the interpretation of the AE-fused data is similar to VF data while capturing key features from OCT measurements. For model training and evaluation, our study included 2504 longitudinal VF and OCT tests from 140 glaucoma patients. VF progression was determined from linear regression slopes of longitudinal mean deviations. Progression detection with AE-fused data was compared to VF-only data (standard clinical method) as well as data from a Bayesian linear regression (BLR) model. In the initial 2-year follow-up period, AE-fused data achieved a detection F1 score of 0.60 (95% CI: 0.57 to 0.62), significantly outperforming ($p < 0.001$) the clinical method (0.45, 95% CI: 0.43 to 0.47) and the BLR model (0.48, 95% CI: 0.45 to 0.51). The capacity of the AEDF model to generate clinically interpretable fused data that improves VF progression detection makes it a promising data integration tool in glaucoma management.

Keywords: autoencoder; data fusion; glaucoma progression; optical coherence tomography; perimetry; visual field

Citation: Li-Han, L.Y.; Eizenman, M.; Shi, R.B.; Buys, Y.M.; Trope, G.E.; Wong, W. Using Fused Data from Perimetry and Optical Coherence Tomography to Improve the Detection of Visual Field Progression in Glaucoma. *Bioengineering* **2024**, *11*, 250. https://doi.org/10.3390/bioengineering11030250

Received: 24 January 2024
Revised: 16 February 2024
Accepted: 27 February 2024
Published: 3 March 2024

Copyright: © 2024 by the authors. Licensee MDPI, Basel, Switzerland. This article is an open access article distributed under the terms and conditions of the Creative Commons Attribution (CC BY) license (https://creativecommons.org/licenses/by/4.0/).

1. Introduction

Glaucoma is a progressive optic neuropathy characterized by irreversible vision loss and abnormal thinning of the retinal nerve fiber layer (RNFL) [1]. Visual field (VF) testing, also known as perimetry, is the primary clinical test for assessing functional vision loss in glaucoma, while optical coherence tomography (OCT) is the standard imaging tool for evaluating the structural integrity of the retinal nerve fiber layer. The subjective and probabilistic nature of VF testing introduces significant variability and can lead to delays in detecting disease progression [2]. While OCT data, such as peripapillary RNFL thickness, is less susceptible to subjective errors, recent research underscores the inadequacy of relying solely on a single modality for glaucoma monitoring [3–5]. This is due to the fact that at different stages of the disease, glaucoma progression is associated with changes in either VF or OCT measurements and these changes can be asynchronous [3–5].

Consequently, there is a pressing need to integrate information from VF and OCT measurements to accurately detect glaucoma progression. However, integration poses challenges owing to substantial differences in scales, dimensions, and variability between data from the two modalities. Clinical perimeters, like the Humphrey Field Analyzer (Carl Zeiss Meditec, Inc.; Dublin, CA, USA), provide combined functional and structural reports yet manual interpretation can be subjective and relies heavily on clinical experience. One potential solution is to use machine-learning techniques as these offer a data-driven approach without the need to assume fixed relationships between the different data streams. However, the black-box nature of machine-learning algorithms presents a challenge in terms of interpretability and explainability, making it difficult for clinicians to trust the results generated from such algorithms.

With respect to data fusion studies in glaucoma, a number of methods have been developed to integrate differential light sensitivity (DLS) measurements from VF testing and RNFL thickness data from OCT. Bizios et al. [6] developed a model to fuse VF and OCT data based on manually defined rules that consider the spatial correspondence between functional and structural measurements. The fused data was then used to train an artificial neural network (ANN) for glaucoma diagnosis. While detecting glaucoma cases with the fused data outperformed models trained with VF data alone (i.e., using pattern deviation maps), the complexity of the fused data limits clinical interpretability of the results. Similarly, in a recent study, Song et al. [7] developed a deep-learning ANN model to integrate global and regional features of VF and OCT measurements. The authors showed that the accuracy of diagnosing glaucoma using the proposed model surpassed that of single-modal approaches. However, the difficulty in tracing through the decision-making process of a deep-learning model can pose challenges in regard to the comprehension of the diagnostic results. Wu and Medeiros [8] proposed a structure-function index to combine information from VF and OCT measurements for glaucoma diagnosis. This method first transforms data from the two testing modalities to the same scale. Subsequently, the transformed data is combined based on predefined rules taking into consideration the difference in measurement ranges of the two data types across varying degrees of glaucoma severity. Although the combined structure-function index exhibited superior diagnostic performance compared to using data from a single modality alone [8], the reliance on hard-coded rules for the data transformation and integration may not account for the variability observed during disease progression. Furthermore, the combined index is challenging to interpret in terms of the established clinical criteria for glaucoma diagnosis.

The above methods are aimed primarily at improving glaucoma diagnosis, and not for the detection of VF progression. A tool for monitoring glaucoma progression should not only indicate whether an eye is deteriorating but also have the capacity to measure the rate of progression [9–11]. Medeiros et al. [11] developed a Bayesian hierarchical linear model that combines changes in visual field index measurements from perimetry and average RNFL thickness data from scanning laser polarimetry. In this method, statistical parameters that describe rates of change in a cohort of glaucoma patients were used as the a priori information to estimate subject-specific progression rates. While the method demonstrated improved performance in detecting progression, the complexity of the model poses challenges in its interpretability. Russell et al. [12] developed a Bayesian linear regression (BLR) model that uses changes in the neuroretinal rim area to improve estimates of VF progression rates. Results from the BLR model are easily interpretable, and the combined progression rates demonstrated reduced error in predicting future VF measurements, which can be further used for progression analysis. However, there are limitations to this approach as the relationship between functional and structural measurements is likely nonlinear.

Inspired by these earlier studies, the objective of this paper is to develop a method that fuses structural RNFL thickness data and functional VF measurements to improve the detection of VF progression. To overcome the nonlinear and changing relationship between structural and functional deteriorations in different glaucoma stages, we employed a trainable ANN model rather than a method based on fixed rules. An ANN is well-suited for describing complex and nonlinear relationships between data sets with substantial differences in scales, dimensions, and variability. Moreover, as fused data should be interpreted in a similar manner to that of data from a standard clinical VF test, we used an autoencoder (AE) to construct the data fusion model. The AE model facilitates a compact representation of input data, referred to as an encoding. This encoding retains essential features required to reconstruct the original input data while removing any redundancies [13], thus making it a widely used technique for multimodality data fusion across diverse fields [14–17]. In what we believe is a key and significant contribution to this area, we introduce an "encoding loss function" to regularize the fused data so that the encoding has a similar structure to that of the VF test. The fused data can then be interpreted in the same way that a visual field test is interpreted, taking a crucial step towards enhancing the clinical interpretability of the machine-learning model's results, thereby addressing a key drawback of conventional AE data fusion models. Finally, to evaluate the efficacy of the AE data fusion (AEDF) model in glaucoma monitoring, we compared VF progression detection performance using only VF data (the standard clinical method), data generated by the BLR model, and data generated by the AEDF model.

2. Materials and Methods:

2.1. VF and OCT Data

In this retrospective study, we evaluated 2504 pairs of longitudinal VF and OCT tests of 140 glaucoma patients who had been followed for a minimum of four years at the glaucoma clinic of the Toronto Western Hospital. Each pair of VF and OCT tests was performed on the same day. All VF tests were conducted using the Humphrey Field Analyzer with the 24-2 SITA Standard algorithm. Each VF consists of 54 differential light sensitivity thresholds at discrete locations extending 30 degrees around the central fixation point. The two VF testing points that are mapped onto the optic nerve head (ONH), i.e., the blind spot, were not used in this study. The ONH, where retinal ganglion cell axons exit the retina and converge to form the optic nerve, represents the primary site of structural damage due to glaucoma. The peripapillary RNFL thickness profile data obtained from OCT encompasses 256 A-scans along a 3.45 mm circle centered at the ONH. This data is typically recorded as a 256-dimensional vector, with each element representing a thickness measurement of the RNFL (in µm) at a particular angular position (0 to 360 degrees) around the ONH (see Figures 1 and 2). All RNFL thickness profile data utilized in this study was acquired using the Cirrus HD-OCT (Carl Zeiss Meditec, Inc.; Dublin, CA, USA). Only pairs of reliable VF and OCT tests were analyzed. In line with prior research [8], the reliability criteria for VF data were false-positive and false-negative rates < 15% and fixation loss rate < 33%. For OCT data, the reliability criterion was defined as signal strength ≥ 7. Patients with severe VF defects, i.e., mean deviation (MD) worse than -20 dB on the first test (visit), were excluded from model training and evaluation, as both VF and OCT changes are notably affected by the floor effect in measurements of advanced glaucoma [18]. Additionally, OCT tests containing missing or corrupted RNFL thickness data points were also excluded.

The study received approval from the Research Ethics Board of the University Health Network, Toronto, Ontario, Canada, and adhered to the tenets of the Declaration of Helsinki.

2.2. Data Fusion Models for Function-Structure Measurements

2.2.1. Autoencoder Data Fusion Model

The AE data fusion (AEDF) model consists of two components: (1) an encoder network $\mathbf{z} = f_\theta(\mathbf{x})$ parameterized by θ, which maps the input data $\mathbf{x} \in \mathbb{R}^{d_x}$ to a low dimensional encoding space $\mathbf{z} \in \mathbb{R}^{d_z}$ ($d_x > d_z$), where the encoder's output is referred to as the AE-fused data, and (2) a decoder network $\mathbf{x}' = g_\phi(\mathbf{z})$ parameterized by ϕ, aiming to reconstruct the input data from the encoding space, such that $\mathbf{x}' \in \mathbb{R}^{d_x}$ [13]. Here, d_x and d_z represent the dimensions of the input and encoding spaces, respectively.

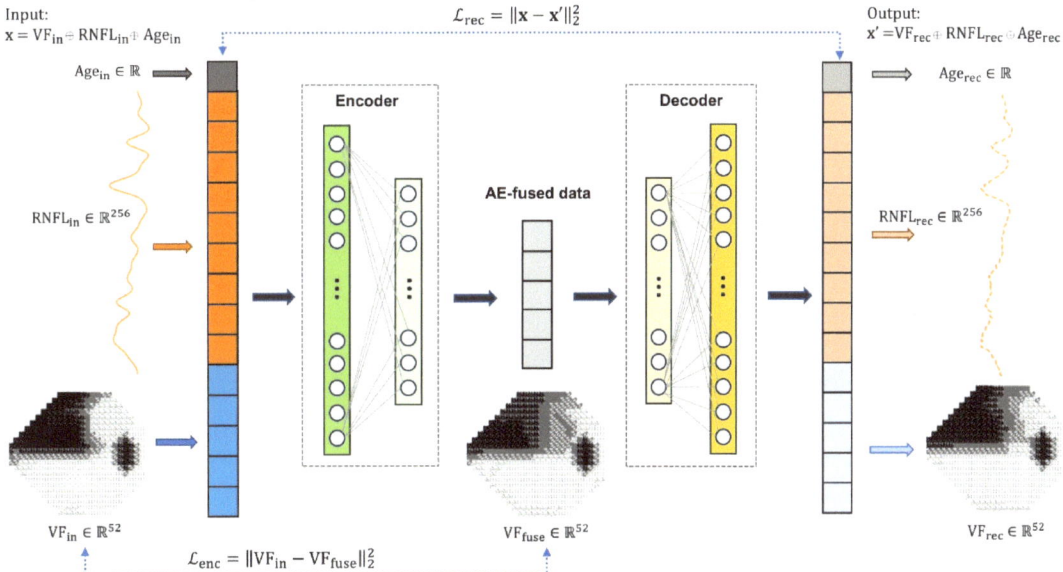

Figure 1. The overall architecture of the autoencoder (AE) data fusion model. The input to the model is a vector that includes pointwise differential light sensitivity thresholds from visual field (VF) testing (52-dimensional vector), retinal nerve fiber layer (RNFL) thickness profile (256-dimensional vector), and patient's age at the time of the test (scalar). The encoder network, constructed with a two-hidden layer multilayer perceptron (MLP) model, processes the input vector and generates a 52-dimensional encoding vector as the AE-fused data. The decoder network, a symmetrically structured MLP model, aims to reconstruct the input data from the encoding vector. The reconstruction loss ($\mathcal{L}_{rec}$) is the mean squared error (MSE) between the input and output vectors of the AE data fusion model. The encoding loss ($\mathcal{L}_{enc}$) is the MSE between the AE-fused data and the measured VF. The training objective is to minimize the convex combination of the reconstruction loss and the encoding loss, weighted by a scalar λ.

Given that similar defects in visual field or retinal structure at different ages may lead to different clinical interpretations, we used the patient's age at the time of testing, a (where $a \in \mathbb{R}$), as one of the input parameters to the AEDF model. The other input parameters are the VF pointwise DLS thresholds, denoted as $\mathbf{v}$ (where $\mathbf{v} \in \mathbb{R}^{52}$), and the RNFL thickness profile, denoted as $\mathbf{r}$ (where $\mathbf{r} \in \mathbb{R}^{256}$). As such, the dimension of the input to the AE data fusion model, denoted as $\mathbf{x} = (\mathbf{v}, \mathbf{r}, a)$, is $d_x = 309$. Note that in Section 3.5, we described sensitivity analysis that investigates the contribution of the input parameters to the detection of VF progression.

In a similar manner, the output of the decoder is expressed as $\mathbf{x}' = (\mathbf{v}', \mathbf{r}', a')$, where $\mathbf{v}' \in \mathbb{R}^{52}$, $\mathbf{r}' \in \mathbb{R}^{256}$, and $a' \in \mathbb{R}$ represent the reconstructed VF, RNFL thickness profile, and patient's age, respectively. Furthermore, the encoding space has the same dimensionality as the input VF data, i.e., $d_z = 52$ for the 24-2 pattern, so that the AE-fused data ($\mathbf{z}$) can be interpreted in the same space as the data from the VF test.

In the training phase, parameters of the encoder and decoder were simultaneously updated to minimize the reconstruction loss $\mathcal{L}_{rec}$, which was defined as the mean squared error (MSE) between the input and reconstructed data (Equation (1)):

$$\mathcal{L}_{rec}(g_\phi(f_\theta(\mathbf{x})), \mathbf{x}) = \frac{1}{N}\sum_{i=1}^{N}(\mathbf{x}'_i - \mathbf{x}_i)^2 \qquad (1)$$

Here, x_i and x'_i represent the input and reconstructed vectors of the i-th sample, respectively, and N is the total number of training examples. Training the AEDF model by solely minimizing the reconstruction loss, as defined in Equation (1), does not ensure that the AE-fused data (the output of the encoder) will retain the appearance of standard VF tests.

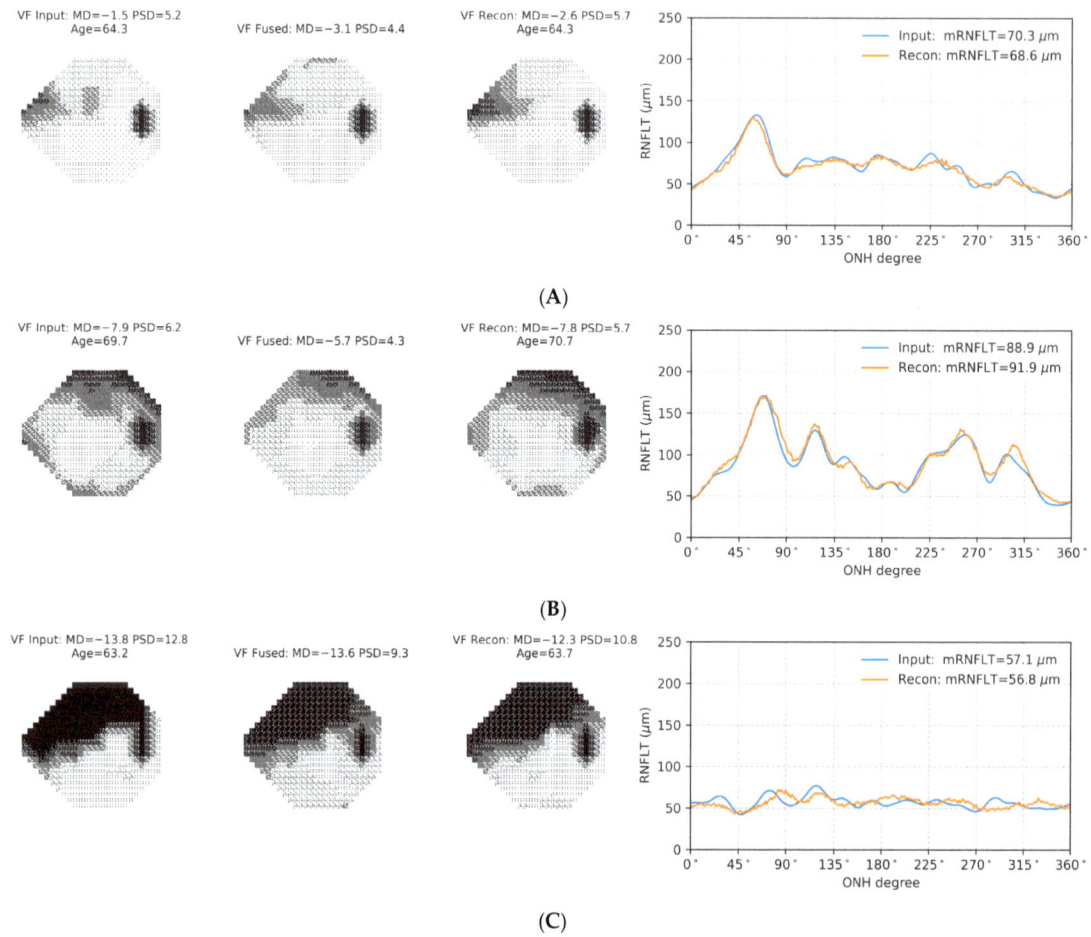

Figure 2. Examples of the autoencoder (AE) data fusion model for eyes with mild (panel **A**), moderate (panel **B**), and severe (panel **C**) VF defects. In each panel, the three visual field (VF) plots represent the input VF to the AE data fusion model (left), the AE-fused data (middle), and the reconstructed VF (right). The right graph in each panel illustrates the input retinal nerve fiber layer thickness (RNFLT) profile data to the AE data fusion model (blue curve) and the reconstructed RNFLT profile data (orange curve) from the AE data fusion model. The RNFLT profile data, a 256-dimensional vector, is visualized as a curve of the RNFLT, where the horizontal axis represents the angular position (0 to 360 degrees) around the optic nerve head (ONH), and the vertical axis represents the RNFL thickness measurement (in μm). These examples provide visualized representations of the way that the AE data fusion model dynamically combines results from VF and OCT tests.

To address this challenge, we introduced an encoding loss, in addition to the reconstruction loss, to the model training. Specifically, the encoding loss function was defined as the MSE between the encoding and the input VF data, formulated as $\mathcal{L}_{\text{enc}} = \frac{1}{N}\sum_i^N (\mathbf{z_i} - \mathbf{v_i})^2$, where z_i and v_i are the encoding and measured VF data of the i-th sample, respectively. With the incorporation of the encoding loss, each point in the AE-fused data represents a modified version of the differential light sensitivity from the input VF, making it interpretable by standard clinical methods for VF progression analysis. Thus, it is unnecessary to develop new criteria for the AE-fused data to detect progression, as it aligns with existing standard clinical techniques to analyze VF data.

To this point, the total loss function ($\mathcal{L}$) for the training of the AEDF model was defined as the convex combination of the reconstruction loss and the encoding loss, formulated as:

$$\mathcal{L} = (1-\lambda)\mathcal{L}_{\text{rec}} + \lambda \mathcal{L}_{\text{enc}} = \frac{1-\lambda}{N}\sum_{i=1}^{N}(\mathbf{x}'_i - \mathbf{x}_i)^2 + \frac{\lambda}{N}\sum_{i=1}^{N}(\mathbf{z}_i - \mathbf{v}_i)^2 \qquad (2)$$

where $\lambda \in [0, 1]$ is a hyperparameter that controls the strength of regularization from the encoding loss. As suggested in Equation (2), the reconstruction and encoding losses work in an adversarial manner, meaning a reduction in one loss results in an increase in the other. In our experiments, a grid search was performed to determine the λ that leads to the best performance in the detection of VF progression (optimal λ). The effects of introducing encoding loss and the selection of λ on the AE-fused data are discussed later.

Figure 1 illustrates the overall architecture of the proposed AEDF model. The encoder and decoder networks were constructed using symmetrical structures of multilayer perceptron models with two hidden layers. Gaussian error linear unit (GELU) activation function [19] and layer normalization technique [20] were adopted to accelerate convergence in model training. The performance of the AE data fusion model was assessed using 10-fold cross-validation (CV). In each CV fold, the data from the same eye was not used simultaneously in both training and validation. Meanwhile, the data distribution in terms of disease severity was held constant across all training and validation sets. All data were normalized to the range between 0 and 1, based on their corresponding minimum and maximum values, and were converted back to the original scales for evaluation and visualization. The AE data fusion model was implemented using the deep learning platform PyTorch v1.13.0 [21] for Python.

2.2.2. Bayesian Linear Regression Model

To serve as a baseline comparison for the AE data fusion model, we implemented a Bayesian linear regression (BLR) model that combines the progression rates of structural and functional measurements. This model is based on the work of Russell et al. [12], who used the linear regression slope of the neuroretinal rim area as a priori information for the posterior progression rate of VF mean-sensitivity (MS). Since our study involved a different type of structural data, we utilized the mean RNFL thickness data from OCT measurements to derive the prior distribution of the progression rate. Additionally, we measured the MD slope rather than the MS slope to maintain consistency with clinical methods to detect VF progression. These modifications do not introduce significant changes to Russell et al.'s work [12], as the MD and MS slopes in our data are comparable in value (-0.21 ± 0.44 dB/year for MD slope vs. -0.26 ± 0.42 dB/year for MS slope). Moreover, the RNFL thickness data used to derive the prior was also suggested as a possible extension in their original study [12].

In general, the posterior progression rate for the BLR model is a weighted average of the likelihood (functional) and prior (structural) progression rates, with the weights determined by the variances of the functional and structural measurements so that data with lower variability receives a higher weight in determining the posterior progression rate. More details on the BLR model can be found in the Appendix A.

2.3. Performance Evaluation

For performance evaluation, we first implemented a data augmentation strategy by dividing each longitudinal VF series for each eye (both measured VF and AE-fused data) into short-term segments using variable-length sliding windows. In this way, long-term gradual VF progression was represented by multiple short-term progressions with different progression rates, thereby enhancing data diversity. Moreover, this segmentation strategy helped mitigate the influence of nonlinearity in long-term measurements, making the clinical linear model a better fit for VF progression detection. Note that the nonlinear trend is commonly associated with the physiological nature of glaucoma and interventions in management [22,23]. As a result, the evaluation metrics obtained with the segmented data are likely to be more representative of the real performance in detecting progression.

A sliding window of 4, 5, 6, 7, and 8 years was applied to both the longitudinal AE-fused data and the measured VF data for each eye to generate increasingly longer data segments. These segments of measured VF data were used to determine the ground truth label of progression via the calculation of a linear regression slope of MD. The segments of AE-fused data were assigned the same ground truth labels as their corresponding measured VF segments. The criterion for VF progression was defined as the MD deteriorating at a rate worse than 0.5 dB per year (i.e., MD linear regression slope <-0.5 dB/year), a common clinical indicator for moderate VF progression [24]. Classification of VF progression was then conducted based on segments of AE-fused and measured VF data from the 1–3 years of data in each segment (in 0.5-year intervals). The classification results were compared to the corresponding ground truth labels to derive the sensitivity, specificity, and F1 score in classifying VF progression. The F1 score is an evaluation metric that uses a single numerical measure to describe the classifier's capacity to correctly identify true positives and avoid false positives, providing a comprehensive assessment of the performance. The evaluation metrics are defined as follows:

$$\text{Sensitivity} = \frac{TP}{TP + FN}$$

$$\text{Specificity} = \frac{TN}{TN + FP}$$

$$F1 = \frac{2TP}{2TP + FP + FN}$$

where TP, TN, FP, and FN represent true positives, true negatives, false positives, and false negatives in the classification, respectively.

Furthermore, we aggregated metrics obtained with different sliding windows at each time point, forming the confidence interval for classification performance with longitudinal VF data of various durations. These aggregated metrics for AE-fused data segments were compared to those from measured VF segments and to results from the BLR model to assess the model's effectiveness in detecting VF progression. In these comparisons, the Wilcoxon signed-rank tests were used to determine statistical significance. Statistical analyses were performed with the SciPy library [25] for Python.

3. Results

3.1. Data Characteristics

A total of 2504 pairs of reliable VF and OCT tests from 253 eyes of 140 glaucoma patients were included in this study. Across all patients, the average age at the first visit was 63.7 ± 11.8 years (mean ± standard deviation), ranging from 29.7 to 88.5 years. The average follow-up length was 7.7 ± 1.7 years (range: 4.2 to 10.6 years), with a mean number of visits of 9.9 ± 3.7. For all eyes, the average mean deviation (MD) in the first VF test was -3.2 ± 5.8 dB, with a mean progression rate (linear regression slope) of -0.21 ± 0.44 dB/year. The mean RNLF thickness for the first OCT test was 78.7 ± 14.4 μm, with an average progression rate of -0.24 ± 0.97 μm/year. Detailed data characteristics of the cohort are summarized in Table 1.

Table 1. Data Characteristics.

Measurements	Mean (Standard Deviation)	Median (Interquartile Range)
Age (years)	63.7 (11.8)	65.7 (56.4 to 71.8)
Follow-up years	7.7 (1.7)	8.1 (6.8 to 8.8)
Number of visits	9.9 (3.7)	10.0 (7.0 to 13.0)
Initial mean deviation [1] (dB)	−3.2 (5.8)	−1.4 (−4.2 to 0.4)
Initial mRNFLT [2] (μm)	78.7 (14.4)	78.3 (66.9 to 89.8)
MD slope [3] (dB/year)	−0.21 (0.44)	−0.15 (−0.33 to 0.02)
mRNFLT slope [4] (μm/year)	−0.24 (0.97)	−0.24 (−0.58 to 0.10)

Note: [1] The mean deviation (MD) in the first visual field testing. [2] The mean retinal nerve fiber layer thickness (mRNFLT) in the first optical coherence tomograph test. [3] The linear regression slope of the longitudinal MD measurements in each eye. [4] The linear regression slope of the longitudinal mRNFLT measurements in each eye.

3.2. Autoencoder Data Fusion Model

We first investigated the reconstruction performance of the AE data fusion (AEDF) model, as it is a crucial factor in determining the model's ability to represent information from both VF and OCT tests. Over the 10-fold cross-validation, the AEDF model achieved an average pointwise mean absolute error (MAE) of 2.0 dB for VF reconstruction in the testing phase, with a 95% confidence interval (CI) ranging from 1.8 to 2.4 dB. For RNFL thickness data, the AEDF model had an average reconstruction MAE of 3.6 μm (95% CI: 2.8 to 4.4 μm) in testing. Additionally, the average pointwise MAE between the input VF data and AE-fused data (representing the encoding loss) was 2.5 dB (95% CI: 2.1 to 2.9 dB). This high-level reconstruction performance demonstrated the model's effectiveness in extracting and integrating representative features from both modalities into the resulting fused data. Meanwhile, the relatively low encoding loss indicated that the AE-fused data maintained good consistency with VF measurements, which assures its clinical interpretability. This will be explained next in more detail.

Figure 2 provides examples of data from the AEDF model for eyes with mild (Figure 2A), moderate (Figure 2B), and severe (Figure 2C) VF defects. Each example shows the input data, AE-fused data, and the output reconstructed data, providing visualized representations of the way that the AE data fusion model combines results from the two testing modalities. For the eye with mild VF defect (Figure 2A), the RNFL thickness measurements (the rightmost plot) exhibit notable thinning in the 225° to 315° region. As such, the mean RNFL thickness (70.3 μm) falls below the normal range of 75.0 μm to 107.2 μm suggested by the Cirrus HD-OCT device [26]. This localized RNFL thinning is reflected in the AE-fused data as a VF defect of more depression in the superior nasal region of the field (the middle VF plot). It should be noted that the region where the VF loss lies in the AE-fused data matches the area of RNFL thinning according to the structure-function map [27]. Since the VF and OCT data describe the same defect, the AE-fused data tend to have a lower MD (−3.1 dB in the middle VF plot) than the MD of measured VF data (−1.5 dB in the left VF plot).

In Figure 2B, the measured VF shows a moderate defect with MD of −7.9 dB. In this field, the defect pattern is a superposition of an actual VF loss in the superior field and lens rim artifact. Considering that the RNFL thickness in this eye is overall normal (mRNLFT = 88.9 μm), the impact of lens rim artifacts is removed in the AE-fused data (the middle VF plot), leading to milder VF loss (MD = −5.7 dB), while maintaining the arcuate defect pattern in the superior field. Note that the reconstructed VF data (the right VF plot) maintains good consistency with the measured VF data in terms of the shape and the depth of the defect (MD = −7.8 dB), showing that the information from the measured VF test has been embedded into the AE-fused data.

In the advanced glaucoma case shown in Figure 2C, the floor effect dominates the RNFL thickness measurements, plateauing at the level of around 50 μm (the rightmost plot). In this case, the resulting AE-fused data (the middle VF plot) is more dependent on data from VF testing and, correspondingly, shows greater consistency (MD = −13.6 dB) with the measured VF data (MD = −13.8 dB).

3.3. Detecting VF Progression

As discussed in the Section 2, each long-term VF and OCT data series was divided into multiple short-term segments to increase the number of cases with progressing and stable VF series. Using the segmented data, we compared the performance of detecting VF progression with AE-fused data, measured VF data (clinical data), and data from the BLR model.

For all three methods, we computed specificity, sensitivity, and F1 scores for VF progression using data collected over the first two years of the follow-up period in each segment. The selection of a two-year period was based on the minimum suggested follow-up duration for reliable estimation of VF progression rate [24]. In the initial 2 year follow-up, the specificity of detecting VF progression using AE-fused data was 0.70 (95% CI: 0.68 to 0.71), representing a 94% improvement ($p < 0.001$) over the detection specificity with the measured VF data (0.36, 95% CI: 0.35 to 0.38) and a 27% improvement ($p < 0.001$) over the detection specificity when using data from the BLR model (0.55, 95% CI: 0.54 to 0.57). In the same period (i.e., the initial 2 years), the sensitivity of detecting VF progression with the AE-fused data was 0.53 (95% CI: 0.47 to 0.58), outperforming that of the BLR model (0.35, 95% CI: 0.31 to 0.38, $p < 0.001$) and insignificantly lower than that of using measured VF data (0.54, 95% CI: 0.51 to 0.58, $p = 0.291$). When considering both specificity and sensitivity, the F1 score for VF progression detection with AE-fused data was 0.60 (95% CI: 0.57 to 0.62), surpassing the F1 scores obtained with the measured VF data (0.45, 95% CI: 0.43 to 0.47, $p < 0.001$) and data from the BLR model (0.48, 95% CI: 0.45 to 0.51, $p < 0.001$).

Figure 3 shows the performance of VF progression detection (specificity, sensitivity, and F1 scores) with the three data models at different time points over the initial three years of the follow-up period. As observed, the F1 scores with the AE-fused data consistently outperformed those with only VF data (clinical method) or data from the BLR model. Moreover, the improved performance with AE-fused data was mainly attributed to a significant increase in the detection specificity compared to the other two methods.

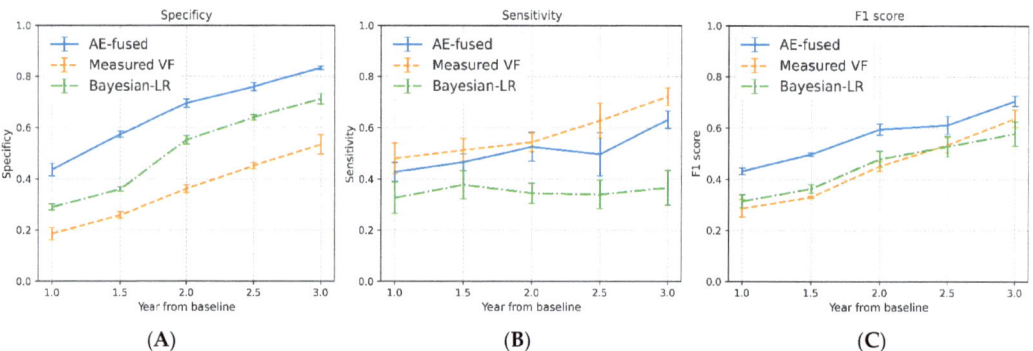

Figure 3. Specificity (panel **A**), sensitivity (panel **B**), and F1 scores (panel **C**) for the detection of visual field (VF) progression using data generated by the autoencoder data fusion model (blue), data of VF measurements (orange), and data from the Bayesian linear regression model (green) at different time points. The x-axis shows the time point, ranging from 1 to 3 years relative to the first test, in which the detection (classification) was made. Each point on the curves is the average performance for the VF time series with various lengths (ranging from 4 to 8 years), with error bars presenting the 95% confidence intervals. As expected, the overall detection performance, measured by F1 scores, for all three data models improved when the number of available data points for the detector increased, i.e., longer time along the x-axis. At different time points, the overall VF progression detection performance (F1 scores) with AE-fused data consistently outperformed the other two methods.

For the results presented above, VF progression was defined as a sequence of VFs in which the MD linear regression slope is worse than -0.5 dB/year. Considering that there is no consensus on the criteria for VF progression, clinics may adopt different thresholds for the detection of VF progression. We investigated the robustness of the detection performance when the criteria for VF progression was either relaxed (MD slope < -0.2 dB/year) or became stricter (MD slope < -1.0 dB/year).

Table 2 shows a summary of the VF progression detection performance (specificity, sensitivity, and F1 scores) for AE-fused data, measured VF data, and the BLR model's data in the initial 2 years, for the above three criteria for VF progression. Detection with AE-fused data achieved the highest F1 scores for all three thresholds, demonstrating that the performance gained by using the AE data fusion model is robust to variation in the VF progression criteria. Moreover, the performance patterns for the AE-fused data compared to the other methods were also consistent across different selections of the progression criteria, i.e., substantial improvement in specificity while keeping sensitivity approximately the same.

Table 2. Performance of visual field (VF) progression detection with different criteria.

Criteria [1]	Metrics	AE-Fused Data [2]	Measured Data	BLR Data
<-0.2 dB/year	Specificity	0.67 ± 0.01	0.34 ± 0.01	0.50 ± 0.01
	Sensitivity	0.53 ± 0.01	0.56 ± 0.01	0.51 ± 0.02
	F1 score	0.62 ± 0.01	0.50 ± 0.01	0.52 ± 0.02
<-0.5 dB/year	Specificity	0.70 ± 0.01	0.36 ± 0.01	0.55 ± 0.01
	Sensitivity	0.53 ± 0.03	0.54 ± 0.02	0.35 ± 0.02
	F1 score	0.60 ± 0.01	0.45 ± 0.01	0.48 ± 0.02
<-1.0 dB/year	Specificity	0.70 ± 0.01	0.36 ± 0.02	0.55 ± 0.01
	Sensitivity	0.41 ± 0.07	0.49 ± 0.06	0.27 ± 0.04
	F1 score	0.50 ± 0.03	0.37 ± 0.02	0.44 ± 0.03

Note: [1] The criteria for visual field (VF) progression are based on the value of mean deviation linear regression slopes. [2] The three data columns show the performance of detecting VF progression with data from the autoencoder data fusion model (AE-fused data), the measured visual field data (Measured data), and the data from the Bayesian linear regression model (BLR data) in the initial 2 years of the follow-up period, respectively. The performance metrics are presented in the form of mean $\pm$ standard error of the mean.

3.4. Selection of λ in the Loss Function

When training the AE data fusion model, the hyperparameter λ was used to balance the contributions of the reconstruction and encoding loss terms. We conducted a grid search for the optimal λ that leads to the best VF progression detection performance by varying λ from 0 to 1, in steps of 0.1. The result showed that when $\lambda = 0.6$ (optimal λ), the best overall VF progression detection performance can be achieved, with the F1 score of 0.60.

Figure 4 presents three examples that demonstrate the interaction between reconstruction and encoding losses with different λ values. When the training objective of the AE data fusion model was to only minimize reconstruction loss (Figure 4A, $\lambda = 0$), the VF defect pattern and DLS thresholds of the AE-fused data (the middle VF plot) are so different from the measured VF that the AE-fused data cannot be interpreted in a manner similar to that of the measured VF data. Note that the MD of the fused data in Figure 4A is -13.7 dB, whereas the MD of the measured VF is -7.7 dB. Consequently, standard clinical VF progression detection techniques cannot be used to analyze the AE-fused data when $\lambda = 0$, even though the AE-fused data retains sufficient information from both testing modalities to reconstruct the input data (low reconstruction errors for both VF and RNFL thickness data). When the training objective is to minimize the encoding loss without considering reconstruction loss (Figure 4C, $\lambda = 1$), the fused data becomes so akin to the input VF data that it fails to adequately represent information from the RNFL thickness measurements. As a result, the detection performance remains the same as that of using only measured VF data. When both reconstruction and encoding losses contribute to the detection performance (Figure 4B, $\lambda = 0.6$), the AE-fused data can be interpreted in the same framework as the measured VF data while incorporating sufficient information from structural OCT measurements to improve the detection of VF progression.

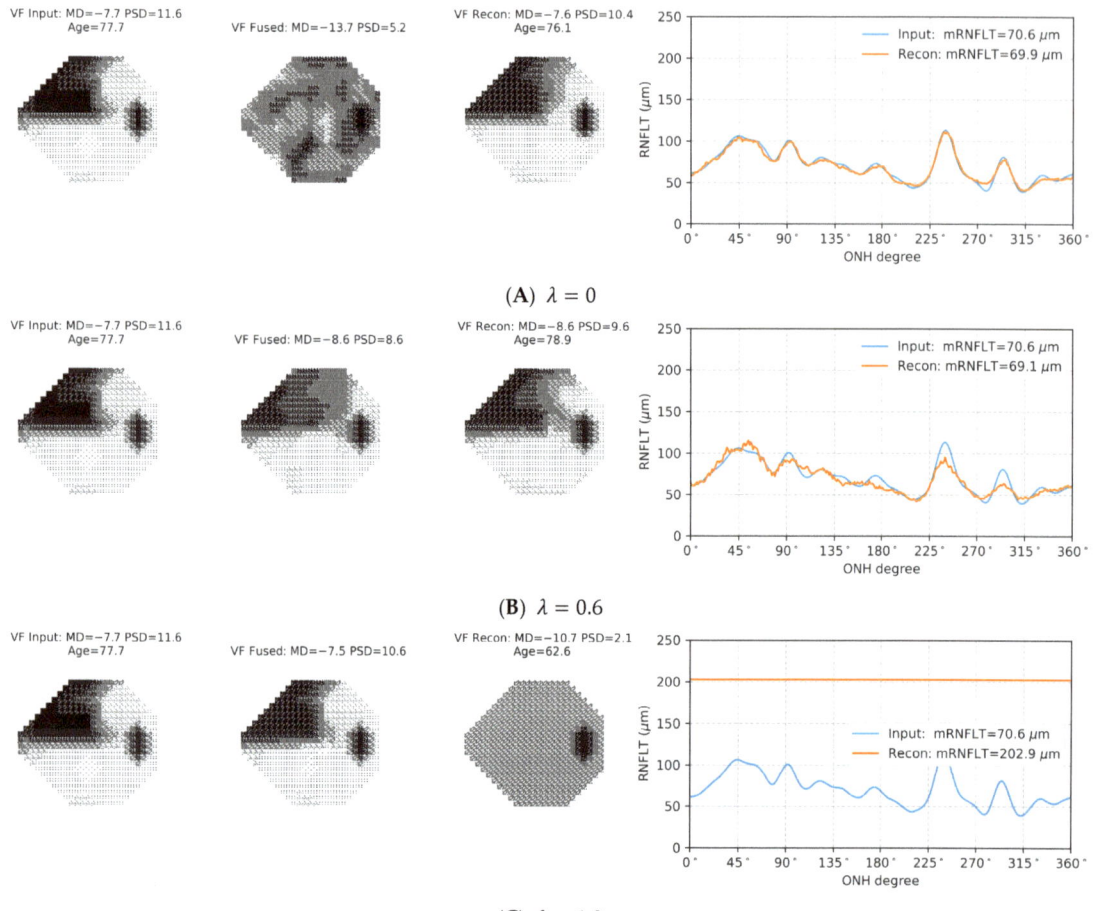

Figure 4. Examples of the autoencoder (AE) data fusion model trained with different λ selections in the loss function. When $\lambda = 0$ (panel **A**), the training objective of the AE data fusion model was to only minimize reconstruction loss. As such, the AE-fused data (the middle visual field [VF] plot) are so different from the input/measured VF (the left VF plot) that they cannot be interpreted and analyzed with clinical methods. When $\lambda = 1$ (panel **C**), the training objective was solely to minimize the encoding loss without considering reconstruction loss. The AE-fused data (the middle VF plot) closely resembles the measured VF (the left VF plot) so that it barely contains additional information from retinal nerve fiber layer thickness measurements. With $\lambda = 0.6$ (panel **B**), both reconstruction and encoding losses contribute to the training of the AE data fusion model. The AE-fused data can be interpreted with clinical knowledge in terms of the VF defect pattern and depth while incorporating sufficient information from both structural and functional tests.

3.5. Sensitivity to Input Parameters

We carried out a sensitivity analysis to examine the contribution of various input parameters to the performance of VF progression detection. When using both VF and

RNFL thickness data as the input to train the AE data fusion model, the detection specificity with the initial two years of AE-fused data significantly outperformed that with measured VF data alone (0.64 vs. 0.36, $p < 0.001$). Meanwhile, the detection sensitivity with the AE-fused data containing both VF and RNFL thickness information showed no substantial difference from that obtained using measured VF data (0.52 vs. 0.54, $p = 0.178$). Moreover, in addition to the VF and RNFL thickness data, when incorporating the patient's age information into the AE data fusion model, the detection specificity of the AE-fused data can be further improved from 0.64 to 0.70 ($p < 0.001$) while maintaining the sensitivity at the same level (0.53 vs. 0.52, $p = 0.313$).

4. Discussion

In this study, we present a method to improve the detection of VF progression by combining differential light sensitivity data from perimetry and RNFL thickness profile data from OCT using an autoencoder data fusion (AEDF) model. Unlike previous methods that rely on statistical or fixed rules for multimodality data integration [6,8,12], the AEDF model learns the function-structure interrelations in glaucoma from patients' perimetry and OCT data. This data-driven approach offers more flexibility and accuracy in describing nonlinear relationship between structural and functional measurements throughout the course of glaucoma progression. Moreover, a key contribution of our approach is the introduction of an encoding loss function that helps structure the fused data similar to the input VF data, allowing for an easy and intuitive interpretation of the model's results.

The overall VF progression detection performance (measured by the F1 score) when using the initial two years of AE-fused data was 33% better than the clinical standard method of using only VF data ($p < 0.001$). The improved detection performance was mainly attributed to a significant increase of 94% in detection specificity. When compared with the Bayesian linear regression model, VF progression detection sensitivity and specificity with AE-fused data were enhanced by 51% and 27%, respectively, leading to 25% improvement ($p < 0.001$) in the F1 score. Furthermore, the performance improvement with the AE-fused data is robust to changes in the criteria used to determine VF progression and to the length of the follow-up period.

The loss function employed in the training of the AE data fusion model comprises reconstruction and encoding loss terms, with a weight factor (λ) that controls the relative contributions of the two loss terms. As shown in Figure 4, λ can be used to change the AE-fused data, i.e., the output of the encoder, by adjusting the effect of structural OCT measurements on the VF data. With the optimal λ for VF progression detection ($\lambda = 0.6$), the appearance of AE-fused data is similar to that of the measured VF data so that the AE-fused data can be interpreted and analyzed by standard methods that are used with VF measurements. At the same time, the combination of reconstruction and encoding losses through λ assures that the AE-fused data incorporate representative features associated with structural measurements from OCT (see Figure 2), which contributes to improved VF progression detection.

The sensitivity analysis in the Section 3 showed that including the patient's age in the input of the AEDF model played a role in enhancing detection of VF progression. This observation aligns with data showing that aging is a major risk factor for glaucoma progression [28]. For that reason, it can be expected that VF progression detection could be further improved by incorporating other parameters that are associated with glaucoma into the AEDF model. Such inputs may include intraocular pressure, cup-to-disc ratio, fundus images, or macular OCT measurements, etc. This extended multimodality data integration can be easily realized by expanding the input to the AEDF model to include these parameters, with minimal adjustments to the model's structure. For instance, fundus image data can be incorporated by reshaping the image to a vector and concatenating with other data types as the input to the AEDF model. Furthermore, by modifying the target of the encoding loss function, one can adapt the AEDF model for different data interpretation and analysis purposes. For example, if the encoding loss function in this study was to compute the difference between the encoding and RNFL thickness data during model training, the resulting AE-fused data would be similar in structure to that of the RNFL thickness data while incorporating features associated with perimetric data. In this case, the AE-fused data will be analyzed by standard clinical methods for interpreting RNFL thickness data. Therefore, the unsupervised nature of the AE model and the flexibility in the design of the encoding loss function can collectively make the AEDF model a promising candidate for generalized data integration approach in glaucoma management.

It should be noted that to accommodate the relatively low-dimensional space of the input data in this study (in contrast to image-based data), we designed the encoder and decoder networks of the AEDF model with a lightweight, simple architecture, i.e., multilayer perceptron with two hidden layers. This approach enhances the robustness and generalizability of the model by avoiding the capture of noise or irrelevant features in the training data, i.e., overfitting, leading to improved performance when applied to unseen data. For a different task with more complex data inputs, a comprehensive investigation of the model architecture and the optimal set of weights for the loss function is imperative. Additionally, in this study, we focused on utilizing the encoder component of the AEDF model for compacted representations of data from VF and OCT tests. As the AE-fused data contains sufficient information from both modalities, the decoder component of the trained AEDF model can be used for simulation purposes, such as generating RNFL thickness profile data based on the corresponding VF measurements. In this case, autoencoder models that excel in generative tasks, such as variation autoencoder [29] and adversarial autoencoder [30], may warrant further investigations.

Constructing the AE fused data in a structure that is similar to VF data provides an intuitive understanding of how OCT data is combined and integrated into perimetric data. The examples in Figure 2 demonstrate that the AEDF model can dynamically combine information from VF and OCT tests in glaucoma patients that are at different stages of the disease. This capacity is particularly important in the context of glaucoma management, as measurements from functional and structural testing modalities may hold distinct clinical significance at different stages. It is typically believed that RNFL thickness measurements are more sensitive to subtle changes in the early stage of glaucoma, while VF measurements have a broader dynamic range that can better support monitoring glaucoma progression in moderate-to-advanced cases [3,4]. For the early-to-moderate glaucoma cases, the RNFL thickness data provides the complementary information to improve the robustness of VF measurements, e.g., to emphasize the depth of VF defects based on corresponding structural damage (Figure 2A) or to remove artifacts in VF measurements (Figure 2B). In advanced glaucoma cases where RNFL thickness data plateaued, the dynamic data integration ability of the AE data fusion model reduces the impact of overly stabilized RNFL thickness data on the AE-fused data. In comparison, the BLR model combines structural and functional progression rates with hard-coded rules that are based only on the uncertainty of the estimates. For eyes with moderate to severe loss (e.g., Figure 2C), the posterior VF progression rate of the BLR model is likely underestimated due to low variability in the plateaued RNFL thickness data. This might explain the lower sensitivity to detect VF progression with data from the BLR model (Figure 3).

This study has several limitations. One of the main limitations is the absence of a reliable and generalized definition of VF progression, especially in early or mild progression. In the study, we coped with this limitation by generating labels based on all longitudinal VF measurements available for each eye, while performance was assessed based on detections made using subsets of the longitudinal data. As VF data are subject to measurements noise, the progression label derived from the entire VF series may be suboptimal and, hence, may affect the performance evaluation. Furthermore, data used for model training and evaluation in this study were sourced from a single glaucoma clinic. As a result, performance evaluation was limited by the number of available longitudinal VF and OCT data, especially for the progressing cases. Testing with data collected from a single clinic with similar clinical management strategies, such as follow-up and treatments, can also introduce bias in the evaluation of the AEDF model. Future evaluations with external datasets containing a greater number of longitudinal data would be essential to comprehensively understanding the generalization of the AEDF model.

5. Conclusions

In this study, we developed an autoencoder data fusion model aimed at learning compact encoding (the AE-fused data) from functional VF data and structural OCT data. In the model training, we introduced an encoding loss to ensure that the AE-fused data can be interpreted in a manner similar to the VF data. Comparisons with the clinical standard method to detect VF progression and the Bayesian linear regression model that integrates structure-functional data showed a significant improvement in the specificity of VF progression detection when using AE-fused data. The unique capability of the AE data fusion model to generate interpretable fused data holds the potential to improve its clinical usability. The flexibility of the autoencoder model makes it as a promising candidate for a generalized data integration model to aid in glaucoma management.

Author Contributions: Conceptualization and Methodology, L.Y.L.-H., M.E., R.B.S. and W.W.; Software, L.Y.L.-H.; Validation, L.Y.L.-H., M.E., R.B.S. and W.W.; Resources and Data Curation, Y.M.B., G.E.T. and L.Y.L.-H.; Writing—Original Draft Preparation, L.Y.L.-H., M.E., R.B.S. and W.W.; Review and Editing, L.Y.L.-H., M.E., R.B.S., Y.M.B., G.E.T. and W.W.; Supervision, M.E. and W.W.; Funding Acquisition, M.E. and W.W. All authors have read and agreed to the published version of the manuscript.

Funding: This study is supported in part by a Natural Sciences and Engineering Research Council of Canada (NSERC) Discovery Grant 458039 to W.W., Vision Science Research Program (VSRP) Fellowship to L.Y.L.-H. and R.B.S., a Bell Graduate Scholarship to L.Y.L.-H., and an NSERC Canada Graduate Scholarships—Doctoral program to R.B.S.

Institutional Review Board Statement: The study is approved by the Research Ethics Board of the University Health Network, Toronto, Ontario, Canada (CAPCR ID: 18-6232.7 approved on 2 December 2022).

Informed Consent Statement: Not applicable.

Data Availability Statement: The data used in this study is not publicly available due to ethical restrictions. The code, trained models, and examples are publicly available on GitHub: https://github.com/lcapacitor/glaucoma-vf-oct-data-fusion.

Conflicts of Interest: The authors declare no conflicts of interest.

Appendix A

Bayesian linear regression model:

The Bayesian linear regression (BLR) model is based on the work of Russell et al. [12]. In this study, the mean retinal nerve fiber layer thickness (mRNFLT) data from optical coherence tomography (OCT), representing the structural changes, were utilized to derive the prior distribution of the progression rates. The mean deviation (MD) data from visual field (VF) tests, representing functional changes, were used for the likelihood of the progression rates. Then, the posterior progression rates, i.e., the combination of structural and functional changes, was derived through a weighted average of the structural and functional progression rates. Note that the progression rate was represented by the slope of the linear regression line.

Following Russell et al.'s method [12], before estimating the prior distribution of progression rates, the mRNFLT data (in μm) in our study were first converted into the same scale as MD measurements (in dB). This transformation was achieved by fitting a Passing–Bablok linear regression model [31] to MD and mRNFLT measurements. The resulting linear model based on our data can be expressed as: $MD_{RNFL} = 0.241 \times mRNFLT - 21.310$, where MD_{RNFL} represents the MD value estimated by the corresponding mRNFLT data.

For computational simplicity, we employed the conjugate prior to derive the posterior distribution of the progression rate [32]. Specifically, we assumed that MD progression rates follows Gaussian distribution, denoted as $N(\mu_1, \sigma_1^2)$, where μ_1 and σ_0^2 are the mean and variance parameters. Here, both μ_1 and σ_1^2 were derived by fitting an ordinary least square linear regression (OLSLR) model to the measured longitudinal MD data for each eye, where μ_1 is the measured slope and σ_1^2 is estimated by MSE/S_{xx} from the OLSLR model. Specifically, MSE is the mean squared error (or average squared residual) of the OLSLR model and can be derived from $MSE = \frac{1}{n-2}\sum_{t=1}^{n}(y_t - \hat{y}_t)^2$, where y_t and $\hat{y}_t$ represent the measured MD value at time t and predicted MD from the OLSLR model at the same time t, respectively, and n is the total number of MD measurements (visits) in this time series of MD. S_{xx} is the sum of squares of x from the OLSLR model and can be calculated through $S_{xx} = \sum_{t=1}^{n}(x_t - \bar{x})^2$, where x_t and $\bar{x}$ represent the patient's age at visit t and mean age over the n visits for this patient, respectively.

When further assuming that σ_1^2 is a constant, the structural progression rates (i.e., progression rates of MD_{RNFL}) follows Gaussian distribution, denoted as $N(\mu_0, \sigma_0^2)$, where μ_0 and σ_0^2 are the mean and variance parameters of the prior distribution. Likewise, μ_0 and σ_0^2 can be derived by fitting an OLSLR model to the longitudinal MD_{RNFL} data for the same eye, following the same method elaborated above. Therefore, the posterior distribution of progression rate also follows Gaussian distribution, denoted as $N(\mu, \sigma^2)$. The parameter μ of the posterior distribution represents estimated progression rate that combines changes in both structural and functional measurements. It can be analytically derived from a weighted average of the functional progression rate (μ_1) and the structural progression rate (μ_0), with the weights determined by the variances of these progression rate distributions (Equation (A1)).

$$\mu = \frac{\sigma_0^2}{\sigma_0^2 + \sigma_1^2}\mu_1 + \frac{\sigma_1^2}{\sigma_0^2 + \sigma_1^2}\mu_0 \tag{A1}$$

In other words, whether functional or structural changes, the distribution with the lower variance receives a higher weight in determining the posterior progression rate.

References

1. Stein, J.D.; Khawaja, A.P.; Weizer, J.S. Glaucoma in adults—Screening, diagnosis, and management: A review. *JAMA* **2021**, *325*, 164–174. [CrossRef] [PubMed]
2. Heijl, A.; Lindgren, A.; Lindgren, G. Test-retest variability in glaucomatous visual fields. *Am. J. Ophthalmol.* **1989**, *108*, 130–135. [CrossRef] [PubMed]
3. Hood, D.C.; Kardon, R.H. A framework for comparing structural and functional measures of glaucomatous damage. *Prog. Retin. Eye Res.* **2007**, *26*, 688–710. [CrossRef] [PubMed]
4. Malik, R.; Swanson, W.H.; Garway-Heath, D.F. Structure–function relationship in glaucoma: Past thinking and current concepts. *Clin. Exp. Ophthalmol.* **2012**, *40*, 369–380. [CrossRef] [PubMed]
5. Denniss, J.; Turpin, A.; McKendrick, A.M. Relating optical coherence tomography to visual fields in glaucoma: Structure–function mapping, limitations and future applications. *Clin. Exp. Optom.* **2019**, *102*, 291–299. [CrossRef] [PubMed]
6. Bizios, D.; Heijl, A.; Bengtsson, B. Integration and fusion of standard automated perimetry and optical coherence tomography data for improved automated glaucoma diagnostics. *BMC Ophthalmol.* **2011**, *11*, 20. [CrossRef] [PubMed]
7. Song, D.; Fu, B.; Li, F.; Xiong, J.; He, J.; Zhang, X.; Qiao, Y. Deep relation transformer for diagnosing glaucoma with optical coherence tomography and visual field function. *IEEE Trans. Med. Imaging* **2021**, *40*, 2392–2402. [CrossRef]
8. Wu, Z.; Medeiros, F.A. A simplified combined index of structure and function for detecting and staging glaucomatous damage. *Sci. Rep.* **2021**, *11*, 3172. [CrossRef]
9. Chauhan, B.C.; Malik, R.; Shuba, L.M.; Rafuse, P.E.; Nicolela, M.T.; Artes, P.H. Rates of glaucomatous visual field change in a large clinical population. *Investig. Ophthalmol. Vis. Sci.* **2014**, *55*, 4135–4143. [CrossRef]
10. Heijl, A.; Buchholz, P.; Norrgren, G.; Bengtsson, B. Rates of visual field progression in clinical glaucoma care. *Acta Ophthalmol.* **2013**, *91*, 406–412. [CrossRef] [PubMed]
11. Medeiros, F.A.; Leite, M.T.; Zangwill, L.M.; Weinreb, R.N. Combining structural and functional measurements to improve detection of glaucoma progression using Bayesian hierarchical models. *Investig. Ophthalmol. Vis. Sci.* **2011**, *52*, 5794–5803. [CrossRef]
12. Russell, R.A.; Malik, R.; Chauhan, B.C.; Crabb, D.P.; Garway-Heath, D.F. Improved estimates of visual field progression using Bayesian linear regression to integrate structural information in patients with ocular hypertension. *Investig. Ophthalmol. Vis. Sci.* **2012**, *53*, 2760–2769. [CrossRef] [PubMed]

13. Goodfellow, I.; Bengio, Y.; Courville, A. *Chapter 14 Autoencoder. Deep Learning*; MIT Press: Cambridge, MA, USA, 2016.
14. Bengio, Y.; Courville, A.; Vincent, P. Representation learning: A review and new perspectives. *IEEE Trans. Pattern Anal. Mach. Intell.* **2013**, *35*, 1798–1828. [CrossRef] [PubMed]
15. Ma, M.; Sun, C.; Chen, X. Deep coupling autoencoder for fault diagnosis with multimodal sensory data. *IEEE Trans. Ind. Inform.* **2018**, *14*, 1137–1145. [CrossRef]
16. Chaudhary, K.; Poirion, O.B.; Lu, L.; Garmire, L.X. Deep learning–based multi-omics integration robustly predicts survival in liver cancer. *Clin. Cancer Res.* **2018**, *24*, 1248–1259. [CrossRef] [PubMed]
17. Hu, D.; Zhang, H.; Wu, Z.; Wang, F.; Wang, L.; Smith, J.K.; Lin, W.; Li, G.; Shen, D. Disentangled-multimodal adversarial autoencoder: Application to infant age prediction with incomplete multimodal neuroimages. *IEEE Trans. Med. Imaging* **2020**, *39*, 4137–4149. [CrossRef]
18. Rao, H.L.; Kumar, A.U.; Babu, J.G.; Senthil, S.; Garudadri, C.S. Relationship between severity of visual field loss at presentation and rate of visual field progression in glaucoma. *Ophthalmology* **2011**, *118*, 249–253. [CrossRef]
19. Hendrycks, D.; Gimpel, K. Gaussian error linear units (gelus). *arXiv* **2016**, arXiv:1606.08415.
20. Ba, J.L.; Kiros, J.R.; Hinton, G.E. Layer normalization. *arXiv* **2016**, arXiv:1607.06450.
21. Paszke, A.; Gross, S.; Massa, F.; Lerer, A.; Bradbury, J.; Chanan, G.; Killeen, T.; Lin, Z.; Gimelshein, N.; Gimelshein, A.; et al. PyTorch: An Imperative Style, High-Performance Deep Learning Library. In *Advances in Neural Information Processing Systems 32*; Curran Associates, Inc.: Nice, France, 2019; pp. 8024–8035. Available online: http://papers.neurips.cc/paper/9015-pytorch-an-imperative-style-high-performance-deep-learning-library.pdf (accessed on 10 January 2023).
22. Medeiros, F.A.; Zangwill, L.M.; Bowd, C.; Mansouri, K.; Weinreb, R.N. The structure and function relationship in glaucoma: Implications for detection of progression and measurement of rates of change. *Investig. Ophthalmol. Vis. Sci.* **2012**, *53*, 6939–6946. [CrossRef]
23. Otarola, F.; Chen, A.; Morales, E.; Yu, F.; Afifi, A.; Caprioli, J. Course of glaucomatous visual field loss across the entire perimetric range. *JAMA Ophthalmol.* **2016**, *134*, 496–502. [CrossRef]
24. Chauhan, B.C.; Garway-Heath, D.F.; Goñi, F.J.; Rossetti, L.; Bengtsson, B.; Viswanathan, A.; Heijl, A. Practical recommendations for measuring rates of visual field change in glaucoma. *Br. J. Ophthalmol.* **2008**, *92*, 569–573. [CrossRef]
25. Virtanen, P.; Gommers, R.; Oliphant, T.E.; Haberland, M.; Reddy, T.; Cournapeau, D.; Burovski, E.; Peterson, P.; Weckesser, W.; Bright, J.; et al. SciPy 1.0: Fundamental algorithms for scientific computing in Python. *Nat. Methods* **2020**, *17*, 261–272. [CrossRef]
26. Available online: https://www.zeiss.com/content/dam/Meditec/us/brochures/cirrus_how_to_read.pdf (accessed on 1 November 2023).
27. Garway-Heath, D.F.; Poinoosawmy, D.; Fitzke, F.W.; Hitchings, R.A. Mapping the visual field to the optic disc in normal tension glaucoma eyes. *Ophthalmology* **2000**, *107*, 1809–1815. [CrossRef] [PubMed]
28. Coleman, A.L.; Miglior, S. Risk factors for glaucoma onset and progression. *Surv. Ophthalmol.* **2008**, *53*, S3–S10. [CrossRef] [PubMed]
29. Kingma, D.P.; Welling, M. Auto-encoding variational bayes. *arXiv* **2013**, arXiv:1312.6114.
30. Makhzani, A.; Shlens, J.; Jaitly, N.; Goodfellow, I.; Frey, B. Adversarial autoencoders. *arXiv* **2015**, arXiv:1511.05644.
31. Passing, H.; Bablok, W. A new biometrical procedure for testing the equality of measurements from two different analytical methods. Application of linear regression procedures for method comparison studies in clinical chemistry, Part I. *Clin. Chem. Lab. Med.* **1983**, *21*, 709–720. [CrossRef] [PubMed]
32. Murphy, K.P. Conjugate Bayesian analysis of the Gaussian distribution. *def* **2007**, *1*, 16.

Disclaimer/Publisher's Note: The statements, opinions and data contained in all publications are solely those of the individual author(s) and contributor(s) and not of MDPI and/or the editor(s). MDPI and/or the editor(s) disclaim responsibility for any injury to people or property resulting from any ideas, methods, instructions or products referred to in the content.

Article

Proactive Decision Support for Glaucoma Treatment: Predicting Surgical Interventions with Clinically Available Data

Mark Christopher [1], Ruben Gonzalez [1], Justin Huynh [1], Evan Walker [1], Bharanidharan Radha Saseendrakumar [1], Christopher Bowd [1], Akram Belghith [1], Michael H. Goldbaum [1], Massimo A. Fazio [2], Christopher A. Girkin [2], Carlos Gustavo De Moraes [3], Jeffrey M. Liebmann [3], Robert N. Weinreb [1], Sally L. Baxter [1] and Linda M. Zangwill [1],*

1. Hamilton Glaucoma Center and Division of Ophthalmology Informatics and Data Science, Shiley Eye Institute, Viterbi Family Department of Ophthalmology, University of California, San Diego, CA 92037, USA; mac157@health.ucsd.edu (M.C.)
2. Department of Ophthalmology and Vision Sciences, Heersink School of Medicine, University of Alabama at Birmingham, Birmingham, AL 35233, USA
3. Bernard and Shirlee Brown Glaucoma Research Laboratory, Department of Ophthalmology, Edward S. Harkness Eye Institute, Columbia University Medical Center, New York, NY 10032, USA
* Correspondence: lzangwill@health.ucsd.edu

Citation: Christopher, M.; Gonzalez, R.; Huynh, J.; Walker, E.; Radha Saseendrakumar, B.; Bowd, C.; Belghith, A.; Goldbaum, M.H.; Fazio, M.A.; Girkin, C.A.; et al. Proactive Decision Support for Glaucoma Treatment: Predicting Surgical Interventions with Clinically Available Data. *Bioengineering* 2024, 11, 140. https://doi.org/10.3390/bioengineering11020140

Academic Editors: Karanjit S. Kooner, Osamah J. Saeedi and Yunfeng Wu

Received: 29 November 2023
Revised: 6 January 2024
Accepted: 27 January 2024
Published: 30 January 2024

Copyright: © 2024 by the authors. Licensee MDPI, Basel, Switzerland. This article is an open access article distributed under the terms and conditions of the Creative Commons Attribution (CC BY) license (https://creativecommons.org/licenses/by/4.0/).

Abstract: A longitudinal ophthalmic dataset was used to investigate multi-modal machine learning (ML) models incorporating patient demographics and history, clinical measurements, optical coherence tomography (OCT), and visual field (VF) testing in predicting glaucoma surgical interventions. The cohort included 369 patients who underwent glaucoma surgery and 592 patients who did not undergo surgery. The data types used for prediction included patient demographics, history of systemic conditions, medication history, ophthalmic measurements, 24-2 VF results, and thickness measurements from OCT imaging. The ML models were trained to predict surgical interventions and evaluated on independent data collected at a separate study site. The models were evaluated based on their ability to predict surgeries at varying lengths of time prior to surgical intervention. The highest performing predictions achieved an AUC of 0.93, 0.92, and 0.93 in predicting surgical intervention at 1 year, 2 years, and 3 years, respectively. The models were also able to achieve high sensitivity (0.89, 0.77, 0.86 at 1, 2, and 3 years, respectively) and specificity (0.85, 0.90, and 0.91 at 1, 2, and 3 years, respectively) at an 0.80 level of precision. The multi-modal models trained on a combination of data types predicted surgical interventions with high accuracy up to three years prior to surgery and could provide an important tool to predict the need for glaucoma intervention.

Keywords: glaucoma; glaucoma progression; glaucoma surgery; OCT; visual field; machine learning; deep learning

1. Introduction

Glaucoma is characterized by a progressive loss of vision and is the leading cause of irreversible blindness in the world [1]. Lowering intraocular pressure (IOP) is the only currently known effective way to slow disease progression, and available treatments focus on lowering IOP through medications, laser therapies, and/or surgical intervention [2–4]. However, it is difficult or impossible to predict the rate of glaucoma progression or identify which patients will require surgical intervention to prevent blindness [5]. The identification of patients with a high risk of progression will help to reduce the risk of vision loss and preserve patients' quality of life.

Advances in artificial intelligence (AI) and machine learning (ML) over the past several years have provided tools that may help to address this need. AI-based approaches have had a large impact on many different prediction tasks across nearly all fields of medicine as well as numerous applications in ophthalmology and glaucoma [6–9]. Glaucoma care is data- and imaging-intensive. The current standard of care includes the collection of ophthalmic and systemic measurements, fundus and optical coherence tomography (OCT) imaging to assess the retinal structure, and visual field (VF) testing to evaluate visual function [10]. These sources of data have been exploited to build tools for glaucoma-related prediction tasks. Our group and others have employed AI techniques to detect disease, segment OCT images, and predict functions from structures, among other tasks [11–16]. These approaches, however, have largely relied on models that use a single data type. Multi-modal approaches that integrate the different types of data collected as part of routine clinical glaucoma management may improve the ability to identify glaucoma predictions. Moreover, they may address unmet needs, such as forecasting the need for surgical intervention in glaucoma.

There is increasing interest in multi-modal models that incorporate data types from multiple different sources into a single predictive model [17]. In the case of glaucoma predictions, potentially informative data sources can include not only ophthalmic measurements, OCT imaging, and VF testing but also patient demographics, medical history, and data regarding systemic conditions and medications [18,19]. The current study investigates the use of multi-modal models that incorporate each of these data types to predict the need for surgical intervention in varying time windows up to 3 years pre-intervention.

2. Materials and Methods

2.1. Datasets

The primary dataset was collected from a cohort of glaucoma participants recruited as part of two longitudinal glaucoma studies: the Diagnostic Innovations in Glaucoma Study (DIGS, clinicaltrials.gov identifier: NCT00221897) and the African Descent and Glaucoma Evaluation Study (ADAGES, clinicaltrials.gov identifier: NCT00221923) [20]. ADAGES is an ongoing, multicenter collaboration between four primary academic medical centers with high-volume glaucoma clinics and one high-volume private practice (delineated in this study as sites one through five for the purpose of preserving patient privacy). Details of the study design for these studies have been described previously [20]. In short, all participants provided written informed consent, and the institutional review boards at all sites approved the study methods. All methods adhere to the tenets of the Declaration of Helsinki and to the Health Insurance Portability and Accountability Act. Inclusion in the DIGS/ADAGES glaucoma cohorts required participants to meet the following criteria at study entry: 20/40 or better best corrected Snellen visual acuity and at least two consecutive reliable standard automated perimetry VF tests. For this analysis, glaucoma was defined as eyes having repeated abnormal VF results.

All clinical, demographic, VF testing, and OCT imaging data were stored in UCSD's HIPAA-compliant Amazon Web Services (AWS)-based data management and analysis system, iDARE (intelligent Design for AI-Readiness in Eye Research).

2.2. Patient Exams and Interviews

An ophthalmological exam was completed at each study visit, which included measurements of IOP, central corneal thickness (CCT), spherical equivalent (SE), and axial length (AL). Patient interviews were also conducted to collect self-reported demographic information (age, sex, race, and ethnicity), systemic medical conditions, history of non-ocular medications, and history of ocular medications and interventions. Medications were mapped to the anatomical therapeutic chemical (ATC) drug classification defined by the World Health Organization using RxNav [21,22]. Complete details on the ADAGES examination and interview procedures have been described previously [20]. Both medical record data and self-reported history of interventions were used as the basis for the surgery ground truth used in the model's training and testing. Self-reported data were used because clinical records were not available for some of the patients in the DIGS/ADAGES cohorts considered here.

2.3. OCT Imaging

OCT imaging consisted of Spectralis (Heidelberg Engineering GmbH, Heidelberg, Germany) circle scans containing single B-scans comprised of 1536 A-scans, each captured in a circular region surrounding the optic nerve head (ONH). The retinal nerve fiber layer (RNFL) was segmented using built-in software provided by the device manufacturer. The segmentation resulted in global and sectoral mean RNFL thickness measurements. The mean RNFL included global, temporal, temporal-superior, temporal-inferior, nasal, nasal-superior, and nasal-inferior. The OCT images and segmentations were evaluated for quality by the UC San Diego Imaging Data Evaluation and Analysis (IDEA) Reading Center according to standard protocols [20]. The OCT images with poor signal quality, imaging artifacts, and/or segmentation errors that could not be manually corrected were excluded from further analysis.

2.4. Visual Field Testing

VF testing consisted of 24-2 testing using the Swedish interactive thresholding algorithm (SITA standard, Humphrey Field Analyzer II; Carl Zeiss Meditec, Inc., Jena, Germany). VF results with more than 33% fixation losses, 33% false-negative errors, or 33% false-positive errors were excluded. The VF results were processed and evaluated to assess their quality according to standard protocols by the University of California San Diego Visual Field Assessment Center [20]. For inclusion in the models, the mean deviation (MD), pattern standard deviation (PSD), visual field index (VFI), and individual test point pattern deviation (PD) measurements were extracted.

2.5. Multi-Modal Models

The model input was constructed by combining the data sources described above into a final set of inputs. This set consisted of patient demographics (age, sex, and race), systemic medical conditions (e.g., hypertension, diabetes, and cancer), history of non-ocular and ocular medications, ophthalmic measurements (IOP, CCT, SE, and AL) along with 24-2 VF results (MD, PSD, VFI, and individual test point PD), and RNFL measurements (mean global and sectoral thickness). A complete list of model inputs and outputs is provided in the supplementary materials (Table S1 in Supplementary Materials). For this analysis, the model input consisted of quantitative (e.g., RNFL thickness, VF MD, age, IOP, etc.), categorical (e.g., self-reported race), and binary (e.g., gender and presence of systemic conditions) variables derived from the multiple modalities described above. Eyes missing VF, OCT, and IOP data were excluded from the final datasets in our study. The primary outcome of interest was whether the patients progressed to requiring glaucoma surgical intervention. The ground truth labels for the outcome (glaucoma surgery or no surgery) were identified based on the self-reported data collected during the patients' exams. For this analysis, any patients with incisional, laser, or minimally invasive glaucoma surgery (MIGS) procedures were included in the surgery group, as these procedures would still typically represent advancing disease. This is also consistent with the methodology employed in prior studies examining glaucoma surgery as a proxy for disease progression [13,14,18]. Figure 1 provides surgery and non-surgery examples.

The longitudinal aspect of our dataset was used to simulate predicting surgeries at various lengths of time in the future. To this end, we defined a time horizon for our models. The time horizon represents the period of time prior to the surgery date during which no data is used as input to the model for training or testing. This is important for establishing applicability to future deployments of these models in real-world clinical environments. For example, as a clinician, one may want to understand a specific patient's risk of progressing to needing surgery within the next year. With this framework in mind, when training models with existing data, it would not be sensible to include all data leading up to surgery. Therefore, a model with a time horizon of 12 months would be trained and tested only on input data that was collected at least 12 months prior to the surgery date.

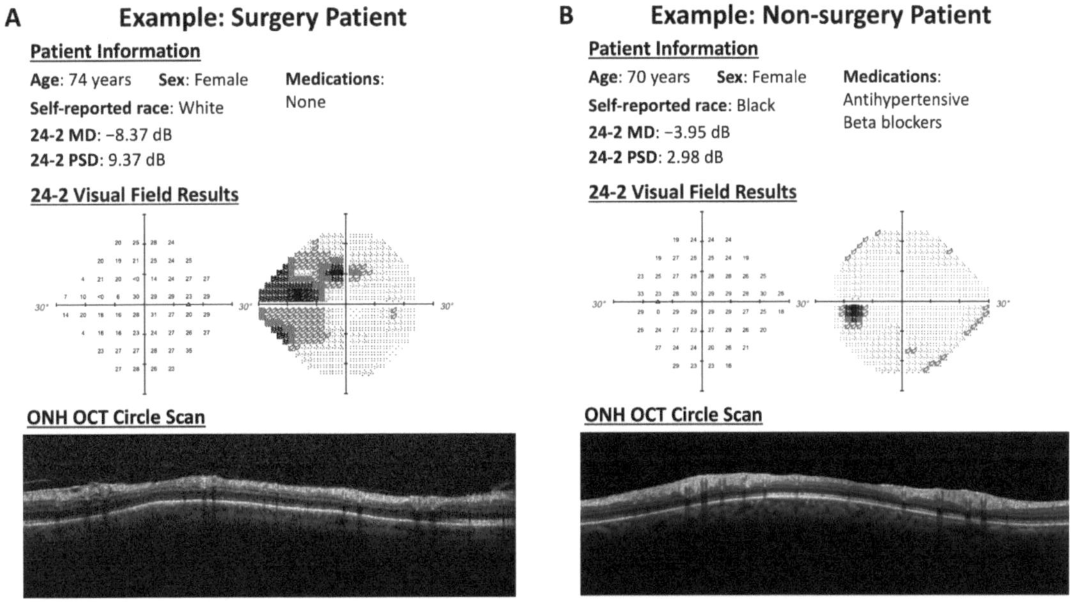

A   Example: Surgery Patient

Patient Information
Age: 74 years   **Sex**: Female   **Medications**:
Self-reported race: White   None
24-2 MD: −8.37 dB
24-2 PSD: 9.37 dB

24-2 Visual Field Results

ONH OCT Circle Scan

B   Example: Non-surgery Patient

Patient Information
Age: 70 years   **Sex**: Female   **Medications**:
Self-reported race: Black   Antihypertensive
24-2 MD: −3.95 dB   Beta blockers
24-2 PSD: 2.98 dB

24-2 Visual Field Results

ONH OCT Circle Scan

Figure 1. Example of a patient who progressed to needing glaucoma surgery and (**A**) another patient who did not end up needing surgery. (**B**) Patient demographics, medication history, visual field results, and optic nerve head circle scans.

To quantify how the accuracy of our models changed as we predicted surgeries at various lengths of time in the future, we trained and evaluated models on several time horizons, including 3 months, 1 year, 2 years, and 3 years. For each patient who did not progress to surgery, one study visit was randomly selected as a "pseudo surgery" date (to establish an analogous observation period) and was used to identify the data eligible for each time horizon. The actual input data to the models consisted of a single measurement for each feature selected from the most recently collected eligible data (based on the time horizon).

Several different model types were evaluated, including logistic regression, random forests, gradient-boosting machines (GBMs), gradient-boosted decision trees (XGBoost), and custom deep neural networks (DNNs) [23–26]. In all cases, the model parameters were selected via empirical testing on the training data.

2.6. Model Evaluation

The data were partitioned into training and testing datasets by selecting data collected at most sites to serve as a training/validation dataset (sites one to four) and one study site (site five) to serve as an external test dataset. This approach ensured that the testing data was collected from an independent, geographically separated population. The training dataset was further split (90/10, according to participant) into training and validation subsets. Models of different types with a range of parameters were trained on the training subset and evaluated on the validation subset. The model type with the best overall performance on the validation subset was selected for evaluation on the testing dataset. The models were evaluated using the area under the receiver operating characteristic (AUC) and precision-recall curves, as well as precision, sensitivity, and specificity. Using these metrics, the models were evaluated on the entire testing dataset and on the subsets stratified by age, self-reported race, and disease severity to estimate the impact of these factors on the model's performance. When evaluating the models, we aimed to optimize the precision (also known as the positive predictive value), which is the proportion of patients who actually progressed to glaucoma surgery among those who were predicted to progress. This was in line with the primary goal of the future clinical deployment of this model. Subsequently, decision thresholds were tuned so that each model achieved a precision score of at least 0.80 while also maximizing the sensitivity. We then also evaluated how the model performance metrics varied at various thresholds for precision (from 0.75 to 0.95 in 0.05 increments).

To enhance model explainability, the decision-making process employed by the best-performing model was explored using Shapley additive explanations (SHAPs) [27]. SHAP is a game theory-based approach to measure the impact of input features on model output. Using this approach, we quantified the contribution of each input to predictions regarding the need for glaucoma surgery.

3. Results

A summary of the surgery and non-surgery cohorts used to train and test the models to predict glaucoma surgical intervention is provided in Table 1. In both the training and testing data, the surgery group was, on average older (68.4 years vs. 63.2 years in the training data, $p < 0.001$), had worse 24-2 MD (-7.17 dB vs. -1.40 dB in the training data, $p < 0.001$), and had a larger proportion of Black/African American patients (51.8% vs. 41.3% in the training data, $p < 0.001$).

Table 1. A summary of the testing and training datasets. The training data consisted of data collected at sites one through four, while site five data was held out as an independent testing dataset.

	Training		Testing	
	Glaucoma Surgery	No Surgery	Glaucoma Surgery	No Surgery
Participants/eyes (n, %)	419/610 (45.8%)	496/830 (54.2%)	137/178 (51.9%)	127/221 (48.1%)
Age (years, 95% CI)	68.4 (67.5–69.4)	63.2 (62.3–64.2)	66.3 (64.5–68.0)	59.8 (58.0–61.6)
24-2 MD (dB, 95% CI)	-7.17 (-7.70–-6.63)	-1.40 (-1.85–-0.94)	-8.39 (-9.74–-7.05)	-1.19 (-2.18–-0.20)
Mean RNFL thickness (μm, 95% CI)	75.2 (72.2–78.1)	86.6 (84.8–88.3)	73.3 (68.1–78.5)	84.4 (79.6–89.2)
Sex (n, % female)	217 (51.8%)	306 (61.7%)	91 (66.4%)	71 (55.9%)
Self-reported race (n, %)				
Black/African American	217 (51.8%)	205 (41.3%)	90 (65.7%)	67 (52.8%)
White	194 (46.3%)	253 (51.0%)	47 (34.3%)	60 (47.2%)
Other/not reported	8 (1.9%)	38 (7.7%)	0 (0.0%)	0 (0.0%)
Surgery type				
Incisional	170 (27.9%)	-	67 (37.6%)	-
MIGS	1 (0.2%)	-	6 (3.4%)	-
Laser	439 (72.0%)	-	105 (59.0%)	-
None	0 (0.0%)	830 (100.0%)	0 (0.0%)	221 (100.0%)

The best-performing surgical intervention prediction model was a GBM. It achieved an AUC (95% CI) of 0.94 (0.91–0.96) at the 3-month time horizon, incorporating all available data preceding surgery, 0.93 (0.90–0.95) at 1 year prior to surgery, 0.92 (0.89–0.95) at 2 years prior to surgery, and 0.93 (0.89–0.97) at 3 years prior to surgery. The decision thresholds were tuned so that each model achieved a precision score of at least 0.80 while also maximizing the sensitivity. For models with the 0.80 precision threshold, the specificity increased with longer time horizons (0.81, 0.85, 0.90, and 0.91 at 3 months, 1 year, 2 years, and 3 years, respectively), while sensitivity/recall was at a maximum at 3 months (0.93, 0.89, 0.77, and 0.86 at 3 months, 1 year, 2 years, and 3 years, respectively). The full results for this model are shown in Table 2 and illustrated as ROCs and precision-recall curves in Figure 2. The results for all the models are summarized in Table S2 in Supplementary Materials.

The best-performing model (GBM) was also evaluated as a function of disease severity, age, and self-reported race using stratified analyses (Table 3). Comparing patients with 24-2 MD > -6.0 dB to those ≤ -6.0 dB, the model achieved AUCs of 0.89 vs. 0.88 at 3 months, 0.90 vs. 0.78 at 1 year, 0.87 vs. 0.86 at 2 years, and 0.88 vs. 0.96 at 3 years. With respect to age, the model had similar AUCs for patients below and above 60 years old at all time horizons (AUCs of 0.94 vs. 0.93 at 3 months, 0.93 vs. 0.91 at 1 year, 0.92 vs. 0.92 at 2 years, and 0.94 vs. 0.92 at 3 years). With respect to self-reported race, only the Black/African American and White groups had sufficient patients to perform the analyses. Similar model performance was found in Black/African American vs. White patients; the AUCs were 0.93 vs. 0.94 at 3 months, 0.91 vs. 0.94 at 1 year, 0.92 vs. 0.94 at 2 years, and 0.94 vs. 0.93 at 3 years. Of note, no statistically significant differences at the 0.05 level were observed in the model performance according to disease severity, age, or race.

The SHAP analysis of the model predictions revealed the features with the greatest impact on the predictions at each time horizon (Figure 3). Some particularly impactful features were common among all the time horizons (age, gender, self-reported race, AL, CCT, IOP, MD, VFI, and PSD). At shorter time horizons, the RNFL measurements were more important. At longer time horizons, the list of the most impactful features was dominated by the VF measurements. The self-reported patient conditions and medications did not appear among the most impactful features.

Table 2. Performance of the best-performing model (GBM) in predicting glaucoma surgical interventions at time horizons up to 3 years. The models used patient demographics, ophthalmic measurements, VF results, OCT measurements, and self-reported history of systemic conditions and medications to predict surgical interventions. Multiple tuning thresholds were used for evaluating the model performance metrics at varying levels of precision.

Precision at Time Horizons	AUC	Precision	Recall/Sensitivity	Specificity
3 months				
0.75 precision		0.75 (0.71–0.79)	0.94 (0.91–0.98)	0.75 (0.69–0.80)
0.80 precision		0.80 (0.76–0.85)	0.93 (0.89–0.96)	0.81 (0.76–0.86)
0.85 precision	0.94 (0.91–0.96)	0.85 (0.80–0.90)	0.83 (0.78–0.88)	0.88 (0.84–0.92)
0.90 precision		0.90 (0.85–0.95)	0.68 (0.61–0.75)	0.94 (0.91–0.97)
0.95 precision		0.95 (0.91–0.99)	0.56 (0.49–0.63)	0.98 (0.96–1.00)
1 year				
0.75 precision		0.75 (0.71–0.80)	0.95 (0.91–0.98)	0.79 (0.73–0.84)
0.80 precision		0.80 (0.75–0.85)	0.89 (0.84–0.93)	0.85 (0.80–0.90)
0.85 precision	0.93 (0.90–0.95)	0.85 (0.80–0.91)	0.66 (0.59–0.74)	0.92 (0.89–0.95)
0.90 precision		0.91 (0.84–0.97)	0.45 (0.37–0.53)	0.97 (0.94–0.99)
0.95 precision		0.96 (0.89–1.00)	0.30 (0.22–0.38)	0.99 (0.98–1.00)
2 years				
0.75 precision		0.75 (0.70–0.82)	0.86 (0.79–0.92)	0.85 (0.80–0.90)
0.80 precision		0.80 (0.74–0.86)	0.77 (0.69–0.85)	0.90 (0.86–0.94)
0.85 precision	0.92 (0.89–0.95)	0.85 (0.78–0.91)	0.70 (0.61–0.78)	0.94 (0.90–0.97)
0.90 precision		0.90 (0.84–0.97)	0.58 (0.48–0.67)	0.97 (0.94–0.99)
0.95 precision		0.96 (0.90–1.00)	0.39 (0.30–0.47)	0.99 (0.98–1.00)
3 years				
0.75 precision		0.75 (0.69–0.82)	0.90 (0.83–0.95)	0.87 (0.82–0.92)
0.80 precision		0.80 (0.74–0.88)	0.86 (0.79–0.93)	0.91 (0.87–0.94)
0.85 precision	0.93 (0.89–0.97)	0.85 (0.78–0.92)	0.75 (0.66–0.83)	0.94 (0.91–0.97)
0.90 precision		0.90 (0.84–0.97)	0.67 (0.58–0.77)	0.97 (0.94–0.99)
0.95 precision		0.97 (0.90–1.00)	0.33 (0.23–0.43)	0.99 (0.99–1.00)

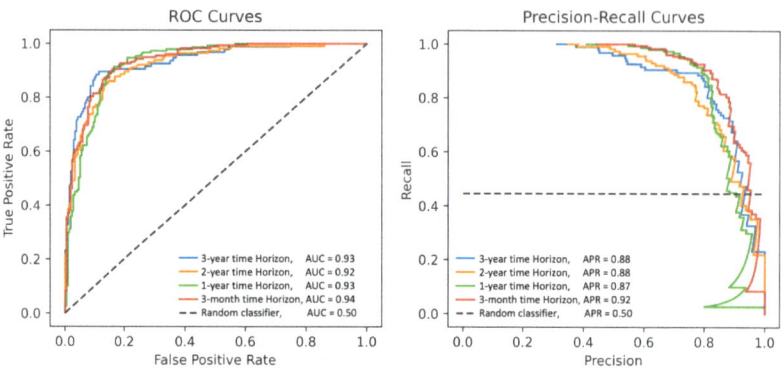

Figure 2. Receiver operating characteristics and precision-recall curves of the best-performing model (GBM) at each time horizon.

Table 3. AUC of the best-performing model (GBM) stratified by age, self-reported race, and glaucoma severity at each time horizon.

	3 Months	1 Year	2 Years	3 Years
Glaucoma severity				
MD > −6.0 dB (n = 153)	0.89 (0.84–0.93)	0.90 (0.86–0.94)	0.87 (0.79–0.92)	0.88 (0.79–0.94)
MD ≤ −6.0 dB (n = 50)	0.88 (0.72–0.97)	0.78 (0.58–0.92)	0.86 (0.70–0.96)	0.96 (0.84–1.00)
Age				
>60 years (n = 113)	0.93 (0.88–0.96)	0.91 (0.86–0.95)	0.92 (0.86–0.96)	0.92 (0.86–0.96)
≤60 years (n = 107)	0.94 (0.89–0.97)	0.93 (0.89–0.96)	0.92 (0.85–0.96)	0.94 (0.87–0.99)
Self-reported race				
Black/African American (n = 134)	0.93 (0.89–0.96)	0.91 (0.87–0.95)	0.92 (0.86–0.96)	0.94 (0.88–0.97)
White (n = 86)	0.94 (0.90–0.97)	0.94 (0.90–0.97)	0.94 (0.88–0.97)	0.93 (0.85–0.98)

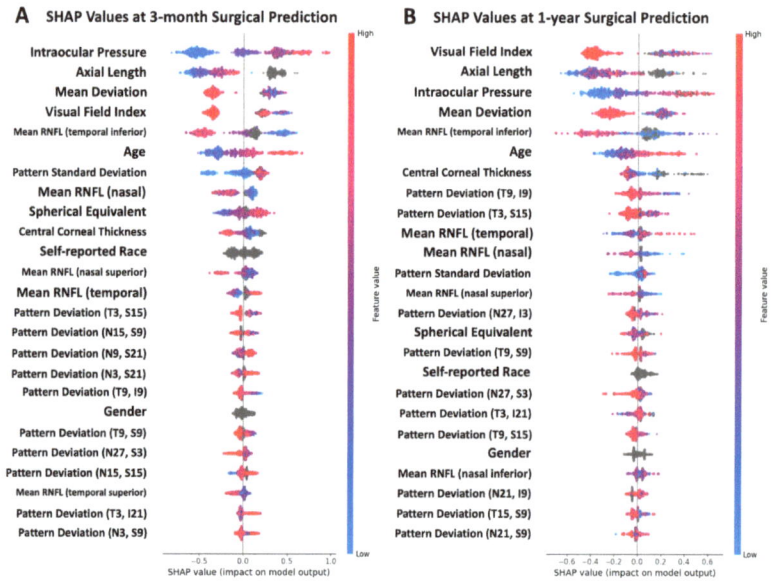

Figure 3. *Cont.*

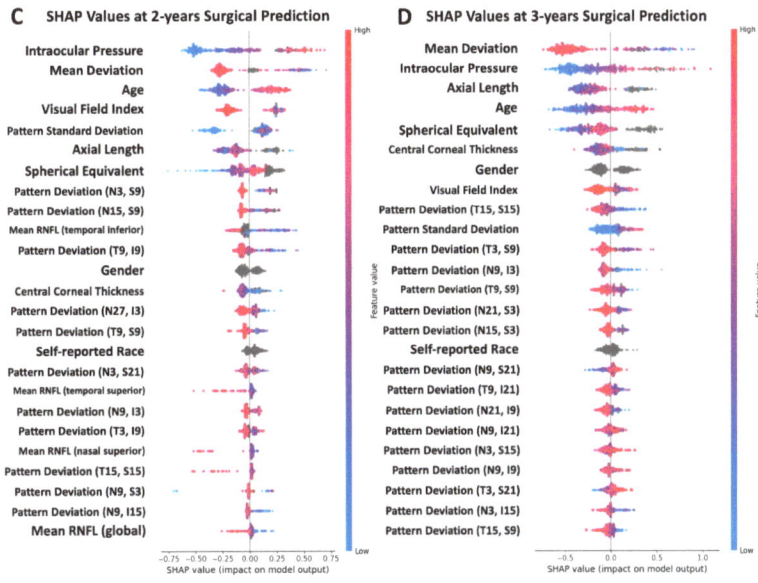

Figure 3. Shapley analysis of the features associated with the greatest impact on predictions for the best-performing model trained with all data leading to surgery. (**A**) The best-performing model trained with data up to 1 year prior to surgery, (**B**) up to 2 years prior to surgery, and (**C**) up to 3 years prior to surgery. (**D**) In these plots, each point corresponds to a single patient, and the color (blue to pink) indicates the normalized value of the indicated feature for that patient. The x-axis represents the SHAP value (the contribution of the indicated feature to the model prediction for each patient). The positive values pushed the model toward a surgery prediction, while the negative values pushed the model toward a non-surgery prediction. The features with all points clustered near zero had relatively little impact on the model decisions, while the features with higher SHAP values had a larger impact.

4. Discussion

4.1. Model Performance

In this study, we developed ML models that achieved high accuracy in predicting surgical intervention in glaucoma up to 3 years prior to the intervention. Even at 3 years prior to surgery (meaning no data were used within 3 years preceding the surgical intervention for training the model), the model achieved a high AUC (0.93) and sensitivity (0.86) as well as high specificity (0.91) and precision (0.80). Accurate methods to identify patients who are at high risk of progression and need glaucoma interventions, like those presented here, are a critical need in glaucoma. For patients, the early identification of risk may help to inform downstream interventions that can preserve vision and quality of life. For clinicians, forecasting patients who are at a high risk of glaucoma progression can help to inform which patients need closer monitoring and enable more efficient allocation of limited clinical time and resources.

Models were evaluated at several time horizons (i.e., predictions at different time periods prior to an intervention) spanning from 3 months up to 3 years prior to the surgery date. This is clinically relevant, as we plan to implement these models for point-of-care decision support in a prospective fashion. Clinicians would be interested in predicting a specific patient's risk of requiring glaucoma surgery in the future and would be limited to whatever data may be currently available to them. Thus, while developing these models with retrospective data, we decided to censor data preceding surgery of varying lengths to simulate the prospective clinical implementation scenario. Developing models to predict the risk of progression to surgery up to 3 years in advance represents an improvement over prior studies that have examined shorter time windows for prediction, such as 6 months [13]. Compared to 3 months, longer time horizons performed better than expected. While the 3-month predictions did achieve the highest sensitivity, it was comparable to or worse than the other time horizons with respect to the AUC, precision, and specificity. Part of the similarity in performance across time horizons may be due to the availability of data during the corresponding time periods. Different time horizon models were restricted in what data they had access to, but the restrictions only limited model access to data nearer to surgery. Shorter horizon models did have access to older data and could use it if newer measurements were not available. For instance, the 1-year model could not use any time collected within the year prior to surgery but could use older data if needed. For a specified patient, if all the data for a particular measurement had been collected more than 3 years prior to surgery, all the time horizon models would have access to that data. In addition, some variables are stable over time and would be the same for all time horizons (e.g., some patient demographics). This means that, in some cases, models at different time horizons were relying on similar input to identify their predictions, possibly leading to greater-than-expected similarity.

In estimating the performance of the models, we relied on the multi-center data collected in the DIGS/ADAGES dataset. Patient recruitment and data collection were performed at five geographically separated locations across the U.S. from demographically diverse populations. This provided the opportunity to withhold data from one study site to use as an independent test set to better estimate the generalizability of the models. The lack of external validation or test datasets is a commonly known challenge in the reproducibility of AI/ML models, so the evaluation of model performance in a completely independent set of patients is an important advancement over some prior clinical prediction models that largely relied on internal validation alone [28–30]. The diversity of our datasets also allowed for estimates of model performance across patient subgroups using stratified analyses based on disease severity, age, and self-reported race. There were some differences in model performance based on severity. The model performed better in patients with MD > -6.0 dB at 1 year and better in patients with MD ≤ -6.0 dB at 3 years, although these differences were not statistically significant. With respect to age and self-reported race, the models performed comparably at all the time horizons in both the over and under 60 years patient groups and the Black/African American and White patient groups. This is an important finding, as several recent studies have found other clinical AI models demonstrating inferior performance in minority populations, potentially creating a source of algorithmic bias with the potential to exacerbate existing health disparities [31–33]. Given that Blacks carry a disproportionate burden of glaucoma and face existing disparities in care, the high performance of our models in this population is important in the context of health equity [34,35]. Regardless of the stratified group, the model AUC remained high (≥ 0.88, with the most substantially higher), suggesting that the models could maintain performance across a wide variety of patient subgroups. The validation of these models in additional datasets, especially in real-world clinical data from diverse patient populations, is a critical next step toward deploying these models for clinical use.

4.2. Feature Importance

The Shapley analysis that was used to interrogate model decision-making may help explain the model performance at different time horizons. The incorporation of this analysis is also helpful for enhancing the explainability of the models, which is important considering that ML models have frequently been criticized for their "black box" nature [36]. Across all time horizons, several features had a moderate-to-large impact on the model predictions, including age, gender, self-reported race, axial length, CCT, IOP, MD, VFI, and PSD. Model reliance on the features that were relatively stable in impact across the time horizons may help to explain the similarity in performance at different time horizons. Other features, however, had large jumps in their impact at later time horizons. In particular, the impact of

OCT-based RNFL thickness measurements had a larger impact during shorter time horizons. At longer time horizons, the relative importance of RNFL measurements decreased compared to the other features. Given the ubiquity of OCT in glaucoma management, this decrease in informativity at longer time horizons warrants further study. Overall, the Shapley analysis provided information about relative feature importance that coincided with known clinical features of glaucoma, instilling confidence in the models.

4.3. Limitations and Future Directions

There are limitations to the current study. First, there are differences in the rates of surgical interventions across clinicians and departments that could impact model performance. Across the three sites with the largest number of patients, the rates of glaucoma surgery cases were 24.7% (site one), 53.6% (site two), and 40.0% (site five). These are comparable to prior studies from both academic medical centers and nationwide cohorts such as the All of Us Research Program and the IRIS registry, which also demonstrate a high level of variability in the rates of glaucoma surgery [13,14,37,38]. The range of rates could be the results of differences in the patient populations, differences in decision-making regarding glaucoma interventions and local practice variations, patients refusing or postponing surgery, or a combination of these factors. Incorporating training and testing data from additional sites could help in training and evaluating models that are more robust to these differences. This additional data could also be useful in evaluating model performance across types of surgical interventions (laser, incisional, and MIGS). The current dataset has a limited number of subtypes (especially MIGS), limiting our ability to evaluate performance across subtypes. In addition, we limited our analyses to patients with available VF, OCT, and IOP data. However, missing data is often an issue in real-world clinical settings and may limit the applicability of our models to those settings. Future work will include developing models that can handle missingness without a large loss of performance, as this is an important consideration for clinical adoption [39]. Finally, the surgery ground truth was based on both the clinical record and self-reported data because clinical records were not available for many patients in the cohort. While this approach allowed us to include a larger number of patients in the analysis, there could be issues with accuracy in patient self-reporting. Future work will incorporate data from clinical records to determine surgical intervention ground truth.

There are several possible future directions to build on the current study, which include (1) incorporating raw OCT and fundus photography data into models, (2) extending the current methods to better take advantage of longitudinal data, and (3) working to incorporate models into clinical workflows. The models reported in the current study only used summary metrics (mean RNFL thickness values) and did not take advantage of the information represented by the raw OCT image data. Incorporating this data (as well as fundus photos) could help to improve performance even further. Additionally, the current study identified predictions largely based on measurements collected at single visits. Extending our methods to use serial data from clinical, imaging, and VF testing visits could also improve accuracy or extend the timespan of surgical intervention prediction. A variety of machine and deep learning methods exist (e.g., recurrent neural networks, transformers, etc.) that explicitly model the longitudinal aspects of our datasets [40,41]. Our longer-term goal is to incorporate these predictive models into clinical settings so that they can be utilized at the point of care to help identify high-risk patients and inform downstream impacts on patient care. To this end, future work will also focus on validating methods in real-world clinical data and developing computational infrastructure to provide clinicians with real-time predictions to support their clinical decision-making.

5. Conclusions

In summary, our ML estimates achieved high accuracy in predicting surgical interventions in glaucoma up to 3 years in advance. The model accuracy was consistently high across age and racial subgroups in the test dataset. These results show that multi-modal ML approaches can achieve high accuracy in a critical glaucoma prediction task and suggest the potential for a large impact on patient care.

Supplementary Materials: The following supporting information can be downloaded at: https://www.mdpi.com/article/10.3390/bioengineering11020140/s1, Table S1: Summary of all model input features and output in predicting surgical intervention; Table S2: Performance of all models in predicting surgical interventions at time horizons up to 3 years.

Author Contributions: Conceptualization, M.C., C.B., A.B., M.H.G., S.L.B. and L.M.Z.; methodology, M.C., R.G., E.W., B.R.S., C.B., A.B., M.H.G., S.L.B. and L.M.Z.; software, R.G., J.H., E.W. and B.R.S.; validation, M.C., R.G., E.W. and B.R.S.; formal analysis, M.C., R.G. and E.W.; investigation, M.C., R.G., E.W., B.R.S., C.B., A.B., M.H.G., S.L.B. and L.M.Z.; resources, M.C., M.A.F., C.A.G., C.G.D.M., J.M.L., R.N.W., S.L.B. and L.M.Z.; data curation, R.G., J.H., E.W. and B.R.S.; writing—original draft preparation, M.C., R.G., S.L.B. and L.M.Z.; writing—review and editing, M.C., R.G., J.H., E.W., B.R.S., C.B., A.B., M.H.G., M.A.F., C.A.G., C.G.D.M., J.M.L., R.N.W., S.L.B. and L.M.Z.; visualization, M.C., R.G., E.W. and S.L.B.; supervision, M.C., C.B., A.B., M.H.G., S.L.B. and L.M.Z.; project administration, M.C., M.A.F., C.A.G., C.G.D.M., J.M.L., R.N.W., S.L.B. and L.M.Z.; funding acquisition, M.C., M.A.F., C.A.G., C.G.D.M., J.M.L., R.N.W., S.L.B. and L.M.Z. All authors have read and agreed to the published version of the manuscript.

Funding: This research was funded by the National Eye Institute grants R01EY029058, R01MD014850, R01EY026574, K12EY024225, R00EY030942, DP5OD029610, R01EY034146, R41EY034424, T35EY033704, Core Grant P30EY022589; The Glaucoma Foundation; unrestricted grant from Research to Prevent Blindness (New York, NY, USA).

Institutional Review Board Statement: The study was conducted in accordance with the Declaration of Helsinki and approved by the Institutional Review Board of UCSD (protocol #140276 and approved on 14 October 2022).

Informed Consent Statement: Informed consent was obtained from all the subjects involved in the study.

Data Availability Statement: The data presented in this study are available on reasonable request from the corresponding author. The data are not publicly available due to potential privacy issues.

Conflicts of Interest: Two authors (M.C. and L.M.Z.) are co-founders of AISight Health. The terms of this arrangement have been reviewed and approved by the University of California, San Diego in accordance with its conflict-of-interest policies. This relationship had no role in the design of the study; in the collection, analyses, or interpretation of data; in the writing of the manuscript; or in the decision to publish the results.

References

1. Tham, Y.C.; Li, X.; Wong, T.Y.; Quigley, H.A.; Aung, T.; Cheng, C.Y. Global prevalence of glaucoma and projections of glaucoma burden through 2040: A systematic review and meta-analysis. *Ophthalmology* **2014**, *121*, 2081–2090. [CrossRef]
2. Kass, M.A.; Heuer, D.K.; Higginbotham, E.J.; Johnson, C.; Keltner, J.L.; Miller, J.P.; Parrish, R.K.; Wilson, M.R.; Gordon, M.O. The Ocular Hypertension Treatment Study: A randomized trial determines that topical ocular hypotensive medication delays or prevents the onset of primary open-angle glaucoma. *Arch. Ophthalmol.* **2002**, *120*, 701–713; discussion 829–830. [CrossRef]
3. Heijl, A.; Leske, M.C.; Hyman, L.; Bengtsson, B.; Hussein, M. Reduction of intraocular pressure and glaucoma progression: Results from the Early Manifest Glaucoma Trial. *Arch. Ophthalmol.* **2002**, *120*, 1268–1279. [CrossRef]
4. Gordon, M.O.; Beiser, J.A.; Brandt, J.D.; Heuer, D.K.; Higginbotham, E.J.; Johnson, C.A.; Keltner, J.L.; Miller, J.P.; Parrish, R.K.; Wilson, M.R.; et al. The Ocular Hypertension Treatment Study: Baseline factors that predict the onset of primary open-angle glaucoma. *Arch. Ophthalmol.* **2002**, *120*, 714–720; discussion 829–830. [CrossRef] [PubMed]
5. Wu, Z.; Saunders, L.J.; Daga, F.B.; Diniz-Filho, A.; Medeiros, F.A. Frequency of Testing to Detect Visual Field Progression Derived Using a Longitudinal Cohort of Glaucoma Patients. *Ophthalmology* **2017**, *124*, 786–792. [CrossRef]
6. Esteva, A.; Robicquet, A.; Ramsundar, B.; Kuleshov, V.; DePristo, M.; Chou, K.; Cui, C.; Corrado, G.; Thrun, S.; Dean, J. A guide to deep learning in healthcare. *Nat. Med.* **2019**, *25*, 25–29. [CrossRef]
7. Ting, D.S.W.; Pasquale, L.R.; Peng, L.; Campbell, J.P.; Lee, A.Y.; Raman, R.; Tan, G.S.W.; Schmetterer, L.; Keane, P.A.; Wong, T.Y. Artificial intelligence and deep learning in ophthalmology. *Br. J. Ophthalmol.* **2019**, *103*, 167–175. [CrossRef] [PubMed]
8. Devalla, S.K.; Liang, Z.; Pham, T.H.; Boote, C.; Strouthidis, N.G.; Thiery, A.H.; Girard, M.J. Glaucoma management in the era of artificial intelligence. *Br. J. Ophthalmol.* **2019**, *104*, 301–311. [CrossRef]

9. Thompson, A.C.; Jammal, A.A.; Medeiros, F.A. A Review of Deep Learning for Screening, Diagnosis, and Detection of Glaucoma Progression. *Transl. Vis. Sci. Technol.* **2020**, *9*, 42. [CrossRef]
10. European Glaucoma Society Terminology and Guidelines for Glaucoma, 5th Edition. *Br. J. Ophthalmol.* **2021**, *105* (Suppl. S1), 1–169. [CrossRef] [PubMed]
11. Christopher, M.; Belghith, A.; Bowd, C.; Proudfoot, J.A.; Goldbaum, M.H.; Weinreb, R.N.; Girkin, C.A.; Liebmann, J.M.; Zangwill, L.M. Performance of Deep Learning Architectures and Transfer Learning for Detecting Glaucomatous Optic Neuropathy in Fundus Photographs. *Sci. Rep.* **2018**, *8*, 16685. [CrossRef]
12. Christopher, M.; Bowd, C.; Proudfoot, J.A.; Belghith, A.; Goldbaum, M.H.; Rezapour, J.; Fazio, M.A.; Girkin, C.A.; De Moraes, G.; Liebmann, J.M. Deep Learning Estimation of 10-2 and 24-2 Visual Field Metrics Based on Thickness Maps from Macula OCT. *Ophthalmology* **2021**, *128*, 1534–1548. [CrossRef]
13. Baxter, S.L.; Marks, C.; Kuo, T.T.; Ohno-Machado, L.; Weinreb, R.N. Machine Learning-Based Predictive Modeling of Surgical Intervention in Glaucoma Using Systemic Data From Electronic Health Records. *Am. J. Ophthalmol.* **2019**, *208*, 30–40. [CrossRef]
14. Wang, R.; Bradley, C.; Herbert, P.; Hou, K.; Ramulu, P.; Breininger, K.; Unberath, M.; Yohannan, J. Deep Learning-Based Identification of Eyes at Risk for Glaucoma Surgery. *Sci. Rep.* **2024**, *14*, 599. [CrossRef]
15. Devalla, S.K.; Chin, K.S.; Mari, J.-M.; Tun, T.A.; Strouthidis, N.G.; Aung, T.; Thiéry, A.H.; Girard, M.J.A. A Deep Learning Approach to Digitally Stain Optical Coherence Tomography Images of the Optic Nerve Head. *Invest. Ophthalmol. Vis. Sci.* **2018**, *59*, 63–74. [CrossRef]
16. Medeiros, F.A.; Jammal, A.A.; Thompson, A.C. From Machine to Machine: An OCT-Trained Deep Learning Algorithm for Objective Quantification of Glaucomatous Damage in Fundus Photographs. *Ophthalmology* **2019**, *126*, 513–521. [CrossRef]
17. Acosta, J.N.; Falcone, G.J.; Rajpurkar, P.; Topol, E.J. Multimodal biomedical AI. *Nat. Med.* **2022**, *28*, 1773–1784. [CrossRef]
18. Zheng, W.; Dryja, T.P.; Wei, Z.; Song, D.; Tian, H.; Kahler, K.H.; Khawaja, A.P. Systemic Medication Associations with Presumed Advanced or Uncontrolled Primary Open-Angle Glaucoma. *Ophthalmology* **2018**, *125*, 984–993. [CrossRef] [PubMed]
19. De Moraes, C.G.; Cioffi, G.A.; Weinreb, R.N.; Liebmann, J.M. New Recommendations for the Treatment of Systemic Hypertension and their Potential Implications for Glaucoma Management. *J. Glaucoma* **2018**, *27*, 567–571. [CrossRef] [PubMed]
20. Sample, P.A.; Girkin, C.A.; Zangwill, L.M.; Jain, S.; Racette, L.; Becerra, L.M.; Weinreb, R.N.; Medeiros, F.A.; Wilson, M.R.; De León-Ortega, J.; et al. The African Descent and Glaucoma Evaluation Study (ADAGES): Design and baseline data. *Arch. Ophthalmol.* **2009**, *127*, 1136–1145. [CrossRef] [PubMed]
21. WHO Collaborating Centre for Drug Statistics Methodology. *Guidelines for ATC Classification and DDD Assignment, 2023*; WHO: Oslo, Norway, 2022.
22. Zeng, K.; Bodenreider, O.; Kilbourne, J.; Nelson, S. RxNav: A web service for standard drug information. *AMIA Annu. Symp. Proc.* **2006**, *2006*, 1156. Available online: https://www.ncbi.nlm.nih.gov/pubmed/17238775 (accessed on 30 September 2023). [PubMed]
23. Breiman, L. Random Forests. *Mach. Learn.* **2001**, *45*, 5–32. [CrossRef]
24. Friedman, J.H. Greedy function approximation: A gradient boosting machine. *Ann. Stat.* **2001**, *29*, 1189–1232. [CrossRef]
25. Chen, T.; Guestrin, C. XGBoost: A Scalable Tree Boosting System. *arXiv* **2016**, arXiv:1603.02754. [CrossRef]
26. h2o: R Interface for H2O. (2022). Available online: https://github.com/h2oai/h2o-3 (accessed on 1 July 2023).
27. Lundberg, S.M.; Lee, S.-I. A Unified Approach to Interpreting Model Predictions. In Proceedings of the Advances in Neural Information Processing Systems 30, Long Beach, CA, USA, 4 December 2017.
28. Haymond, S.; Master, S.R. How Can We Ensure Reproducibility and Clinical Translation of Machine Learning Applications in Laboratory Medicine? *Clin. Chem.* **2022**, *68*, 392–395. [CrossRef]
29. Haibe-Kains, B.; Adam, G.A.; Hosny, A.; Khodakarami, F.; Shraddha, T.; Kusko, R.; Sansone, S.-A.; Tong, W.; Wolfinger, R.D.; Mason, C.E.; et al. Transparency and reproducibility in artificial intelligence. *Nature* **2020**, *586*, E14–E16. [CrossRef]
30. Bleeker, S.E.; Moll, H.A.; Steyerberg, E.W.; Donders, A.R.T.; Derksen-Lubsen, G.; Grobbee, D.; Moons, K.G.M. External validation is necessary in prediction research: A clinical example. *J. Clin. Epidemiol.* **2003**, *56*, 826–832. [CrossRef] [PubMed]
31. Obermeyer, Z.; Powers, B.; Vogeli, C.; Mullainathan, S. Dissecting racial bias in an algorithm used to manage the health of populations. *Science* **2019**, *366*, 447–453. [CrossRef]
32. Hong, C.; Pencina, M.J.; Wojdyla, D.M.; Hall, J.L.; Judd, S.E.; Cary, M.; Engelhard, M.M.; Berchuck, S.; Xian, Y.; D'agostino, R. Predictive Accuracy of Stroke Risk Prediction Models Across Black and White Race, Sex, and Age Groups. *JAMA* **2023**, *329*, 306–317. [CrossRef]
33. Coley, R.Y.; Johnson, E.; Simon, G.E.; Cruz, M.; Shortreed, S.M. Racial/Ethnic Disparities in the Performance of Prediction Models for Death by Suicide After Mental Health Visits. *JAMA Psychiatry* **2021**, *78*, 726–734. [CrossRef]
34. Delavar, A.; Saseendrakumar, B.R.; Weinreb, R.N.; Baxter, S.L. Racial and Ethnic Disparities in Cost-Related Barriers to Medication Adherence Among Patients With Glaucoma Enrolled in the National Institutes of Health All of Us Research Program. *JAMA Ophthalmol.* **2022**, *140*, 354–361. [CrossRef]
35. Melchior, B.; Valenzuela, I.A.; De Moraes, C.G.; Paula, J.S.; Fazio, M.A.; Girkin, C.A.; Proudfoot, J.; Cioffi, G.A.; Weinreb, R.N.; Zangwill, L.M.; et al. Glaucomatous Visual Field Progression in the African Descent and Glaucoma Evaluation Study (ADAGES): Eleven Years of Follow-up. *Am. J. Ophthalmol.* **2022**, *239*, 122–129. [CrossRef]
36. Gu, B.; Sidhu, S.; Weinreb, R.N.; Christopher, M.; Zangwill, L.M.; Baxter, S.L. Review of Visualization Approaches in Deep Learning Models of Glaucoma. *Asia Pac. J. Ophthalmol.* **2023**, *12*, 392–401. [CrossRef]

37. Yang, S.A.; Mitchell, W.; Hall, N.; Elze, T.; Lorch, A.C.; Miller, J.W.; Zebardast, N.; Pershing, S.; Hyman, L.; Haller, J.A.; et al. Trends and Usage Patterns of Minimally Invasive Glaucoma Surgery in the United States: IRIS(R) Registry Analysis 2013–2018. *Ophthalmol. Glaucoma* **2021**, *4*, 558–568. [CrossRef] [PubMed]
38. Wang, S.Y.; Tseng, B.; Hernandez-Boussard, T. Deep Learning Approaches for Predicting Glaucoma Progression Using Electronic Health Records and Natural Language Processing. *Ophthalmol. Sci.* **2022**, *2*, 100127. [CrossRef] [PubMed]
39. Marino, M.; Lucas, J.; Latour, E.; Heintzman, J.D. Missing data in primary care research: Importance, implications and approaches. *Fam. Pract.* **2021**, *38*, 200–203. [CrossRef] [PubMed]
40. Hochreiter, S.; Schmidhuber, J. Long Short-term Memory. *Neural Comput.* **1997**, *9*, 1735–1780. [CrossRef] [PubMed]
41. Dosovitskiy, A.; Beyer, L.; Kolesnikov, A.; Weissenborn, D.; Zhai, X.; Unterthiner, T.; Dehghani, M.; Minderer, M.; Heigold, G.; Gelly, S.; et al. An Image is Worth 16x16 Words: Transformers for Image Recognition at Scale. *arXiv* **2020**, arXiv:2010.11929. [CrossRef]

Disclaimer/Publisher's Note: The statements, opinions and data contained in all publications are solely those of the individual author(s) and contributor(s) and not of MDPI and/or the editor(s). MDPI and/or the editor(s) disclaim responsibility for any injury to people or property resulting from any ideas, methods, instructions or products referred to in the content.

Brief Report

Multi-Dataset Comparison of Vision Transformers and Convolutional Neural Networks for Detecting Glaucomatous Optic Neuropathy from Fundus Photographs

Elizabeth E. Hwang [1,2,†], Dake Chen [1,†], Ying Han [1], Lin Jia [3,*] and Jing Shan [1,*]

1. Department of Ophthalmology, University of California, San Francisco, San Francisco, CA 94143, USA
2. Medical Scientist Training Program, University of California, San Francisco, San Francisco, CA 94143, USA
3. Digillect LLC, San Francisco, CA 94158, USA
* Correspondence: linjia@digillect.xyz (L.J.); jing.shan@ucsf.edu (J.S.)
† These authors contributed equally to this work.

Abstract: Glaucomatous optic neuropathy (GON) can be diagnosed and monitored using fundus photography, a widely available and low-cost approach already adopted for automated screening of ophthalmic diseases such as diabetic retinopathy. Despite this, the lack of validated early screening approaches remains a major obstacle in the prevention of glaucoma-related blindness. Deep learning models have gained significant interest as potential solutions, as these models offer objective and high-throughput methods for processing image-based medical data. While convolutional neural networks (CNN) have been widely utilized for these purposes, more recent advances in the application of Transformer architectures have led to new models, including Vision Transformer (ViT,) that have shown promise in many domains of image analysis. However, previous comparisons of these two architectures have not sufficiently compared models side-by-side with more than a single dataset, making it unclear which model is more generalizable or performs better in different clinical contexts. Our purpose is to investigate comparable ViT and CNN models tasked with GON detection from fundus photos and highlight their respective strengths and weaknesses. We train CNN and ViT models on six unrelated, publicly available databases and compare their performance using well-established statistics including AUC, sensitivity, and specificity. Our results indicate that ViT models often show superior performance when compared with a similarly trained CNN model, particularly when non-glaucomatous images are over-represented in a given dataset. We discuss the clinical implications of these findings and suggest that ViT can further the development of accurate and scalable GON detection for this leading cause of irreversible blindness worldwide.

Keywords: glaucoma; deep learning; vision transformer; fundus photography

1. Introduction

Glaucoma is a group of chronic, progressive optic neuropathies are a leading cause of vision loss worldwide [1]. Primary open-angle glaucoma (POAG) is the most common type of glaucoma, with cases estimated to rise from 2.7 million in 2011 to 7.3 million by 2050 in the United States alone [2]. While most often associated with increased intraocular pressure (IOP), the disease process can also occur with normal or low IOP and is often referred to as the "silent thief of sight" because it typically progresses slowly and without noticeable symptoms in its early stages. Thus, early detection, close monitoring, and timely interventions are key to preserving vision in glaucoma patients, especially among minority populations such as Hispanics/Latinos and African Americans, who are disproportionately affected relative to non-Hispanic Whites [3]. However, currently the United States Preventive Services Task Force (USPSTF) does not recommend screening for primary open-angle glaucoma in asymptomatic adults 40 years or older. In their updated 2022 review, the USPSTF cited the need for targeted screening among high-risk populations (such as individuals with a family history of glaucoma or from disproportionately affected minority groups), optimizing contemporary screening approaches and modalities to improve both efficiency and cost-effectiveness, and clinical trials demonstrating the utility of such screening approaches in vision-related patient outcomes [4].

Deep learning-aided diagnostic interpretation has received significant interest for its potential to improve the accuracy of diagnosing glaucoma and deliver high-throughput screening tools optimized for early diagnosis in at-risk patients [5]. Glaucoma diagnosis often requires complex medical imaging of the optic nerve and retina in a specialist setting, and even then, is subject to inter-observer variability. Deep learning models have the potential to detect subtle structural changes missed by the eye, provide consistent results, and improve efficiency by reducing the burden on glaucoma specialists. Application of deep learning models to glaucoma diagnosis would also allow for high-throughput screening to identify asymptomatic disease and improve patient outreach, particularly in resource-limited settings.

Among the imaging modalities, fundus photography is a widely available, relatively low-cost approach already employed for clinical use in diabetic retinopathy tele-screening. In glaucoma, fundus photos provide the vertical optic nerve cup-to-disc ratio (vCDR), which quantifies the relationship between the cup (the central depression on the optic nerve head) and the disc (the entire optic nerve head) which enlarges as the disease progresses. Interpretation of these photos, however, can be difficult to reproduce among even expert specialists, and exhibit high rates of inter-observer variability [6–8], as well as being subject to observer bias (e.g., the tendency to under-call optic neuropathy in small optic discs while overcalling disease in physiologically large discs [9]). Therefore, the development and application of an AI tool to classify GON could greatly enhance fundus photography's utility as a population-based screening tool.

Previous studies have shown that deep learning models individually trained on color fundus photos [10], visual field analysis [11–14], and optical coherence tomography (OCT) [15–19] are able to identify glaucomatous optic neuropathy (GON) with robust performance (comparisons of specific deep learning models developed for glaucoma diagnosis and discussions of the different approaches are thoroughly covered in excellent reviews from Thompson et al. [5] and Yousefi [20]). Indeed, a recent meta-analysis of 17 deep-learning models trained on diagnosing GON from fundus photographs reported an overall AUC of 0.93 (95% CI 0.92–0.94), slightly lower than the AUC reported for studies using OCT (overall AUC 0.96, 95% CI 0.94–0.98) [21]. Several of these studies included external validation sets of up to six cohorts, suggesting that their models may generalize to unseen outside data. However, because these models are large and require intensive computational resources to train, they have been trained on datasets that are most often inaccessible to the public, thus making it difficult to compare whether the models themselves show differences during the training process.

To date, many AI models for glaucoma classification have utilized convolutional neural networks (CNNs). CNNs provide a scalable approach to object recognition within images by processing spatial patterns and extracting relevant features [22]. This architecture enables CNNs to automatically learn hierarchical representations of the features in an image. In supervised learning, CNNs are trained on labeled datasets, while in unsupervised learning, unsupervised methods like autoencoders are utilized for feature extraction. Semi-supervised learning, such as transfer learning, are also commonly described, as pre-trained models can be fine-tuned with smaller labeled datasets to improve performance [5,23]. However, a well-known attribute of CNNs is their inherent bias towards translation-invariant object recognition [24] which permits the interpretation of features outside of their spatial context [25], leaving models vulnerable to artifactual errors. Current models attempt to alleviate this by strict standardization of inputs, which, unfortunately, further restricts the ability of CNN algorithms to generalize to new, and even related, tasks without labor-intensive preprocessing.

In the last decade, Vision Transformer (ViT) [26], among other transformer architectures [27], has taken advantage of the self-attention mechanisms used in natural language processing to improve upon these limitations of CNNs. In contrast to CNNs, ViT models process the entire image as a sequence of patches, thus allowing for the capture of global relationships. ViTs have also been shown to generalize from smaller datasets than CNNs, which are heavily reliant upon pre-training and fine-tuning for optimal performance [26]. ViT models have now been applied to the analysis and interpretation of a wide range of clinical data ranging from electrocardiograms [28] to intraoperative surgical techniques [29]. In ophthalmology, there are increasing reports of ViT models trained to classify retinal pathologies from fundus photography [30–32] and OCT imaging [33–36], including several assessing their performance relative to CNNs [31,34,35]. Given that glaucoma diagnosis often requires multimodal imaging that correlates structural and functional data, it has been theorized that the global attention mechanisms utilized by ViTs offer an advantage over CNNs' dependency upon local features. However, few such reports in the ophthalmic literature benchmark one model against the other, and even fewer compare the outcomes from more than one training dataset. This represents a knowledge gap for AI-guided GON detection, since an optimal architecture should be able to generalize across variables that vary by clinical setting, such as patient population, image format, and disease prevalence.

In this report, we describe the training of ViT and CNN models on six publicly available, independent datasets, compare the two models' accuracies, and discuss the potential clinical applications for each type of model. We propose that the choice between these two model architectures may depend upon the specific clinical setting, labeled data availability, and computational resources. Ultimately, we hope that our results provide insight into model selection for specific clinical tasks as well as effective database construction.

2. Materials and Methods
2.1. Datasets

A total of six public datasets were included for analysis in this paper (Figure 1, Table 1) [37–42]. Complete dataset sizes varied between 101 images (Drishti-GS1) to 720 images (REFUGE2). Though representation of non-glaucomatous (control) and glaucomatous classes varied between the datasets, no obvious correlation existed between total dataset size and class ratios (Table 1). When provided by the original authors, the patient selection criteria and instrument cameras are also noted in Table 1.

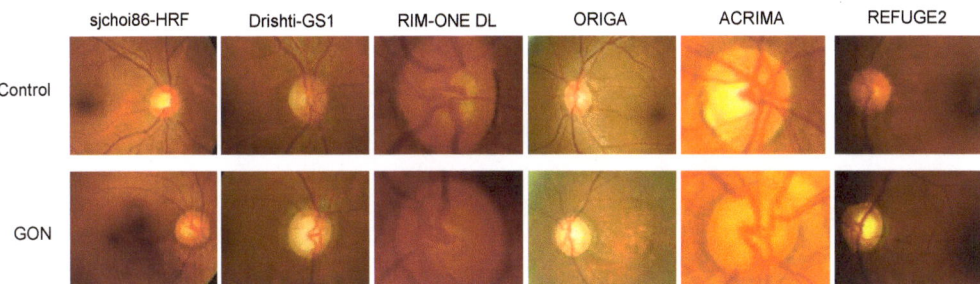

Figure 1. Representative fundus photographs from the datasets used in this study. GON: glaucomatous optic neuropathy.

All datasets included ground truth labels indicating GON or control. Most datasets derived ground truth from expert labeling and clinical annotations with the exceptions of ORIGA (algorithm-based) and sjchoi86-HRF (unknown). Three datasets (sjchoi86-HRF, ORIGA, and REFUGE2) provided whole fundus images, one provided OD-centered images (Drishti-GS1), and two provided OD-cropped images (RIM-ONE DL and ACRIMA) (Figure 1). Sources accessed for each of the datasets are provided in the references. No photographs were excluded from our analysis.

Table 1. Characteristics of publicly available datasets used in this study, as ordered by total set size.

Study	Patient Selection	Instrument	Ground Truth	Non-GC (Control)	GC	Total Size	Class Ratio *
Drishti-GS1 [37]	Glaucomatous and routine refraction images selected by experts from patients between ages 40 and 80 at Aravind Eye Hospital in India	Not noted	4 experts	31	70	101	0.44
sfchoi86-HRF [38]	Unknown	Unknown	Unknown	300	101	401	2.97
RIM-ONE DL [39]	Curated extraction from RIM-ONE V1, V2, and V3 of glaucomatous and healthy patients from 3 hospitals in Spain	V1/V2: Nidek AFC-210 camera V3: Kowa WX 3D non-stereo camera	2 experts with tiebreaker	312	173	485	1.80
ORIGA [40]	Glaucomatous and randomly selected non-GC images from cross-sectional population Singaporean study (SiMES) of Malay adults between ages 40 and 80	Canon CR-DGi	ORIGA-GT	482	168	650	2.87
ACRIMA [41]	Glaucomatous and normal images selected by experts in Spain based on clinical findings	Topcon TRC retinal camera	2 experts	309	396	705	0.78
REFUGE2 [42]	Random selection from glaucoma and myopia study cohorts in China (Zongshan Ophthalmic Center)	KOWA, TOPCON	7 experts	720	80	800	9.00

* Calculated as a ratio of non-GC: GC images. GC = glaucomatous.

2.2. Image Preprocessing

To minimize the presence of redundant information which could potentially impact deep learning model performance, we conducted pre-processing of all images to extract the region around the optic nerve head from each fundus image as shown in the depicted model (Figure 2). This was achieved using deeplabv3plus [43], a semantic segmentation model. Once the region of interest was extracted, we automatically cropped a square area centered around the disc. These extracted images were then utilized to train the CNN and ViT models described below. By focusing on specific areas, we aimed to improve the model's ability to identify glaucoma-related features and enhance the accuracy of the automated detection system.

2.3. Vision Transformer (ViT) and ResNet Training and Evaluation

Each one of the public databases was split into a training set (80%) and a testing set (20%) (Figure 2). This ensured a consistent and fair evaluation of both models using identical testing datasets. An overview of our method is shown in Figure 2. For the CNN model, we leveraged the standard ResNet-50 which has 50 layers with incorporated residual connections with no further tuning [43]. For the ViT architecture [25], we used 12 attention layers and a patch size of 16, hidden size of 768, and 12 heads. Following the practices established by [44] and [25], we pretrained the ViT on the ImageNet dataset [45]. The images were resized to a uniform size of 224 × 224 pixels. Additionally, we normalized the pixel values to a range between 0 and 1. During training, we used a batch size of 16 and employed the AdamW optimizer with a learning rate of $6e^{-4}$ and regularization of $6e^{-2}$. These hyperparameters (included in Figure 2) were chosen to optimize the model's convergence and performance. To compute the loss during training, we employed cross-entropy loss with 0.1 label smoothing.

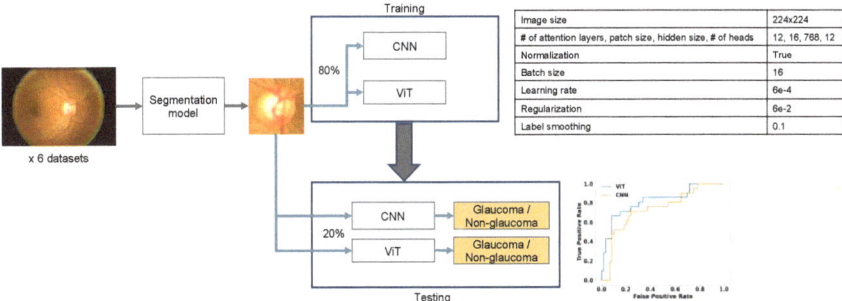

Figure 2. Workflow of ViT vs. CNN model training (with hyperparameters) and validation.

Performance metrics, including area under the receiver operating characteristic curve (AUC), sensitivity, specificity, accuracy, F1 score, and mAP (mean Average Precision), were calculated from models evaluated on the held-out test sets (Table 2). Specificities were calculated at a fixed sensitivity threshold of 50%. Confidence intervals (CI) were determined by bootstrap resampling of the test sets with replacement (n = 10 times) while the training sets and models remained fixed.

Table 2. Performance statistics for ViT and CNN models evaluated on held-out test sets. Confidence intervals of 95% are reported in parentheses.

	ViT			CNN		
	AUC	*Sensitivity*	*Specificity*	*AUC*	*Sensitivity*	*Specificity*
Drishti-GS1	0.67 (0.44, 0.97)	0.93 (0.79, 1.00)	0.40 (0.00, 1.00)	0.67 (0.38, 0.91)	0.73 (0.50, 0.93)	0.91 (0.00, 1.00)
sjchoi86-HRF	0.79 (0.67, 90)	0.67 (0.46, 0.82)	0.92 (0.84, 0.98)	0.71 (0.59, 82)	0.52 (0.31, 0.75)	0.90 (0.81, 0.97)
RIM-ONE DL	0.88 (0.81, 0.94)	0.91 (0.79, 1.00)	0.85 (0.76, 0.93)	0.86 (0.78, 0.93)	0.81 (0.67, 0.94)	0.91 (0.83, 0.97)
ORIGA	0.69 (0.60, 0.77)	0.52 (0.37, 0.67)	0.85 (0.77, 0.92)	0.62 (0.54, 0.70)	0.36 (0.21, 0.52)	0.88 (0.81, 0.95)
ACRIMA	0.94 (0.90, 0.97)	1.00 (1.00, 1.00)	0.88 (0.79, 0.95)	0.92 (0.88, 0.96)	0.84 (0.76, 0.92)	1.00 (1.00, 1.00)
REFUGE2	0.95 (0.88, 1.00)	0.94 (0.80, 1.00)	0.97 (0.94, 0.99)	0.89 (0.78, 0.99)	0.81 (0.60, 1.00)	0.97 (0.94, 0.99)
	ViT			CNN		
	Accuracy	*F1 Score*	*mAP*	*Accuracy*	*F1 Score*	*mAP*
Drishti-GS1	0.80 (0.60, 0.95)	0.87 (0.73, 0.97)	0.82 (0.63, 0.99)	0.70 (0.50, 0.90)	0.79 (0.58, 0.93)	0.82 (0.61, 0.98)
sjchoi86-HRF	0.81 (0.72, 0.89)	0.68 (0.51, 0.82)	0.53 (0.35, 0.72)	0.80 (0.71, 0.89)	0.58 (0.37, 0.76)	0.46 (0.28, 0.65)
RIM-ONE DL	0.87 (0.79, 0.93)	0.82 (0.71, 0.90)	0.70 (0.56, 0.85)	0.88 (0.80, 0.94)	0.81 (0.69, 0.91)	0.72 (0.56, 0.86)
ORIGA	0.74 (0.66, 0.82)	0.57 (0.44, 0.69)	0.50 (0.37, 0.63)	0.71 (0.63, 0.78)	0.46 (0.31, 0.60)	0.44 (0.32, 0.56)
ACRIMA	0.94 (0.91, 0.98)	0.95 (0.91, 0.98)	0.91 (0.84, 0.96)	0.91 (0.87, 0.96)	0.92 (0.86, 0.96)	0.93 (0.89, 0.97)
REFUGE2	0.97 (0.94, 0.99)	0.83 (0.64, 0.95)	0.72 (0.47, 0.92)	0.96 (0.93, 0.99)	0.79 (0.60, 0.93)	0.64 (0.39, 0.86)

3. Results

In Table 2 and Figure 3, we present the performance statistics and contingency tables of the CNN and ViT models trained to classify non-glaucomatous (non-GC) from glaucomatous (GC) eyes on each of the six public datasets. When compared using relative AUC, the ViT models were non-inferior to the CNN models and appeared to outperform the CNN models on five of the six datasets, though this was not statistically significant given the overlapping confidence intervals. The greatest differences were observed among the sjchoi86-HRF (0.79 ViT vs. 0.71 CNN), ORIGA (0.69 ViT vs. 0.62 CNN) and REFUGE2 (0.95 ViT vs. 0.89 CNN) datasets. No difference in mean AUC was observed for only one dataset, Drishti-GS1 (both 0.67). The performance on the remaining two datasets were also consistently, if marginally, higher for the ViT models (RIM-ONE: 0.88 ViT vs. 0.86 CNN; ACRIMA 0.94 ViT vs. 0.92 CNN). Similar observations were made for the accuracies, F1 scores, mAP, as reflected in the average statistic and the 95% confidence intervals.

The recall or sensitivity of the ViT models surpassed those of the CNN models among the six datasets by an average of 0.14, with the largest difference observed in the Dristhi-GS1 (0.93 ViT vs. 0.73 CNN) and the smallest in REFUGE2 (0.94 ViT vs. 0.81 CNN). By contrast, the specificities were more varied between the two methods, ranging from comparable (sjchoi86-HRF: ViT 0.92 vs. CNN 0.90; ORIGA: 0.85 ViT vs. 0.88 CNN; REFUGE2: ViT 0.97 vs. CNN 0.97) to favoring the CNN model (RIM-ONE: 0.85 ViT vs. 0.91 CNN; ACRIMA: 0.88 ViT vs. 1.00 CNN). For Drishti-GS1, the small sample size of non-GC images in the held-out test set ($n = 5$ images) resulted in inconclusive specificity statistics as reflected by the 95% confidence intervals of (0,1) for both models.

We noted that ViT tended call more false positives (i.e., label control images as GON) than CNN models in several datasets, including RIM-ONE (10 ViT false positives (FP) vs. 6 CNN FP), ORIGA (13 ViT FP vs. 10 CNN FP), and ACRIMA (8 ViT FP vs. 0 CNN FP). Accordingly, two of these ViT models demonstrated lower specificities than their CNN equivalents (i.e., RIM-ONE: 0.74 ViT vs. 0.81 CNN; ACRIMA: 0.91 ViT vs. 1.00 CNN).

Interestingly, ViT outperformed CNN on datasets with higher ratios of non-GC to GC photos (Figure 4a), though not with total dataset size (Figure 4b). This was most clearly evidenced by the delta AUC of the Drishti-GS1 and REFUGE2 models, whose datasets harbored the lowest (0.44) and highest (9.0) ratios of non-GC to GC images, respectively. Furthermore, the differences in specificity between the ViT and CNN models diminished as the ratio of non-GC to GC images increased (Figure 4c): when trained on the REFUGE2 dataset, the specificity of the ViT model overlapped with that of the CNN model (0.97, CI 0.94–0.99).

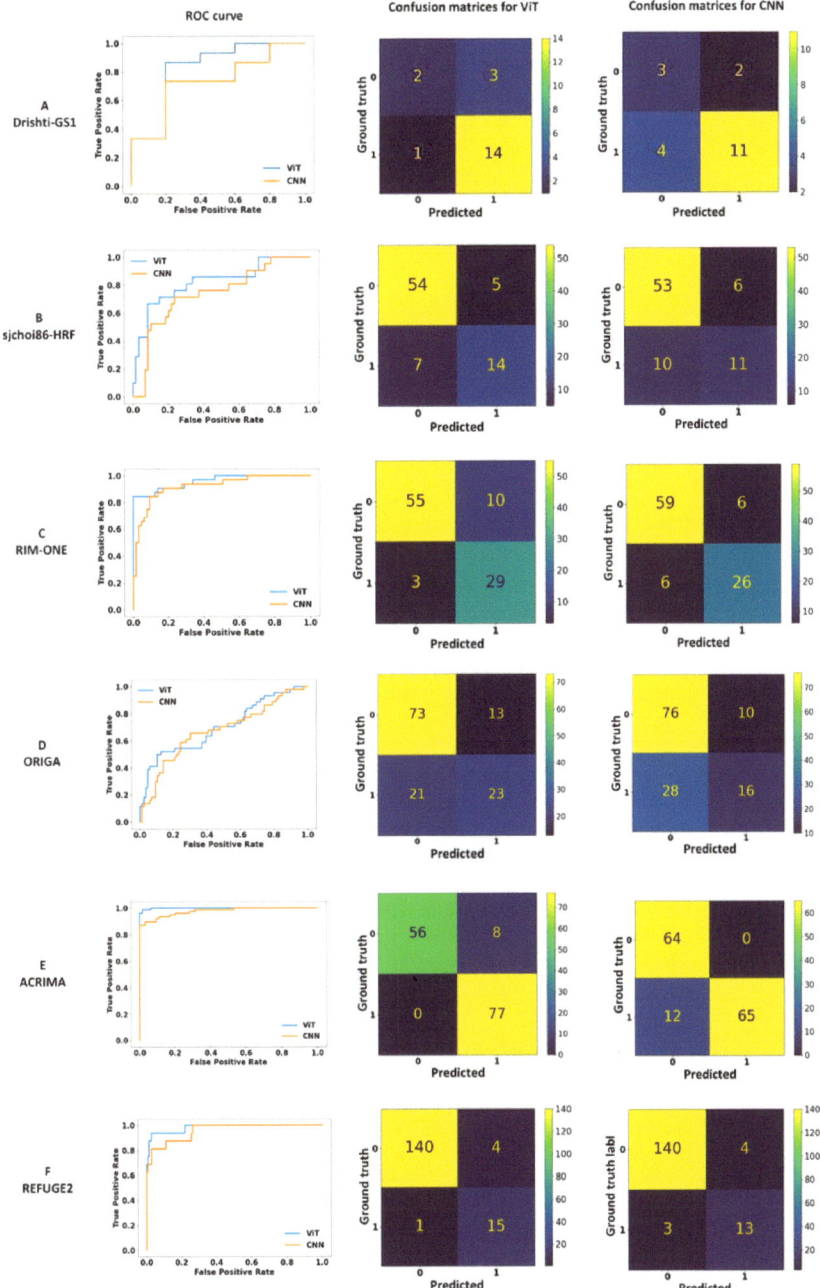

Figure 3. ROC curves and confusion matrices for ViT and CNN models trained on individual datasets (**A**–**F**). For the confusion matrices, a classification of 0 refers to control/non-glaucomatous, whereas a classification of 1 refers to glaucomatous. Ground truth labels were used as provided by the original datasets (ref. Table 1).

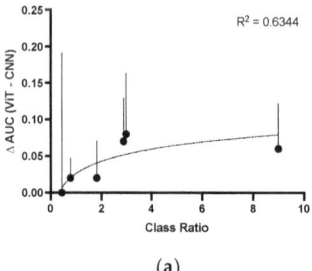 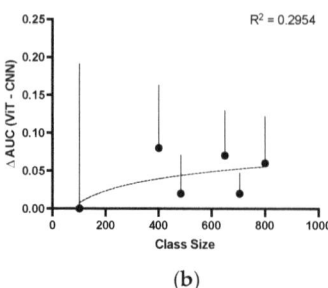 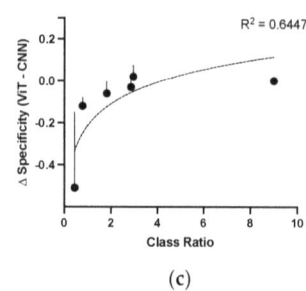

Figure 4. ViT outperforms CNN models in datasets with greater class imbalance but not class size. (Δ = ViT − CNN, where ViT outperforms CNN when $\Delta > 0$, and CNN outperforms ViT when $\Delta < 0$) Log-linear regression models (dotted lines) are included with coefficients of determination as indicated. (**a**) ΔAUC as a function of class ratio. (**b**) ΔAUC as a function of class size. (**c**) ΔSpecificity as a function of class ratio. See Table 1 for class sizes and ratios.

4. Discussion

Here we focus the performance of ViT and CNN models trained on glaucoma detection from a single imaging modality, fundus photography. We take advantage of ImageNet pre-trained models to test and train each architecture on six publicly available annotated datasets, which were collected from at least four countries (India, Spain, Singapore, and China) and varied in size from 101 to 800 total images. Class imbalances between control and glaucomatous labeled images were present to varying degrees among the datasets, from the most evenly matched (Drishti-GS1, class ratio of 0.44 or 69% glaucoma prevalence) to the least (REFUGE2, class ratio of 9.0 or 10% glaucoma prevalence). A survey of 14 US-based studies found all glaucoma prevalence rates ranging from 2.1% to 25.5%, and POAG prevalence rates between 1.86% and 13.8% [44]. Thus, though these datasets represent selected cohorts rather than a population survey, it is likely that the "lower" prevalence cohorts more accurately reflect the dataset composition that would be expected from a moderate to high-risk screening population. We had two objectives from comparing the models trained upon multiple, rather than pooled, datasets: first, to ask whether the ViT model could perform equal to, or better than, a widely accepted CNN model, ResNet-50, and second, to determine whether ViT or CNN models, when trained on datasets of different sizes and class representations, demonstrated any trends in performance metrics including AUC, sensitivity, and specificity, that might inform future clinical application of the two architectures.

As predicted, we found that the pre-trained ViT model matched or outperformed the equivalent CNN model on all six datasets by AUC and accuracy measures. We also observed that the ViT models increasingly outperformed CNN models, as measured by AUC and specificity, on datasets with greater representations of controls (i.e., higher class ratio). We suggest that this difference may reflect that, when presented with insufficient control representation, ViT struggles with the greater variability present among non-glaucomatous optic nerve discs due to the wider array of potential relationships when using a global attention mechanism. However, as the ViT algorithm is presented with an increasing number of control examples, it can better assign global relationships to a given class, even when it is as varied as "non-glaucoma".

Given our observations, we would recommend CNN models for GON detection in tasks with uniform data collection where high test specificity outweighs other considerations. In contrast, we would nominate ViT models for tasks requiring collaborative data collections (e.g., clinical trials, multi-site tele-screening), whereby different operators, patient demographics, camera models, and data processing standards are likely to result in datasets with levels of heterogeneity beyond that which CNN models can accommodate. ViT performance could be enhanced further by targeted deployment to patients with identifiable risk factors, such as a family history of glaucoma, advanced age, or predisposing conditions such as steroid use. Such at-risk populations exhibit higher pre-test probability and would thus benefit from ViT's greater sensitivity.

While CNN and ViT models are widely utilized for high-throughput image analysis and classification, their differences in feature detection and training requirements have led to the suggestion that ViT models may improve upon CNN performance. CNN architecture utilizes a sliding window method to extract features in a local fashion and thus has strict input requirements [45]. Previous strategies for improving CNN performance of photographic GON detection have focused on the optimization of pre-processing techniques like data augmentation [46] and feature extraction [47], as well as more clinically motivated strategies such as structure-function correlation between multiple testing modalities [48,49]. More recently, transformer architectures have gained interest for their ability to use global attention mechanisms to identify long-range interactions [50] and their flexibility in allowing for non-uniform inputs [26]. Therefore, while the prevalence of inductive biases in CNNs relative to transformers may enable ResNet models to outperform ViT models when classification relies upon the presence or absence of locally identifiable features (e.g., optic nerve thinning in defined superior-temporal or inferior-nasal patterns) [26,51]. ViT may ultimately offer superior performance when diagnostic features are distributed in a disconnected manner (e.g., identifying glaucomatous features such as bayoneting). This would be particularly applicable in the setting of multimodal imaging datasets that could potentially rely upon global features, such as the correlation of functional visual field testing with structural changes in the OCT, which so far have required a multi-algorithmic approach [52].

Yet, despite the potential for transformers to incorporate long-range feature detection from multimodal datasets, the literature comparing ViT models to CNN models have generally focused on single modalities due to the challenges of multimodal data integration as an input into a single algorithm [31,32,35,51–55]. One outstanding report compares ViT to CNN models trained on Diabetic Retinopathy (DR) detection from multiple independent datasets consisting of either fundus photos or OCT imaging, and finds that ViT models are superior in both cases; however, no multimodal datasets are used [31]. To the best of our knowledge, only two publications so far have compared the performance of published ViT and CNN models on glaucoma detection from fundus photos [51,56]. In one report, the authors found that Data-efficient Image Transformer (DeIT) models outperformed similarly trained ResNet-50 models [51]. They further compared the DeIT attention maps with ResNet-50 average saliency maps to demonstrate that the transformer model more precisely focused upon the borders of the optic disc where glaucomatous features are most often identified, whereas the CNN saliency maps highlighted the entire optic nerve. Intriguingly, the more recent report found that the ViT model underperformed the CNN models (VGG, ResNet, Inception, MobileNet, DenseNet) on an external validation set [56]. While not directly comparable to our results, we note that their training set was also comprised of three nearly equally represented classes (GON, non-GON, and normal optic discs), perhaps resembling our "lower" class ratio datasets, such as ACRIMA.

Our work builds upon these studies by incorporating the use of independent training sets similar to [31] as well as avoiding the use of fine-tuning between datasets, thus allowing for observations on the baseline performance of the two architectures in multiple settings. Within the constraints of the public datasets utilized by our models, our results suggest that simply switching to ViT-based architecture alone will not significantly improve model performance. This is for two reasons: First, though there is an appreciable trend of higher mean AUCs across the ViT models, the differences between the individual ViT and CNN models were not statistically significant. Second, while ViT models uniformly demonstrated greater sensitivities than the CNN models, we observed that the under-representation of non-GC images during training may have led to lower model specificities. This implies that one trade-off of ViT's global attention mechanism may result in increased dependence upon sufficient class representation during training, which aligns with previous observations that, for smaller datasets, ViT-based architectures are more dependent upon training set representation than CNN-based architectures [26]. Thus, improving model performance may not rely only upon optimizing the model itself, but also the training data and processes involved. Here we utilized pre-trained models, but other techniques to improve model performance have included transfer learning [57], artifact-tolerant feature representation [10], cross-teaching between CNN and transformer models [58], and hybrid CNN-ViT architectures which extract local features in a patch-based manner [55]. While not addressed here, many of these strategies appear promising and merit further investigation.

We acknowledge a couple of limitations in our study. First, our comparisons of the two models were limited to datasets containing only fundus photography, while in practice, the gold standard diagnosis of glaucomatous optic neuropathy requires the correlation of structural findings (optic nerve thinning) with functional ones (visual field defects) [45]. Secondly, we pre-processed the fundus photos with optic nerve head segmentation to avoid biasing the models with non-disc-related information. Given that ViT uses a global mechanism, we anticipate that the performance of the ViT models may have been disproportionately affected relative to the CNN models. However, given that real-world application of these models often incorporates similar pre-processing for a variety of reasons [59,60], we suggest that this approach remains relevant to clinical practice.

Future works based on these findings may benefit from comparisons of CNN-based vs. ViT-based models on larger cohorts, more rigorous investigations of whether non-glaucomatous representation impacts model performance, and ideally, side-by-side comparisons of both models in various clinical contexts (i.e., screening vs. specialty visits) to determine their efficacies and practicalities in different settings.

5. Conclusions

Overall, our findings suggest that ViT-based algorithms show excellent results regarding glaucoma detection in line with previous studies. However, our results indicate that, despite recent publication trends, CNN models may offer advantages over Transformer models for training datasets with more equal representation of both non-glaucomatous and glaucomatous images. For high-risk populations or other situations where the importance of detecting any disease outweighs the risk of false positives, we propose that ViT models should be considered superior to the more widely utilized CNN-based architectures established within the field.

Automated image processing algorithms for the detection of glaucomatous optic neuropathy can empower population-based screening towards preventing irreversible vision loss. We hope our findings here can further the development of accurate and scalable high-throughput methods for this leading cause of blindness worldwide.

Author Contributions: Conceptualization, Methodology, and Software, J.S., L.J. and D.C.; Investigation, Formal Analysis, and Visualization, E.E.H. and D.C.; Writing—Methods, D.C.; Writing—Original Draft Preparation, E.E.H.; Writing—Review and Editing, E.E.H., J.S., D.C., Y.H. and L.J.; Supervision, Project Administration, and Funding Acquisition, J.S. All authors have read and agreed to the published version of the manuscript.

Funding: This research was funded by Computational Innovator Faculty Research Award to Dr. Jing Shan from the UCSF Initiative for Digital Transformation in Computational Biology & Health, All May See Foundation and Think Forward Foundation to Dr. Jing Shan, UCSF Irene Perstein Award to Dr. Jing Shan, and National Institutes of Health under NCI Award Number F30CA250157 and NIGMS T32GM007618 MSTP training grant to Elizabeth Hwang.

Institutional Review Board Statement: Not applicable.

Informed Consent Statement: Not applicable.

Data Availability Statement: Publicly available datasets were analyzed in this study.

Conflicts of Interest: The authors declare no conflict of interest.

References

1. Tham, Y.-C.; Li, X.; Wong, T.Y.; Quigley, H.A.; Aung, T.; Cheng, C.-Y. Global prevalence of glaucoma and projections of glaucoma burden through 2040: A systematic review and meta-analysis. *Ophthalmology* 2014, *121*, 2081–2090. [CrossRef] [PubMed]
2. Vajaranant, T.S.; Wu, S.; Torres, M.; Varma, R. The changing face of primary open-angle glaucoma in the United States: Demographic and geographic changes from 2011 to 2050. *Arch. Ophthalmol.* 2012, *154*, 303–314.e3. [CrossRef] [PubMed]
3. Stein, J.D.; Khawaja, A.P.; Weizer, J.S. Glaucoma in Adults—Screening, Diagnosis, and Management: A Review. *JAMA* 2021, *325*, 164–174. [CrossRef] [PubMed]
4. Chou, R.; Selph, S.; Blazina, I.; Bougatsos, C.; Jungbauer, R.; Fu, R.; Grusing, S.; Jonas, D.E.; Tehrani, S. Screening for Glaucoma in Adults: Updated Evidence Report and Systematic Review for the US Preventive Services Task Force. *JAMA* 2022, *327*, 1998–2012. [CrossRef]
5. Thompson, A.C.; Jammal, A.A.; Medeiros, F.A. A Review of Deep Learning for Screening, Diagnosis, and Detection of Glaucoma Progression. *Transl. Vis. Sci. Technol.* 2020, *9*, 42. [CrossRef]
6. Chan, H.H.; Ong, D.N.; Kong, Y.X.; O'Neill, E.C.; Pandav, S.S.; Coote, M.A.; Crowston, J.G. Glaucomatous optic neuropathy evaluation (gone) project: The effect of monoscopic versus stereoscopic viewing conditions on optic nerve evaluation. *Am. J. Ophthalmol.* 2014, *157*, 936–944. [CrossRef]
7. Denniss, J.; Echendu, D.; Henson, D.B.; Artes, P.H. Discus: Investigating subjective judgment of optic disc damage. *Optom. Vis. Sci.* 2011, *88*, E93–E101. [CrossRef]
8. Jampel, H.D.; Friedman, D.; Quigley, H.; Vitale, S.; Miller, R.; Knezevich, F.; Ding, Y. Agreement Among Glaucoma Specialists in Assessing Progressive Disc Changes From Photographs in Open-Angle Glaucoma Patients. *Arch. Ophthalmol.* 2009, *147*, 39–44.e1. [CrossRef]
9. Nixon, G.J.; Watanabe, R.K.; Sullivan-Mee, M.; DeWilde, A.; Young, L.; Mitchell, G.L. Influence of Optic Disc Size on Identifying Glaucomatous Optic Neuropathy. *Optom. Vis. Sci.* 2017, *94*, 654–663. [CrossRef]
10. Shi, M.; Lokhande, A.; Fazli, M.S.; Sharma, V.; Tian, Y.; Luo, Y.; Pasquale, L.R.; Elze, T.; Boland, M.V.; Zebardast, N.; et al. Artifact-Tolerant Clustering-Guided Contrastive Embedding Learning for Ophthalmic Images in Glaucoma. *IEEE J. Biomed. Health Inform.* 2023, *27*, 4329–4340. [CrossRef]
11. Datta, S.; Mariottoni, E.B.; Dov, D.; Jammal, A.A.; Carin, L.; Medeiros, F.A. RetiNerveNet: Using recursive deep learning to estimate pointwise 24-2 visual field data based on retinal structure. *Sci. Rep.* 2021, *11*, 12562. [CrossRef] [PubMed]
12. Kang, J.H.; Wang, M.; Frueh, L.; Rosner, B.; Wiggs, J.L.; Elze, T.; Pasquale, L.R. Cohort Study of Race/Ethnicity and Incident Primary Open-Angle Glaucoma Characterized by Autonomously Determined Visual Field Loss Patterns. *Transl. Vis. Sci. Technol.* 2022, *11*, 21. [CrossRef]
13. Saini, C.; Shen, L.Q.; Pasquale, L.R.; Boland, M.V.; Friedman, D.S.; Zebardast, N.; Fazli, M.; Li, Y.; Eslami, M.; Elze, T.; et al. Assessing Surface Shapes of the Optic Nerve Head and Peripapillary Retinal Nerve Fiber Layer in Glaucoma with Artificial Intelligence. *Ophthalmol. Sci.* 2022, *2*, 100161. [CrossRef]
14. Yousefi, S.; Pasquale, L.R.; Boland, M.V.; Johnson, C.A. Machine-Identified Patterns of Visual Field Loss and an Association with Rapid Progression in the Ocular Hypertension Treatment Study. *Ophthalmology* 2022, *129*, 1402–1411. [CrossRef] [PubMed]
15. Mariottoni, E.B.; Datta, S.; Shigueoka, L.S.; Jammal, A.A.; Tavares, I.M.; Henao, R.; Carin, L.; Medeiros, F.A. Deep Learning–Assisted Detection of Glaucoma Progression in Spectral-Domain OCT. *Ophthalmol. Glaucoma* 2023, *6*, 228–238. [CrossRef] [PubMed]
16. Mariottoni, E.B.; Jammal, A.A.; Urata, C.N.; Berchuck, S.I.; Thompson, A.C.; Estrela, T.; Medeiros, F.A. Quantification of Retinal Nerve Fibre Layer Thickness on Optical Coherence Tomography with a Deep Learning Segmentation-Free Approach. *Sci. Rep.* 2020, *10*, 402. [CrossRef]
17. Medeiros, F.A.; Jammal, A.A.; Mariottoni, E.B. Detection of Progressive Glaucomatous Optic Nerve Damage on Fundus Photographs with Deep Learning. *Ophthalmology* 2021, *128*, 383–392. [CrossRef]
18. Shigueoka, L.S.; Mariottoni, E.B.; Thompson, A.C.; Jammal, A.A.; Costa, V.P.; Medeiros, F.A. Predicting Age From Optical Coherence Tomography Scans with Deep Learning. *Transl. Vis. Sci. Technol.* 2021, *10*, 12. [CrossRef]
19. Xiong, J.; Li, F.; Song, D.; Tang, G.; He, J.; Gao, K.; Zhang, H.; Cheng, W.; Song, Y.; Lin, F.; et al. Multimodal Machine Learning Using Visual Fields and Peripapillary Circular OCT Scans in Detection of Glaucomatous Optic Neuropathy. *Ophthalmology* 2022, *129*, 171–180. [CrossRef]
20. Yousefi, S. Clinical Applications of Artificial Intelligence in Glaucoma. *J. Ophthalmic. Vis. Res.* 2023, *18*, 97–112. [CrossRef]
21. Aggarwal, R.; Sounderajah, V.; Martin, G.; Ting, D.S.W.; Karthikesalingam, A.; King, D.; Ashrafian, H.; Darzi, A. Diagnostic accuracy of deep learning in medical imaging: A systematic review and meta-analysis. *npj Digit. Med.* 2021, *4*, 65. [CrossRef]
22. Krichen, M. Convolutional Neural Networks: A Survey. *Computers* 2023, *12*, 151. [CrossRef]
23. Shan, J.; Li, Z.; Ma, P.; Tun, T.A.; Yonamine, S.; Wu, Y.; Baskaran, M.; Nongpiur, M.E.; Chen, D.; Aung, T.; et al. Deep Learning Classification of Angle Closure based on Anterior Segment OCT. *Ophthalmol. Glaucoma* 2023. [CrossRef] [PubMed]
24. Myburgh, J.C.; Mouton, C.; Davel, M.H. Tracking Translation Invariance in CNNs. In *Southern African Conference for Artificial Intelligence Research*; Springer International Publishing: Cham, Switzerland, 2020; pp. 282–295.
25. Sadeghzadeh, H.; Koohi, S. Translation-invariant optical neural network for image classification. *Sci. Rep.* 2022, *12*, 17232. [CrossRef] [PubMed]

26. Dosovitskiy, A.; Beyer, L.; Kolesnikov, A.; Weissenborn, D.; Zhai, X.; Unterthiner, T.; Dehghani, M.; Minderer, M.; Heigold, G.; Gelly, S. An image is worth 16x16 words: Transformers for image recognition at scale. *arXiv* **2020**, arXiv:2010.11929.
27. Touvron, H.; Cord, M.; Douze, M.; Massa, F.; Sablayrolles, A.; Jegou, H. Training data-efficient image transformers & distillation through attention. *Pr. Mach. Learn. Res.* **2021**, *139*, 7358–7367.
28. Vaid, A.; Jiang, J.; Sawant, A.; Lerakis, S.; Argulian, E.; Ahuja, Y.; Lampert, J.; Charney, A.; Greenspan, H.; Narula, J.; et al. A foundational vision transformer improves diagnostic performance for electrocardiograms. *NPJ Digit. Med.* **2023**, *6*, 108. [CrossRef]
29. Kiyasseh, D.; Ma, R.; Haque, T.F.; Miles, B.J.; Wagner, C.; Donoho, D.A.; Anandkumar, A.; Hung, A.J. A vision transformer for decoding surgeon activity from surgical videos. *Nat. Biomed. Eng.* **2023**, *7*, 780–796. [CrossRef]
30. Liu, H.; Teng, L.; Fan, L.; Sun, Y.; Li, H. A new ultra-wide-field fundus dataset to diabetic retinopathy grading using hybrid preprocessing methods. *Comput. Biol. Med.* **2023**, *157*, 106750. [CrossRef]
31. Playout, C.; Duval, R.; Boucher, M.C.; Cheriet, F. Focused Attention in Transformers for interpretable classification of retinal images. *Med. Image Anal.* **2022**, *82*, 102608. [CrossRef]
32. Yu, S.; Ma, K.; Bi, Q.; Bian, C.; Ning, M.; He, N.; Li, Y.; Liu, H.; Zheng, Y. MIL-VT: Multiple Instance Learning Enhanced Vision Transformer for Fundus Image Classification. In *Medical Image Computing and Computer Assisted Intervention–MICCAI 2021: 24th International Conference, Strasbourg, France, 27 September–1 October 2021, Proceedings, Part VIII 24*; Springer International Publishing: Cham, Switzerland, 2021; pp. 45–54.
33. Kihara, Y.; Shen, M.; Shi, Y.; Jiang, X.; Wang, L.; Laiginhas, R.; Lyu, C.; Yang, J.; Liu, J.; Morin, R.; et al. Detection of Nonexudative Macular Neovascularization on Structural OCT Images Using Vision Transformers. *Ophthalmol. Sci.* **2022**, *2*, 100197. [CrossRef] [PubMed]
34. Li, A.L.; Feng, M.; Wang, Z.; Baxter, S.L.; Huang, L.; Arnett, J.; Bartsch, D.-U.G.; Kuo, D.E.; Saseendrakumar, B.R.; Guo, J.; et al. Automated Detection of Posterior Vitreous Detachment on OCT Using Computer Vision and Deep Learning Algorithms. *Ophthalmol. Sci.* **2023**, *3*, 100254. [CrossRef] [PubMed]
35. Philippi, D.; Rothaus, K.; Castelli, M. A vision transformer architecture for the automated segmentation of retinal lesions in spectral domain optical coherence tomography images. *Sci. Rep.* **2023**, *13*, 517. [CrossRef]
36. Xuan, M.; Wang, W.; Shi, D.; Tong, J.; Zhu, Z.; Jiang, Y.; Ge, Z.; Zhang, J.; Bulloch, G.; Peng, G.; et al. A Deep Learning–Based Fully Automated Program for Choroidal Structure Analysis Within the Region of Interest in Myopic Children. *Transl. Vis. Sci. Technol.* **2023**, *12*, 22. [CrossRef] [PubMed]
37. Sivaswamy, J.; Krishnadas, S.R.; Joshi, G.D.; Jain, M.; Tabish, A.U.S. Drishti-Gs: Retinal Image Dataset for Optic Nerve Head(ONH) Segmentation. In Proceedings of the 2014 IEEE 11th International Symposium on Biomedical Imaging (ISBI), Beijing, China, 29 April–2 May 2014; pp. 53–56. Available online: https://cvit.iiit.ac.in/projects/mip/drishti-gs/mip-dataset2/Home.php (accessed on 16 October 2023).
38. sjchoi86. sjchoi86-HRF. Available online: https://github.com/yiweichen04/retina_dataset (accessed on 20 October 2016).
39. Fumero, F.; Diaz-Aleman, T.; Sigut, J.; Alayon, S.; Arnay, R.; Angel-Pereira, D. Rim-One Dl: A Unified Retinal Image Database for Assessing Glaucoma Using Deep Learning. *Image Anal. Ster.* **2020**, *39*, 161–167. [CrossRef]
40. Zhang, Z.; Yin, F.S.; Liu, J.; Wong, W.K.; Tan, N.M.; Lee, B.H.; Cheng, J.; Wong, T.Y. ORIGA(-light): An online retinal fundus image database for glaucoma analysis and research. *Annu. Int. Conf. IEEE Eng. Med. Biol. Soc.* **2010**, *2010*, 3065–3068. Available online: https://www.kaggle.com/datasets/sshikamaru/glaucoma-detection (accessed on 16 October 2023). [CrossRef]
41. Diaz-Pinto, A.; Morales, S.; Naranjo, V.; Köhler, T.; Mossi, J.M.; Navea, A. CNNs for automatic glaucoma assessment using fundus images: An extensive validation. *Biomed. Eng. Online* **2019**, *18*, 29. [CrossRef]
42. Fang, H.; Li, F.; Wu, J.; Fu, H.; Sun, X.; Son, J.; Yu, S.; Zhang, M.; Yuan, C.; Bian, C. REFUGE2 Challenge: A Treasure Trove for Multi-Dimension Analysis and Evaluation in Glaucoma Screening. *arXiv* **2022**, arXiv:2202.08994. Available online: https://ai.baidu.com/broad/download (accessed on 16 October 2023).
43. Chen, L.-C.; Zhu, Y.; Papandreou, G.; Schroff, F.; Adam, H. Encoder-Decoder with Atrous Separable Convolution for Semantic Image Segmentation. In Proceedings of the European Conference on Computer Vision (ECCV), Cham, Switzerland, 8–14 September 2018; pp. 833–851.
44. Vision & Eye Health Surveillance System. Review: Glaucoma. Available online: https://www.cdc.gov/visionhealth/vehss/data/studies/glaucoma.html (accessed on 12 October 2023).
45. Dhillon, A.; Verma, G.K. Convolutional neural network: A review of models, methodologies and applications to object detection. *Prog. Artif. Intell.* **2019**, *9*, 85–112. [CrossRef]
46. Wang, P.; Yuan, M.; He, Y.; Sun, J. 3D augmented fundus images for identifying glaucoma via transferred convolutional neural networks. *Int. Ophthalmol.* **2021**, *41*, 2065–2072. [CrossRef]
47. Rogers, T.W.; Jaccard, N.; Carbonaro, F.; Lemij, H.G.; Vermeer, K.A.; Reus, N.J.; Trikha, S. Evaluation of an AI system for the automated detection of glaucoma from stereoscopic optic disc photographs: The European Optic Disc Assessment Study. *Eye* **2019**, *33*, 1791–1797. [CrossRef] [PubMed]
48. Jammal, A.A.; Thompson, A.C.; Mariottoni, E.B.; Berchuck, S.I.; Urata, C.N.; Estrela, T.; Wakil, S.M.; Costa, V.P.; Medeiros, F.A. Human Versus Machine: Comparing a Deep Learning Algorithm to Human Gradings for Detecting Glaucoma on Fundus Photographs. *Arch. Ophthalmol.* **2020**, *211*, 123–131. [CrossRef] [PubMed]
49. Thompson, A.C.; Jammal, A.A.; Medeiros, F.A. A Deep Learning Algorithm to Quantify Neuroretinal Rim Loss From Optic Disc Photographs. *Arch. Ophthalmol.* **2019**, *201*, 9–18. [CrossRef]

50. Ma, J.; Bai, Y.; Zhong, B.; Zhang, W.; Yao, T.; Mei, T. Visualizing and Understanding Patch Interactions in Vision Transformer. In *IEEE Transactions on Neural Networks and Learning Systems*; IEEE: New York, NY, USA, 2023; pp. 1–10. [CrossRef]
51. Fan, R.; Alipour, K.; Bowd, C.; Christopher, M.; Brye, N.; Proudfoot, J.A.; Goldbaum, M.H.; Belghith, A.; Girkin, C.A.; Fazio, M.A.; et al. Detecting Glaucoma from Fundus Photographs Using Deep Learning without Convolutions: Transformer for Improved Generalization. *Ophthalmol. Sci.* **2023**, *3*, 100233. [CrossRef]
52. Song, D.; Fu, B.; Li, F.; Xiong, J.; He, J.; Zhang, X.; Qiao, Y. Deep Relation Transformer for Diagnosing Glaucoma With Optical Coherence Tomography and Visual Field Function. *IEEE Trans. Med. Imaging* **2021**, *40*, 2392–2402. [CrossRef] [PubMed]
53. Hou, K.; Bradley, C.; Herbert, P.; Johnson, C.; Wall, M.; Ramulu, P.Y.; Unberath, M.; Yohannan, J. Predicting Visual Field Worsening with Longitudinal OCT Data Using a Gated Transformer Network. *Ophthalmology* **2023**, *130*, 854–862. [CrossRef]
54. Yi, Y.; Jiang, Y.; Zhou, B.; Zhang, N.; Dai, J.; Huang, X.; Zeng, Q.; Zhou, W. C2FTFNet: Coarse-to-fine transformer network for joint optic disc and cup segmentation. *Comput. Biol. Med.* **2023**, *164*, 107215. [CrossRef]
55. Zhang, Y.; Li, Z.; Nan, N.; Wang, X. TranSegNet: Hybrid CNN-Vision Transformers Encoder for Retina Segmentation of Optical Coherence Tomography. *Life* **2023**, *13*, 976. [CrossRef]
56. Vali, M.; Mohammadi, M.; Zarei, N.; Samadi, M.; Atapour-Abarghouei, A.; Supakontanasan, W.; Suwan, Y.; Subramanian, P.S.; Miller, N.R.; Kafieh, R.; et al. Differentiating Glaucomatous Optic Neuropathy From Non-glaucomatous Optic Neuropathies Using Deep Learning Algorithms. *Arch. Ophthalmol.* **2023**, *252*, 1–8. [CrossRef]
57. Christopher, M.; Belghith, A.; Bowd, C.; Proudfoot, J.A.; Goldbaum, M.H.; Weinreb, R.N.; Girkin, C.A.; Liebmann, J.M.; Zangwill, L.M. Performance of deep learning architectures and transfer learning for detecting glaucomatous optic neuropathy in fundus photographs. *Sci. Rep.* **2018**, *8*, 16685. [CrossRef]
58. Luo, X.; Hu, M.; Song, T.; Wang, G.; Zhang, S. Semi-Supervised Medical Image Segmentation via Cross Teaching between CNN and Transformer. In Proceedings of the 5th International Conference on Medical Imaging with Deep Learning, Zurich, Switzerland, 6–8 July 2022; pp. 820–833.
59. Li, A.; Cheng, J.; Wong, D.W.K.; Liu, J. Integrating holistic and local deep features for glaucoma classification. In Proceedings of the 2016 38th Annual International Conference of the IEEE Engineering in Medicine and Biology Society (EMBC), Orlando, FL, USA, 16–20 August 2016; pp. 1328–1331.
60. Li, L.; Xu, M.; Liu, H.; Li, Y.; Wang, X.; Jiang, L.; Wang, Z.; Fan, X.; Wang, N. A Large-Scale Database and a CNN Model for Attention-Based Glaucoma Detection. *IEEE Trans. Med. Imaging* **2020**, *39*, 413–424. [CrossRef] [PubMed]

Disclaimer/Publisher's Note: The statements, opinions and data contained in all publications are solely those of the individual author(s) and contributor(s) and not of MDPI and/or the editor(s). MDPI and/or the editor(s) disclaim responsibility for any injury to people or property resulting from any ideas, methods, instructions or products referred to in the content.

Review

Advances and Challenges in Wearable Glaucoma Diagnostics and Therapeutics

Ryan Shean [1,†], Ning Yu [2,†], Sourish Guntipally [3], Van Nguyen [4], Ximin He [5], Sidi Duan [5], Kimberly Gokoffski [4], Yangzhi Zhu [3] and Benjamin Xu [4,*]

1. Keck School of Medicine, University of Southern California, 1975 Zonal Avenue, Los Angeles, CA 90033, USA; rshean@usc.edu
2. Department of Chemical Engineering, Stanford University, Stanford, CA 94305, USA; ning.yu@email.ucr.edu
3. Terasaki Institute for Biomedical Innovation, 21100 Erwin Street, Los Angeles, CA 90064, USA; s.guntipally@yahoo.com (S.G.); yzhu@terasaki.org (Y.Z.)
4. Roski Eye Institute, Keck School of Medicine, University of Southern California, 1450 San Pablo Street, Los Angeles, CA 90033, USA; van.nguyen2@med.usc.edu (V.N.); kimberly.gokoffski@med.usc.edu (K.G.)
5. Department of Materials Science and Engineering, University of California, 410 Westwood Plaza, Los Angeles, CA 90095, USA; ximinhe@ucla.edu (X.H.); sididuan@ucla.edu (S.D.)
* Correspondence: benjamin.xu@med.usc.edu; Tel.: +1-232-442-6780
† These authors contributed equally to this work.

Citation: Shean, R.; Yu, N.; Guntipally, S.; Nguyen, V.; He, X.; Duan, S.; Gokoffski, K.; Zhu, Y.; Xu, B. Advances and Challenges in Wearable Glaucoma Diagnostics and Therapeutics. *Bioengineering* 2024, 11, 138. https://doi.org/10.3390/bioengineering11020138

Academic Editor: Dimitrios Karamichos

Received: 30 December 2023
Revised: 24 January 2024
Accepted: 26 January 2024
Published: 30 January 2024

Copyright: © 2024 by the authors. Licensee MDPI, Basel, Switzerland. This article is an open access article distributed under the terms and conditions of the Creative Commons Attribution (CC BY) license (https://creativecommons.org/licenses/by/4.0/).

Abstract: Glaucoma is a leading cause of irreversible blindness, and early detection and treatment are crucial for preventing vision loss. This review aims to provide an overview of current diagnostic and treatment standards, recent medical and technological advances, and current challenges and future outlook for wearable glaucoma diagnostics and therapeutics. Conventional diagnostic techniques, including the rebound tonometer and Goldmann Applanation Tonometer, provide reliable intraocular pressure (IOP) measurement data at single-interval visits. The Sensimed Triggerfish and other emerging contact lenses provide continuous IOP tracking, which can improve diagnostic IOP monitoring for glaucoma. Conventional therapeutic techniques include eye drops and laser therapies, while emerging drug-eluting contact lenses can solve patient noncompliance with eye medications. Theranostic platforms combine diagnostic and therapeutic capabilities into a single device. Advantages of these platforms include real-time monitoring and personalized medication dosing. While there are many challenges to the development of wearable glaucoma diagnostics and therapeutics, wearable technologies hold great potential for enhancing glaucoma management by providing continuous monitoring, improving medication adherence, and reducing the disease burden on patients and healthcare systems. Further research and development of these technologies will be essential to optimizing patient outcomes.

Keywords: glaucoma; theranostics; diagnostics; therapeutics; smart contact lens

1. Introduction

Glaucoma is the leading cause of irreversible blindness worldwide, affecting an estimated 3.5% of people aged from 40 to 80 years globally [1,2]. The majority of glaucoma cases go undetected and untreated, leading to irreversible vision loss and, ultimately, blindness. The World Health Organization estimates that 7.7 million people with glaucoma experience preventable visual impairment or blindness globally [3]. By 2040, the number of people with glaucoma is expected to increase to 111.8 million, highlighting the urgent need for innovative diagnostic and therapeutic strategies [1].

Glaucoma is a chronic and slowly progressive disease of the optic nerve that tends to be asymptomatic in its early stages. When left untreated, glaucoma can progress to irreversible tunnel vision or even complete visual field loss [4]. Elevated intraocular pressure (IOP) is a major risk factor for the development and progression of glaucoma. Several landmark clinical trials, including the Ocular Hypertension Treatment Study (OHTS) and the Early Manifest Glaucoma Trial (EMGT), have demonstrated that reducing IOP can delay or prevent the onset of glaucoma in individuals with ocular hypertension or early glaucoma [5,6]. These studies provided evidence for the importance of early treatment and IOP control in the management of glaucoma.

Current diagnostic and therapeutic strategies for glaucoma have limitations. As early-stage glaucoma is often asymptomatic, patients may experience delays in receiving diagnostic tests to detect glaucoma, thereby leading to delays in treatment. Even when patients with early glaucoma are evaluated, IOP fluctuations throughout each day may not be accurately represented by a single measurement in the clinic. Therefore, continuous IOP monitoring could advance our understanding of the relationship between IOP and glaucoma onset or progression [7]. Poor patient adherence to medication regimens is another significant problem that contributes to preventable disease progression and vision loss. Additionally, fluctuations and spikes in IOP are considered detrimental, similar to elevated average IOP. Consequently, both conditions benefit from regular monitoring and prompt intervention [8,9]. Therefore, the development of convenient drug delivery systems for glaucoma management has the potential to significantly improve treatment standards. This review aims to provide an overview of current diagnostic and treatment standards, recent medical and technological advances, and current challenges and future outlook for wearable glaucoma diagnostics and therapeutics.

Advancements in wearable devices for the management of glaucoma mirror broader progress observed across numerous medical fields. A recent review article by Kazanskiy et al. highlighted the rapidly growing demand for wearable devices in monitoring various medical conditions, such as diabetes, heart disease, and hypertension [10]. The worldwide market for wearables was USD 978.86 million in 2022 and is expected to grow to USD 4336.7 million by 2032 [10]. Another recent review by Wu et al. outlined numerous advancements in intraocular pressure biosensor engineering [11]. This review aims to expand upon these articles by focusing on extraocular IOP sensors with integrated drug delivery capabilities. By focusing on diagnostic devices with therapeutic capabilities, we present a forward-looking perspective on theranostics—a convergence of diagnostics and therapeutics in an all-in-one device that could reshape future glaucoma management.

2. Diagnostics of Glaucoma

2.1. Current Methods for Measuring IOP

2.1.1. Goldmann Applanation Tonometer

Goldmann applanation tonometry (GAT) stands as the benchmark for measuring IOP in clinical practice and research and is widely recognized as the gold standard technique [12]. GAT is a slit-lamp-mounted device that uses a calibrated prism to apply a known force to the cornea (Figure 1). The amount of force needed to indent the cornea by a fixed amount is directly proportional to the IOP as long as the cornea has a standard thickness and curvature. GAT has high reproducibility in measuring IOP, demonstrating less variability in IOP measurement than other methods of tonometry [13]. The American Academy of Ophthalmology (AAO) recommends GAT as the standard for IOP measurements in clinical practice. In addition, GAT is the standard in scientific research, including clinical trials to assess the efficacy of glaucoma treatments [14].

The accuracy of GAT measurements can be affected by corneal thickness and curvature, which may lead to an over- or underestimation of IOP [15,16]. Accurate and consistent measurement of IOP with GAT requires a trained operator. GAT involves a patient being seated in a slit-lamp biomicroscope, a setup that may not be practical in situations where patients are bedridden or unable to sit upright. Furthermore, as the tonometer makes direct contact with the cornea, it may cause discomfort for some patients. Additionally, GAT may be affected by ocular surface conditions, including dry eye or corneal edema, which can obscure the true IOP measurement [17]. Finally, as GAT requires specialized equipment and a trained operator, it can only be used in-office and cannot be used to monitor IOP fluctuations continuously.

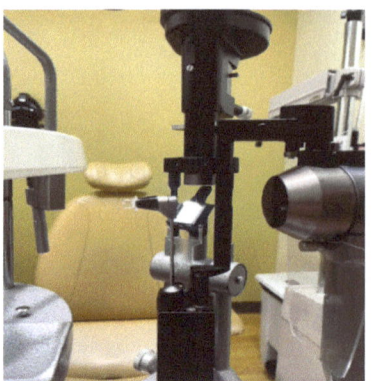 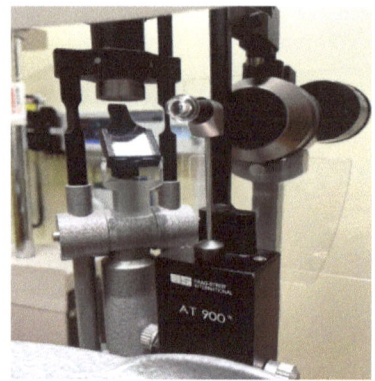

Figure 1. Photographs of a Goldmann Applanation Tonometer.

2.1.2. Handheld Tonometers

A rebound tonometer, an alternative to GAT, is a handheld device that measures the deceleration of a small probe rebounding off of the cornea to calculate IOP. These devices are non-invasive, portable, and can be easily used in clinics or at home, especially by patients who do not tolerate GAT. Rebound tonometers show a high degree of concordance with GAT and yield reproducible IOP measurements [18–20]. They also require less patient cooperation and little to no topical anesthesia. A study showed that 73.7% of patients rated rebound tonometry more comfortable than GAT [21]. Therefore, rebound tonometers are well-suited for use with children, elderly patients, and patients with cognitive or physical disabilities who cannot tolerate other IOP measurement methods. Rebound tonometers enable a rapid measuring process, taking just a few seconds per eye, and their compact, portable design allows for use in a wider range of settings compared to GAT.

Similar to GAT, the accuracy of rebound tonometry can be affected by corneal thickness and curvature, overestimating IOP in patients with thicker corneas [22–24]. The accuracy and consistency of IOP reading also depend on patient cooperation, blinking during measurements, and corneal hysteresis [25].

The iCare HOME Tonometer (Figure 2) is a portable rebound tonometer device that allows patients to monitor their IOP levels at home without the need for assistance from a healthcare professional. The iCare HOME Tonometer is an effective tool for monitoring IOP, with 80% of the measurements falling within 3.0 mmHg of IOP measured with GAT [26]. The device is especially useful for patients who have difficulty visiting clinics due to age or mobility, or for those who live in remote areas with limited access to healthcare facilities. The iCare HOME Tonometer can remind patients to check their IOP and alert patients when IOP measurement results are outside of a target range. The users can store and track their IOP data over time using a mobile phone application. Survey results suggest that 78.5% of patients found the iCare Home Tonometer easy to use [27]. The primary limitation of the iCare HOME Tonometer is its cost; the device is relatively expensive and is not covered by most insurances.

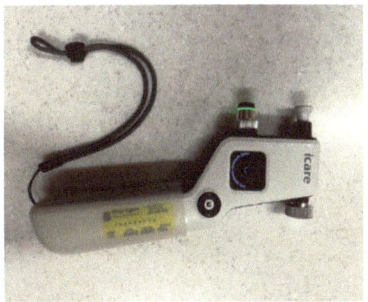

Figure 2. Photographs of an iCare Home Tonometer.

2.1.3. Sensimed Triggerfish and GlakoLens

The Sensimed Triggerfish (Figure 3) is a soft contact lens equipped with a microsensor that continuously measures changes in corneal curvature, which can be used to calculate relative changes in IOP [28]. The device is worn on the eye like a regular contact lens and can be worn during normal daily activities, providing continuous IOP monitoring without the need for manual measurement at specific times. Conventional techniques for measuring IOP, as discussed in Section 2.1, only measure IOP at a single point in time. Therefore, they may not capture diurnal or nocturnal fluctuations in IOP [29]. The Sensimed Triggerfish is the only FDA-approved device for continuous IOP monitoring, offering invaluable data that can enhance the precision of diagnosis and treatment [30,31]. Research about the Triggerfish reported a high level of patient safety and tolerability [32,33].

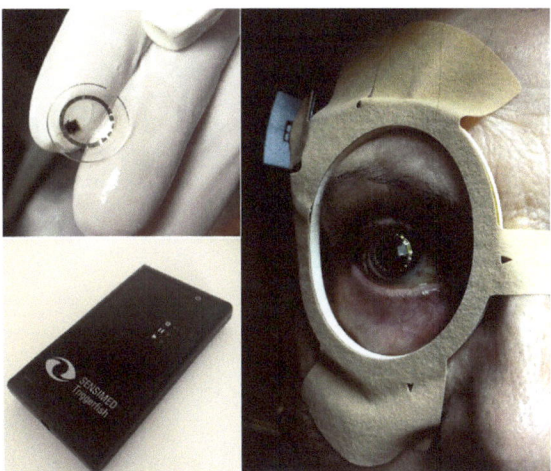

Figure 3. Sensimed Triggerfish contact lens. Adjusted with permission [34]. Copyright 2017, Elsevier.

The Sensimed Triggerfish has several limitations that have hindered its widespread clinical use despite its IOP-monitoring capabilities. First, the Triggerfish measures relative changes in IOP, so another device is still necessary to establish baseline IOP. It can also cause irritation and discomfort because it is an electronic device that is worn on the eye for extended periods of time. Additionally, changes among certain factors, such as air temperature, may add noise to the electronic signal, which can affect the accuracy of intraocular measurements [35,36]. Another major limitation is that the device is expensive but not currently covered by many insurances, which limits its accessibility. Finally, the device cannot be used in patients who are unable to tolerate wearing contact lenses.

In addition to the Triggerfish, the GlakoLens (Istanbul, Turkey) is an IOP-sensing smart contact lens that has not yet received FDA approval. The design of the GlakoLens system involves an electrically passive sensor embedded in a disposable soft contact lens which utilizes resonant frequency shift of a metallic resonator for 24-h IOP monitoring. The sensor's meta-material properties enable accurate measurement of IOP fluctuations by detecting changes in corneal geometry. Data from the sensor can be collected wirelessly via a low-power RF signal generated by a Holter-monitor-like device with an electronic circuit and wearable antenna. GlakoLens has taken proactive steps at commercialization, developing a user-friendly website and intellectual property protection (US Patent No: US10067075B2 and pending PCT applications). By prioritizing ease of use, design innovation, and intellectual property protection, GlakoLens may achieve successful commercialization in the coming years.

2.2. Recent Advances in Wearable Diagnostics

Wearable diagnostics are a relatively new development in ophthalmology, offering the ability to monitor IOP using contact lenses and implants. Wearable IOP sensors can measure IOP continuously or at predefined intervals, enabling real-time monitoring and timely detection of IOP fluctuations. In addition to contact lens-based sensors, there has been research into implantable IOP sensors in sites such as the anterior chamber, capsular bags, vitreous, and choroid [37]. Wearable IOP sensors and other diagnostic tools can be used for diagnosing, monitoring, and predicting the progression of glaucoma. Continuous IOP monitoring can provide valuable information on IOP fluctuations, which is an important risk factor for glaucoma progression [37,38]. In addition to IOP, these sensors have the potential to monitor glucose, lactate, and cortisol levels as well as the humidity of the ocular surface, which enables the possibility of monitoring for and detecting additional conditions, such as diabetes and liver disease [37,39–41].

Wang et al. developed a smart contact lens with a dual-sensing platform for real-time monitoring of IOP and detecting matrix metalloproteinase-9 (MMP-9) in tears, which is a biomarker in eye-related diseases including glaucoma [42]. The unique design (Figure 4) contained surface-enhanced Raman spectroscopy (SERS) substrates. IOP could be monitored by observing structural color changes, and glaucoma could be predicted using quantitative SERS measurement of MMP-9 levels in tear film. Tests using porcine eyes found that the contact lens provided accurate measurements of IOP in the range from 0 to 30 mmHg (R2 = 0.98), covering the normal physiological IOP range (10–20 mmHg). Meanwhile, the SERS analysis can effectively detect MMP-9 down to concentrations of 1.29 ng/mL.

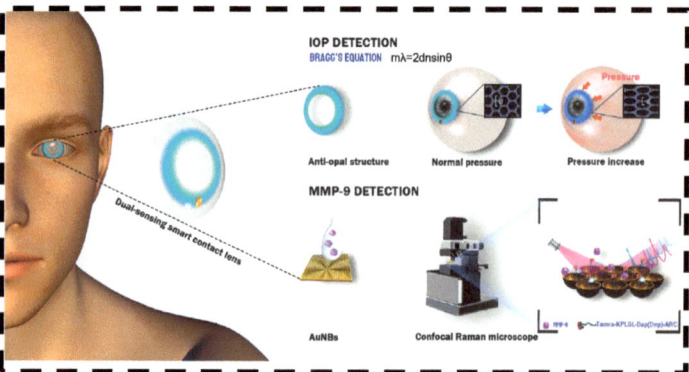

Figure 4. Schematic illustration of dual-functional contact lens for IOP detection and MMP-9 detection. Open access [42]. Copyright 2022, Wiley-VCH GmbH.

Electronic smart contact lenses that monitor IOP changes usually have relatively simple structural designs and can be constructed from easily accessible materials. However, these devices sometimes have difficulty differentiating small changes in IOP from signal noise induced by activities such as blinking. In order to reduce noise, Hu et al. created a contact lens with a self-lubricating layer that reduces the coefficient of friction to remove interference from the tangential forces [43]. The contact lens is essentially composed of three distinct layers (Figure 5), including a substrate layer, a flexible reinforced sensing layer, and a self-lubricating layer. The IOP sensor maintains the same level of sensitivity to normal forces with the addition of the self-lubricating layer, while significantly reducing the effects induced by blinking and eye movement.

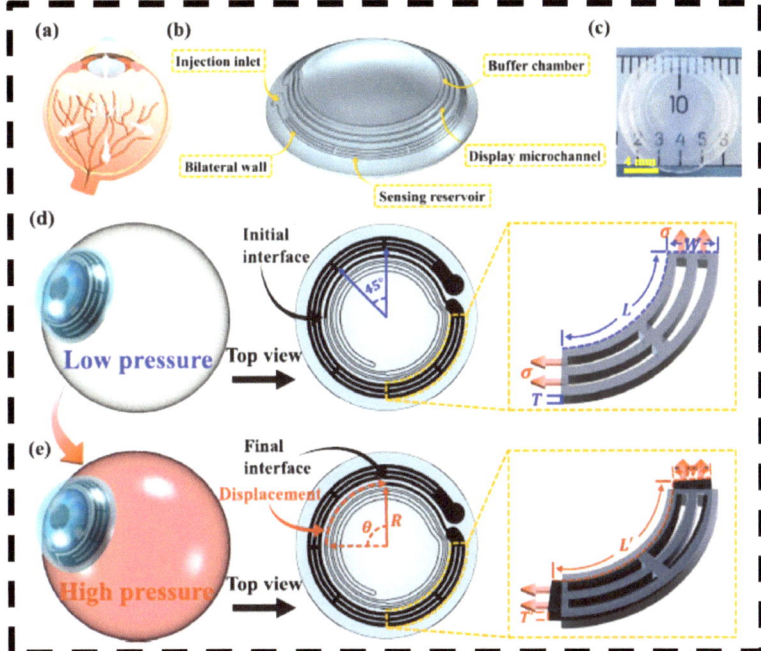

Figure 5. Schematic diagrams, photograph, and operation principle of the microfluidic contact lens. (a) Schematic diagram of the human eye with the contact lens. (b) Schematic diagram of the contact lens. (c) Photograph of the contact lens. (d,e) The operation principle of the contact lens in the setting of elevated IOP. The mechanical mechanism of the sensor is represented on the right. Adapted with permission [44]. Copyright 2023, Elsevier.

While electronic wearable devices can provide consistent real-time IOP monitoring, the inclusion of electronic components in contact lenses reduces water permeability and may cause corneal damage with long-term use. A structurally colored contact lens that changes color in response to changes in IOP can effectively avoid this problem. These contact lenses are composed of flexible materials and are free of complex electronic components [45–48]. Chen et al. designed a high-sensitivity microfluidic contact lens sensor composed of a sensing reservoir filled with dyed fish oil, a display microchannel, and a buffer chamber [43]. When IOP increases, the corneal radius of curvature increases and the enlarged volume of the sensing reservoir under the surface tension of the tear film pushes liquid in the display microchannel into the sensing reservoir. The displacement of the liquid interface reflects the IOP change. The contact lens can reach a sensitivity of 660 μm/mmHg in the IOP fluctuation range of 10–30 mmHg with a linear regression coefficient R^2 up to 0.99.

However, one limitation of this device is that small IOP changes are indistinguishable by the naked eye and are only detectable with specialized reflection spectrometers.

Continuous monitoring of IOP during sleep has been challenging in glaucoma care. In an attempt to address this issue, Lee et al. created a smart soft contact lens (SSCL) with an ocular tonometer built into its structure [49]. This device uses a circuit with a resistor, inductor, and capacitor in series (RLC) to produce a distinct electrical resonance frequency based on the characteristics of the inductor and capacitor (Figure 6). The electrical properties of the RLC circuit in this ocular tonometer vary depending on the radial and axial deformations by the contact lens. As a result, the RLC circuit produces a different resonance frequency when IOP increases, followed by curvature changes of both the eye and the contact lens. The device offers overnight wearability, superior signal quality compared to existing wearable ocular tonometers, and comfort on par with the Triggerfish. Additionally, the lens has no internal power source; rather, its IOP measurements can be read using glasses or a sleep mask fitted with a reader coil inductively paired with the ocular tonometer. This SSCL was fabricated to mirror commercial brands, preserving its inherent characteristics such as great biocompatibility, softness, transparency, and oxygen transmissibility.

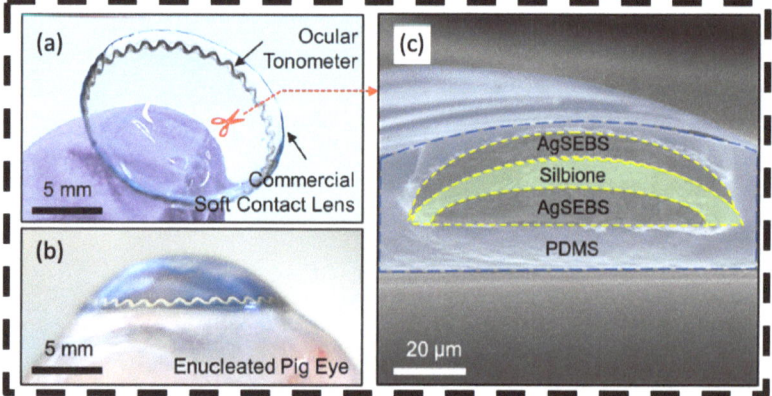

Figure 6. Schematic and optical images of smart soft contact lens for 24 h IOP monitoring. (**a**) Photograph of the contact lens. (**b**) Cross-sectional scanning electron microscope (SEM) image of the contact lens. (**c**) Photograph of the contact lens in an enucleated pig eye. Open access [49]. Copyright 2022, Springer Nature.

Variations in oxygen and water content, as well as the stiffness of the contact lenses, may lead to discomfort for certain users and introduce signal noise. To address these issues, Zolfaghari et al. created wearable glasses with a laser source, lenses, mirrors, mask, and a camera for personalized real-time IOP monitoring [50]. Using the lens, mirrors, and mask, it creates a grid on the cornea which it measures with the camera to detect changes in corneal curvature. This device was justified with analytical modeling, ray tracing, and FEM simulations [50]. It produced a pressure measurement resolution of 2.4 mmHg between 0 and 55 mmHg pressure. Zolfaghari et al. produced a separate glasses-based solution with an implantable diffraction grating MEMS sensor [51]. The readout glasses were embedded with a laser dioxide, miniaturized aspheric lenses, and a complementary-metal-oxide semiconductor (CMOS) camera [51].

While wearable IOP sensors and other glaucoma diagnostics offer several advantages, they also have limitations. Some materials used in contact lens sensors, such as polymerized hydroxyethyl methacrylate (pHEMA) and silicone hydrogel (SiH), may be affected by local hydration levels, which can introduce noise and inaccuracies to IOP measurements [37]. Due to signal noise and variations in corneal parameters, these wearable diagnostics are not as accurate or stable as GAT, which is considered the clinical standard for measuring IOP [14,37]. Furthermore, the cost of wearable glaucoma diagnostics may prevent some patients from accessing them in the future. Further research is needed to establish the efficacy, safety, and cost-effectiveness of wearable glaucoma diagnostics.

3. Therapeutics

3.1. Current Methods for Lowering IOP

3.1.1. Eye Drops

Eye drops are a common form of glaucoma treatment that can be prescribed in different classes, including prostaglandin analogs, beta blockers, alpha agonists, and carbonic anhydrase inhibitors (CAIs). The primary goal of using eye drops is to prevent or slow down the progression of glaucoma by lowering IOP, either by decreasing the amount of aqueous humor produced by the eye or improving its outflow from the eye [52].

Prostaglandin analogs are the most commonly prescribed eye drops due to their effectiveness in reducing IOP and convenient once-a-day dosing [52]. They work by increasing the outflow of aqueous from the eye via the uveoscleral pathway. They have greater efficacy, lowering IOP by 25 to 30%, and fewer systemic side effects compared to other eye drop medication classes. However, like many typical glaucoma medications, prostaglandin analogs may induce eye redness and other localized side effects in the eye and periocular area, such as pigmentary changes, lengthening of eyelashes, fat atrophy, and an increased risk of uveitis and cystoid macular edema [53].

Latanoprostene bunod, marketed under the name Vyzulta, is a novel glaucoma treatment that combines latanoprost, a prostaglandin analog, with a nitric oxide donor, offering a dual mechanism to reduce IOP [54]. Latanoprost increases uveoscleral outflow, while nitric oxide increases conventional outflow through the trabecular meshwork, resulting in an average IOP reduction of 27% within 24 h with a generally favorable safety profile [55,56]. Beta-blockers reduce IOP by inhibiting aqueous humor production, but they can adversely affect cardiovascular and respiratory systems, potentially reducing lung function and increasing asthma morbidity [57–59]. Alpha-adrenergic agonists decrease aqueous humor production and increase outflow, achieving a 20–25% IOP reduction. Despite effectiveness, they may cause allergic conjunctivitis and systemic side effects, making them less common as first-choice monotherapy but widely used in multi-drug glaucoma treatment [60,61].

Carbonic anhydrase inhibitors reduce IOP by decreasing aqueous humor production but are less commonly used as monotherapy due to potential side effects like metallic taste, ocular irritation, and corneal edema [6,62]. Netarsudil, a novel eye drop, lowers IOP by inhibiting Rho kinase and norepinephrine transporter, enhancing trabecular outflow and reducing aqueous production [54]. Clinical studies show a 20–25% IOP reduction, with ocular side effects like conjunctival hyperemia, subconjunctival hemorrhage, and blurred vision, but relatively rare systemic side effects compared to beta-blockers and alpha agonists [54,63].

Despite the benefits of using eye drops as a therapeutic option for glaucoma, there are numerous limitations associated with their use. Eye drops require careful administration and may have side effects such as ocular surface irritation and systemic effects [58,59]. Moreover, some patients may not be able to tolerate the side effects associated with different classes of eye drops, resulting in nonadherence and insufficient IOP management, particularly among those patients with limited health literacy [64]. Finally, patients who have difficulty administering the drops may need support from a caregiver or healthcare professional, potentially leading to financial difficulties and increased noncompliance [65].

3.1.2. Intracameral Implants

Intracameral implants emerge as a promising new approach for treating glaucoma. These are small devices that can be implanted into the eye to deliver medications. Several varieties of intracameral implants, such as DURYSTA (Allergan, Irvine, CA, USA), OTX-TIC (Ocular Therapeutix, Bedford, MA, USA), and iDose (Glaukos, Aliso Viejo, CA, USA), offer direct and sustained medication release into the anterior chamber of the eye. DURYSTA is an intracameral implant that was recently approved by the FDA for the treatment of open-angle glaucoma or ocular hypertension [66]. It is a biodegradable implant that slowly releases bimatoprost, a prostaglandin analog. DURYSTA is the only FDA-approved intracameral implant, providing a sustained reduction in IOP of 20 to 25% [67]. However, DURYSTA is only approved for a single dose due to the risk of corneal edema [66,68]. iDose is another example of an intracameral implant that has demonstrated success in several clinical trials [68]. It is a small titanium implant that is surgically inserted into the trabecular meshwork to deliver a formulation of travoprost over a period of six to twelve months [68]. After all the travoprost has been depleted, this implant can be removed and exchanged [69]. OTX-TIC is another intracameral implant that has shown promising results in clinical trials [68]. Similar to DURYSTA, it is a biodegradable implant that is designed to release a formulation of micronized travoprost over an extended period, typically spanning four to six months. Once the travoprost supply is depleted, the implant is bioresorbable.

The main disadvantage of intracameral implants is that they are invasive and have short-term risks, such as infection or bleeding, and long-term complications, such as endothelial damage and persistent corneal edema. Intracameral implants can also be more expensive than other treatment options [68]. Finally, given that intracameral implants are a relatively recent therapeutic innovation, further research is required to ascertain the safety of repeated injections [70].

3.1.3. Lasers and Surgery

Laser and surgical treatments serve as adjuncts to medications in the comprehensive management of glaucoma. Laser therapies are generally considered low-risk and can decrease aqueous production or enhance aqueous outflow via the trabecular meshwork, efficiently lowering IOP and managing glaucoma. Common laser therapies used to treat glaucoma include trabeculoplasty, peripheral iridotomy, and cyclophotocoagulation [71]. Surgery remains the definitive standard for lowering IOP when glaucoma proves resistant to medical or laser treatment [72,73]. Glaucoma surgeries range from minimally invasive glaucoma surgery (MIGS), offering reduced risk with modest efficacy, to traditional glaucoma surgeries like trabeculectomy and tube shunts, which yield higher efficacy but come with increased risk. Nonetheless, glaucoma surgery carries significant risks, including infection, bleeding, and permanent vision loss [74]. Surgery may also be inaccessible to some patients due to the high costs of surgical treatment.

3.2. Recent Advances in Wearable Therapeutics

Drug-eluting contact lenses, soft contact lenses designed to release medication into the eye over an extended period of time, can be used to treat a wide range of ocular conditions [75,76]. These contact lenses offer the advantage of sustained drug release, which improves medication adherence and reduces side effects. Drug-eluting contact lenses can release therapeutic levels of medication for up to several months, demonstrating their potential as a sustained drug delivery system [77]. Some studies have reported that drug-eluting contact lenses result in greater IOP reduction than eye drops [78]. As patient nonadherence to conventional eye drops poses a significant obstacle to effective glaucoma care, drug-eluting contact lenses are seen as a promising alternative treatment option [79].

A novel type of contact lens was recently developed by Que et al. with embedded microtubes as drug containers (Figure 7) [80]. This type of contact lens can be fabricated by combining a ball-mold fabrication process and soft lithography. Drug release is based on diffusion and adaptive mechanisms; when IOP rises, the contact lens is stretched and the deformation of the microtubes triggers the diffusion of drugs. By tuning the tube size, density, and drug loading, the contact lens can achieve an extended drug release of up to 45 days.

While the immersion method is the simplest and most cost-effective way to prepare drug-loaded contact lenses, it has the disadvantages of low drug loading and fast release speed [81,82]. Shiau et al. resolved these issues by incorporating large-pore mesoporous silica nanoparticles (LPMSNs) with high surface area, large pore volume, and tunable pore size into drug-eluting contact lenses (DCLs) [83]. LPMSN-laden DCLs do not require drug preloading and can be fabricated with current contact lens manufacturing processes. Compared with standard DCLs, LPMSN-laden DCLs exhibited enhanced maximum loading and release capacities for glaucoma drugs and slower release rates, significantly extending the duration of drug release.

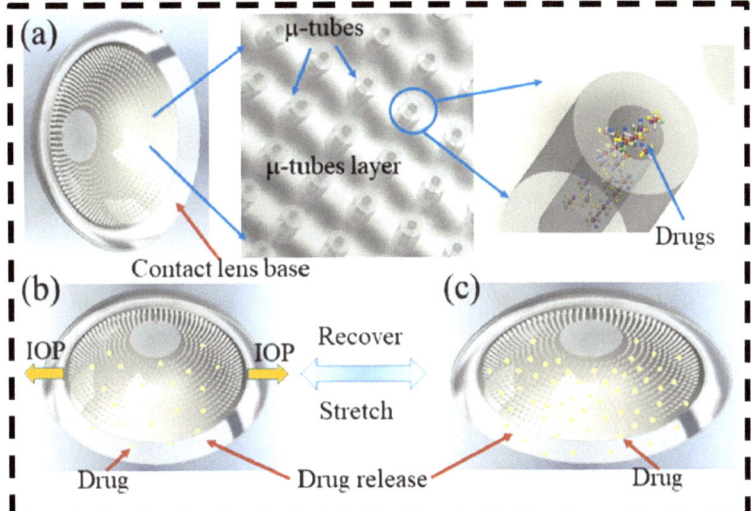

Figure 7. Schematic illustration of contact lens embedded with microtubes as drug containers. (a) Schematic diagram of the contact lens with embedded microtubes as drug containers for diffusion-based drug delivery and adaptive drug delivery. (b) Schematic of the contact lens device under a neutral state for diffusion-based drug release. (c) Illustration of the mechanically stretched contact lens device with more drug being released. Adjusted with permission [80]. Copyright 2020, American Chemical Society.

In another study, a type of flat microfluidic contact lens was fabricated, integrating a microchannel and a micropump (Figure 8) [47]. Drugs confined in the microchannels can be released by applying pressure on the pump chamber, such as through blinking, making liquid release controllable and adjustable. Different types of drugs can be loaded in different microchambers for multi-drug treatment applications. The contact lens exhibited good flexibility, light transmission, and biocompatibility without the need for electronic components, providing a safe, convenient, and effective method to treat ocular diseases.

The bimatoprost periocular ring (Allergan, Irvine, CA, USA) is a ring-shaped drug-eluting device that rests under the eyelids in the fornix. The device has passed Phase 2 trials, with the ability to elute bimatoprost and reduce IOP for up to 6 months [84]. Similar to contact lenses, periocular rings allow eye care providers to address the issue of medication nonadherence. However, to date, no periocular rings have received FDA approval for the treatment of glaucoma.

There are currently no FDA-approved drug-eluting contact lenses to treat glaucoma. However, drug-eluting contact lenses have been approved to treat other eye conditions, including myopia [85]. One issue with drug-eluting contact lenses is that they can deliver only small amounts of medication to patients, and the rate of release may not be linear [77]. In addition to potential safety concerns, patient and practitioner acceptance, and storage considerations, fit, comfort, and cost present further challenges [86]. Further research and development are essential to overcome these limitations before the full potential of drug-eluting contact lenses as a wearable therapeutic for glaucoma can be harnessed.

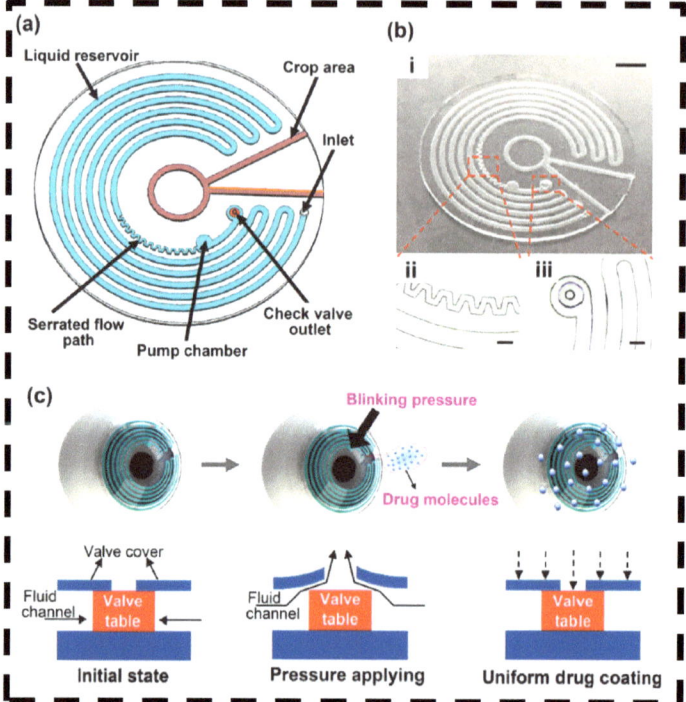

Figure 8. Schematic diagram and work principle of flat microfluidic contact lens. (**a**) Schematic diagram of the microfluidic chip. (**b**) (**i**) Picture of a flat microfluidic chip, scale bar: 2 mm; (**ii**) Enlarged view of the serrated flow path, scale bar: 300 μm; (**iii**) Enlarged view of the check valve outlet, scale bar: 300 μm. (**c**) Schematic illustration of drug release process realized by the pressure-triggered microfluidic contact lens. Adjusted with permission [47]. Copyright 2022, American Chemical Society.

4. Emerging Theranostic Platform

4.1. Description of Theranostics

Theranostics are a category of medical devices that integrate therapeutic drugs and diagnostic modalities into a united system [87]. Instead of using a "one size fits all" strategy, theragnostic adopts a personalized approach to patient management by customizing care to individual disease profiles and treatment responses [87,88]. Consequently, theranostic devices can automatically modify medication dosages in response to disease condition fluctuations [87]. In the context of glaucoma care, an example of a theranostic device is one that can monitor IOP and use this information to dynamically modulate the release of IOP-lowering drugs. Moreover, theranostic devices have the potential to monitor and image diseased tissue, analyze delivery kinetics, and maximize drug efficacy [87].

The interest in theranostics has grown significantly since the beginning of the century. Between 2000 and 2011, the annual publication count for research on theranostics and multifunctional therapies increased from none to 120 and 160 papers, respectively. The growing interest in theranostics, including dedicated journals on the topic, indicates its rising importance in medicine [89].

4.2. Clinical Uses of Theranostics

One of the challenges in glaucoma management is the accurate and continuous monitoring of IOP. As previously mentioned, traditional methods only provide intermittent IOP measurements during clinic visits. However, theranostic smart soft contact lenses (SSCLs) can provide continuous and non-invasive IOP monitoring. These lenses incorporate biosensing components that measure and transmit real-time IOP data wirelessly. By providing a comprehensive understanding of IOP dynamics, theranostic SSCLs facilitate personalized treatment strategies and timely interventions.

Glaucoma often requires long-term medication to control IOP and prevent disease progression. Conventional treatment methods, such as eye drops, have limitations in terms of patient adherence and drug bioavailability [64]. Theranostic SSCLs address these challenges by offering personalized drug delivery directly to the ocular surface. These lenses facilitate the controlled release of therapeutic agents, maintaining consistent drug levels over an extended duration. Theragnostic SSCLs hold the potential to circumvent the drawbacks of eye drops, thereby enhancing medication adherence, improving drug efficacy, and reducing patient discomfort [78]. Moreover, theranostic SSCLs can incorporate feedback mechanisms that optimize treatment based on individual patient needs. By integrating biosensing capabilities with drug delivery systems, these lenses can monitor IOP levels and adjust drug release accordingly. This dynamic approach ensures precise medication administration, minimizing IOP fluctuations and optimizing therapeutic outcomes. Theranostic SSCLs offer a minimally invasive and patient-friendly approach to provide continuous IOP monitoring and personalized drug delivery, which have the potential to revolutionize glaucoma management. By improving the treatment of glaucoma, they may even reduce the number of clinic visits and invasive procedures. Further research is needed to fully explore the capabilities of these technologies and their integration into routine clinical practice.

4.3. Emerging Technologies

Recent advances in smart contact lenses for glaucoma diagnosis and drug delivery have inspired growing research interest in the integration of both types of functionalities to enable continuous IOP monitoring and effective on-demand drug delivery to treat glaucoma. This closed-loop feedback system features a sensor in the contact lens that detects elevated IOP and an on-board signal processor that triggers the immediate release of preloaded drugs to lower IOP. When the IOP drops below a predetermined threshold, drug release would then be halted.

Hahn et al. recently developed a theranostic device that integrates electrical circuits on a contact lens for IOP monitoring, wireless data transmission, and coordinated drug delivery (Figure 9) [90]. The key attribute of this lens is its feedback mechanism, equipped with a highly sensitive gold hollow nanowire sensor for real-time IOP monitoring, and an adaptive drug delivery system that releases the glaucoma medication timolol on demand to modulate IOP. The IOP sensor can attain measurements on par with commercial tonometers, with a correlation coefficient of $R = 0.94$. Another significant benefit is that the technology can be personalized to react to a patient's unique IOP levels and treatment sensitivity. A wireless board receives the IOP measurements from the lens, which are subsequently transferred to and interpreted by a computer using low-energy Bluetooth.

Xie et al. developed another closed-loop theranostic SSCL that demonstrated IOP sensing and drug delivery capabilities called the wireless theragnostic contact lens (WTCL) [91]. This lens utilizes an iontophoresis drug delivery system, which allows for electrically controlled medication release and improved drug permeation. A double-layer structure was adopted to overcome the challenge of integrating multiple modules onto the space-limited, curved contact lens (Figure 9). Sensors and wireless power transfer circuits were embedded inside the contact lens between the double layers to avoid direct contact with the ocular surface. An ultra-soft air dielectric film between the layers provides high sensitivity to IOP fluctuations. Drug release from the hydrogel layer coated on an iontophoretic electrode is activated when IOP is higher than 21 mmHg. Using iontophoresis drug delivery, the WTCL can achieve an IOP reduction of over 20%, surpassing the 6.9 ± 14.7% lowering seen with slow diffusion over extended periods (~2 h) in rabbits. The WTCL IOP sensor maintains an error rate of <42%, compared to 10–14% error associated with a Tonopen.

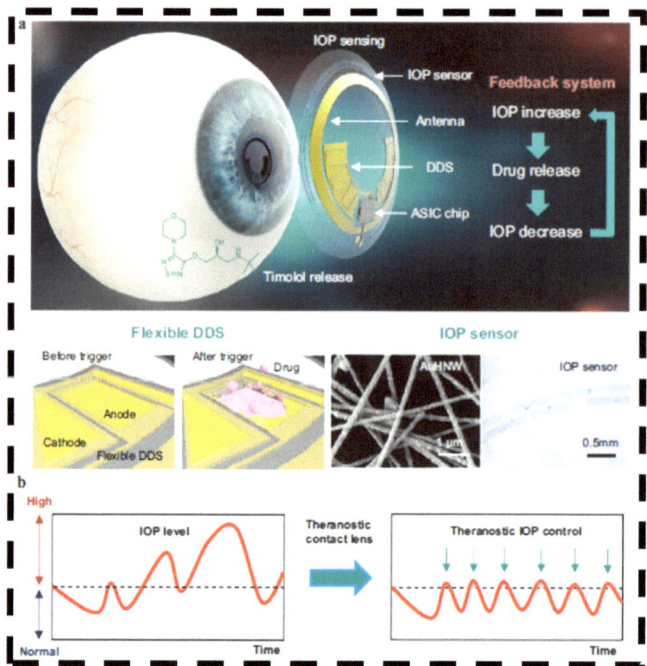

Figure 9. Schematic illustration of AuHNW-based theranostic contact lens for glaucoma treatment. (a) The structure of theranostic smart contact lens with a fully integrated AuHNW-based IOP senor, a DDS, and wireless circuits for wireless glaucoma treatments with a feedback system for IOP sensing and timolol release. (b) Schematic representation of the conventional continuous IOP monitoring and the IOP control by IOP monitoring and on-demand drug delivery for the treatment of glaucoma. Open access [90]. Copyright 2022, Springer Nature.

5. Current Challenges and Clinical Outlook

Smart contact lenses with diagnostic and therapeutic capabilities have the potential to revolutionize glaucoma management by providing continuous monitoring and targeted treatments. Diagnostic lenses offer a non-invasive and user-friendly alternative to traditional diagnostic tools like GAT, enabling the continuous monitoring of IOP fluctuations, facilitating early glaucoma detection and allowing for timely treatment before disease progression. Despite the advancements in the field of diagnostic lenses, current iterations are not without their limitations. Challenges such as the bulkiness and rigidity of embedded circuits, the failure to detect minute fluctuations in IOP, and the complexities involved in their manufacture, persist. Hydrogel-based colorimetric sensors emerge as a promising solution to these issues thanks to their inherent flexibility, straightforward production process, and the elimination of the need for an external power source [92]. These sensors employ various mechanisms for color or transparency alteration, including interference [93], scattering [94], and diffraction [95]. Some mechanisms have shown exceptional sensitivity to even the slightest deformations [96]. The potential for integrating such hydrogel sensors into contact lenses to monitor IOP changes is significant, bringing new possibilities to non-invasive tracking of ocular health. Therapeutic contact lenses can be personalized based on each patient's specific needs for controlled drug delivery. Advancements in smart contact lens technology could significantly improve patient adherence to glaucoma medications and reduce the need for frequent visits to the doctor's office for monitoring. The development of closed-loop theranostic contact lenses combines the advantages of diagnostic and therapeutic lenses by simultaneously diagnosing and treating glaucoma.

While the potential benefits of smart contact lenses are exciting, it is important to note that the technology is still in the early stages and numerous challenges remain. First, the integration of sensors, drug delivery, and communication components into a contact lens presents a complex challenge, as ensuring the reliability, stability, and compatibility of all these components is demanding. Second, the complicated designs of the smart contact lenses and their use of electronics could make them uncomfortable to wear and increase their cost of production, which would hinder adoption. In addition, smart contact lenses come into direct contact with the eye; therefore, materials must be safe and biocompatible to avoid irritation and inflammation. Finally, closed-loop smart contact lenses require a power source that is stable and compact enough to operate the sensors, deliver the drugs, and wirelessly communicate with external systems.

While smart contact lenses have the potential to transform glaucoma diagnosis and treatment, their development and commercialization might take several more years due to the aforementioned challenges. As technology continues to advance, smart contact lenses could shift glaucoma practice paradigms by providing a lower-cost, closed-loop theranostic system that addresses the urgent needs of both eye care providers and patients. However, further research, validation, and regulatory clearances are necessary before smart contact lenses can become a standard component of routine glaucoma management.

Author Contributions: Conceptualization, B.X. and Y.Z.; validation, R.S., N.Y., S.G.,V.N., X.H., S.D., K.G., Y.Z. and B.X.; writing—original draft preparation, R.S., N.Y., S.G. and B.X.; writing—review and editing, R.S., V.N., X.H., S.D., K.G., Y.Z. and B.X.; visualization, R.S., N.Y., S.G. and B.X.; supervision, B.X. and Y.Z.; project administration, R.S., Y.Z. and B.X.; All authors have read and agreed to the published version of the manuscript.

Funding: This work was supported by grant K23 5K23EY029763-05 from the National Institute of Health, Bethesda, Maryland.

Institutional Review Board Statement: Not applicable.

Informed Consent Statement: Not applicable.

Data Availability Statement: Not applicable.

Conflicts of Interest: The authors declare no conflicts of interest.

References

1. Tham, Y.C.; Li, X.; Wong, T.Y.; Quigley, H.A.; Aung, T.; Cheng, C.Y. Global prevalence of glaucoma and projections of glaucoma burden through 2040: A systematic review and meta-analysis. *Ophthalmology* **2014**, *121*, 2081–2090. [CrossRef] [PubMed]
2. Kang, J.M.; Tanna, A.P. Glaucoma. *Med. Clin. N. Am.* **2021**, *105*, 493–510. [CrossRef] [PubMed]
3. Blindness and Visual Impairment. Available online: https://www.who.int/news-room/fact-sheets/detail/blindness-and-visual-impairment (accessed on 26 April 2023).
4. Gupta, D.; Chen, P.P. Glaucoma. *Am. Fam. Physician* **2016**, *93*, 668.
5. Gordon, M.O.; Beiser, J.A.; Brandt, J.D.; Heuer, D.K.; Higginbotham, E.J.; Johnson, C.A.; Keltner, J.L.; Miller, J.P.; Parrish, R.K.; Wilson, M.R.; et al. The Ocular Hypertension Treatment Study: Baseline factors that predict the onset of primary open-angle glaucoma. *Arch. Ophthalmol.* **2002**, *120*, 714–720. [CrossRef] [PubMed]
6. Leske, M.C.; Heijl, A.; Hyman, L.; Bengtsson, B.; Early Manifest Glaucoma Trial Group. Early Manifest Glaucoma Trial: Design and baseline data. *Ophthalmology* **1999**, *106*, 2144–2153. [CrossRef]
7. Grippo, T.M.; Liu, J.H.K.; Zebardast, N.; Arnold, T.B.; Moore, G.H.; Weinreb, R.N. Twenty-four–hour pattern of intraocular pressure in untreated patients with ocular hypertension. *Investig. Ophthalmology Vis. Sci.* **2013**, *54*, 512–517. [CrossRef]
8. Matlach, J.; Bender, S.; König, J.; Binder, H.; Pfeiffer, N.; Hoffmann, E.M. Investigation of intraocular pressure fluctuation as a risk factor of glaucoma progression. *Clin. Ophthalmol.* **2019**, *13*, 9. [CrossRef] [PubMed]
9. Kim, J.H.; Caprioli, J. Intraocular pressure fluctuation: Is it important? *J. Ophthalmic Vis. Res.* **2018**, *13*, 170. [CrossRef]

10. Kazanskiy, N.L.; Khonina, S.N.; Butt, M.A. A review on flexible wearables-Recent developments in non-invasive continuous health monitoring. *Sens. Actuators A Phys.* **2024**, *366*, 114993. [CrossRef]
11. Wu, K.Y.; Mina, M.; Carbonneau, M.; Marchand, M.; Tran, S.D. Advancements in Wearable and Implantable Intraocular Pressure Biosensors for Ophthalmology: A Comprehensive Review. *Micromachines* **2023**, *14*, 1915. [CrossRef]
12. Zeppieri, M.; Gurnani, B. Applanation Tonometry. Available online: https://www.ncbi.nlm.nih.gov/books/NBK582132/ (accessed on 13 July 2023).
13. Tonnu, P.-A.; Ho, T.; Sharma, K.; White, E.; Bunce, C.; Garway-Heath, D. A comparison of four methods of tonometry: Method agreement and interobserver variability. *Br. J. Ophthalmol.* **2005**, *89*, 847–850. [CrossRef] [PubMed]
14. Gazzard, G.; Konstantakopoulou, E.; Garway-Heath, D.; Garg, A.; Vickerstaff, V.; Hunter, R.; Ambler, G.; Bunce, C.; Wormald, R.; Nathwani, N.; et al. Selective laser trabeculoplasty versus eye drops for first-line treatment of ocular hypertension and glaucoma (LiGHT): A multicentre randomised controlled trial. *Lancet* **2019**, *393*, 1505–1516. [CrossRef]
15. Zhang, Y.; Zhao, J.-L.; Bian, A.-L.; Liu, X.-L.; Jin, Y.-M. Effects of central corneal thickness and corneal curvature on measurement of intraocular pressure with Goldmann applanation tonometer and non-contact tonometer. *Zhonghua Yan Ke Za Zhi Chin. J. Ophthalmol.* **2009**, *45*, 713–718.
16. Stodtmeister, R. Applanation tonometry and correction according to corneal thickness. *Acta Ophthalmol. Scand.* **1998**, *76*, 319–324. [CrossRef] [PubMed]
17. Hamilton, K.E.B.; Pye, D.C.M.; Hali, A.B.; Lin, C.B.; Kam, P.B.; Ngyuen, T.B. The effect of contact lens induced corneal edema on Goldmann applanation tonometry measurements. *J. Glaucoma* **2007**, *16*, 153–158. [CrossRef]
18. Stamper, R.L. A history of intraocular pressure and its measurement. *Optom. Vis. Sci.* **2011**, *88*, E16–E28. [CrossRef] [PubMed]
19. Arribas-Pardo, P.; Mendez-Hernández, C.; Valls-Ferran, I.; Puertas-Bordallo, D. Icare-Pro rebound tonometer versus hand-held applanation tonometer for pediatric screening. *J. Pediatr. Ophthalmol. Strabismus* **2018**, *55*, 382–386. [CrossRef] [PubMed]
20. Suman, S.; Agrawal, A.; Pal, V.K.; Pratap, V.B. Rebound tonometer: Ideal tonometer for measurement of accurate intraocular pressure. *J. Glaucoma* **2014**, *23*, 633–637. [CrossRef]
21. Pakrou, N.; Gray, T.; Mills, R.; Landers, J.; Craig, J. Clinical comparison of the Icare tonometer and Goldmann applanation tonometry. *J. Glaucoma* **2008**, *17*, 43–47. [CrossRef]
22. Ting, S.L.; Lim, L.T.; Ooi, C.Y.; Rahman, M.M. Comparison of icare rebound tonometer and Perkins applanation tonometer in community eye screening. *Asia-Pacific J. Ophthalmol.* **2019**, *8*, 229–232. [CrossRef]
23. Galgauskas, S.; Strupaite, R.; Strelkauskaite, E.; Asoklis, R. Comparison of intraocular pressure measurements with different contact tonometers in young healthy persons. *Int. J. Ophthalmol.* **2016**, *9*, 76. [CrossRef]
24. Poostchi, A.; Mitchell, R.; Nicholas, S.; Purdie, G.; Wells, A. The iCare rebound tonometer: Comparisons with Goldmann tonometry, and influence of central corneal thickness. *Clin. Exp. Ophthalmol.* **2009**, *37*, 687–691. [CrossRef]
25. Chui, W.-S.; Lam, A.; Chen, D.; Chiu, R. The influence of corneal properties on rebound tonometry. *Ophthalmology* **2008**, *115*, 80–84. [CrossRef]
26. Fernandes, P.; Díaz-Rey, J.A.; Queirós, A.; Gonzalez-Meijome, J.M.; Jorge, J. Comparison of the ICare® rebound tonometer with the Goldmann tonometer in a normal population. *Ophthalmic Physiol. Opt.* **2005**, *25*, 436–440. [CrossRef]
27. Cvenkel, B.; Velkovska, M.A.; Jordanova, V.D. Self-measurement with Icare HOME tonometer, patients' feasibility and acceptability. *Eur. J. Ophthalmol.* **2020**, *30*, 258–263. [CrossRef]
28. Leonardi, M.; Leuenberger, P.; Bertrand, D.; Bertsch, A.; Renaud, P. First steps toward noninvasive intraocular pressure monitoring with a sensing contact lens. *Investig. Opthalmol. Vis. Sci.* **2004**, *45*, 3113–3117. [CrossRef]
29. Tong, J.B.; Huang, J.P.; Kalloniatis, M.P.; Coroneo, M.M.; Zangerl, B.P. Clinical trial: Diurnal IOP fluctuations in glaucoma using Latanoprost and Timolol with self-tonometry. *Optom. Vis. Sci.* **2021**, *98*, 901–913. [CrossRef] [PubMed]
30. De Novo Classification Request for Sensimed Triggerfish Contact Lens Sensor. Available online: https://www.fda.gov/news-events/press-announcements/fda-permits-marketing-device-senses-optimal-time-check-patients-eye-pressure (accessed on 26 April 2023).
31. Mansouri, K.; Weinreb, R.N.; Liu, J.H.K. Efficacy of a contact lens sensor for monitoring 24-h intraocular pressure related patterns. *PLoS ONE* **2015**, *10*, e0125530. [CrossRef]
32. Dunbar, G.E.; Shen, B.Y.; Aref, A. The Sensimed Triggerfish contact lens sensor: Efficacy, safety, and patient perspectives. *Clin. Ophthalmol.* **2017**, *11*, 875–882. [CrossRef] [PubMed]
33. Lorenz, K.; Korb, C.; Herzog, N.; Vetter, J.M.; Elflein, H.; Keilani, M.M.; Pfeiffer, N. Tolerability of 24-hour intraocular pressure monitoring of a pressure-sensitive contact lens. *J. Glaucoma* **2013**, *22*, 311–316. [CrossRef] [PubMed]
34. Osorio-Alayo, V.; Pérez-Torregrosa, V.T.; Clemente-Tomás, R.; Olate-Pérez, Á.; Cerdà-Ibáñez, M.; Gargallo-Benedicto, A.; Barreiro-Rego, A.; Duch-Samper, A. Efficacy of the SENSIMED Triggerfish® in the postoperative follow-up of PHACO-ExPRESS combined surgery. *Arch. Soc. Española Oftalmol.* **2017**, *92*, 372–378. (In English) [CrossRef]
35. Vitish-Sharma, P.; Acheson, A.G.; Stead, R.; Sharp, J.; Abbas, A.; Hovan, M.; Maxwell-Armstrong, C.; Guo, B.; King, A.J. Can the SENSIMED Triggerfish® lens data be used as an accurate measure of intraocular pressure? *Acta Ophthalmol.* **2018**, *96*, e242–e246. [CrossRef] [PubMed]
36. Rabensteiner, D.F.; Rabensteiner, J.; Faschinger, C. The influence of electromagnetic radiation on the measurement behaviour of the triggerfish® contact lens sensor. *BMC Ophthalmol.* **2018**, *18*, 338. [CrossRef] [PubMed]

37. Yang, C.; Huang, X.; Li, X.; Yang, C.; Zhang, T.; Wu, Q.; Liu, D.; Lin, H.; Chen, W.; Hu, N.; et al. Wearable and implantable intraocular pressure biosensors: Recent progress and future prospects. *Adv. Sci.* **2021**, *8*, 2002971. [CrossRef] [PubMed]
38. Zhang, W.; Huang, L.; Weinreb, R.N.; Cheng, H. Wearable electronic devices for glaucoma monitoring and therapy. *Mater. Des.* **2021**, *212*, 110183. [CrossRef]
39. Kim, J.; Kim, M.; Lee, M.-S.; Kim, K.; Ji, S.; Kim, Y.-T.; Park, J.; Na, K.; Bae, K.-H.; Kim, H.K.; et al. Wearable smart sensor systems integrated on soft contact lenses for wireless ocular diagnostics. *Nat. Commun.* **2017**, *8*, 14997. [CrossRef]
40. Ku, M.; Kim, J.; Won, J.-E.; Kang, W.; Park, Y.-G.; Park, J.; Lee, J.-H.; Cheon, J.; Lee, H.H.; Park, J.-U. Smart, soft contact lens for wireless immunosensing of cortisol. *Sci. Adv.* **2020**, *6*, eabb2891. [CrossRef]
41. Wang, Y.; Zhao, Q.; Du, X. Structurally coloured contact lens sensor for point-of-care ophthalmic health monitoring. *J. Mater. Chem. B* **2020**, *8*, 3519–3526. [CrossRef]
42. Ye, Y.; Ge, Y.; Zhang, Q.; Yuan, M.; Cai, Y.; Li, K.; Li, Y.; Xie, R.; Xu, C.; Jiang, D.; et al. Smart contact lens with dual-sensing platform for monitoring intraocular pressure and matrix metalloproteinase-9. *Adv. Sci.* **2022**, *9*, e2104738. [CrossRef]
43. Ren, X.; Zhou, Y.; Lu, F.; Zhai, L.; Wu, H.; Chen, Z.; Wang, C.; Zhu, X.; Xie, Y.; Cai, P.; et al. Contact Lens Sensor with Anti-jamming Capability and High Sensitivity for Intraocular Pressure Monitoring. *ACS Sens.* **2023**, *8*, 2691–2701. [CrossRef]
44. Yuan, M.; Liu, Z.; Wu, X.; Gou, H.; Zhang, Y.; Ning, X.; Li, W.; Yao, Z.; Wang, Y.; Pei, W.; et al. High-sensitive microfluidic contact lens sensor for intraocular pressure visualized monitoring. *Sens. Actuators A Phys.* **2023**, *354*, 114250. [CrossRef]
45. Jiang, N.; Montelongo, Y.; Butt, H.; Yetisen, A.K. Microfluidic contact lenses. *Small* **2018**, *14*, e1704363. [CrossRef] [PubMed]
46. Yang, X.; Yao, H.; Zhao, G.; Ameer, G.A.; Sun, W.; Yang, J.; Mi, S. Flexible, wearable microfluidic contact lens with capillary networks for tear diagnostics. *J. Mater. Sci.* **2020**, *55*, 9551–9561. [CrossRef]
47. Du, Z.; Zhao, G.; Wang, A.; Sun, W.; Mi, S. Pressure-Triggered Microfluidic Contact Lens for Ocular Drug Delivery. *ACS Appl. Polym. Mater.* **2022**, *4*, 7290–7299. [CrossRef]
48. Moreddu, R.; Elsherif, M.; Adams, H.; Moschou, D.; Cordeiro, M.F.; Wolffsohn, J.S.; Vigolo, D.; Butt, H.; Cooper, J.M.; Yetisen, A.K. Integration of paper microfluidic sensors into contact lenses for tear fluid analysis. *Lab Chip* **2020**, *20*, 3970–3979. [CrossRef] [PubMed]
49. Zhang, J.; Kim, K.; Kim, H.J.; Meyer, D.; Park, W.; Lee, S.A.; Dai, Y.; Kim, B.; Moon, H.; Shah, J.V.; et al. Smart soft contact lenses for continuous 24-hour monitoring of intraocular pressure in glaucoma care. *Nat. Commun.* **2022**, *13*, 5518. [CrossRef] [PubMed]
50. Zolfaghari, P.; Yalcinkaya, A.D.; Ferhanoglu, O. Smart glasses to monitor intraocular pressure using optical triangulation. *Opt. Commun.* **2023**, *546*, 129752. [CrossRef]
51. Zolfaghari, P.; Yalcinkaya, A.D.; Ferhanoglu, O. MEMS Sensor-Glasses Pair for Real-Time Monitoring of Intraocular Pressure. *IEEE Photonics-Technol. Lett.* **2023**, *35*, 887–890. [CrossRef]
52. Marshall, L.L.; Hayslett, R.L.; Stevens, G.A. Therapy for open-angle glaucoma. *Consult. Pharm.® * **2018**, *33*, 432–445. [CrossRef]
53. Digiuni, M.; Fogagnolo, P.; Rossetti, L. A review of the use of latanoprost for glaucoma since its launch. *Expert Opin. Pharmacother.* **2012**, *13*, 723–745. [CrossRef]
54. Mehran, N.A.; Sinha, S.; Razeghinejad, R. New glaucoma medications: Latanoprostene bunod, netarsudil, and fixed combination netarsudil-latanoprost. *Eye* **2020**, *34*, 72–88. [CrossRef]
55. Araie, M.; Sforzolini, B.S.; Vittitow, J.; Weinreb, R.N. Evaluation of the effect of latanoprostene bunod ophthalmic solution, 0.024% in lowering intraocular pressure over 24 h in healthy Japanese subjects. *Adv. Ther.* **2015**, *32*, 1128–1139. [CrossRef]
56. Kawase, K.; Vittitow, J.L.; Weinreb, R.N.; Araie, M.; JUPITER Study Group Shigeru Hoshiai Setsuko Hashida Miki Iwasaki Kiyoshi Kano Kazuhide Kawase Takuji Kato Yasuaki Kuwayama Tomoyuki Muramatsu Masatada Mitsuhashi Sakae Matsuzaki Toru Nakajima Isao Sato Yuzuru Yoshimura. Long-term safety and efficacy of latanoprostene bunod 0.024% in Japanese subjects with open-angle glaucoma or ocular hypertension: The JUPITER study. *Adv. Ther.* **2016**, *33*, 1612–1627. [CrossRef] [PubMed]
57. Demailly, P.; Allaire, C.; Bron, V.; Trinquand, C. Effectiveness and Tolerance of beta-Blocker/Pilocarpine Combination Eye Drops in Primary Open-Angle Glaucoma and High Intraocular Pressure. *J. Glaucoma* **1995**, *4*, 235–241. [CrossRef] [PubMed]
58. Zimmerman, T.J. Topical ophthalmic beta blockers: A comparative review. *J. Ocul. Pharmacol. Ther.* **1993**, *9*, 373–384. [CrossRef] [PubMed]
59. Morales, D.R.; Dreischulte, T.; Lipworth, B.J.; Donnan, P.T.; Jackson, C.; Guthrie, B. Respiratory effect of beta-blocker eye drops in asthma: Population-based study and meta-analysis of clinical trials. *Br. J. Clin. Pharmacol.* **2016**, *82*, 814–822. [CrossRef]
60. Reis, R.; Queiroz, C.F.; Santos, L.C.; Avila, M.P.; Magacho, L. A randomized, investigator-masked, 4-week study comparing timolol maleate 0.5%, brinzolamide 1%, and brimonidine tartrate 0.2% as adjunctive therapies to travoprost 0.004% in adults with primary open-angle glaucoma or ocular hypertension. *Clin. Ther.* **2006**, *28*, 552–559. [CrossRef]
61. Nocentini, A.; Supuran, C.T. Adrenergic agonists and antagonists as antiglaucoma agents: A literature and patent review (2013–2019). *Expert Opin. Ther. Patents* **2019**, *29*, 805–815. [CrossRef] [PubMed]
62. Stoner, A.; Harris, A.; Oddone, F.; Belamkar, A.; Verkicharla, A.C.V.; Shin, J.; Januleviciene, I.; Siesky, B. Topical carbonic anhydrase inhibitors and glaucoma in 2021: Where do we stand? *Br. J. Ophthalmol.* **2022**, *106*, 1332–1337. [CrossRef]
63. Bacharach, J.; Dubiner, H.B.; Levy, B.; Kopczynski, C.C.; Novack, G.D.; AR-13324-CS202 Study Group. Double-masked, randomized, dose–response study of AR-13324 versus latanoprost in patients with elevated intraocular pressure. *Ophthalmology* **2015**, *122*, 302–307. [CrossRef]

64. Slota, C.; Sayner, R.; Vitko, M.; Carpenter, D.M.; Blalock, S.J.; Robin, A.L.; Muir, K.W.; Hartnett, M.E.; Sleath, B. Glaucoma patient expression of medication problems and nonadherence. *Optom. Vis. Sci. Off. Publ. Am. Acad. Optom.* **2015**, *92*, 537. [CrossRef] [PubMed]
65. Cooper, J. Improving compliance with glaucoma eye-drop treatment. *Nurs. Times* **1996**, *92*, 36–37. [PubMed]
66. Shirley, M. Bimatoprost implant: First approval. *Drugs Aging* **2020**, *37*, 457–462. [CrossRef] [PubMed]
67. Brandt, J.D.; Sall, K.; DuBiner, H.; Benza, R.; Alster, Y.; Walker, G.; Semba, C.P.; Budenz, D.; Day, D.; Flowers, B.; et al. Six-month intraocular pressure reduction with a topical bimatoprost ocular insert: Results of a phase II randomized controlled study. *Ophthalmology* **2016**, *123*, 1685–1694. [CrossRef]
68. Cvenkel, B.; Kolko, M. Devices and treatments to address low adherence in glaucoma patients: A narrative review. *J. Clin. Med.* **2022**, *12*, 151. [CrossRef] [PubMed]
69. Kesav, N.P.; Young, C.E.C.; Ertel, M.K.; Seibold, L.K.; Kahook, M.Y. Sustained-release drug delivery systems for the treatment of glaucoma. *Int. J. Ophthalmol.* **2021**, *14*, 148. [CrossRef] [PubMed]
70. Kompella, U.B.; Hartman, R.R.; Patil, M.A. Extraocular, periocular, and intraocular routes for sustained drug delivery for glaucoma. *Prog. Retin. Eye Res.* **2021**, *82*, 100901. [CrossRef] [PubMed]
71. Kumar, H.; Mansoori, T.; Warjri, G.B.; Somarajan, B.I.; Bandil, S.; Gupta, V. Lasers in glaucoma. *Indian J. Ophthalmol.* **2018**, *66*, 1539. [CrossRef]
72. Ha, J.Y.; Lee, T.H.; Sung, M.S.; Park, S.W. Efficacy and safety of intracameral bevacizumab for treatment of neovascular glaucoma. *Korean J. Ophthalmol.* **2017**, *31*, 538–547. [CrossRef]
73. Lusthaus, J.; Goldberg, I. Current management of glaucoma. *Med. J. Aust.* **2019**, *210*, 180–187. [CrossRef]
74. Stefan, C.; Batras, M.; De Simone, A.; Hosseini-Ramhormozi, J. Current Options for Surgical Treatment of Glaucoma. *Rom. J. Ophthalmol.* **2016**, *59*, 194–201.
75. Zhu, Y.; Li, S.; Li, J.; Falcone, N.; Cui, Q.; Shah, S.; Hartel, M.C.; Yu, N.; Young, P.; de Barros, N.R.; et al. Lab-on-a-Contact Lens. Recent Advances and Future Opportunities in Diagnostics and Therapeutics. *Adv. Mater.* **2022**, *34*, 2108389. [CrossRef]
76. Zhu, Y.; Haghniaz, R.; Hartel, M.C.; Mou, L.; Tian, X.; Garrido, P.R.; Wu, Z.; Hao, T.; Guan, S.; Ahadian, S.; et al. Recent Advances in Bioinspired Hydrogels: Materials, Devices, and Biosignal Computing. *ACS Biomater. Sci. Eng.* **2021**, *9*, 2048–2069. [CrossRef] [PubMed]
77. Ciolino, J.B.; Hoare, T.R.; Iwata, N.G.; Behlau, I.; Dohlman, C.H.; Langer, R.; Kohane, D.S. A drug-eluting contact lens. *Investig. Ophthalmol. Vis. Sci.* **2009**, *50*, 3346–3352. [CrossRef] [PubMed]
78. Hsu, K.-H.; Carbia, B.E.; Plummer, C.; Chauhan, A. Dual drug delivery from vitamin E loaded contact lenses for glaucoma therapy. *Eur. J. Pharm. Biopharm.* **2015**, *94*, 312–321. [CrossRef] [PubMed]
79. Taniguchi, E.V.; Kalout, P.; Pasquale, L.R.; Kohane, D.S.; Ciolino, J.B. Clinicians' perspectives on the use of drug-eluting contact lenses for the treatment of glaucoma. *Ther. Deliv.* **2014**, *5*, 1077–1083. [CrossRef]
80. Ding, X.; Ben-Shlomo, G.; Que, L. Soft contact lens with embedded microtubes for sustained and self-adaptive drug delivery for glaucoma treatment. *ACS Appl. Mater. Interfaces* **2020**, *12*, 45789–45795. [CrossRef]
81. Soluri, A.; Hui, A.; Jones, L. Delivery of ketotifen fumarate by commercial contact lens materials. *Optom. Vis. Sci.* **2012**, *89*, 1140–1149. [CrossRef]
82. Chaudhari, P.; Ghate, V.M.; Lewis, S.A. Next-generation contact lenses: Towards bioresponsive drug delivery and smart technologies in ocular therapeutics. *Eur. J. Pharm. Biopharm.* **2021**, *161*, 80–99. [CrossRef]
83. Lai, C.F.; Shiau, F.J. Enhanced and Extended Ophthalmic Drug Delivery by pH-Triggered Drug-Eluting Contact Lenses with Large-Pore Mesoporous Silica Nanoparticles. *ACS Appl. Mater. Interfaces* **2023**, *15*, 18630–18638. [CrossRef]
84. Brandt, J.D.; DuBiner, H.B.; Benza, R.; Sall, K.N.; Walker, G.A.; Semba, C.P.; Budenz, D.; Day, D.; Flowers, B.; Lee, S.; et al. Long-term safety and efficacy of a sustained-release bimatoprost ocular ring. *Ophthalmology* **2017**, *124*, 1565–1566. [CrossRef]
85. Ruiz-Pomeda, A.; Villa-Collar, C. Slowing the progression of myopia in children with the MiSight contact lens: A narrative review of the evidence. *Ophthalmol. Ther.* **2020**, *9*, 783–795. [CrossRef]
86. Lanier, O.L.; Christopher, K.G.; Macoon, R.M.; Yu, Y.; Sekar, P.; Chauhan, A. Commercialization challenges for drug eluting contact lenses. *Expert Opin. Drug Deliv.* **2020**, *17*, 1133–1149. [CrossRef] [PubMed]
87. Kelkar, S.S.; Reineke, T.M. Theranostics: Combining imaging and therapy. *Bioconjug. Chem.* **2011**, *22*, 1879–1903. [CrossRef] [PubMed]
88. Jeelani, S.; Reddy, R.J.; Maheswaran, T.; Asokan, G.S.; Dany, A.; Anand, B. Theranostics: A treasured tailor for tomorrow. *J. Pharm. Bioallied Sci.* **2014**, *6* (Suppl. S1), S6. [CrossRef] [PubMed]
89. Svenson, S. Theranostics: Are we there yet? *Mol. Pharm.* **2013**, *10*, 848–856. [CrossRef] [PubMed]
90. Kim, T.Y.; Mok, J.W.; Hong, S.H.; Jeong, S.H.; Choi, H.; Shin, S.; Joo, C.-K.; Hahn, S.K. Wireless theranostic smart contact lens for monitoring and control of intraocular pressure in glaucoma. *Nat. Commun.* **2022**, *13*, 6801. [CrossRef] [PubMed]
91. Yang, C.; Wu, Q.; Liu, J.; Mo, J.; Li, X.; Yang, C.; Liu, Z.; Yang, J.; Jiang, L.; Chen, W.; et al. Intelligent wireless theranostic contact lens for electrical sensing and regulation of intraocular pressure. *Nat. Commun.* **2022**, *13*, 2556. [CrossRef] [PubMed]
92. Qin, M.; Sun, M.; Hua, M.; He, X. Bioinspired structural color sensors based on responsive soft materials. *Curr. Opin. Solid State Mater. Sci.* **2019**, *23*, 13–27. [CrossRef]
93. Qin, M.; Sun, M.; Bai, R.; Mao, Y.; Qian, X.; Sikka, D.; Zhao, Y.; Qi, H.J.; Suo, Z.; He, X. Bioinspired hydrogel interferometer for adaptive coloration and chemical sensing. *Adv. Mater.* **2018**, *30*, e1800468. [CrossRef]

94. Frenkel, I.; Hua, M.; Alsaid, Y.; He, X. Self-Reporting Hydrogel Sensors Based on Surface Instability-Induced Optical Scattering. *Adv. Photonics Res.* **2021**, *2*, 2100058. [CrossRef]
95. Choi, J.; Hua, M.; Lee, S.Y.; Jo, W.; Lo, C.; Kim, S.; Kim, H.; He, X. Hydrocipher: Bioinspired dynamic structural color-based cryptographic surface. *Adv. Opt. Mater.* **2020**, *8*, 1901259. [CrossRef]
96. Sun, M.; Bai, R.; Yang, X.; Song, J.; Qin, M.; Suo, Z.; He, X. Hydrogel interferometry for ultrasensitive and highly selective chemical detection. *Adv. Mater.* **2018**, *30*, 1804916. [CrossRef] [PubMed]

Disclaimer/Publisher's Note: The statements, opinions and data contained in all publications are solely those of the individual author(s) and contributor(s) and not of MDPI and/or the editor(s). MDPI and/or the editor(s) disclaim responsibility for any injury to people or property resulting from any ideas, methods, instructions or products referred to in the content.

Review

Recent Advancements in Glaucoma Surgery—A Review

Bryan Chin Hou Ang [1,2,*], Sheng Yang Lim [1], Bjorn Kaijun Betzler [3,4], Hon Jen Wong [4], Michael W. Stewart [5] and Syril Dorairaj [5]

1. Department of Ophthalmology, National Healthcare Group Eye Institute, Tan Tock Seng Hospital, Singapore 308433, Singapore
2. Department of Ophthalmology, National Healthcare Group Eye Institute, Woodlands Health Campus, Singapore 737628, Singapore
3. Department of Surgery, Tan Tock Seng Hospital, National Healthcare Group, Singapore 308433, Singapore
4. Yong Loo Lin School of Medicine, National University of Singapore, Singapore 119077, Singapore
5. Department of Ophthalmology, Mayo Clinic, Jacksonville, FL 32224, USA; stewart.michael@mayo.edu (M.W.S.)
* Correspondence: drbryanang@gmail.com

Abstract: Surgery has long been an important treatment for limiting optic nerve damage and minimising visual loss in patients with glaucoma. Numerous improvements, modifications, and innovations in glaucoma surgery over recent decades have improved surgical safety, and have led to earlier and more frequent surgical intervention in glaucoma patients at risk of vision loss. This review summarises the latest advancements in trabeculectomy surgery, glaucoma drainage device (GDD) implantation, and minimally invasive glaucoma surgery (MIGS). A comprehensive search of MEDLINE, EMBASE, and CENTRAL databases, alongside subsequent hand searches—limited to the past 10 years for trabeculectomy and GDDs, and the past 5 years for MIGS—yielded 2283 results, 58 of which were included in the final review (8 trabeculectomy, 27 GDD, and 23 MIGS). Advancements in trabeculectomy are described in terms of adjunctive incisions, Tenon's layer management, and novel suturing techniques. Advancements in GDD implantation pertain to modifications of surgical techniques and devices, novel methods to deal with postoperative complications and surgical failure, and the invention of new GDDs. Finally, the popularity of MIGS has recently promoted modifications to current surgical techniques and the development of novel MIGS devices.

Keywords: glaucoma; trabeculectomy; glaucoma tube shunts; minimally invasive glaucoma surgery; device; eye

Citation: Ang, B.C.H.; Lim, S.Y.; Betzler, B.K.; Wong, H.J.; Stewart, M.W.; Dorairaj, S. Recent Advancements in Glaucoma Surgery—A Review. *Bioengineering* 2023, 10, 1096. https://doi.org/10.3390/bioengineering10091096

Academic Editors: Karanjit S. Kooner, Osamah J. Saeedi and Hiroshi Ohguro

Received: 6 July 2023
Revised: 10 September 2023
Accepted: 11 September 2023
Published: 19 September 2023

Copyright: © 2023 by the authors. Licensee MDPI, Basel, Switzerland. This article is an open access article distributed under the terms and conditions of the Creative Commons Attribution (CC BY) license (https://creativecommons.org/licenses/by/4.0/).

1. Introduction

Glaucoma is the leading cause of irreversible blindness worldwide, with little variability according to race, ethnicity, or location [1]. Elevated intraocular pressure (IOP) remains the primary modifiable risk factor for glaucoma progression, thereby mandating that treatments lower the IOP. This is the only therapeutic strategy that prevents damage to the optic nerve and the progression of visual field defects [2]. Anti-glaucoma medications, most of which work by lowering aqueous production or increasing outflow, as well as laser procedures, such as peripheral iridotomy or trabeculoplasty, are generally considered to be first-line therapy [3]. Surgery is usually indicated when glaucoma medications and lasers are unable to reduce IOP sufficiently to halt visual field loss [4].

Trabeculectomy has long been the gold standard for the surgical management of glaucoma, but new surgical techniques and devices, including glaucoma drainage devices (GDD) and minimally invasive glaucoma surgery (MIGS), have been recently developed [5]. Because the field of glaucoma surgery has changed so significantly over the past decade, the authors believe a comprehensive review that consolidates and summarises recent developments and innovations in trabeculectomy surgery, GDDs, and MIGS is warranted.

2. Materials and Methods

A comprehensive search of PubMed, EMBASE, and the Cochrane Central Register of Controlled Trials (CENTRAL) was performed on 29th August 2022. Combinations of the following keywords and MeSH terms were used: "Glaucoma", "Trabeculectomy", "Glaucoma Drainage Implants", "Tube", "Tube Shunt", "Tube Shunts", "Ahmed", "Baerveldt", "Clearpath", "Molteno", "Paul", "Minimally Invasive Surgical Procedures", "MIGS", "Minimally Invasive Glaucoma Surgery", "Trabectome", "Trabeculectomy", "GATT", "Gonioscopy-assisted transluminal trabeculotomy", "Trab360", "iStent", "Hydrus", "XEN", "Preserflo", "Canaloplasty", "ABiC", "iTrack", "Kahook", "KDB", "Omni", "Visco360", "Visiplate", "Cypass", and "Durysta". The search was restricted to only adult studies (>19 years of age) and studies published in English. The literature searches for trabeculectomy and GDDs dated back to 29th August 2012 (10 years) and for MIGS dated from 29th August 2017 (5 years). Identified studies were evaluated and manually searched to identify other eligible studies, which were added as hand searches.

Advancements were defined as developments in the following predefined areas: trabeculectomy—"incisional technique" and "closure technique"; GDDs—"GDD surgical technique", "Existing GDDs", and "New GDDs"; MIGS—"MIGS technique", "combination MIGS", and "new MIGS devices". Any surgical advancement or development that fell under the predefined categories and was described within the specified time frame was included. A supplementary manual search was conducted if the included study was deemed not to be the original/first description of the advancement. Criteria for inclusion did not include consideration of the significance or extent of real-life adoption of the particular advancement. Only studies involving United States Food and Drug Administration (FDA) or European Conformité Européenne (CE)-approved MIGS devices were included for final review under the MIGS section.

Manuscripts were assessed by three reviewers (S.Y.L., B.K.B., and H.J.W.) for inclusion. Disagreements were resolved through discussion and consensus, and when unsuccessful, a senior reviewer (B.C.H.A.) was consulted.

The database searches yielded 2261 results—844 (PubMed), 459 (EMBASE), and 958 (CENTRAL). An additional 6 (Trabeculectomy), 14 (GDD), and 2 (MIGS) studies were added from hand searches, and 517 duplicates were subsequently removed. After the initial title–abstract sieve, 98 (out of 1766 articles) remained—16 (Trabeculectomy), 38 (GDDs), and 44 (MIGS). Following a full-text review, 58 articles—8 (Trabeculectomy), 27 (GDD), and 23 (MIGS)—were included in the final review on advancements. Results of the literature searches and reviews are presented in a PRISMA flowchart [6] (Figure 1). A summary list of all included studies is presented in Table A1 (Appendix A).

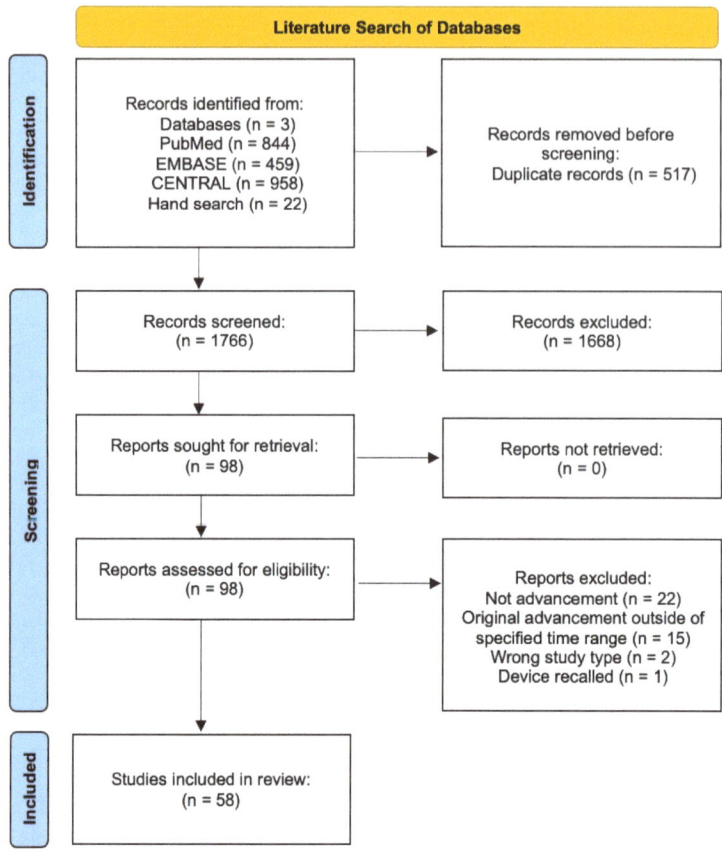

Figure 1. PRISMA flowchart [6].

3. Trabeculectomy

As a technique to divert aqueous from the anterior chamber into the subconjunctival space, conventional trabeculectomy was first described by Cairns in 1968 [7]. Surgery includes the creation of a fornix or limbal-based conjunctival flap, dissection of the underlying Tenon's layer, the creation of a partial-thickness scleral flap, the formation of an ostium into the anterior chamber, and finally, a surgical iridectomy to prevent postoperative occlusion of the ostium [8]. Aqueous flows down a pressure gradient from the anterior chamber into the subconjunctival space, resulting in the formation of a filtering bleb with a reduction in the IOP. Trabeculectomy remains the gold standard, first-line, subconjunctival filtration surgery for the treatment of vision-threatening glaucoma.

The performance of an "ideal" trabeculectomy is said to follow the "10-10-10" rule—a surgical time of 10 min, the achievement of a postoperative IOP of 10 mmHg, and an effect that lasts for 10 years or longer [9]. The Moorfields Safer Surgery System, adopted by many trabeculectomy surgeons worldwide, has been designed to facilitate trabeculectomy outcomes following the "10-10-10" rule [10]. Despite the long history of trabeculectomy, challenges, including bleb failure from postoperative fibrosis, concerns regarding long-term IOP-lowering efficacy, bleb complications, such as leaks, hypotony, and endophthalmitis, and the ongoing need for vigilant postoperative monitoring and interventions to sustain surgical efficacy, persist [11]. This section will explore advancements in trabeculectomy surgery that are meant to overcome these challenges and improve outcomes, with a special focus on incisional and closure techniques.

3.1. Incisional Technique

Trabeculectomy augmented by limited deep sclerectomy (LDS) was first described in 2017 by Dada et al. [8]. LDS involves elevating and excising a 3×3 mm block of deep scleral tissue below the initial scleral flap, thereby creating a crater in the scleral bed [8]. This augmentative intra-operative procedure is intended to surgically thin the remaining sclera, thereby enhancing permeability and increasing aqueous drainage. The pooling of aqueous within the inner scleral layers is intended to promote supraciliary and suprachoroidal outflow of aqueous [8,12] with the size of the pressure difference between the anterior chamber and suprachoroidal spaces driving uveoscleral outflow [8]. These additional filtration pathways reduce reliance on subconjunctival filtration and theoretically allow for greater reductions in IOP. The intrascleral lake supports the scleral flap, preventing its collapse, and scleral flap elevation reduces the risk of local episcleral and intrascleral fibrosis [8]. LDS-augmented trabeculectomy appears to be a potential alternative to conventional trabeculectomy. In a randomised controlled trial of 68 patients with primary open-angle glaucoma or primary angle closure glaucoma with pseudophakia, LDS-augmented trabeculectomy reduced IOP from a baseline of 29 ± 4.6 mmHg to 12.54 ± 1.67 mmHg at 12 months whilst conventional trabeculectomy reduced IOP from a baseline of 30 ± 5.2 mmHg to 13.45 ± 1.83 mmHg at 12 months [8]. None of the eyes in the LDS group required postoperative bleb needling [8], a procedure often performed to revive a non-functional fibrotic bleb. Both LDS-augmented and conventional trabeculectomy reduced the need for postoperative glaucoma medications (3.36 ± 0.48 to 0.46 ± 0.76 vs. 3.34 ± 0.48 to 0.9 ± 0.76) and LDS-augmented trabeculectomy had a lower rate of surgical failure [8].

In 2022, Dada et al. [13] further enhanced the LDS-augmented trabeculectomy by creating a cyclodialysis in two patients who had high IOP after vitrectomy. In this surgical modification, a controlled separation of the ciliary body from the scleral spur is performed at the site of the scleral flap. The excised deep scleral tissue during LDS is used as a spacer in the cyclodialysis cleft (Figure 2) [13], which prevents closure and fibrosis of the cleft, and ensures suprachoroidal drainage of aqueous [13]. The spacer also inhibits the excessive aqueous outflow that commonly occurs with standard cyclodialysis procedures. These alternative drainage pathways alleviate the IOP-lowering burden on the subconjunctival bleb and minimise the risk of bleb-related complications, such as fibrosis and bleb leak. Good short-term IOP outcomes, from 38 mmHg to 12 mmHg and 44 mmHg to 10 mmHg, were seen in both patients, without leaking blebs at 6 months [13]. Unfortunately, the long-term efficacy and safety of this procedure have yet to be reported.

The Extended Subscleral Technique (ESST) was first described in 2014 by Saeed et al. [14] as an adjunct to trabeculectomy. In ESST, a narrow longitudinal strip of deep sclera posterior to the scleral flap is removed to create an extended scleral tunnel, approximately 6 mm in length from the limbus, that allows aqueous passage into the posterior subconjunctival space. The pressure gradient between the original filtering ostium and the additional subscleral tunnel produces a regulated posterior flow [14,15]. Consistent with Bernoulli's principle, the different channel diameters produce variable aqueous velocities, with regions of low and high pressures. The resultant force balances out the pressure difference, thereby encouraging a posteriorly directed, controlled flow of aqueous. By acting as another outlet for aqueous, the ESST limits aqueous outflow velocity, minimises the development of a shallow anterior chamber post-operatively, and promotes the formation of a more widely-distributed posterior bleb [15]. The enhanced diffusion of aqueous into the wider adjacent subconjunctival space may inhibit the formation of a ring of scar tissue—the "Ring of Steel"—that often forms at the junction between bleb and normal conjunctiva, leading to a localised, elevated, and thin bleb that increases the risk of postoperative leaks [15]. ESST may reduce the need for postoperative bleb needling, a procedure that is often used to re-establish drainage after fibrosis and encapsulation [16]. A randomised controlled trial that examined outcomes of ESST-augmented trabeculectomy vs. conventional trabeculectomy in 40 eyes with primary open angle glaucoma found no bleb-related complications in the ESST group. ESST also produced a greater reduction in IOP from baseline compared to that following conventional trabeculectomy with a superior reduction in IOP that was statistically significant at 7 days (80.0% vs. 56.0%) and 180 days (66% vs. 53.6%), but non-significant at 1 day (66.7% vs. 57.2%) and 1 year post-operatively (67.5% vs. 53.1%) [15]. A statistically greater reduction in the need for postoperative glaucoma medications was observed in the ESST group (2.53 ± 0.9 to 0.052 ± 0.2) compared to conventional trabeculectomy (2.85 ± 0.59 to 0.65 ± 0.2) at 1 year [15].

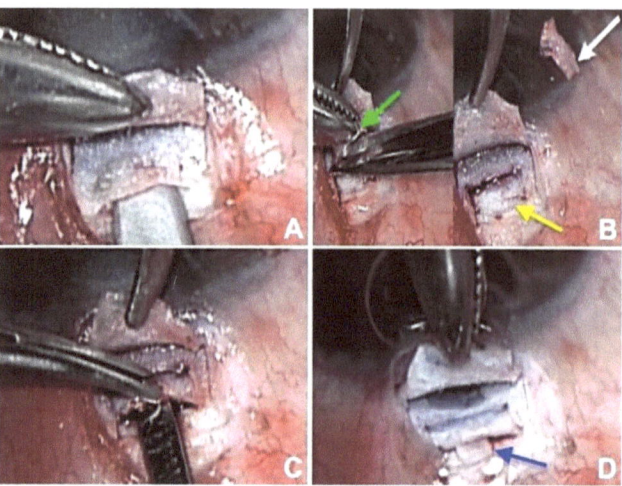

Figure 2. (**A**) The making of a partial-thickness scleral flap to create a deep crater; (**B**) further dissection of the scleral block to create an even deeper crater; (**C**) cyclodialysis cleft made using cyclodialysis spatula; and (**D**) deep scleral tissue inserted at the cyclodialysis cleft. Courtesy of Dada et al. [13].

3.2. Closure Technique

Trabeculectomy surgery concludes with suturing of the scleral flap, adjustment of suture tension, and watertight closure of the conjunctiva and Tenon's layers. These steps are critical for preventing bleb leaks and ensuring adequate postoperative IOP reduction. In recent years, new closure techniques have been developed to improve both the efficacy and safety profile of trabeculectomy.

Chan et al. [17] reported a novel, modified conjunctival closure technique that repositions and separates Tenon's layer from the conjunctiva, which differs from the traditional simultaneous closure of conjunctiva and the Tenon. The Tenon is dissected from the conjunctiva and anchored close to and overlying the scleral flap; the conjunctiva is then closed separately [17]. This technique positions the inner surface of Tenon's layer further from the anterior sclera, which lowers the risk of postoperative fibrosis and enhances aqueous flow into the sub-Tenon's space alongside the intentional misalignment of the Tenon and sclera. This helps maintain space patency and encourages posterior aqueous flow. Approximating the Tenon on top of the partial-thickness scleral flap creates a tensional force that may prevent over-drainage, thereby reducing the risk of hypotony. Anchoring the Tenon allows conjunctival closure with minimal tension, thereby reducing the risk of a buttonhole, dehiscence, and bleb leak. Finally, this closure method reduces the risk of Tenon's layer retraction and encourages the development of a thicker bleb wall, which may reduce the risk of a cystic bleb and bleb leaks [17]. In this non-comparative case series, 30 Chinese patients underwent fornix-based trabeculectomy with mitomycin C and experienced a reduction in mean IOP from 28.5 ± 9.6 mmHg to 15.5 ± 2.6 mmHg and a decrease in the need for postoperative glaucoma medications (4.4 ± 0.9 to 0.8 ± 0.12). No wound leaks were observed.

The literature describes better long-term results with fornix-based trabeculectomy than with limbal-based trabeculectomy but with greater risks of conjunctival wound leakage [18]. In 2015, Olawoye et al. described a new closure technique for fornix-based trabeculectomy that uses a horizontal conjunctival suture [19]. This ensures a watertight limbal conjunctival wound and mechanically separates the conjunctiva from the cornea. A non-comparative case series of 79 eyes with primary open angle glaucoma or secondary glaucoma, such as exfoliative and pigmentary glaucoma, that were at high risk of surgical failure reported low rates of bleb leakage (7 (8.8%) eyes), as compared to other forms of previously described conjunctival closure techniques in fornix-based trabeculectomy. Significant reductions in mean IOP (31.5 ± 8.1 mmHg to 14.2 ± 6.0 mmHg) and postoperative glaucoma medications (3.7 ± 0.8 to 0.6 ± 0.12) were also observed [19].

Kirk et al. in 2014 [18] modified the original Wise closure technique [20] by creating a limbal lip of conjunctiva. The firm adhesion between the anterior and posterior conjunctival edges promotes healing and minimises leakage at the wound site. The closing suture, which is secured to both the conjunctiva and sclera peripheral to the original conjunctival incision, evenly distributes mechanical traction across the wound [18]. This contrasts with traditional closure techniques that rely heavily on the sutures at both ends of the conjunctival flap [18,21]. A retrospective comparative study (313 patients) that investigated the efficacy and safety profile of the modified Wise closure [21] found that the incidence of bleb leaks was lower after the modified Wise closure than with winged sutures (6.4% vs. 16.6%). The modified Wise closure exhibited a stronger protective effect against bleb leaks compared to techniques employing winged sutures for closure (odds ratio of 0.345; 95% CI 0.16–0.74; $p = 0.007$) [21]. The postoperative IOP reduction from baseline in six months was significantly greater following the modified Wise closure compared to closure with winged sutures (-14.5 ± 10.8 mmHg vs. -11.6 ± 9.1 mmHg) [21] though no significant difference in the need for postoperative glaucoma medications between groups was found.

Figus M et al. first described the use of a scleral flap everting suture for anterior filtering procedures with a scleral flap in 2016 [22]. This technique involved passing an everting 10-0 nylon suture through the distal margin of the flap, then through the limbus twice before knotting and forming a closed ellipse with a loop on the cornea. If IOP reduction is required postoperatively, traction can be applied to the exposed loop to increase aqueous outflow and restore the bleb. In 92 eyes that underwent filtering surgery, the authors reported the need to traction the everting suture by 4 months postoperatively in 26 out of 92 eyes, of which the procedure was successful in reopening the scleral flap in 25 eyes. However, IOP results were not reported, as the abovementioned study was still undergoing approval by the institutional ethics committee. Baykara M. et al. in 2017 reported a modification of the scleral flap everting suture—the accordion suture—in a population of eight eyes with neovascular glaucoma [23]. The technique involved first passing the suture through the mid-distal edge of the scleral flap, internal to external, and then through the mid-left edge of the flap, external to internal. Next, the suture is passed through the clear cornea at the limbus and again through the clear cornea, creating a U-shaped loop. Subsequently, the suture is passed through the mid-right edge of the scleral flap, internal to external, and finally through the mid-distal edge of the flap, external to internal. Lastly, both ends of the suture are placed underneath the scleral flap and finely tied with a 3-1-1 slip knot manner after adjustment of desired tension. The use of the accordion suture has been postulated by the authors to result in an even lifting pressure applied to both edges of the flap, which delivers a more substantive decrease in IOP. The mean removal time of the accordion suture was reported to be 3.5 ± 0 weeks post-operatively, with the mean IOP before and after the procedure at 22.63 ± 2.06 mmHg and 11.12 ± 2.64 mmHg, respectively.

4. Glaucoma Drainage Devices

Glaucoma drainage devices (GDDs) have become a mainstay in the surgical management of advanced, refractory glaucoma, particularly in those eyes with a prior history of failed filtering surgery. GDDs divert aqueous humour (AH) from the anterior chamber to an external reservoir, over which a fibrous capsule forms at 4–6 weeks after surgery. AH diffuses between the collagenous fibres of the capsule and is absorbed by capillaries and lymphatic vessels within the Tenon and conjunctiva. The base plate prevents conjunctival adhesion to the sclera and maintains the AH reservoir [24], though the fibrous capsule encapsulating the base plate is the site most resistant to AH flow [24,25]. Overall, GDDs successfully control IOP in eyes with previously failed trabeculectomy [26], and in eyes with prior conjunctiva scarring that precludes other forms of subconjunctival filtering surgery. GDDs are usually classified as flow-restrictive (valved) or non-flow-restrictive (non-valved) types, with devices varying according to size and base plate material. Since the Molteno drainage implant device [27] was first introduced into clinical practice, attempts have been made to improve the safety and efficacy of GDDs by modifying intra-operative techniques, exploring ways to manage postoperative complications and surgical failure, modifying existing devices and creating new GDDs.

4.1. Modifications to Existing Techniques of GDD Implantation

Assessing the function and patency of GDDs is important for ensuring good and predictable outcomes after GDD implantation. Grover et al. [28] used Trypan blue to assess adequate GDD flow in three situations: (1) when completing the second stage of Baerveldt tube implantation; (2) when blockage of a valved implant is suspected after it had previously functioned well; and (3) when the valve mechanism of an implant seems to have failed. In the first situation, Trypan blue injection through the drainage tube stained the capsule, signifying that the dye successfully reached the plate. In the second situation, elevated IOP was observed in the first postoperative week after Ahmed Glaucoma Valve (AGV) implantation. To assess for GDD function, diluted Trypan blue was flushed into the tube with blue staining of the capsule and plate, confirming GDD function. In the third situation, high IOP occurred during the second postoperative week after the implantation of an AGV. After initial irrigation attempts had failed, the tube was externalised and flushed aggressively with diluted Trypan blue, thereby re-establishing good flow. In this case, the dye served both to confirm and re-establish flow in the GDD, but the authors also advised against the overuse of Trypan blue because of its association with endothelial toxicity at high concentrations and prolonged exposure [29]. In this study, the authors minimised the risk of toxicity by diluting three drops of the dye with 3 mL of balanced salt solution.

Another modification to the GDD implantation procedure involves the placement of the GDD tube through a sclerotomy port during vitreoretinal surgery. In eyes with compromised anterior segments due to previous surgeries or disease processes, as well as in post-corneal transplant eyes, GDDs have been implanted in the sulcus or vitreous cavity. Gupta et al. [30] described pars plana placement of the AGV through a sclerotomy port in the only-seeing eye of an aphakic patient with post-penetrating keratoplasty refractory glaucoma and a history of trabeculectomy. In this case, the AGV tube was trimmed to an intravitreal length of 6mm and inserted through a superotemporal 25 G vitrectomy port to minimise the number of entry wounds and, hopefully, to limit postoperative fibro-vascular proliferation and exaggerated wound healing [31].

While GDDs usually drain into the subconjunctival space, Maldonado-Junyent et al. [32] followed the principles of a ventriculoperitoneal shunt used in the treatment of hydrocephalus, to drain aqueous humour into the peritoneal cavity. A hydrocephalus valve (Medtronic PS Medical Strata NSC) was used, regulated at level 2.5 to operate at pressures between 14 and 16 mmHg. Good IOP was maintained for the first four weeks, but longer-term results have yet to be published. This unique modification to GDD implantation raises the possibility of diverting aqueous to other spaces outside of the eye.

4.2. Novel Techniques to Manage Surgical Complications and Failure following GDD Implantation

Tube exposure is a well-known complication of GDD implantation [33] that may result from the eye's immunologic response, repeated mechanical irritation caused by blinking, outward pressure against the tube from the eye, or vaulting of the tube due to intrinsic tube elasticity. Various strategies, including the creation of an overlying scleral flap and the application of patch grafts, have been used during surgery to reduce the incidence. In a retrospective series of 36 eyes with refractory glaucoma, Ma et al. [34] used a modified scleral tunnel technique. After the tube was inserted into the scleral tunnel, it was covered by both the tunnel and an overlying scleral flap at the point of intersections. Through 21 months of follow-up (mean), no conjunctival tube exposure was reported.

In a retrospective series of 30 eyes, Eslami et al. [35] reported the use of a single long tunnel to prevent tube exposure. The authors proposed that preventing tube-conjunctiva contact would reduce the risk of tube exposure, and through a 37.2-month follow-up (mean), no cases of tube exposure were reported. The surgical technique (Figure 3) begins with an 8 mm half-thickness scleral tunnel, after which the plate of the shunt device is secured to the sclera. The silicone tube is trimmed, threaded through the scleral tunnel, and inserted into the anterior chamber under a scleral flap and through a partial paracentesis. The limbal scleral flap is closed to prevent leakage.

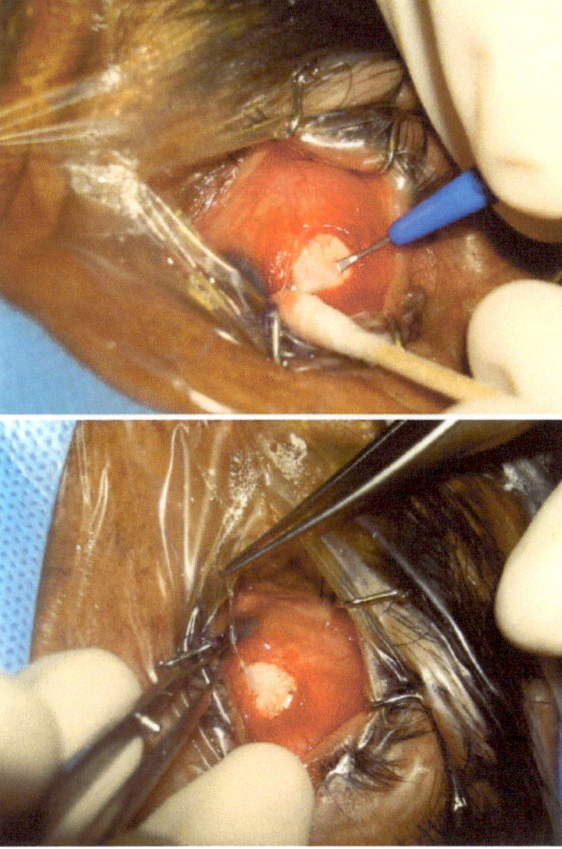

Figure 3. Creation of an 8 mm half-thickness scleral tunnel (**top**); passage of shunt tube through the scleral tunnel (**bottom**). Courtesy of Eslami et al. [35].

Brouzas et al. [36] developed a 'double scleral tunnel in tandem' technique to exceed the maximum length of a single tunnel. Two scleral incisions are made parallel to the limbus at 4 and 12 mm. A half-thickness scleral tunnel is dissected between the two incisions (Figure 4), and a second tunnel is made from the proximal incision to the limbus. After injection of a viscoelastic into the anterior chamber and the creation of a paracentesis, the tube is inserted through both the distal and proximal tunnels, and then into the anterior chamber. The proximal incisions are sutured, and the tube and conjunctiva are secured. In a series of 28 eyes, only two (7.1%) cases had tube exposure after a mean follow-up of 60 months.

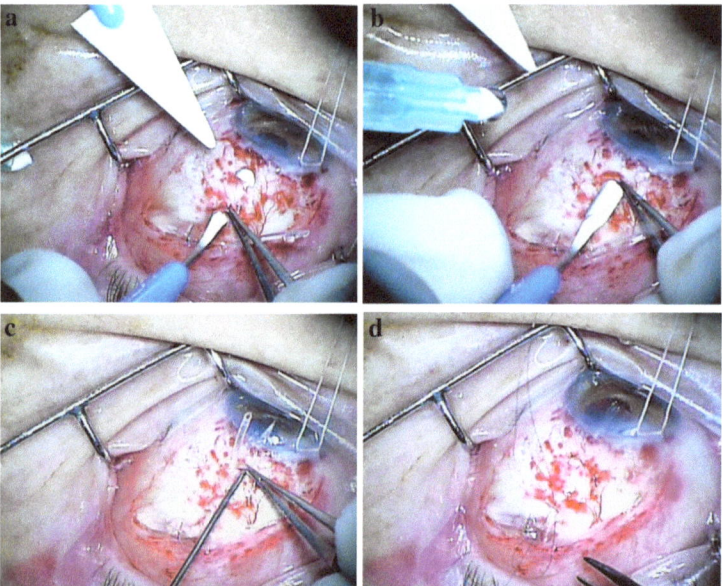

Figure 4. (a) The distal-to-limbus tunnel is prepared with a bevel-up lancet between the two scleral incisions; (b) the proximal-to-limbus tunnel is fashioned from the proximal-to-limbus incisions to the limbus; (c) a paracentesis is created with a 23-gauge needle through the proximal-to-limbus tunnel into the anterior chamber; (d) the tube is secured with a 10-0 nylon suture (distal incision-sclera, sclera-distal incision). Courtesy of Brouzas et al. [36].

Alternative techniques to cover the GDD implant during surgery have also been described. In a randomised clinical trial, Pakravan et al. [37] reported the use of a graft-free, short tunnel, small flap method of AGV implantation and compared it with a scleral patch graft. Comparable success rates, including postoperative IOPs, glaucoma medication burden, and complication rates through 1 year, were found with each approach. These data suggest that the graft-free, short tunnel, small flap technique may be a viable way to reduce the risks associated with scleral patch grafts. Gupta et al. [38] also used a graft-free scleral sleeve technique (Figure 5) in a single patient during the COVID-19 pandemic to reduce the risk of viral transmission through a donor scleral graft and reported no tube exposure through 6 months.

GDD efficacy is often limited by bleb fibrosis, with no clear consensus on the effectiveness of intra-operative antimetabolite use in reducing the rate of surgical failure [39]. Alternate adjuncts have been used with the hope of preventing bleb fibrosis following GDD implantation. In a randomised prospective multicentre clinical trial with 58 patients, Sastre-Ibanez et al. [40] used the Ologen collagen matrix, but could not demonstrate an efficacy or safety benefit over traditional AGV implantation surgery after 12 months postoperatively.

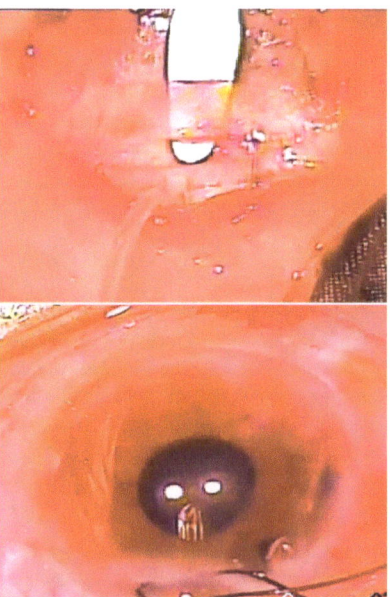

Figure 5. (**top**) The lamellar scleral tunnel was created with a crescent blade for the passage of the AGV tube. (**bottom**) Insertion of the tube in the sulcus. Courtesy of Prakavan et al. [37].

New surgical techniques have been developed to better manage intra-operative complications. Mungale et al. [41] described a novel method to manage inadvertent tube-cut by a ligature that sometimes occurs during aurolab aqueous drainage (AADI) implant surgery. The authors removed the short end of the tube attached to the implant and reinserted the long, transected end into the back plate of the implant. Management options, in this case, were limited by the absence of spares and other materials for tube extension, and the authors cautioned that similar techniques might not be applicable to valved implants like the AGV, where the tube fits tightly into the base plate.

Early GDD failure may occur because a blood clot obstructs the tube, particularly in eyes with neovascular or inflammatory glaucoma. At the end of surgery, Hwang et al. [42] injected filtered air into the anterior chamber through a 30-gauge needle. The authors hypothesised that a large air bubble would keep blood from entering the tube opening and prevent an obstructive clot from forming.

GDDs may also become occluded by iris tissue. In a single case, Kataria et al. [43] used a single trans-corneal suture to manage iris tuck in an AADI tube. A trans-corneal sling suture was passed through the cornea and behind the tube, approximately 2 mm from the limbus. The suture tension was adjusted to lift the tube away from the iris while keeping it a safe distance from the corneal endothelium. The authors acknowledged, however, the risks of suture-related infection, corneal astigmatism, and persistent tube-iris or tube-cornea touch.

Surgical options, including the implantation of additional GDDs and "piggyback" drainage devices, have been developed to treat primary GDD failure [44]. In a series of 8 eyes, Lee et al. [45] reported the implantation of an additional AGV device in patients with IOP persistently ≥ 30 mmHg, despite having a GDD and receiving maximally tolerated medical therapy. Seven (of eight) patients had a statistically significant decrease in glaucoma medications 1 year post-operatively. No cases of diplopia or corneal decompensation were observed. In 16 eyes of 14 patients with uncontrolled glaucoma, Valimaki et al. [46] inserted a second glaucoma drainage implant in a piggyback manner. The sequential implant was rotated so that the tube of the 'piggyback' implant was directed towards the quadrant containing the original implant and inserted into the bleb, thereby converting a one-plate into a two-plate implant. The mean IOP was reduced from 29.2 mmHg to 17.3 mmHg, suggesting that a piggyback approach may be a viable option in patients with a failed GDD. In a series of 18 eyes, Dervan et al. [47] sutured a Baerveldt (250 or 350 mm) or Molteno3 GDD into an unused scleral quadrant and connected the silicone tube to the primary plate bleb. Mean IOP was reduced from 27.1 mmHg to 18.4 mmHg at the last follow-up. Several studies [44,48,49] suggest that piggyback GDD placement may be a viable surgical option for primary tube failure, without risking the corneal decompensation that may occur when inserting a second GDD into the anterior chamber [45,46].

Tube retraction, a complication of GDD implantation, often requires surgical revision to maintain drainage. Chiang et al. [50] reported successful outcomes in three patients with a 'tube-in-tube' technique that extended the existing tube of the Baerveldt GDD. The anterior portion of the drainage tube was exposed, and its patency was assessed. A tube segment from either a new GDD or a Tube Extender was inserted into the original tube, or vice versa. Advantages of this technique include the need for only minimal surgical dissection and disruption of the pre-existing GDD bleb, having a low risk of joined tube migration due to the high tensile strength, not requiring fixation sutures at the 'tube-in-tube' interface, not requiring additional scleral grafting, and the ease with which this technique can be learned. No tube migration occurred during follow-up periods of 1 month to 3 years.

The EX-PRESS Glaucoma Filtration Device (Alcon Laboratories, Fort Worth, TX, USA) has also been implanted in different locations when required by a unique clinical situation. Yen et al. [51] described an eye that had previously undergone pars plana vitrectomy (PPV) with prolonged silicone oil tamponade (22 months) for a rhegmatogenous retinal detachment and had developed neovascular glaucoma (NVG). Trabeculectomy with EX-PRESS implantation was performed, but bleb failure developed three times in four years, and the IOP reached 40 mmHg despite topical anti-glaucoma medications and oral acetazolamide. The existing EX-PRESS device was re-implanted into the posterior segment, and the IOP remained at 8 mmHg for more than 8 months after surgery and without medications [51].

4.3. Modifications to Existing GDDs

Over the past few years, GDDs have been repeatedly modified to enhance safety and efficacy. In 42 patients with neovascular glaucoma, Gil-Carrasco et al. [52] compared the safety and efficacy of the AGV model M4 (high-density porous polyethylene plate) and the model S2 (polypropylene plate). The AGV model M4, because of its porous polyethylene plate, was believed to increase aqueous outflow, but no differences in efficacy were seen at 1 year.

4.4. Invention of New GDDs

The Paul Glaucoma Implant (PGI) was created to reduce complications while preserving efficacy [53]. The PGI differs by having a smaller tube diameter—the external tube diameter is 467 μm, and the internal tube diameter is 127 μm. By occupying less space in the anterior chamber and preserving a large endplate surface area for aqueous absorption, damage to the corneal endothelium and risk of tube erosion are theoretically lowered [53]. The smaller tube calibre makes intraoperative surgical occlusion easier. At 24 months [54], complete success was achieved in 71.1% of patients, and the mean number of glaucoma medications decreased from 3.2 to 0.29. Complications included a self-limiting shallow anterior chamber, hypotony that required intervention, and tube occlusion.

The Ahmed ClearPath GDD (ACP, New World Medical, Rancho Cucamonga, CA, USA) [55] was introduced in 2019 as a valveless device, available in both 250 and 350 mm^2 sizes, and with a flexible plate that conforms to the curvature of the globe. Anteriorly located suture fixation points make implantation easier, the posteriorly positioned plate on the 350 model avoids muscle insertions, and an optional pre-threaded 4-0 polypropylene rip cord and a co-packaged 23-gauge needle simplify the creation of a sclerostomy.

The lower profile of the plate purportedly reduces the risk of conjunctival erosion and produces a low, diffuse bleb [56]. In a multicentre retrospective analysis of 104 eyes with medically and/or surgically uncontrolled glaucoma, Grover et al. [55] reported good IOP outcomes with both the 250 or 350 mm^2 devices. Significant reductions in mean IOP (13.6 to 16.7 mmHg) and medications (3.9 to 1.9) were seen at 6 months [55].

A series of newly designed GDDs can be adjusted post-operatively to reduce the incidences of hypotony and hypertension. These devices include the eyeWatch (eW, Rheon Medical, Lausanne, Switzerland) and others currently undergoing animal testing [57,58]. The eW has a deformable silicone tube that can undergo targeted compression to alter its cross-sectional area and thereby change fluidic resistance. Post-operatively, IOP may be changed non-invasively by moving the position of an internal magnetic rotor with an external control unit (the "eyeWatch Pen"). A pilot study found fewer postoperative episodes of hypotony and IOP spikes, with a complete success rate of 40% [59]. Subsequent studies produced outcomes comparable to those with the AGV [60]. Adjustable GDDs may reduce the need for intra-operative measures, such as tube ligation, and enable better postoperative IOP control.

The primary objective of GDDs has been to improve the drainage of aqueous, but GDDs are now being developed as extended-release drug reservoirs [61,62]. No US FDA-approved GDD drug delivery systems have reached the market, but base plates are being redesigned as reservoirs for drug storage. The tube would deliver the drug into either the anterior or posterior chamber through a one-way pressure-dependent valve. A wireless programming system is being developed to control drug delivery [63,64] with challenges that include the creation of a micro-delivery system and the need to resupply the reservoir.

Base plates may be replaced with tube shunt devices that have expanded membranes. In a study of 43 eyes, Ahn et al. [65] reported that the MicroMT (Figure 6), a membrane-tube shunt device, significantly reduced IOP from 22.5 mmHg to 11.1 mmHg after 3 years. The MicroMT has a reduced device profile, which decreases the risk of diplopia and conjunctival erosion.

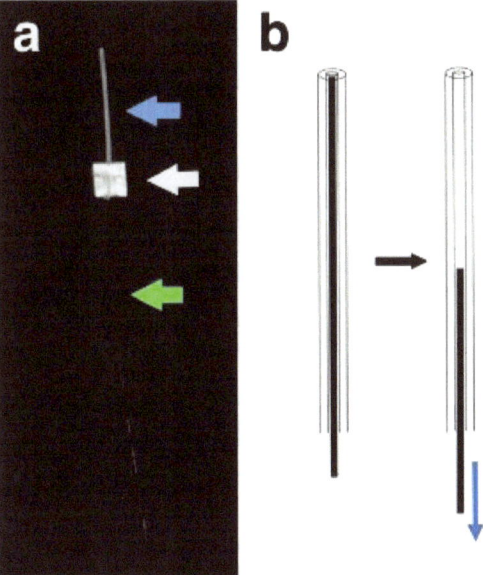

Figure 6. (**a**) The MicroMT consists of an expanded polytetrafluoroethylene membrane (white arrow) and silicone tube (blue arrow) with an intraluminal stent (green arrow); (**b**) the stent can be retracted after the operation. Courtesy of Ahn et al. [65].

5. Minimally Invasive Glaucoma Surgery (MIGS)

Minimally Invasive Glaucoma Surgery (MIGS) refers to a group of IOP-lowering surgical procedures that have emerged during the last decade. MIGS generally cause minimal trauma with little or no scleral dissection or conjunctival manipulation [66], incorporate either an ab interno or ab externo approach, and have good safety profiles and rapid recovery times [66]. MIGS are broadly classified into the following three categories according to the site of implantation or augmentation [67]: (1) angle-based MIGS, which enhance trabecular outflow by bypassing or manipulating angle structures, such as the trabecular meshwork and Schlemm's canal; (2) suprachoroidal MIGS, which increases uveoscleral outflow through a suprachoroidal drainage shunt; and (3) subconjunctival MIGS, which creates an aqueous outflow pathway into the subconjunctival or sub-Tenon's space.

MIGS procedures have evolved rapidly over the past decade, with continuing, robust research and development into new techniques and devices [68]. As with any recently developed surgical device or technique, various challenges have emerged in the performance of surgery and the management of complications. Many of these challenges are common to all MIGS procedures, and they can be broadly classified as follows: (1) perioperative challenges (e.g., difficulty with intra-operative handling, visualisation, and implantation of the device, or bleeding and hypotony in the immediate postoperative period); and (2) long-term postoperative problems (e.g., bleb fibrosis, scarring, stent occlusion, and insufficient long-term IOP lowering). In general, these perioperative and long-term postoperative problems tend to be mild [69,70], and serious sight-threatening complications, such as retinal detachment or endophthalmitis following MIGS, are rare [69,70]. Areas for improvement remain, and since MIGS are becoming an increasingly important option for the management of glaucoma, they are being continuously evaluated [5].

Recent advancements in MIGS have attempted to address current limitations in surgical success rates and ease of use in the following ways: (1) modifications to existing MIGS techniques, (2) combination MIGS, and (3) development of new MIGS. The next section will explore recent advancements in MIGS procedures and devices, and provide examples as to how they attempt to address existing limitations.

5.1. Recent Modifications to MIGS Techniques

The XEN45 gel stent (Allergan, Dublin, Ireland), a subconjunctival MIGS device, has demonstrated good safety and efficacy in the management of open-angle glaucoma [71], but many investigators have reported the need for postoperative interventions, such as bleb needling, with or without antifibrotic usage, to maintain the long-term patency of the device and sustain its IOP-lowering effect [72,73]. These additional interventions impose additional cost, risk, and inconvenience to both the patient and surgeon. The XEN45 was originally approved by the US FDA to be implanted with an ab interno, closed conjunctiva technique [74], but glaucoma surgeons have adopted an ab externo approach (with either opened or closed conjunctiva) in an attempt to improve safety, efficacy, and ease of implantation [75–77]. Some studies have reported higher rates of surgical success and IOP-lowering and lower rates of bleb interventions in the open conjunctiva ab externo approach. A retrospective case series by Tan et al. [75] showed a greater mean IOP reduction in the ab externo open conjunctiva group compared to the ab interno closed conjunctiva group (12.8 ± 3.0 mmHg (40.1% decrease) vs. 8.4 ± 1.7 mmHg (28.6% decrease); $p = 0.208$) at the 12-month follow-up. Needling was required in fewer ab externo than ab interno cases (26.7% vs. 42%; $p = 0.231$), but the superiority of the ab externo open conjunctiva technique has not been consistently demonstrated across studies [75–77].

The distal end of the XEN Gel Stent can become obstructed by Tenon's, so a transconjunctival ab externo implantation approach [78] has been developed to produce a similar lowering of IOP and medication dependency as the ab interno closed conjunctiva approach but with shorter surgical times and quicker postoperative visual recovery [78]. Another technique to improve XEN implantation in the subconjunctival space is the XEN 'Air' Technique [79]. Prior to placement of the XEN gel stent, air and viscoelastic is injected into the subconjunctival space to create a mixed pneumatic/viscoelastic dissection, thus preparing a subconjunctival pocket for subsequent XEN insertion with a larger bleb to reduce rates of postoperative fibrosis.

The Preserflo Microshunt (Santen, Osaka, Japan) is a similar subconjunctival MIGS device but is meant to be implanted via an ab externo approach into the anterior chamber through an opened conjunctiva. The Preserflo Microshunt has a significantly smaller diameter than other drainage devices, but it still may damage the corneal endothelium [80], particularly if the implant extends far into the anterior chamber or close to the endothelium. Martinez-de-la-Casa et al. [81] reported a patient with open-angle glaucoma refractory to medical therapy (with an IOP of 26 mmHg on maximal medical therapy) and concomitant granular corneal dystrophy with incipient stromal folds and an endothelial count of 700 cells/mm^2. The Preserflo Microshunt was implanted into the posterior chamber to minimise the possibility of further endothelial damage, and to avoid iris incarceration, the bevel was directed downward as it is when posterior chamber drainage devices are implanted (Figure 7). Six months after surgery, the implant remained functional, with an IOP of 9 mmHg and without additional medical treatment [81].

Figure 7. Preserflo Microshunt implanted into the posterior chamber. Note the orientation of the bevel to avoid incarceration of the iris. Courtesy of Martinez-de-la-Casa et al. [81].

Poor visualisation may prevent the implantation of MIGS devices. Extensive anterior synechiae or significant corneal opacities may prevent visualisation of the angle through conventional gonioscopy, which increases the risk of implantation failure or precludes MIGS usage entirely. To overcome this challenge, glaucoma surgeons have used intraoperative optical coherence tomography (iOCT) [82,83]. Junker et al. [84] reported the use of iOCT to accurately visualise a Trabectome within iridocorneal structures and facilitate the removal of the trabecular meshwork (TM). Ishida et al. [85] used the iOCT to visualise angle structures during ab interno trabeculotomy with the Tanito microhook (M-2215, Inami, Tokyo, Japan). Further research into the outcomes of iOCT-assisted MIGS procedures may improve the overall safety and success of MIGS while enabling patients who were previously ineligible to undergo these surgeries successfully.

Deep learning can create three-dimensional images of iridocorneal structures during angle-based MIGS surgeries to augment direct microscope visualisation [86]. A recent publication [87] by the Artificial Intelligence in Gonioscopy (AIG) Study Group described a convolutional neural network (CNN) that had been trained on videos of gonioscopic ab interno trabeculotomy with the Trabectome to accurately identify the TM in real-time. The CNN developed by Lin et al. [87] managed to consistently identify the TM from surgical videos, outperforming the human experts against which it was tested. Since accurate identification of iridocorneal structures on gonioscopy may be difficult, and errors can lead to surgical complications or suboptimal outcomes, a real-time assistive deep learning model could have applications to MIGS training and intraoperative guidance [87]. Deep learning could also be useful for other MIGS and non-MIGS glaucoma surgeries.

Existing MIGS devices have also been modified to improve device delivery and facilitate surgical handling. The iStent inject (Glaukos Corporation, San Clemente, CA, USA) consists of two trabecular-bypass flange devices designed to facilitate aqueous outflow into Schlemm's canal by bypassing the trabecular meshwork. Randomised controlled trials [88,89] showed a good lowering of IOP and a substantial reduction in postoperative medication use. Additional improvements resulted in the iStent infinite—consisting of three wider-flange devices (increased from 230 μm to 360 μm) on a single preloaded injector. The widened flanges optimise stent visualisation and improve placement, while possibly reducing the risk of stent occlusion by the iris. Additional iStent devices further lower IOP, with an incremental benefit of 3 stents over 2 [90]. The new iStent infinite allows the surgeon to inject three devices while entering the eye only once, thus reducing surgical time and risk. A 12-month multicentre clinical trial showed that the iStent infinite [91] significantly and safely reduces IOP in patients with uncontrolled open-angle glaucoma. The original iTrack microcatheter circumferentially viscodilates and intubates the Schlemm's canal [92], whereas the new iTrack Advance utilises the same microcatheter with a new and improved handheld injector to increase predictability and control during device advancement or retraction. This may reduce complications related to inappropriate device handling or insertion.

5.2. Combination MIGS Procedures

The different but complementary mechanisms of action of MIGS procedures have been combined to effectively lower IOP. The OMNI surgical system (Sight Sciences Inc., Menlo Park, CA, USA) was US FDA-approved in 2021 [93] to perform both canaloplasty (microcatheterisation and transluminal viscodilation of Schlemm's Canal) and trabeculotomy (cutting of TM). This procedure targets the three main sites of outflow resistance in the conventional aqueous outflow pathway—the TM, Schlemm's canal, and the distal collector channels [94]. Three-hundred-and-sixty-degree catheterisation and pressurised viscodilation enlarge Schlemm's canal and dilate distal collector channels, thereby removing distal blockages to aqueous outflow and reducing distal outflow resistance. By addressing both proximal and distal areas of outflow resistance, the OMNI surgical system has the potential to increase the IOP-lowering efficacy of a single-setting procedure [94–96].

MIGS has been used in combination with traditional glaucoma filtering surgery. To mitigate hypotony and corneal endothelial cell loss [97] after placement of the Baerveldt tube (Advanced Medical Optics, Inc., Santa Ana, CA, USA), D'Alessandro et al. [98] placed (ab externo) an XEN implant into the anterior chamber and inserted the Baerveldt tube more posteriorly. The newly formed double tube was sutured and covered by the scleral flap [98]. In eyes with refractory open-angle glaucoma, Bravetti et al. [99] reported a significant IOP decrease from baseline to 12 months (29.9 $\pm$ 13.2 to 15.2 $\pm$ 6.6 mmHg (-49.2%); $p < 0.0001$) and medication use decreased from 3.0 $\pm$ 1.3 to 1.3 $\pm$ 0.9. However, 41.5% of patients required revision surgery or transscleral cyclodestruction, ocular hypotony (under 6 mmHg for >4 weeks) occurred in 24.4% of eyes, and blockage of the XEN gel stent occurred in 17.1%; no cases of corneal endothelial damage were reported.

5.3. Recent Development of New MIGS

More MIGS devices have been proposed to overcome the limitations of existing devices, provide new mechanisms for aqueous outflow, facilitate ease of use, or improve device efficacy. The following six MIGS devices will be discussed: (1) MINIject DO627 (iStar Medical, Wavre, Belgium); (2) Intra-Scleral Ciliary Sulcus Suprachoroidal Microtube; (3) iDose TR (Glaukos Corporation, California, USA); (4) Beacon Aqueous Microshunt (MicroOptx, Maple Grove, MN, USA); (5) Minimally Invasive Micro Sclerostomy (MIMS; Sanoculis Ltd., Israel); and (6) STREAMLINE® Surgical System (New World Medical, Rancho Cucamonga, CA, USA).

The MINIject, a suprachoroidal device inserted ab interno into the supraciliary space [100], has garnered significant interest among glaucomatologists. Compared to angle-based MIGS, supraciliary stents are not limited by downstream episcleral venous pressure, which theoretically allows them to produce greater IOP lowering. Supraciliary stents do not form blebs, thereby eliminating bleb-related risks and interventions, though they are prone to postoperative scarring, tissue reaction, and implant failure [101]. Different suprachoroidal shunts have been introduced over the past decade, with varying degrees of success. Despite initial success, the CyPass Micro-Stent (Alcon Laboratories, Inc., Fort Worth, TX, USA) was withdrawn from the global market in August 2018 due to long-term safety concerns over endothelial cell loss [102]. The SOLX gold shunt (SOLX, Inc., Waltham, MA, USA) did not receive US FDA approval due to high fibrosis-related failure rates [103]. The MINIject DO627 [100] aims to overcome the limitations of previous suprachoroidal MIGS devices by using a biocompatible, medical-grade silicone (STAR material NuSil med-6215) that is soft, flexible, and inherently antifibrotic [104]. The 5 mm long implant does not have a patent lumen but rather consists of a meshwork of porous microspheres that allows aqueous to drain down the pressure gradient at a steady state via a sponge effect (Figure 8). In addition, the silicone demonstrates good biointegration, as surrounding tissue colonises the porous structure while preserving drainage and minimising fibrosis and scarring, thereby eliminating the risk of a blocked lumen [100]. Three clinical trials (STAR-I [100], STAR-II [105], and STAR-III [106]) across 11 sites in Central and South America, Asia, and Europe, showed promising IOP-lowering results and medication reduction over 24 months with few adverse events [107]. The ongoing STAR-V [108] trial aims to enrol 350 patients with primary open-angle glaucoma in the US, and the STAR-VI trial will evaluate the MINIject DO627 in patients undergoing concurrent phacoemulsification.

The 'Intrascleral Ciliary Sulcus-Suprachoroidal Microtube' [109] consists of a sterile medical grade silicone tube (Tube extender, New World Medical) with a 300 μm internal diameter and 600 μm external diameter. During insertion, the tube is custom cut, inserted through an inferotemporal conjunctival peritomy to preserve the superior conjunctiva for future surgery, sutured to the sclera to prevent migration, and covered by a partial thickness scleral flap. In a 12-month trial of 36 pseudophakic Black and Afro-Latin patients with glaucoma refractory to topical ocular antihypertensive medications, IOP decreased (21 ± 8.2 to 13.5 ± 4.4 mmHg; $p = 0.032$), as did the mean number of medications (4.2 ± 1.0 to 2.4 ± 1.7; $p = 0.021$), with five patients being medication free. This technique avoids bleb-related complications from traditional trabeculectomy or subconjunctival filtering devices, but there are no data regarding rates of suprachoroidal space scarring or corneal endothelial damage. A larger sample size with longer follow-up is needed.

The iDose TR is a drug-eluting MIGS device that aims to overcome barriers to long-term topical therapy, including patient non-compliance, ocular surface irritation, and difficulty with instilling eye drops [110]. The 1.8×0.5 mm biocompatible titanium implant has three main parts—a scleral anchor that affixes to the TM, the body that serves as a reservoir for the drug (travoprost), and a membrane that elutes the drug intracamerally for a target duration of 6–12 months [111]. The iDose TR is implanted similarly to the iStent inject, another MIGS device that is located in the TM. Two phase III randomised controlled trials [112,113] are ongoing, with preliminary results showing that the iDose TR arms will achieve the primary efficacy endpoint of non-inferiority to the active comparator arm (twice-daily topical timolol 0.5%) at 3 months [114]. A favourable safety profile with no clinically significant corneal endothelial cell loss through 12 months was reported [114]. IOP-lowering is likely to diminish after 12 months when the reservoir empties, which will prompt the question of whether the empty implant should be left in place, refilled, or removed.

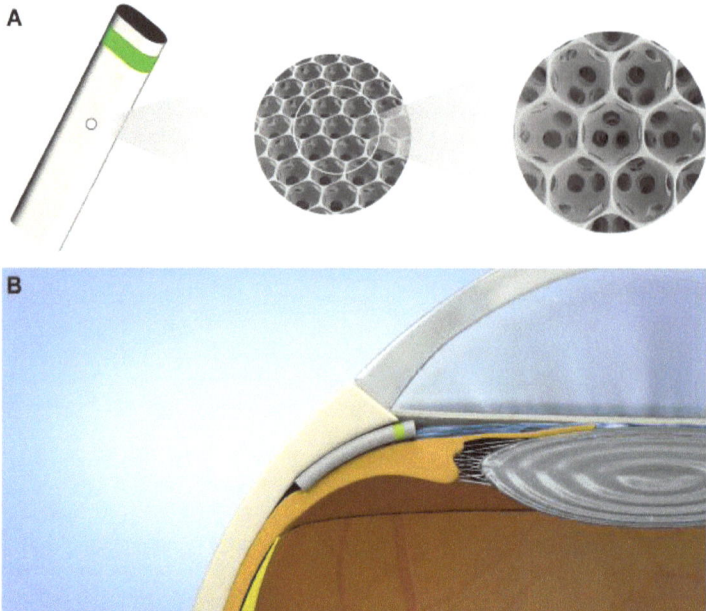

Figure 8. MINIject glaucoma drainage device (iSTAR Medical SA, Wavre, Belgium): (**A**) implant made of STAR material; (**B**) schematic of the device in situ. Courtesy of Denis et al. [100].

The Beacon Aqueous Microshunt [115] is a new class of ab externo MIGS that is implanted at the superior limbus to allow aqueous outflow into the tear film. The microshunt measures 1.70 mm wide by 3.30 mm long, with a 0.03 mm × 0.048 mm internal hydrogel channel. Controlled-outflow resistance depends on the channel diameter, and the shunt has been engineered to produce IOP reductions of 8 to 12 mmHg regardless of baseline [115]. To reduce retrograde bacterial movement and mitigate the risk of endophthalmitis, the polyethylene glycol (PEG) hydrogel channel is composed of anti-biofouling polymers that only allow a one-way laminar flow of aqueous humour towards the ocular surface. In a five-patient safety trial [116], no short-term corneal or infectious complications were seen. In a separate, single-patient case report, a significant IOP reduction from baseline (33 mmHg to 12 mmHg) was achieved. Long-term safety and efficacy need to be further investigated [115].

Minimally Invasive Micro Sclerostomy (MIMS) is an ab interno, stent-free, subconjunctival filtration procedure [117]. The MIMS handpiece consists of a 600 μm needle that rotates around its longitudinal axis and has been designed to carve a permanent tunnel near the corneoscleral junction to connect the AC with the subconjunctival space. MIMS is being touted as a MIGS procedure without foreign body-related complications, such as conjunctival erosions, corneal endothelial cell loss, stent migration, or extrusion, while delivering an IOP reduction that resembles existing subconjunctival MIGS. In an early clinical trial with 31 eyes, short-term IOP was lowered, similar to that expected with subconjunctival filtering MIGS [117]. Iris clogging of the internal sclerostomy causing high IOP spikes was the most common and concerning complication, and some of these could not be cleared with laser [117].

The STREAMLINE® Surgical System (New World Medical, Rancho Cucamonga, CA, USA) [118] is a handheld MIGS device for incisional goniotomies and Schlemm's canal viscodilation. A stainless-steel cannula tip with a retractable outer sleeve is used to make up to eight incisional goniotomies (150 μm diameter each) in the TM, while simultaneously delivering approximately 7 μL of viscoelastic per incision into the Schlemm's canal. In a series of 19 eyes [118], mean IOP reduction was 8.8 mmHg (36.9%) at 6 months, 57.9% (11/19) of subjects were using fewer medications than at screening, and 42.1% (8/19) were medication-free. A prospective randomised study comparing the safety and efficacy of the STREAMLINE® Surgical System to the iStent inject is ongoing [119].

6. Limitations

While this review aims to be a comprehensive one, several limitations are acknowledged. First, to ensure recency of the reviewed surgical procedures and modifications, the scope of this study was limited to trabeculectomy and GDD studies in the last 10 years, and MIGS studies in the last 5 years. Important modifications with significant impact on surgical outcomes may have been introduced outside this timeframe. Second, the emphasis on recent, novel procedures and modifications resulted in the inclusion of case reports and small case studies. This may limit the applicability of this review to the general population. Finally, while this review highlights individual surgeries and procedures, it does not suggest any particular approach to procedure selection in different disease contexts and, hence, may be limited in its clinical applicability.

7. Conclusions

There have been significant advancements in all major types of glaucoma surgery—trabeculectomy, GDD implantation, and MIGS. The increasing armamentarium of available surgical procedures and modified techniques will allow glaucoma surgeons to further personalise a patient's surgical treatment based on the desired magnitude of IOP reduction and anatomical and disease characteristics of the eye, whilst considering the risk-benefit ratio of various techniques. Despite its long history, trabeculectomy surgery continues to be improved with adjunctive incisions, Tenon's layer positioning, and novel suturing techniques. GDD implantation has also been the subject of several surgical and design modifications. The rapid development of MIGS procedures and their widespread adoption appears to be fuelling further development, including novel modifications to surgical techniques, the development of new MIGS devices, and the emergence of combination MIGS with multiple mechanisms of action to lower IOP.

Author Contributions: Conceptualization, B.C.H.A., S.Y.L. and B.K.B.; writing—original draft preparation, B.C.H.A., S.Y.L., B.K.B. and H.J.W.; writing—review and editing, B.C.H.A., S.Y.L., B.K.B., H.J.W., M.W.S. and S.D., results interpretation, B.C.H.A., S.Y.L., B.K.B., H.J.W., M.W.S. and S.D. All authors have read and agreed to the published version of the manuscript.

Funding: This research received no external funding.

Institutional Review Board Statement: Not applicable.

Informed Consent Statement: Not applicable.

Data Availability Statement: This manuscript makes use of publicly available data from published studies. As such, no original data are available for sharing.

Conflicts of Interest: B.C.H.A has received speaker's honoraria from Alcon, Inc., research support and speaker's honoraria from Glaukos Corporation, and speaker's honoraria from Santen Pharmaceutical Asia Pte. Ltd. S.D. has received travel support from New World Medical.

Appendix A

Table A1. Summary table of all included studies.

Trabeculectomy Studies		
Author/Year	Title	Study Type
Dada 2022 [13]	Trabeculectomy Augmented with Limited Deep Sclerectomy and Cyclodialysis with Use of Scleral Tissue as a Spacer	Case Report
Dada 2021 [8]	Efficacy of Trabeculectomy Combined with Limited Deep Sclerectomy Versus Trabeculectomy Alone A Randomised-controlled Trial	Randomised Controlled Trial
Chan 2020 [17]	The Tenons' Layer Reposition Approach of Trabeculectomy: A Longitudinal Case Series of a Mixed Group of Glaucoma Patients	Non-comparative case series
Olawoye 2015 [19]	Fornix-based Trabeculectomy with Mitomycin C Using the Horizontal Conjunctival Suture Technique	Non-comparative case series
Allam R 2020 [15]	Trabeculectomy With Extended Subscleral Tunnel Versus Conventional Trabeculectomy in the Management of POAG: A 1-Year Randomised-controlled Trial	Randomised Controlled Trial
Kirk 2014 [18]	Modified Wise Closure of the Conjunctival Fornix-Based Trabeculectomy Flap	Retrospective Comparative Study
Figus M 2016 [22]	Scleral Flap-Everting Suture for Glaucoma-Filtering Surgery	Non-comparative Case Series
Baykara M 2017 [23]	A Novel Suturing Technique for Filtering Glaucoma Surgery: The Accordion Suture	Non-comparative Case Series
Glaucoma Drainage Device Studies		
Grover 2022 [55]	Clinical Outcomes of Ahmed ClearPath Implantation in Glaucomatous Eyes: A Novel Valveless Glaucoma Drainage Device	Retrospective Case Series
Nakamura 2022 [120]	Tissue Reactivity to, and Stability of, Glaucoma Drainage Device Materials Placed Under Rabbit Conjunctiva	Animal In Vivo Study
Gupta 2021 [38]	A Graft-Free Scleral Sleeve Technique of Ahmed Glaucoma Valve Implantation In Refractory Glaucoma—Rising to the Challenge of COVID-19 Pandemic	Case Report
Gupta 2020 [30]	Pars Plana Placement of Ahmed Glaucoma Valve Tube Through Sclerotomy Port In Refractory Glaucoma: A Novel Surgical Technique	Case Report
Koh 2020 [53]	Treatment Outcomes Using the PAUL Glaucoma Implant to Control Intraocular Pressure in Eyes with Refractory Glaucoma	Interventional Cohort Study
Mungale 2019 [41]	A Novel Simplified Method for Managing Inadvertent Tube Cut During Aurolab Aqueous Drainage Implant Surgery For Refractory Glaucoma	Case Report

Table A1. *Cont.*

	Trabeculectomy Studies	
Author/Year	Title	Study Type
Roy 2019 [59]	Initial Clinical Results of the eyeWatch: A New Adjustable Glaucoma Drainage Device Used in Refractory Glaucoma Surgery	Prospective Non-comparative Clinical Trial
Sastre-Ibanez 2019 [40]	Efficacy of Ologen Matrix Implant in Ahmed Glaucoma Valve Implantation	Prospective Randomised Clinical Trial
Eslami 2019 [35]	Single Long Scleral Tunnel Technique for Prevention of Ahmed Valve Tube Exposure	Retrospective Case Series
Vergados 2019 [121]	Ab Interno Tube Ligation for Refractory Hypotony Following Non-valved Glaucoma Drainage Device Implantation	Retrospective Case Series
Pakravan 2018 [37]	Ahmed Glaucoma Valve Implantation: Graft-Free Short Tunnel Small Flap versus Scleral Patch Graft After 1-Year Follow-up: A Randomised Clinical Trial	Randomised Controlled Trial
Chiang 2017 [50]	A Novel Method of Extending Glaucoma Drainage Tube: "Tube-in-Tube" Technique	Retrospective Non-comparative Case Series
Hwang 2017 [42]	Intracameral Air Injection During Ahmed Glaucoma Valve Implantation In Neovascular Glaucoma for the Prevention of Tube Obstruction with Blood Clot: Case Report	Case Report
Brouzas 2017 [36]	Double Scleral Tunnel In Tandem Technique for Glaucoma Drainage Tube Implants	Case Series
Dervan 2017 [47]	Intermediate-Term and Long-Term Outcome of Piggyback Drainage: Connecting Glaucoma Drainage Device to a Device In Situ for Improved Intraocular Pressure Control	Retrospective Interventional Cohort Study
Park 2016 [122]	Polymeric Check Valve With an Elevated Pedestal for Precise Cracking Pressure In a Glaucoma Drainage Device	In Vitro Study
Kataria 2016 [43]	A Novel Technique of a Transcorneal Suture to Manage an Iris Tuck into the Tube of a Glaucoma Drainage Device	Case Report
Ahn 2016 [65]	Novel Membrane-Tube Type Glaucoma Shunt Device for Glaucoma Surgery	Retrospective Non-comparative Interventional Case Series
Gil-Carrasco 2016 [52]	Comparative Study of the Safety and Efficacy of The Ahmed Glaucoma Valve Model M4 (High-Density Po-Rous Polyethene) And the Model S2 (Polypropylene) In Patients With Neovascular Glaucoma	Prospective Comparative Randomised Study
Ma 2016 [34]	Modified Scleral Tunnel to Prevent Tube Exposure In Patients With Refractory Glaucoma	Retrospective Case Series
Maldonado-Junyent 2015 [32]	Oculo-Peritoneal Shunt: Draining Aqueous Humour To The Peritoneum	Case Report

Table A1. Cont.

Author/Year	Title	Study Type
Trabeculectomy Studies		
Martino 2015 [123]	Surgical Outcomes of Superior Versus Inferior Glauco-Ma Drainage Device Implantation	Retrospective Case Series
Schaefer 2015 [44]	Failed Glaucoma Drainage Implant: Long-Term Out-Comes of a Second Glaucoma Drainage Device Versus Cyclophotocoagulation	Non-randomised Retrospective Cohort Study
Välimäki 2015 [46]	Insertion of Sequential Glaucoma Drainage Implant in a Piggyback Manner	Retrospective Case Series
Lee 2014 [45]	Efficacy of Additional Glaucoma Drainage Device Insertion in Refractory Glaucoma: Case Series with a Systematic Literature Review and Meta-Analysis	Non-comparative Retrospective Case Series
Luong 2014 [124]	A New Design and Application of Bioelastomers for Better Control of Intraocular Pressure In a Glaucoma Drainage Device	In Vitro Study
Grover 2013 [28]	Confirming and Establishing Patency of Glaucoma Drainage Devices Using Trypan Blue	Case Report
Minimally Invasive Glaucoma Surgery Studies		
Geffen 2022 [118]	Minimally Invasive Micro Sclerostomy (MIMS) Procedure: A Novel Glaucoma Filtration Procedure	Prospective Clinical Trial
Martinez-de-la-casa 2022 [81]	Posterior Chamber Implantation of a Preserflo Microshunt In a Patient With a Compromised Endothelium	Case Report
New World Medical 2022 [120]	STREAMLINE®SURGICAL SYSTEM Compared to iStent Inject W®in Patients with Open-Angle Glaucoma	Prospective Randomised Controlled Trial
Lin 2022 [87]	Accurate Identification of the Trabecular Meshwork Under Gonioscopic View in Real Time Using Deep Learning.	Cross-Sectional Study
Bleeker 2022 [95]	Short-Term Efficacy of Combined ab Interno Canaloplasty and Trabeculotomy in Pseudophakic Eyes with Open-Angle Glaucoma	Retrospective Case Series
Lazcano-Gomez 2022 [119]	Interim Analysis of STREAMLINE®Surgical System Clinical Outcomes in Eyes with Glaucoma	Prospective Case Series
Gallardo 2022 [77]	Comparison of Clinical Outcomes Following Gel Stent Implantation via Ab externo and Ab interno Approaches in Patients with Refractory Glaucoma.	Retrospective Case Series
Tan 2021 [75]	Comparison of Safety and Efficacy Between Ab Interno and Ab Externo Approaches to XEN Gel Stent Placement	Retrospective Case Series
Do 2021 [76]	Clinical Outcomes with Open Versus Closed Conjunctiva Implantation of the XEN45 Gel Stent	Retrospective Case Series
Feijoo 2020 [105]	A European Study of the Performance and Safety of MINIject in Patients with Medically Uncontrolled Open-angle Glaucoma (STAR-II)	Prospective Clinical Trial

Table A1. *Cont.*

Trabeculectomy Studies		
Author/Year	Title	Study Type
Ucar 2020 [78]	Xen Implantation in Patients With Primary Open-Angle Glaucoma: Comparison of Two Different Techniques	Retrospective Comparative Interventional Study
Vera 2020 [79]	Surgical Approaches for Implanting Xen Gel Stent without Conjunctival Dissection	Expert Opinion
Ishida 2020 [85]	Observation of Gonio Structures during Microhook Ab Interno Trabeculotomy Using a Novel Digital Microscope with Integrated Intraoperative Optical Coherence Tomography	Retrospective Observational Study
Bravetti 2020 [99]	Xen-Augmented Baerveldt Drainage Device Implantation in Refractory Glaucoma: 1-Year Outcomes	Retrospective Case Series
Denis 2019 [100]	A First-in-Human Study of the Efficacy and Safety of MINIject in Patients with Medically Uncontrolled Open-Angle Glaucoma (STAR-I)	Randomised Controlled Trial
Laroche 2019 [109]	Intra-Scleral Ciliary Sulcus Suprachoroidal Microtube: Making Supraciliary Glaucoma Surgery Affordable	Case Report
Valimaki 2018 [46]	Xen Gel Stent to Resolve Late Hypotony After Glaucoma Drainage Implant Surgery: A Novel Technique	Case Report
Yen 2018 [51]	Pars Plana Insertion of Glaucoma Shunt in Eyes With Refractory Neovascular Glaucoma: Case Report	Case Report
Fili 2018 [104]	The Starflo Glaucoma Implant: Preliminary 12 Months Results	Prospective Case Series

References

1. Quigley, H.A.; Broman, A.T. The number of people with glaucoma worldwide in 2010 and 2020. *Br. J. Ophthalmol.* **2006**, *90*, 262–267. [CrossRef] [PubMed]
2. Pillunat, L.E.; Erb, C.; Jünemann, A.G.; Kimmich, F. Micro-invasive glaucoma surgery (MIGS): A review of surgical procedures using stents. *Clin. Ophthalmol.* **2017**, *11*, 1583–1600. [CrossRef] [PubMed]
3. Saha, B.C.; Kumari, R.; Sinha, B.P.; Ambasta, A.; Kumar, S. Lasers in Glaucoma: An Overview. *Int. Ophthalmol.* **2021**, *41*, 1111–1128. [CrossRef] [PubMed]
4. Sharaawy, T.; Bhartiya, S. Surgical management of glaucoma: Evolving paradigms. *Indian J. Ophthalmol.* **2011**, *59* (Suppl. S1), S123–S130. [CrossRef] [PubMed]
5. Lim, R. The surgical management of glaucoma: A review. *Clin. Exp. Ophthalmol.* **2022**, *50*, 213–231. [CrossRef]
6. Haddaway, N.R.; Page, M.J.; Pritchard, C.C.; McGuinness, L.A. PRISMA2020: An R package and Shiny app for producing PRISMA 2020-compliant flow diagrams, with interactivity for optimised digital transparency and Open Synthesis. *Campbell Syst. Rev.* **2022**, *18*, e1230. [CrossRef]
7. Cairns, J.E. Trabeculectomy. Preliminary report of a new method. *Am. J. Ophthalmol.* **1968**, *66*, 673–679. [CrossRef]
8. Dada, T.; Sharma, A.; Midha, N.; Angmo, D.; Gupta, S.; Sihota, R. Efficacy of Trabeculectomy Combined With Limited Deep Sclerectomy Versus Trabeculectomy Alone: A Randomized-controlled Trial. *J. Glaucoma* **2021**, *30*, 1065–1073. [CrossRef]
9. Yu-Wai-Man, C.; Khaw, P.T. Developing novel anti-fibrotic therapeutics to modulate post-surgical wound healing in glaucoma: Big potential for small molecules. *Expert Rev. Ophthalmol.* **2015**, *10*, 65–76. [CrossRef]
10. Khaw, P.T.; Chiang, M.; Shah, P.; Sii, F.; Lockwood, A.; Khalili, A. Enhanced Trabeculectomy: The Moorfields Safer Surgery System. *Dev. Ophthalmol.* **2017**, *59*, 15–35.
11. Vijaya, L.; Manish, P.; Ronnie, G.; Shantha, B. Management of complications in glaucoma surgery. *Indian J. Ophthalmol.* **2011**, *59* (Suppl. S1), S131–S140. [CrossRef] [PubMed]
12. Figus, M.; Posarelli, C.; Passani, A.; Albert, T.G.; Oddone, F.; Sframeli, A.T.; Nardi, M. The supraciliary space as a suitable pathway for glaucoma surgery: Ho-hum or home run? *Surv. Ophthalmol.* **2017**, *62*, 828–837. [CrossRef] [PubMed]

13. Dada, T.; Shakrawal, J.; Ramesh, P.; Sethi, A. Trabeculectomy Augmented with Limited Deep Sclerectomy and Cyclodialysis with Use of Scleral Tissue as a Spacer. *J. Ophthalmic. Vis. Res.* **2022**, *17*, 596–600. [CrossRef] [PubMed]
14. Saeed, A.; Saleh, S. Modified trabeculectomy with an extended subscleral tunnel: Could it be a secure way toward successful glaucoma surgery? *J. Egypt. Ophthalmol. Soc.* **2014**, *107*, 97–105. [CrossRef]
15. Allam, R.; Raafat, K.A.; Abdel-Hamid, R.M. Trabeculectomy With Extended Subscleral Tunnel Versus Conventional Trabeculectomy in the Management of POAG: A 1-Year Randomized-controlled Trial. *J. Glaucoma* **2020**, *29*, 473–478. [CrossRef]
16. Feyi-Waboso, A.; Ejere, H.O. Needling for encapsulated trabeculectomy filtering blebs. *Cochrane Database Syst. Rev.* **2012**, *2012*, Cd003658. [CrossRef]
17. Chan, P.P.; Wong, L.Y.N.; Chan, T.C.Y.; Lai, G.; Baig, N. The Tenons' Layer Reposition Approach of Trabeculectomy: A Longitudinal Case Series of a Mixed Group of Glaucoma Patients. *J. Glaucoma* **2020**, *29*, 386–392. [CrossRef]
18. Kirk, T.Q.; Condon, G.P. Modified Wise closure of the conjunctival fornix-based trabeculectomy flap. *J. Cataract. Refract. Surg.* **2014**, *40*, 349–353. [CrossRef]
19. Olawoye, O.; Lee, M.; Kuwayama, Y.; Kee, C. Fornix-based Trabeculectomy With Mitomycin C Using the Horizontal Conjunctival Suture Technique. *J. Glaucoma* **2015**, *24*, 455–459. [CrossRef]
20. Wise, J.B. Mitomycin-Compatible Suture Technique for Fornix-Based Conjunctival Flaps in Glaucoma Filtration Surgery. *Arch. Ophthalmol.* **1993**, *111*, 992–997. [CrossRef]
21. Wang, Q.; Zhang, Q.E.; Nauheim, J.; Nayak Kolomeyer, N.; Pro, M.J. Fornix-Based Trabeculectomy Conjunctival Closure: Winged Sutures versus Modified Wise Closure. *Ophthalmol. Glaucoma* **2019**, *2*, 251–257. [CrossRef] [PubMed]
22. Figus, M.; Posarelli, C.; Nasini, F.; Casini, G.; Martinelli, P.; Nardi, M. Scleral Flap-Everting Suture for Glaucoma-filtering Surgery. *J. Glaucoma* **2016**, *25*, 128–131. [CrossRef] [PubMed]
23. Baykara, M.; Can Ermerak, B.; Sabur, H.; Genc, S. A novel suturing technique for filtering glaucoma surgery: The accordion suture. *Int. J. Ophthalmol.* **2017**, *10*, 1931–1934.
24. Aref, A.A.; Gedde, S.J.; Budenz, D.L. Glaucoma Drainage Implant Surgery. *Dev. Ophthalmol.* **2017**, *59*, 43–52.
25. Schwartz, K.S.; Lee, R.K.; Gedde, S.J. Glaucoma drainage implants: A critical comparison of types. *Curr. Opin. Ophthalmol.* **2006**, *17*, 181–189. [CrossRef] [PubMed]
26. Gedde, S.J.; Schiffman, J.C.; Feuer, W.J.; Herndon, L.W.; Brandt, J.D.; Budenz, D.L. Treatment outcomes in the Tube Versus Trabeculectomy (TVT) study after five years of follow-up. *Am. J. Ophthalmol.* **2012**, *153*, 789–803.e782. [CrossRef]
27. Ashburn, F.S.; Netland, P.A. The Evolution of Glaucoma Drainage Implants. *J. Ophthalmic. Vis. Res.* **2018**, *13*, 498–500.
28. Grover, D.S.; Fellman, R.L. Confirming and establishing patency of glaucoma drainage devices using trypan blue. *J. Glaucoma* **2013**, *22*, e1–e2. [CrossRef]
29. van Dooren, B.T.; Beekhuis, W.H.; Pels, E. Biocompatibility of trypan blue with human corneal cells. *Arch. Ophthalmol.* **2004**, *122*, 736–742. [CrossRef]
30. Gupta, R.; Varshney, A. Pars plana placement of Ahmed glaucoma valve tube through sclerotomy port in refractory glaucoma: A novel surgical technique. *Indian J. Ophthalmol.* **2020**, *68*, 234–236. [CrossRef]
31. Schlote, T.; Ziemssen, F.; Bartz-Schmidt, K.U. Pars plana-modified Ahmed Glaucoma Valve for treatment of refractory glaucoma: A pilot study. *Graefes. Arch. Clin. Exp. Ophthalmol.* **2006**, *244*, 336–341. [CrossRef] [PubMed]
32. Maldonado-Junyent, A.; Maldonado-Bas, A.; Gonzalez, A.; Pueyrredón, F.; Maldonado-Junyent, M.; Maldonado-Junyent, A.; Rodriguez, D.; Bulacio, M. Oculo-peritoneal shunt: Draining aqueous humor to the peritoneum. *Arq. Bras. Oftalmol.* **2015**, *78*, 123–125. [CrossRef] [PubMed]
33. Chaku, M.; Netland, P.A.; Ishida, K.; Rhee, D.J. Risk factors for tube exposure as a late complication of glaucoma drainage implant surgery. *Clin. Ophthalmol.* **2016**, *10*, 547–553. [PubMed]
34. Ma, X.-h.; Du, X.-j.; Liu, B.; Bi, H.-S. Modified scleral tunnel to prevent tube exposure in patients with refractory glaucoma. *J. Glaucoma* **2016**, *25*, 883–885. [CrossRef] [PubMed]
35. Eslami, Y.; Azaripour, E.; Mohammadi, M.; Kiarudi, M.Y.; Fakhraie, G.; Zarei, R.; Alizadeh, Y.; Moghimi, S. Single long scleral tunnel technique for prevention of Ahmed valve tube exposure. *Eur. J. Ophthalmol.* **2019**, *29*, 52–56. [CrossRef]
36. Brouzas, D.; Dettoraki, M.; Andreanos, K.; Nomikarios, N.; Koutsandrea, C.; Moschos, M.M. "Double scleral tunnel in tandem" technique for glaucoma drainage tube implants. *Int. Ophthalmol.* **2018**, *38*, 2349–2356.
37. Pakravan, M.; Hatami, M.; Esfandiari, H.; Yazdani, S.; Doozandeh, A.; Samaeili, A.; Kheiri, B.; Conner, I. Ahmed glaucoma valve implantation: Graft-free short tunnel small flap versus scleral patch graft after 1-Year follow-up: A randomized clinical trial. *Ophthalmol. Glaucoma* **2018**, *1*, 206–212. [CrossRef]
38. Gupta, R. A graft-free scleral sleeve technique of Ahmed Glaucoma Valve implantation in refractory glaucoma- Rising to the challenge of COVID-19 pandemic. *Indian J. Ophthalmol.* **2021**, *69*, 1623–1625. [CrossRef]
39. Amoozgar, B.; Lin, S.C.; Han, Y.; Kuo, J. A role for antimetabolites in glaucoma tube surgery: Current evidence and future directions. *Curr. Opin. Ophthalmol.* **2016**, *27*, 164–169. [CrossRef]
40. Sastre-Ibáñez, M.; Cabarga, C.; Canut, M.I.; Pérez-Bartolomé, F.; Urcelay-Segura, J.L.; Cordero-Ros, R.; García-Feijóo, J.; Martínez-de-la-Casa, J.M. Efficacy of Ologen matrix implant in Ahmed Glaucoma Valve Implantation. *Sci. Rep.* **2019**, *9*, 3178. [CrossRef]
41. Mungale, S.; Dave, P. A novel simplified method for managing inadvertent tube cut during aurolab aqueous drainage implant surgery for refractory glaucoma. *Indian J. Ophthalmol.* **2019**, *67*, 694. [PubMed]

42. Hwang, S.H.; Yoo, C.; Kim, Y.Y.; Lee, D.Y.; Nam, D.H.; Lee, J.Y. Intracameral air injection during Ahmed glaucoma valve implantation in neovascular glaucoma for the prevention of tube obstruction with blood clot: Case Report. *Medicine* **2017**, *96*, e9092. [CrossRef] [PubMed]
43. Kataria, P.; Kaushik, S.; Singh, S.R.; Pandav, S.S. A Novel Technique of a Transcorneal Suture to Manage an Iris Tuck into the Tube of a Glaucoma Drainage Device. *J. Glaucoma* **2016**, *25*, e731–e733. [CrossRef] [PubMed]
44. Schaefer, J.L.; Levine, M.A.; Martorana, G.; Koenigsman, H.; Smith, M.F.; Sherwood, M.B. Failed glaucoma drainage implant: Long-term outcomes of a second glaucoma drainage device versus cyclophotocoagulation. *Br. J. Ophthalmol.* **2015**, *99*, 1718–1724. [CrossRef] [PubMed]
45. Lee, N.Y.; Hwang, H.B.; Oh, S.H.; Park, C.K. Efficacy of Additional Glaucoma Drainage Device Insertion in Refractory Glaucoma: Case Series with a Systematic Literature Review and Meta-Analysis. *Semin. Ophthalmol.* **2015**, *30*, 345–351. [CrossRef] [PubMed]
46. Välimäki, J. Insertion of sequential glaucoma drainage implant in a piggyback manner. *Eye* **2015**, *29*, 1329–1334. [CrossRef]
47. Dervan, E.; Lee, E.; Giubilato, A.; Khanam, T.; Maghsoudlou, P.; Morgan, W.H. Intermediate-term and long-term outcome of piggyback drainage: Connecting glaucoma drainage device to a device in-situ for improved intraocular pressure control. *Clin. Exp. Ophthalmol.* **2017**, *45*, 803–811. [CrossRef]
48. Burgoyne, J.K.; WuDunn, D.; Lakhani, V.; Cantor, L.B. Outcomes of sequential tube shunts in complicated glaucoma. *Ophthalmology* **2000**, *107*, 309–314. [CrossRef]
49. Shah, A.A.; WuDunn, D.; Cantor, L.B. Shunt revision versus additional tube shunt implantation after failed tube shunt surgery in refractory glaucoma. *Am. J. Ophthalmol.* **2000**, *129*, 455–460. [CrossRef]
50. Chiang, M.Y.-M.; Camuglia, J.E.; Khaw, P.T. A novel method of extending glaucoma drainage tube:"Tube-in-Tube" technique. *J. Glaucoma* **2017**, *26*, 93–95. [CrossRef]
51. Yen, C.Y.; Tseng, G.L. Pars plana insertion of glaucoma shunt in eyes with refractory neovascular glaucoma: Case report. *Medicine* **2018**, *97*, e10977. [CrossRef] [PubMed]
52. Gil-Carrasco, F.; Jiménez-Román, J.; Turati-Acosta, M.; Portillo, H.B.-L.; Isida-Llerandi, C. Comparative study of the safety and efficacy of the Ahmed glaucoma valve model M4 (high density porous polyethylene) and the model S2 (polypropylene) in patients with neovascular glaucoma. *Arch. Soc. Española Oftalmol.* **2016**, *91*, 409–414. [CrossRef]
53. Koh, V.; Chew, P.; Triolo, G.; Lim, K.S.; Barton, K. Treatment Outcomes Using the PAUL Glaucoma Implant to Control Intraocular Pressure in Eyes with Refractory Glaucoma. *Ophthalmol. Glaucoma* **2020**, *3*, 350–359. [CrossRef] [PubMed]
54. Tan, M.C.J.; Choy, H.Y.C.; Koh Teck Chang, V.; Aquino, M.C.; Sng, C.C.A.; Lim, D.K.A.; Loon, S.C.; Chew Tec Kuan, P. Two-Year Outcomes of the Paul Glaucoma Implant for Treatment of Glaucoma. *J. Glaucoma* **2022**, *31*, 449–455. [CrossRef] [PubMed]
55. Grover, D.S.; Kahook, M.Y.; Seibold, L.K.; Singh, I.P.; Ansari, H.; Butler, M.R.; Smith, O.U.; Sawhney, G.K.; Van Tassel, S.H.; Dorairaj, S. Clinical Outcomes of Ahmed ClearPath Implantation in Glaucomatous Eyes: A Novel Valveless Glaucoma Drainage Device. *J. Glaucoma* **2022**, *31*, 335–339. [CrossRef] [PubMed]
56. Chang, P. Early surgeon experience with a new valveless glaucoma drainage device. In Proceedings of the American Glaucoma Society Annual Meeting, National Harbor, MD, USA, 27 February–1 March 2020.
57. Olson, J.L.; Groman-Lupa, S. Design and performance of a large lumen glaucoma drainage device. *Eye* **2017**, *31*, 152–156. [CrossRef]
58. Villamarin, A.; Roy, S.; Bigler, S.; Stergiopulos, N. A New Adjustable Glaucoma Drainage Device. *Investig. Ophthalmol. Vis. Sci.* **2014**, *55*, 1848–1852. [CrossRef]
59. Roy, S.; Villamarin, A.; Stergiopulos, C.; Bigler, S.; Guidotti, J.; Stergiopulos, N.; Kniestedt, C.; Mermoud, A. Initial Clinical Results of the eyeWatch: A New Adjustable Glaucoma Drainage Device Used in Refractory Glaucoma Surgery. *J. Glaucoma* **2019**, *28*, 452–458. [CrossRef]
60. Roy, S.; Villamarin, A.; Stergiopulos, C.; Bigler, S.; Stergiopulos, N.; Wachtl, J.; Mermoud, A.; Kniestedt, C. Comparison Between the eyeWatch Device and the Ahmed Valve in Refractory Glaucoma. *J. Glaucoma* **2020**, *29*, 401–405. [CrossRef]
61. Sahiner, N.; Kravitz, D.J.; Qadir, R.; Blake, D.A.; Haque, S.; John, V.T.; Margo, C.E.; Ayyala, R.S. Creation of a drug-coated glaucoma drainage device using polymer technology: In vitro and in vivo studies. *Arch. Ophthalmol.* **2009**, *127*, 448–453. [CrossRef]
62. Hovakimyan, M.; Siewert, S.; Schmidt, W.; Sternberg, K.; Reske, T.; Stachs, O.; Guthoff, R.; Wree, A.; Witt, M.; Schmitz, K.P.; et al. Development of an Experimental Drug Eluting Suprachoroidal Microstent as Glaucoma Drainage Device. *Transl. Vis. Sci. Technol.* **2015**, *4*, 14. [CrossRef] [PubMed]
63. Zhang, W.; Huang, L.; Weinreb, R.N.; Cheng, H. Wearable electronic devices for glaucoma monitoring and therapy. *Mater. Des.* **2021**, *212*, 110183. [CrossRef]
64. Cvenkel, B.; Kolko, M. Devices and Treatments to Address Low Adherence in Glaucoma Patients: A Narrative Review. *J. Clin. Med.* **2022**, *12*, 151. [CrossRef] [PubMed]
65. Ahn, B.H.; Hwang, Y.H.; Han, J.C. Novel membrane-tube type glaucoma shunt device for glaucoma surgery. *Clin. Exp. Ophthalmol.* **2016**, *44*, 776–782. [CrossRef]
66. Saheb, H.; Ahmed, I.I. Micro-invasive glaucoma surgery: Current perspectives and future directions. *Curr. Opin. Ophthalmol.* **2012**, *23*, 96–104. [CrossRef]
67. Pereira, I.C.F.; van de Wijdeven, R.; Wyss, H.M.; Beckers, H.J.M.; den Toonder, J.M.J. Conventional glaucoma implants and the new MIGS devices: A comprehensive review of current options and future directions. *Eye* **2021**, *35*, 3202–3221. [CrossRef]

68. Xin, C.; Wang, H.; Wang, N. Minimally Invasive Glaucoma Surgery: What Do We Know? Where Should We Go? *Transl. Vis. Sci. Technol.* **2020**, *9*, 15. [CrossRef]
69. Yook, E.; Vinod, K.; Panarelli, J.F. Complications of micro-invasive glaucoma surgery. *Curr. Opin. Ophthalmol.* **2018**, *29*, 147–154. [CrossRef]
70. Vinod, K.; Gedde, S.J. Safety profile of minimally invasive glaucoma surgery. *Curr. Opin. Ophthalmol.* **2021**, *32*, 160–168. [CrossRef]
71. Grover, D.S.; Flynn, W.J.; Bashford, K.P.; Lewis, R.A.; Duh, Y.J.; Nangia, R.S.; Niksch, B. Performance and Safety of a New Ab Interno Gelatin Stent in Refractory Glaucoma at 12 Months. *Am. J. Ophthalmol.* **2017**, *183*, 25–36. [CrossRef]
72. Betzler, B.K.; Lim, S.Y.; Lim, B.A.; Yip, V.C.H.; Ang, B.C.H. Complications and post-operative interventions in XEN45 gel stent implantation in the treatment of open angle glaucoma-a systematic review and meta-analysis. *Eye* **2022**, *37*, 1047–1060. [CrossRef] [PubMed]
73. Arnljots, T.S.; Kasina, R.; Bykov, V.J.N.; Economou, M.A. Needling With 5-Fluorouracil (5-FU) After XEN Gel Stent Implantation: 6-Month Outcomes. *J. Glaucoma* **2018**, *27*, 893–899. [CrossRef] [PubMed]
74. U.S. Food and Drug Administration. 510(k) Premarket Notification—XEN Glaucoma Treatment System. Available online: https://www.accessdata.fda.gov/cdrh_docs/pdf16/k161457.pdf (accessed on 21 November 2016).
75. Tan, N.E.; Tracer, N.; Terraciano, A.; Parikh, H.A.; Panarelli, J.F.; Radcliffe, N.M. Comparison of safety and efficacy between ab interno and ab externo approaches to XEN gel stent placement. *Clin. Ophthalmol.* **2021**, *15*, 299. [CrossRef] [PubMed]
76. Do, A.; McGlumphy, E.; Shukla, A.; Dangda, S.; Schuman, J.S.; Boland, M.V.; Yohannan, J.; Panarelli, J.F.; Craven, E.R. Comparison of Clinical Outcomes with Open Versus Closed Conjunctiva Implantation of the XEN45 Gel Stent. *Ophthalmol. Glaucoma* **2021**, *4*, 343–349. [CrossRef]
77. Gallardo, M.J.; Vincent, L.R.; Porter, M. Comparison of Clinical Outcomes Following Gel Stent Implantation via Ab-Externo and Ab Interno Approaches in Patients with Refractory Glaucoma. *Clin. Ophthalmol.* **2022**, *16*, 2187–2197. [CrossRef]
78. Ucar, F.; Cetinkaya, S. Xen implantation in patients with primary open-angle glaucoma: Comparison of two different techniques. *Int. Ophthalmol.* **2020**, *40*, 2487–2494. [CrossRef]
79. Vera, V.; Gagne, S.; Myers, J.S.; Ahmed, I.I.K. Surgical Approaches for Implanting Xen Gel Stent without Conjunctival Dissection. *Clin. Ophthalmol.* **2020**, *14*, 2361–2371. [CrossRef]
80. Ibarz-Barberá, M.; Morales-Fernández, L.; Corroto-Cuadrado, A.; Martinez-Galdón, F.; Tañá-Rivero, P.; Gómez de Liaño, R.; Teus, M.A. Corneal Endothelial Cell Loss After PRESERFLO™ MicroShunt Implantation in the Anterior Chamber: Anterior Segment OCT Tube Location as a Risk Factor. *Ophthalmol. Ther.* **2022**, *11*, 293–310. [CrossRef]
81. Martinez-de-la-Casa, J.M.; Saenz-Frances, F.; Morales Fernandez, L.; García-Feijoo, J. Posterior chamber implantation of a Preserflo Microshunt in a patient with a compromised endothelium. *Arch. Soc. Esp. Oftalmol.* **2022**, *97*, 161–164. [CrossRef]
82. Titiyal, J.S.; Kaur, M.; Nair, S.; Sharma, N. Intraoperative optical coherence tomography in anterior segment surgery. *Surv. Ophthalmol.* **2021**, *66*, 308–326. [CrossRef]
83. Ang, B.C.H.; Lim, S.Y.; Dorairaj, S. Intra-operative optical coherence tomography in glaucoma surgery—A systematic review. *Eye* **2020**, *34*, 168–177. [CrossRef] [PubMed]
84. Junker, B.; Jordan, J.F.; Framme, C.; Pielen, A. Intraoperative optical coherence tomography and ab interno trabecular meshwork surgery with the Trabectome. *Clin. Ophthalmol.* **2017**, *11*, 1755–1760. [CrossRef]
85. Ishida, A.; Sugihara, K.; Shirakami, T.; Tsutsui, A.; Manabe, K.; Tanito, M. Observation of Gonio Structures during Microhook Ab Interno Trabeculotomy Using a Novel Digital Microscope with Integrated Intraoperative Optical Coherence Tomography. *J. Ophthalmol.* **2020**, *2020*, 9024241. [CrossRef]
86. Thakur, A.; Goldbaum, M.; Yousefi, S. Predicting Glaucoma before Onset Using Deep Learning. *Ophthalmol. Glaucoma* **2020**, *3*, 262–268. [CrossRef] [PubMed]
87. Lin, K.Y.; Urban, G.; Yang, M.C.; Lee, L.C.; Lu, D.W.; Alward, W.L.M.; Baldi, P. Accurate Identification of the Trabecular Meshwork under Gonioscopic View in Real Time Using Deep Learning. *Ophthalmol. Glaucoma* **2022**, *5*, 402–412. [CrossRef] [PubMed]
88. Samuelson, T.W.; Sarkisian Jr, S.R.; Lubeck, D.M.; Stiles, M.C.; Duh, Y.-J.; Romo, E.A.; Giamporcaro, J.E.; Hornbeak, D.M.; Katz, L.J.; Bartlett, W. Prospective, randomized, controlled pivotal trial of an ab interno implanted trabecular micro-bypass in primary open-angle glaucoma and cataract: Two-year results. *Ophthalmology* **2019**, *126*, 811–821. [CrossRef]
89. Samuelson, T.W.; Singh, I.P.; Williamson, B.K.; Falvey, H.; Lee, W.C.; Odom, D.; McSorley, D.; Katz, L.J. Quality of Life in Primary Open-Angle Glaucoma and Cataract: An Analysis of VFQ-25 and OSDI From the iStent inject® Pivotal Trial. *Am. J. Ophthalmol.* **2021**, *229*, 220–229. [CrossRef]
90. Paletta Guedes, R.A.; Gravina, D.M.; Paletta Guedes, V.M.; Chaoubah, A. Standalone Implantation of 2-3 Trabecular Micro-Bypass Stents (iStent inject ± iStent) as an Alternative to Trabeculectomy for Moderate-to-Severe Glaucoma. *Ophthalmol. Ther.* **2022**, *11*, 271–292. [CrossRef]
91. Sarkisian, S.R., Jr.; Grover, D.S.; Gallardo, M.; Brubaker, J.W.; Giamporcaro, J.E.; Hornbeak, D.M.; Katz, L.J.; Navratil, T. Effectiveness and Safety of iStent infinite Trabecular Micro-Bypass For Uncontrolled Glaucoma. *J. Glaucoma* **2022**, *32*, 9–18. [CrossRef]
92. Khaimi, M.A.; Dvorak, J.D.; Ding, K. An Analysis of 3-Year Outcomes Following Canaloplasty for the Treatment of Open-Angle Glaucoma. *J. Ophthalmol.* **2017**, *2017*, 2904272. [CrossRef]
93. U.S. Food and Drug Administration. 510(k) Premarket Notification—OMNI Surgical System. Available online: https://www.accessdata.fda.gov/cdrh_docs/pdf20/K202678.pdf (accessed on 1 March 2021).

94. Vold, S.D.; Williamson, B.K.; Hirsch, L.; Aminlari, A.E.; Cho, A.S.; Nelson, C.; Dickerson, J.E., Jr. Canaloplasty and Trabeculotomy with the OMNI System in Pseudophakic Patients with Open-Angle Glaucoma: The ROMEO Study. *Ophthalmol. Glaucoma* **2021**, *4*, 173–181. [CrossRef] [PubMed]
95. Bleeker, A.R.; Litchfield, W.R.; Ibach, M.J.; Greenwood, M.D.; Ristvedt, D.; Berdahl, J.P.; Terveen, D.C. Short-Term Efficacy of Combined ab Interno Canaloplasty and Trabeculotomy in Pseudophakic Eyes with Open-Angle Glaucoma. *Clin. Ophthalmol.* **2022**, *16*, 2295–2303. [CrossRef] [PubMed]
96. Klabe, K.; Kaymak, H. Standalone Trabeculotomy and Viscodilation of Schlemm's Canal and Collector Channels in Open-Angle Glaucoma Using the OMNI Surgical System: 24-Month Outcomes. *Clin. Ophthalmol.* **2021**, *15*, 3121–3129. [CrossRef] [PubMed]
97. Iwasaki, K.; Arimura, S.; Takihara, Y.; Takamura, Y.; Inatani, M. Prospective cohort study of corneal endothelial cell loss after Baerveldt glaucoma implantation. *PLoS ONE* **2018**, *13*, e0201342. [CrossRef]
98. D'Alessandro, E.; Guidotti, J.M.; Mansouri, K.; Mermoud, A. XEN-augmented Baerveldt: A New Surgical Technique for Refractory Glaucoma. *J. Glaucoma* **2017**, *26*, e90–e92. [CrossRef]
99. Bravetti, G.E.; Mansouri, K.; Gillmann, K.; Rao, H.L.; Mermoud, A. XEN-augmented Baerveldt drainage device implantation in refractory glaucoma: 1-year outcomes. *Graefes. Arch. Clin. Exp. Ophthalmol.* **2020**, *258*, 1787–1794. [CrossRef]
100. Denis, P.; Hirneiß, C.; Reddy, K.P.; Kamarthy, A.; Calvo, E.; Hussain, Z.; Ahmed, I.I.K. A First-in-Human Study of the Efficacy and Safety of MINIject in Patients with Medically Uncontrolled Open-Angle Glaucoma (STAR-I). *Ophthalmol. Glaucoma* **2019**, *2*, 290–297. [CrossRef]
101. Gigon, A.; Shaarawy, T. The suprachoroidal route in glaucoma surgery. *J. Curr. Glaucoma Pract.* **2016**, *10*, 13.
102. Reiss, G.; Clifford, B.; Vold, S.; He, J.; Hamilton, C.; Dickerson, J.; Lane, S. Safety and Effectiveness of CyPass Supraciliary Micro-Stent in Primary Open-Angle Glaucoma: 5-Year Results from the COMPASS XT Study. *Am. J. Ophthalmol.* **2019**, *208*, 219–225. [CrossRef]
103. Hueber, A.; Roters, S.; Jordan, J.F.; Konen, W. Retrospective analysis of the success and safety of Gold Micro Shunt Implantation in glaucoma. *BMC Ophthalmol.* **2013**, *13*, 35. [CrossRef]
104. Fili, S.; Wölfelschneider, P.; Kohlhaas, M. The STARflo glaucoma implant: Preliminary 12 months results. *Graefes. Arch. Clin. Exp. Ophthalmol.* **2018**, *256*, 773–781. [CrossRef] [PubMed]
105. García Feijoó, J.; Denis, P.; Hirneiß, C.; Aptel, F.; Perucho González, L.; Hussain, Z.; Lorenz, K.; Pfeiffer, N. A European Study of the Performance and Safety of MINIject in Patients With Medically Uncontrolled Open-angle Glaucoma (STAR-II). *J. Glaucoma* **2020**, *29*, 864–871. [CrossRef] [PubMed]
106. MINIject in Patients With Open Angle Glaucoma Using Single Operator Delivery Tool. Available online: https://ClinicalTrials.gov/show/NCT03996200 (accessed on 10 March 2023).
107. Denis, P.; Hirneiß, C.; Durr, G.M.; Reddy, K.P.; Kamarthy, A.; Calvo, E.; Hussain, Z.; Ahmed, I.K. Two-year outcomes of the MINIject drainage system for uncontrolled glaucoma from the STAR-I first-in-human trial. *Br. J. Ophthalmol.* **2022**, *106*, 65–70. [CrossRef] [PubMed]
108. Evaluate the Safety and Effectiveness of iSTAR Medical's MINIject™ Implant for Lowering Intraocular Pressure (IOP) in Subjects With Primary Open-angle Glaucoma. Available online: https://ClinicalTrials.gov/show/NCT05024695 (accessed on 10 March 2023).
109. Laroche, D.; Anugo, D.; Ng, C.; Ishikawa, H. Intra-Scleral Ciliary Sulcus Suprachoroidal Microtube: Making Supraciliary Glaucoma Surgery Affordable. *J. Natl. Med. Assoc.* **2019**, *111*, 427–435. [CrossRef]
110. Newman-Casey, P.A.; Robin, A.L.; Blachley, T.; Farris, K.; Heisler, M.; Resnicow, K.; Lee, P.P. The Most Common Barriers to Glaucoma Medication Adherence: A Cross-Sectional Survey. *Ophthalmology* **2015**, *122*, 1308–1316. [CrossRef]
111. Belamkar, A.; Harris, A.; Zukerman, R.; Siesky, B.; Oddone, F.; Verticchio Vercellin, A.; Ciulla, T.A. Sustained release glaucoma therapies: Novel modalities for overcoming key treatment barriers associated with topical medications. *Ann. Med.* **2022**, *54*, 343–358. [CrossRef]
112. Clinical Study Comparing Two Models of a Travoprost Intraocular Implant. Available online: https://ClinicalTrials.gov/show/NCT03868124 (accessed on 10 March 2023).
113. Randomized Study Comparing Two Models of a Travoprost Intraocular Implant to Timolol Maleate Ophthalmic Solution, 0.5%. Available online: https://ClinicalTrials.gov/show/NCT03519386 (accessed on 10 March 2023).
114. Lewis, C. *Glaukos Announces Positive Topline Outcomes for Both Phase 3 Pivotal Trials of iDose TR, Achieving Primary Efficacy Endpoints and Demonstrating Favorable Tolerability and Safety Profiles*; Business Wire: San Francisco, CA, USA, 2022.
115. Schultz, T.; Schojai, M.; Kersten-Gomez, I.; Matthias, E.; Boecker, J.; Dick, H.B. Ab externo device for the treatment of glaucoma: Direct flow from the anterior chamber to the ocular surface. *J. Cataract. Refract. Surg.* **2020**, *46*, 941–943. [CrossRef]
116. Early Evaluation of the Beacon Aqueous Microshunt in Patients Refractory to Drug Therapy in the European Union. Available online: https://ClinicalTrials.gov/show/NCT03634319 (accessed on 10 March 2023).
117. Geffen, N.; Kumar, D.A.; Barayev, E.; Gershoni, A.; Rotenberg, M.; Zahavi, A.; Glovinsky, Y.; Agarwal, A. Minimally Invasive Micro Sclerostomy (MIMS) Procedure: A Novel Glaucoma Filtration Procedure. *J. Glaucoma* **2022**, *31*, 191–200. [CrossRef]
118. Lazcano-Gomez, G.; Garg, S.J.; Yeu, E.; Kahook, M.Y. Interim Analysis of STREAMLINE(®) Surgical System Clinical Outcomes in Eyes with Glaucoma. *Clin. Ophthalmol.* **2022**, *16*, 1313–1320. [CrossRef]
119. STREAMLINE®SURGICAL SYSTEM Compared to iStent Inject W®in Patients With Open-Angle Glaucoma (VENICE). Available online: https://clinicaltrials.gov/ct2/show/NCT05280366 (accessed on 10 March 2023).

120. Nakamura, K.; Fujimoto, T.; Okada, M.; Maki, K.; Shimazaki, A.; Kato, M.; Inoue, T. Tissue Reactivity to, and Stability of, Glaucoma Drainage Device Materials Placed Under Rabbit Conjunctiva. *Transl. Vis. Sci. Technol.* **2022**, *11*, 9. [CrossRef]
121. Vergados, A.; Mohite, A.A.; Sung, V.C.T. Ab interno tube ligation for refractory hypotony following non-valved glaucoma drainage device implantation. *Graefes Arch. Clin. Exp. Ophthalmol.* **2019**, *257*, 2271–2278. [CrossRef]
122. Park, C.J.; Yang, D.-S.; Cha, J.-J.; Lee, J.-H. Polymeric check valve with an elevated pedestal for precise cracking pressure in a glaucoma drainage device. *Biomed. Microdevices* **2016**, *18*, 20. [CrossRef] [PubMed]
123. Martino, A.Z.; Iverson, S.; Feuer, W.J.; Greenfield, D.S. Surgical outcomes of superior versus inferior glaucoma drainage device implantation. *J. Glaucoma* **2015**, *24*, 32–36. [CrossRef] [PubMed]
124. Luong, Q.M.; Shang, L.; Ang, M.; Kong, J.F.; Peng, Y.; Wong, T.T.; Venkatraman, S.S. A new design and application of bioelastomers for better control of intraocular pressure in a glaucoma drainage device. *Adv. Healthc. Mater.* **2014**, *3*, 205–213. [CrossRef] [PubMed]

Disclaimer/Publisher's Note: The statements, opinions and data contained in all publications are solely those of the individual author(s) and contributor(s) and not of MDPI and/or the editor(s). MDPI and/or the editor(s) disclaim responsibility for any injury to people or property resulting from any ideas, methods, instructions or products referred to in the content.

Article

Swept-Source Anterior Segment Optical Coherence Tomography Imaging and Quantification of Bleb Parameters in Glaucoma Filtration Surgery

Jeremy C.K. Tan [1,2], Hussameddin Muntasser [1,3], Anshoo Choudhary [1], Mark Batterbury [1] and Neeru A. Vallabh [1,3,*]

1. St. Paul's Eye Unit, Royal Liverpool University Hospital, Liverpool L7 8YA, UK; jeremy.c.tan@unsw.edu.au (J.C.K.T.)
2. Faculty of Medicine and Health, University of New South Wales, Kensington, NSW 2032, Australia
3. Department of Eye and Vision Sciences, Institute of Life Course and Medical Sciences, University of Liverpool, Liverpool L69 3BX, UK
* Correspondence: vallabh@liverpool.ac.uk

Citation: Tan, J.C.K.; Muntasser, H.; Choudhary, A.; Batterbury, M.; Vallabh, N.A. Swept-Source Anterior Segment Optical Coherence Tomography Imaging and Quantification of Bleb Parameters in Glaucoma Filtration Surgery. *Bioengineering* 2023, *10*, 1186. https://doi.org/10.3390/bioengineering10101186

Academic Editor: Hiroshi Ohguro

Received: 13 September 2023
Revised: 10 October 2023
Accepted: 11 October 2023
Published: 13 October 2023

Copyright: © 2023 by the authors. Licensee MDPI, Basel, Switzerland. This article is an open access article distributed under the terms and conditions of the Creative Commons Attribution (CC BY) license (https://creativecommons.org/licenses/by/4.0/).

Abstract: This paper describes a technique for using swept-source anterior segment optical coherence tomography (AS-OCT) to visualize internal bleb microstructure and objectively quantify dimensions of the scleral flap and trabeculo-Descemet window (TDW) in non-penetrating glaucoma filtration surgery (GFS). This was a cross-sectional study of 107 filtering blebs of 67 patients who had undergone deep sclerectomy surgery at least 12 months prior. The mean post-operative follow-up duration was 6.5 years +/− 4.1 [standard deviation (SD)]. The maximal bleb height was significantly greater in the complete success (CS) blebs compared to the qualified success (QS) and failed (F) blebs (1.48 vs. 1.17 vs. 1.10 mm in CS vs. QS vs. F, one-way ANOVA, $p < 0.0001$). In a subcohort of deep sclerectomy blebs augmented by intraoperative Mitomycin-C, the trabeculo-Descemet window was significantly longer in the complete success compared to the qualified success group (613.7 vs. 378.1 vs. 450.8 μm in CS vs. QS vs. F, $p = 0.004$). The scleral flap length, thickness, and width were otherwise similar across the three outcome groups. The quantification of surgical parameters that influence aqueous outflow in non-penetrating GFS can help surgeons better understand the influence of these structures on aqueous outflow and improve surgical outcomes.

Keywords: glaucoma surgery; trabeculectomy; deep sclerectomy; mitomycin-C; scleral flap; surgical technique; surgical success

1. Introduction

Glaucoma is a neurodegenerative disease caused by progressive loss of retinal ganglion cells, with a global prevalence of 3.5% in individuals aged 40 to 80 years old [1]. The reduction of intraocular pressure (IOP) is the only known modifiable risk factor for the treatment of glaucoma [2]. This can be done using topical eye drops, laser treatment, microinvasive glaucoma surgery, or glaucoma filtration surgery (GFS). Glaucoma filtration surgery is often performed to achieve greater intraocular pressure reduction or when other earlier interventions fail to halt glaucoma progression. In GFS, an artificial passage is created for the drainage of aqueous humor from the anterior chamber of the eye into the space below the conjunctiva (the subconjunctival space) and Tenon's capsule (subtenons), with subsequent formation of a structure known as a bleb [3].

Non-penetrating procedures such as the deep sclerectomy (DS) have been developed to reduce the incidence of post-operative complications associated with penetrating forms of GFS like trabeculectomy, with intraoperative dissection down to the trabeculo-Descemet window (TDW)—the juxtacanalicular meshwork and inner wall of Schlemm's canal across which aqueous egresses, followed by removal of a block of deep sclera [4]. The TDW functions as a site of controlled aqueous outflow and also forms an intrascleral space to act as a reservoir [5]. In addition, the natural drainage pathway via the Schlemm's canal is also augmented [4]. Successful filtering surgeries (blebs) typically have non-adherent sub-conjunctival tissue without scarring and with healthy vasculature, which allows continuous aqueous outflow. A major cause of failure in GFS is fibrosis and scarring of the sub-tenon's space, which restricts aqueous flow and is caused by fibroblast proliferation and extracellular matrix deposition.

The post-operative evaluation of filtering blebs such as trabeculectomy blebs has traditionally relied on clinical grading systems performed at the slit-lamp, such as the Indiana Bleb Appearance Grading Scale (IBAGS) and the Moorfields Bleb Grading System (MBGS), which document factors associated with surgical success such as bleb area, height, and vascularity. Deep sclerectomy blebs are also assessed clinically at the slit-lamp, although it is uncertain if bleb morphology is correlated with successful IOP reduction given the different aqueous outflow egress pathways compared to in trabeculectomy. This is because while subconjunctival outflow represents an important pathway for egress, augmentation of natural channels such as the Schlemm canal is also an important consequence of non-penetrating glaucoma surgery [4].

Studies have evaluated the microstructure of filtering blebs using non-contact imaging techniques such as anterior segment Optical Coherence Tomography (AS-OCT) and ultrasound biomicroscopy. These modalities can provide quantitative data on the internal structure of blebs, such as bleb wall thickness, presence of microcysts, and measurements of the internal ostium, bleb cavity, and sub-flap space [6–9]. The purpose of this cross-sectional study was to optimize and develop a technique for using AS-OCT to image and subsequently quantify the internal bleb microstructure after glaucoma filtration surgery. Specifically, we quantified the dimensions of surgical parameters such as the scleral flap and the trabeculo-Descemet window of filtering blebs following deep sclerectomy surgery.

2. Methods

This was a cross-sectional study conducted at the St Paul's Eye Unit, Royal Liverpool University Hospital, a tertiary referral eye unit in Liverpool, United Kingdom. The study had the approval of the clinical governance department of the Royal Liverpool University Hospital Trust and adhered to the tenets of the Declaration of Helsinki. Consecutive patients attending the glaucoma clinics of the St. Paul's Eye Unit between November 2022 and February 2023 who had previously undergone deep sclerectomy surgery at least one year prior were recruited. The inclusion criteria were adult patients with a diagnosis of primary open-angle glaucoma, pseudoexfoliative glaucoma, pigment dispersion syndrome glaucoma, or primary angle closure glaucoma. Exclusion criteria were secondary open-angle glaucomas such as uveitic and neovascular glaucoma. All surgeries were performed by consultants and clinical fellows affiliated with the St Paul's Eye Unit.

2.1. Surgical Technique

The deep sclerectomy surgeries were performed by consultants and clinical fellows affiliated with the St. Paul's Eye Unit over the years. The main steps shared in common by the surgeons were the creation of a superior fornix-based conjunctival incision and peritomy, the creation of a superficial scleral flap, the creation of a deep scleral flap, exposure of the trabeculo-Descemet window (TDW), excision of the deep scleral flap, suturing of the superficial scleral flap, and conjunctival closure. Intraoperative mitomycin-C was used in a proportion of cases and was performed following the conjunctival peritomy. The variations in dimensions and thickness of the scleral flap and size of the TDW by the different surgeons provided the variations in quantifiable dimensions which were used in the subsequent analysis

2.2. Anterior-Segment Optical Coherence Tomography Imaging of Filtering Bleb

The bleb surgical site was first examined with the slit lamp by one of the authors (JT). Anterior-segment Optical coherence tomography of the filtering bleb was then performed by the same author using the Anterion® (Heidelberg Engineering GmbH, Heidelberg, Germany) swept-source OCT device. The Anterion® uses a laser light source with a wavelength of 1300 nm and at 50,000 Hz to obtain B-scans with an axial resolution of 10 microns and transverse resolution of 45 microns. Each scan was performed via the imaging module of the Anterion® device using a standardized raster scan measuring 7.5 mm in width and 12 mm in length and comprising 19 slices. The raster block was first oriented parallel to the long axis of the scleral flap (the sagittal plane in relation to the bleb. (Figure 1) The anterior limit of the image window was positioned just anterior to the limbus at the peripheral superior cornea, which allows the trabeculo-Descemet window within the anterior chamber, iridocorneal angle, and the entire length of the scleral flap to be visualized and captured within the raster slices. The sagittal slice overlying the TDW was identified from the 19 raster slices) and exported for analysis. The raster block was then oriented perpendicular to the initial sagittal plane to visualize bleb structures in the coronal plane. The anterior limit of the raster block was placed anterior to the TDW to allow the full width of the TDW and scleral flap to be visualized and captured within the raster. The coronal slices overlying the TDW and mid-point of the scleral flap were identified from the 19 raster slices and exported for analysis.

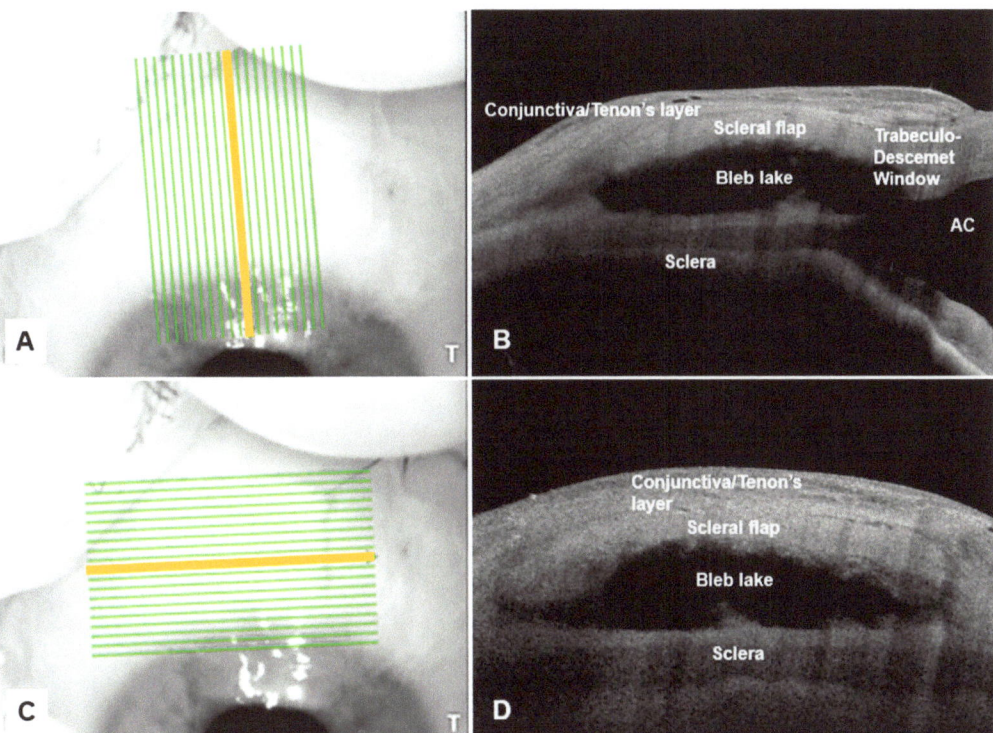

Figure 1. Representative sagittal (**B**) and coronal (**D**) images of a well-functioning deep sclerectomy bleb, with en face oct images in the left column (**A**,**C**). The yellow lines in the left column images represent the sagittal slice overlying the trabeculo-descemet window (TDW) and the coronal slice overlying the mid-point of the flap, which were chosen to produce the adjacent OCT images. Abbreviations: anterior chamber (AC), temporal (T).

2.3. Image Preprocessing

The filtering bleb images were exported to MATLAB for image preprocessing (Mathworks). The following functions were performed on each bleb image sequentially in a standardized manner to improve the visualization of the internal bleb microstructure: conversion to grayscale, contrast enhancement, thresholding, active contouring, and morphological opening (Figure 2).

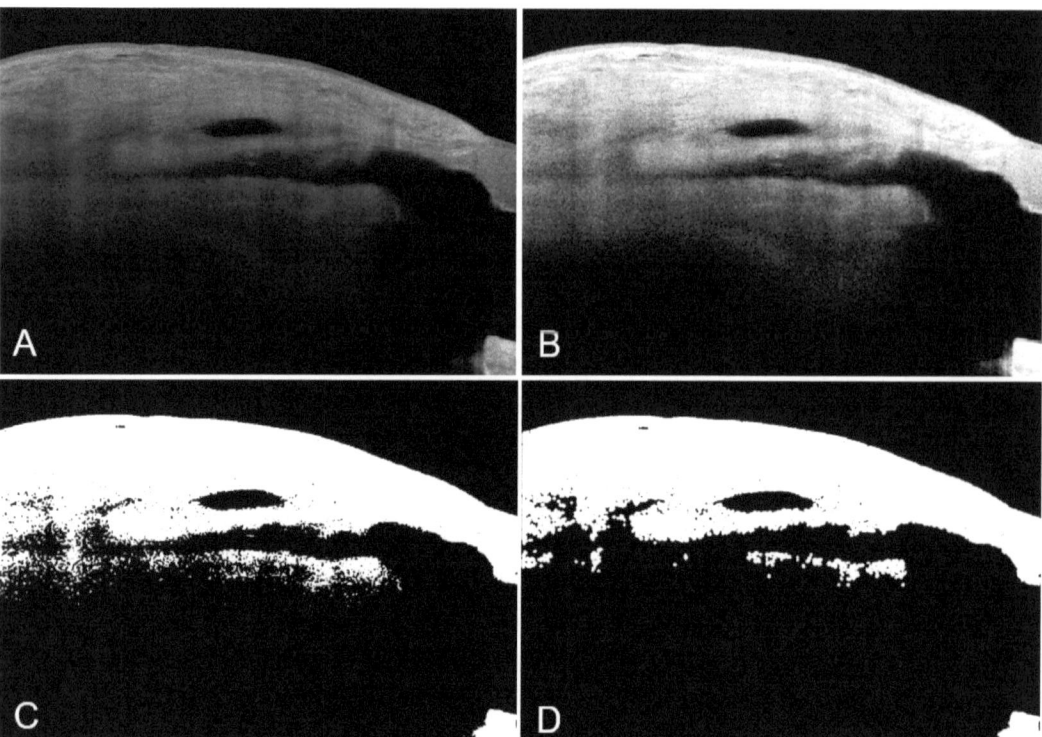

Figure 2. Image processing functions performed on each bleb image—in this representative example using a sagittal trabeculectomy bleb AS-OCT image, to improve visualization of internal bleb microstructure. Original image (**A**), contrast enhancement (**B**), thresholding (**C**), active contouring followed by morphological opening (**D**).

2.4. Visualisation and Quantification of Surgical Parameters

The dimensions of surgical parameters of interest were then quantified using a measurement tool with the superimposed measurements saved onto a separate file for each image. The dimensions of each surgical parameter were converted from image pixel values to true values using a standardized and verified conversion factor. Table 1 displays the surgical parameters of interest and their standardized anatomical reference points captured on the sagittal and coronal AS-OCT slices.

The following surgical parameters were visualized on the sagittal slice of each filtering bleb in DS patients: maximal bleb height, scleral flap length, and TDW length (Figure 3). The scleral flap length was measured using the iridocorneal angle as a reference point, as the anterior limit of the scleral flap was indistinct. The scleral flap length is, therefore, not a true flap length given the orientation of the plane of measurement but rather served as a standardized measure across different blebs. The following surgical parameters were obtained from the coronal slice of each filtering bleb: scleral flap width and thickness.

Table 1. Surgical parameters of interest and associated standardized anatomical reference landmarks captured on the sagittal and coronal AS-OCT imaging of DS blebs.

Imaging Plane	Surgical Parameters of Interest	Anatomical Reference Points of Surgical Parameter
Sagittal	Scleral flap length	Posterior edge of scleral flap to iridocorneal angle
	Trabeculo-Descemet window length	Anterior edge of TDW to posterior edge of TDW
Coronal	Scleral flap width	Nasal edge of scleral flap to temporal edge of scleral flap at midpoint of flap
	Scleral flap thickness	Superior edge of scleral flap to inferior edge of scleral flap at midpoint of flap

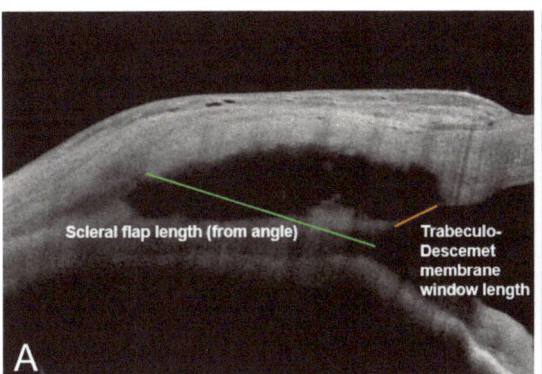
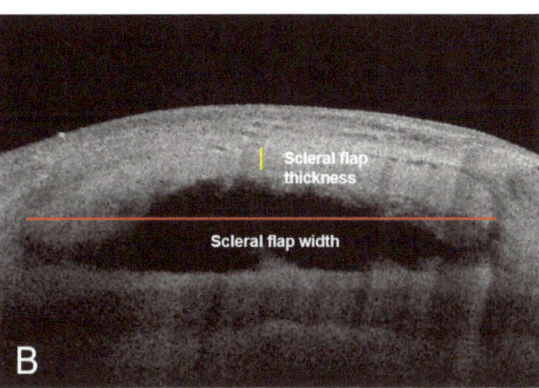
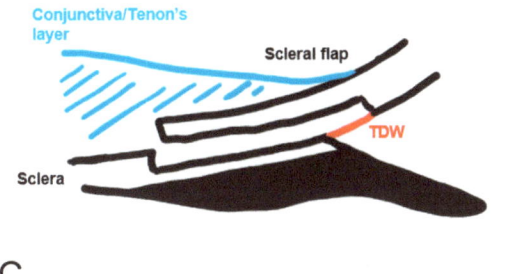
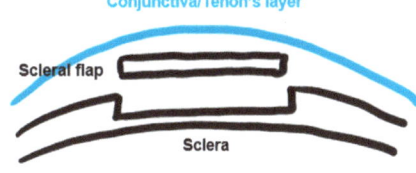

Figure 3. Sagittal (**A**) and coronal (**B**) AS-OCT images of a well-functioning deep sclerectomy bleb with annotated surgical parameters of interest. The relevant anatomical structures are shown in the (**C**) (sagittal) and (**D**) (coronal). TDW: Trabeculo-Descemet window.

2.5. Definitions of Surgical Success

The surgical outcome of each deep sclerectomy case at the index visit was classified into complete success (CS; IOP $\leq$ 18 mmHg with no medications), qualified success (QS; IOP $\leq$ 18 with medications), and failure (F; IOP > 18 mmHg or subsequent filtration surgery procedure performed), as per the World Glaucoma Association consensus on definitions of success 2018 [10].

2.6. Statistical Analysis

Descriptive statistics were used to analyze the demographic characteristics of the cohorts. The distributions of quantitative data were first assessed for normality using a

D'Agostino and Pearson test of normality, with parametric and non-parametric statistics then applied as appropriate. Analyses were conducted using GraphPad Prism version 9 (GraphPad, La Jolla, CA, USA).

3. Results

A total of 107 filtering blebs from 67 patients were included in the study. The mean patient age was 73.0 years (SD 11.2, median 74.1, range 32.6 to 91.0). Sixty patients (89.5%) were of white/Caucasian ethnicity, 4 (6.0%) were Asian, and 3 (4.5%) of afro-Caribbean. The diagnoses were primary open-angle glaucoma (88; 82.2%), primary angle closure glaucoma (9; 8.4%), pigment dispersion glaucoma (4; 3.7%), pseudoexfoliation glaucoma (2; 1.8%), and other/glaucoma not defined (4; 3.7%). The mean best-corrected visual acuity was 0.21 (Snellen equivalent of 20/30. SD 0.23), and the mean deviation was -11.5 dB (SD 7.8) at the index clinic visit. The median post-operative follow-up duration was 6.5 years (IQR 3.1–8.1, SD 4.1 years). The proportion of complete success (CS; IOP $\leq$ 18 mmHg with no medications), qualified success (QS; IOP $\leq$ 18 with medications), and failure (F; IOP > 18 mmHg) was 36.8%, 29.1%, and 34.2%, respectively. The mean intraocular pressure and number of medications were 13.3 mmHg (SD 4.9, range 3–27) and 1.1 (SD 1.2, range 0–4), respectively.

3.1. Bleb Height, Scleral Flap and Trabeculo Descemet Window Dimensions and Surgical Outcomes

The maximum bleb height was significantly greater in the complete success blebs compared to the qualified success and failed blebs (1.48 vs. 1.17 vs. 1.10 mm in CS vs. QS vs. F, one-way ANOVA, $p < 0.0001$). Figure 4 and Table 2 display the dimensions of the surgical parameters of interest across the outcome groups of complete success, qualified success, and failure. There was no significant difference in scleral flap length as measured from the iridocorneal angle across the three groups (2.71 vs. 2.73 vs. 2.96 mm in CS vs. QS vs. F, one-way ANOVA, $p = 0.18$). There was no significant difference in scleral flap thickness across the three groups (270.4 vs. 249.9 vs. 267.6 μm in CS vs. QS vs. F, one-way ANOVA, $p = 0.44$). Scleral flap width was also similar across the CS, QS, and F groups (3.85 vs. 3.56 vs. 3.71 mm in CS vs. QS vs. F, $p = 0.09$). The trabeculo-descemet window length was also similar across the three groups (519.0 vs. 432.9 vs. 441.3 μm, one-way ANOVA, $p = 0.09$).

Table 2. Dimensions (mean and standard deviation below) of dimensions of maximal bleb height, surgical flap, and trabeculo-Descemet window length across the outcome groups of complete success (CS), qualified success (QS), and failure (F). p value of one-way ANOVA of each parameter is reported.

Parameter	CS	QS	F	p Value
Bleb height (mm)	1.48	1.17	1.10	0.001
	0.44	0.44	0.40	
Scleral flap length (mm)	2.71	2.73	2.96	0.178
	0.60	0.54	0.53	
Scleral flap thickness (μm)	270.40	249.90	267.60	0.437
	53.54	53.15	57.26	
Scleral flap width (mm)	3.85	3.56	3.71	0.088
	0.44	0.47	0.58	
Window length (μm)	519.00	432.90	441.30	0.094
	206.40	164.30	146.60	
Scleral flap length, MMC (mm)	2.77	3.11	3.15	0.150
	0.54	0.45	0.51	
Scleral flap thickness, MMC (μm)	298.30	264.40	236.60	0.027
	53.48	34.79	57.81	
Scleral flap width, MMC, (mm)	3.90	3.58	3.97	0.273
	0.40	0.46	0.62	
Window length, MMC (μm)	613.70	378.10	450.80	0.004
	200.20	127.80	152.40	

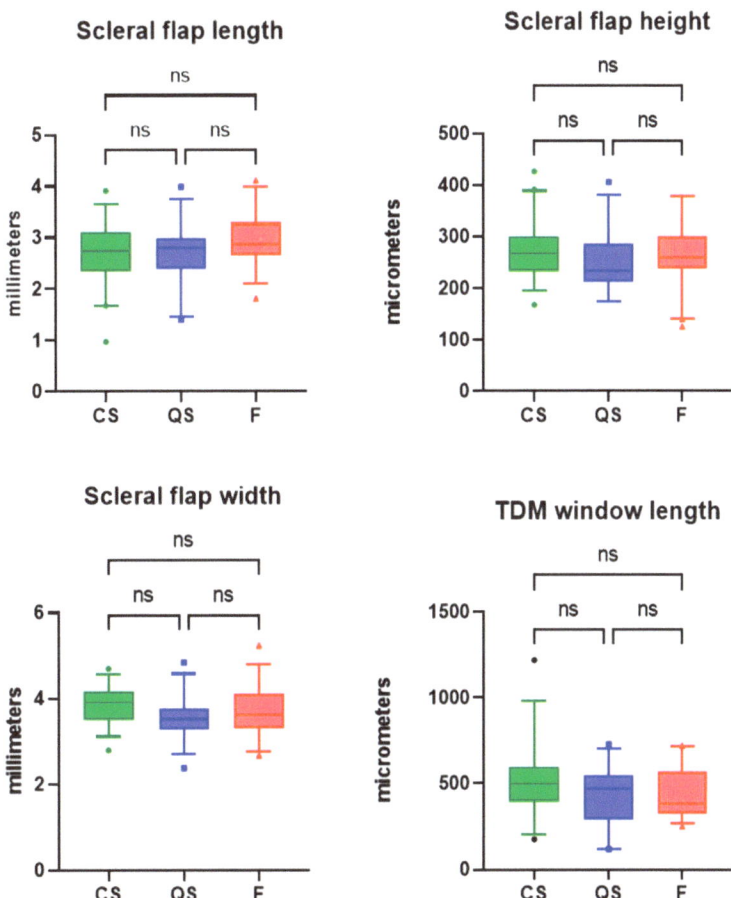

Figure 4. Box and Whisker plots (median, interquartile range, and 5th to 95th percentile) and results of one-way ANOVA of scleral flap dimensions and TDW dimensions across the outcome groups of complete success (CS), qualified success (QS) and failure (F). ns = non-significant.

3.2. Subanalysis in Deep Sclerectomy Cases Augmented with Intraoperative Mitomycin-C

Of the 107 eyes, 28 (26.2%) underwent scleral application of MMC intraoperatively. Given that intraoperative MMC use may represent a confounder in the results of surgical parameters analysis, we performed a subanalysis of cases of deep sclerectomy augmented with intraoperative Mitomycin-C. In this subcohort, the trabeculo-Descemet window was significantly longer in the complete success compared to the qualified success group (613.7 vs. 378.1 vs. 450.8 µm in CS vs. QS vs. F, $p = 0.004$). There was otherwise similarly no significant difference in scleral flap length, thickness, or width across the three groups (Figure 5).

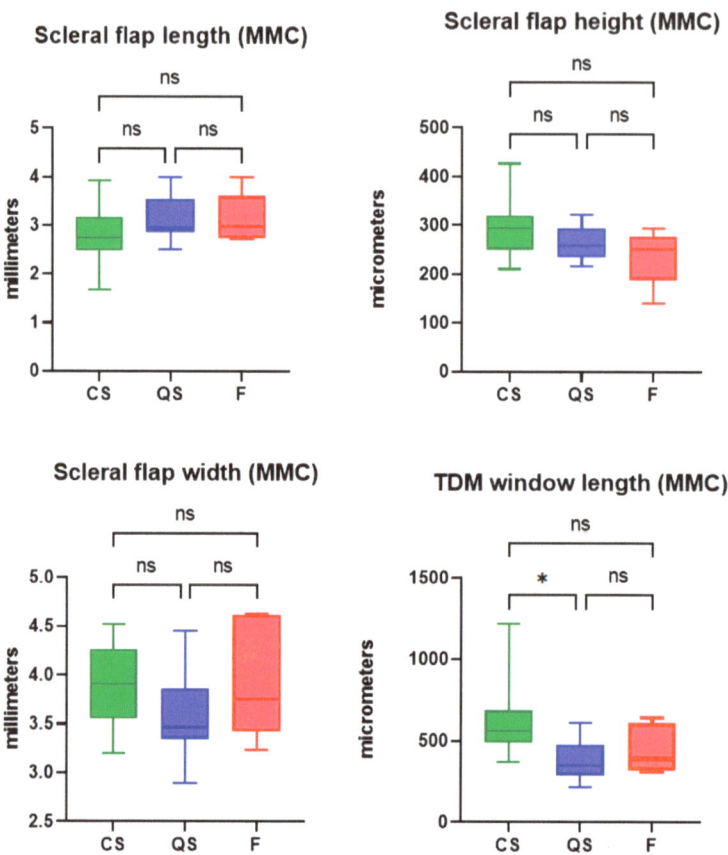

Figure 5. Box and Whisker plots (median, interquartile range, and 5th to 95th percentile) and results of one-way ANOVA of scleral flap dimensions and TDW dimensions across the outcome groups of complete success (CS), qualified success (QS) and failure (F). ns = non-significant. The asterisk denotes statistical significance, i.e., $p < 0.05$.

3.3. Post-Operative Duration and Surgical Outcomes

The prevalence of failure in glaucoma filtration surgery increases over time [11]. We, therefore, analyzed the relationship between post-operative duration (filtering bleb age) and surgical outcomes. In the overall cohort, filtering blebs with complete success and qualified success had significantly shorter post-operative duration compared to blebs with failure (56.2 vs. 72.1 vs. 98.0 months in CS vs. QS vs. F, one-way ANOVA, $p = 0.001$). This was, however, not statistically significant in cases of deep sclerectomy augmented by intra-operative mitomycin-C only. (Figure 6).

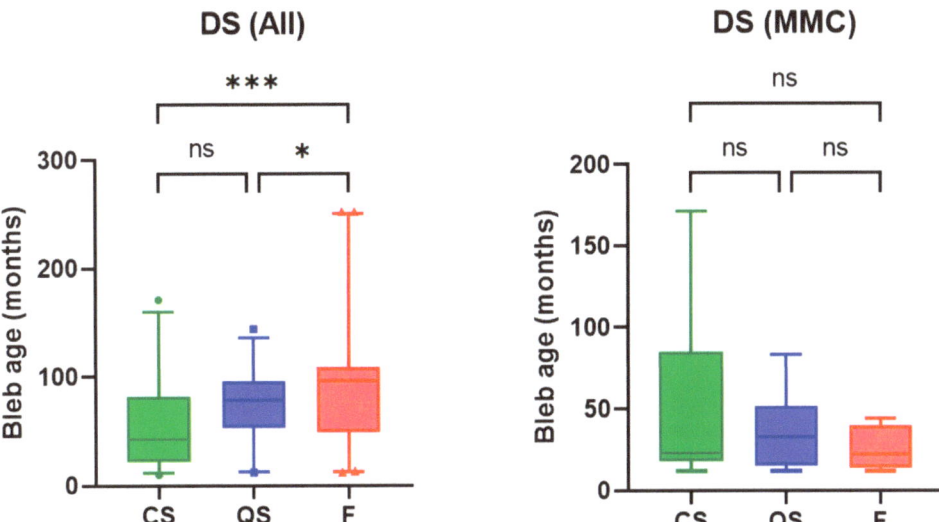

Figure 6. Box and Whisker plots (median, interquartile range, and 5th to 95th percentile) and results of one-way ANOVA of post-operative duration (filtering bleb age) and outcome groups of Complete success (CS), Qualified success (QS), and failure (F). Ns = non-significant. Asterisks denote statistical significance. [i.e., $p < 0.05$ (*), $p < 0.001$ (***)].

4. Discussion

The post-operative evaluation of GFS filtering blebs has traditionally relied on clinical grading systems performed at the slit-lamp, which document factors associated with surgical success such as bleb area, height, and vascularity. In this study, we developed and optimized a technique utilizing swept-source anterior segment-optical coherence tomography to visualize the internal microstructure in the post-operative period of filtering blebs. We then deployed this technique in a cross-sectional cohort of patients who underwent deep sclerectomy surgery with long-term follow-up.

4.1. Post-Operative Evaluation of Filtering Blebs

Glaucoma filtration surgery, such as trabeculectomy and deep sclerectomy, is associated with high rates of short- and long-term complications [12]. Long-term complications are often bleb-related, such as bleb fibrosis, leak, and infections or hypotony [13]. Morphological bleb configuration in the post-operative period can influence the planning of follow-up visits in glaucoma patients [14]. Multiple types of clinical bleb grading systems exist, such as the Moorfields Bleb Grading System or Indiana Bleb Appearance Grading Scale, which document factors associated with surgical success such as bleb area, height, and vascularity [6]. These subjective grading systems have varying levels of interobserver agreement on indices and are influenced by the experience of the observer [14]. These subjective grading systems are, however, often used to guide pharmacological and/or surgical intervention in the early post-operative period to decrease the risk of bleb fibrosis and failure. Post-operative evaluation of deep sclerectomy blebs also involves examination of bleb morphology at the slit-lamp; however, it is unclear how bleb morphology correlates with successful IOP reduction. This is because while subconjunctival outflow represents an important pathway for egress, augmentation of natural channels such as the Schlemm canal is also an important consequence of non-penetrating glaucoma surgery [4]. Nevertheless, an objective and quantifiable method for bleb evaluation may also be beneficial in DS surgery.

4.2. Swept-Source AS-OCT Technology for More Precise Visualization of Glaucoma Surgeries during the Post-Operative Course

The ANTERION (Heidelberg Engineering, Heidelberg, Germany) and CASIAII (Tomey, Nagoya, Japan) are two commercially available SS-OCT systems specifically designed for the evaluation of the anterior segment. Swept-source technology enables more precise visualization of thicker OCT structures than older OCT modalities, such as spectral-domain OCT, due to greater penetration from longer wavelengths used and higher scan speeds. In our study, we used the precise localization of OCT raster slices to visualize bleb morphology and anatomical details of the drainage outflow channel in a cohort of DS patients. We found that the maximal bleb height was significantly greater in complete success compared to qualified success and failure. We also found that the trabeculo-Descemet window was significantly greater in the complete success compared to the qualified success group in the subcohort of patients who had surgery augmented using intraoperative MMC. To our knowledge, the quantification of these microstructural parameters has not been previously reported and may help surgeons better understand the influence of their surgical technique on outcomes.

4.3. Use of Anterior Segment-OCT in Post-Operative Bleb Evaluation

Previous studies have evaluated the microstructure of glaucoma surgeries using non-contact imaging techniques such as anterior segment optical coherence tomography and ultrasound biomicroscopy [6,15]. These modalities can provide quantitative data on the internal structure of blebs, such as bleb wall thickness, presence of microcysts, and measurements of the internal ostium, bleb cavity, and sub-flap space [6–9,15]. These parameters are important as they may predict the success or failure of surgery [15]. For instance, Lenzhofer et al. examined 78 eyes of 60 patients post-XEN® gel stent implantation and found that the prevalence of small diffuse cysts was directly associated with lower IOPs, while cystic encapsulation at three months predicted higher surgical failure [16]. Konstantopoulos et al. examined 50 eyes of 50 patients following trabeculectomy, deep-sclerectomy, or no surgery and found that a tall intrascleral lake and a thick conjunctival/tenon's layer were associated with good post-operative outcomes as defined by intraocular pressure and medication use [17]. Ibarz Barbera et al. used AS-OCT to analyze the morphological evolution of filtering blebs after Preserflo micro shunt implantation and found a progressive horizontal and vertical expansion of the blebs in the sub-Tenon space from baseline to the third month [18]. Gambini et al. similarly characterized Preserflo bleb morphology post-operatively and reported various appearances such as "multiple internal layers" and "microcystic multiform" [19]. In our study, we utilized the high definition provided by swept-source OCT to quantify dimensions of surgical parameters, including the length, width, and thickness of the scleral flap and the length of the TDW. These parameters are important as they are modifiable by the surgeon and also have an effect on aqueous outflow and, therefore, subsequent outcomes.

The routine use of AS-OCT in the early post-operative period may be of particular clinical relevance following trabeculectomy surgery, especially given the frequency of surveillance and bleb manipulation to manage complications and maximize success during this period. Evaluation of bleb morphology may also be beneficial in deep sclerectomy surgery, such as in capturing the patency of the TDW prior to further laser or surgical intervention, such as a Nd:Yag goniopuncture procedure. The latter has been demonstrated to effectively lower IOP further [20,21] but may be dependent on the patency of the TDW. Other potential use cases for AS-OCT in the early post-operative period include the evaluation of causes of hypotony or, conversely, the evaluation of suspected obstruction of sclerostomy/TDW by the iris or haem which result in elevated IOP. Interventions which may benefit from pre- and post-procedure visualization of bleb internal microstructure include scleral flap suture lysis, the removal of releasable sutures, or bleb massage.

4.4. Limitations

We acknowledge several important limitations of our study. Firstly, there are other crucial determinants of surgical success that were not assessed in our analysis, predominantly due to the cross-sectional nature of our study of long follow-up duration. The pre-operative and intra-operative details of surgery were unable to be located in the archived medical records for many patients and had to be excluded from the analysis. These factors include the intraoperative dimensions of the scleral flap, number and type of suture placement on the scleral flap, concentration and duration of intraoperative mitomycin-C use [22], pre-operative intraocular pressure and duration and number of prior topical medications [23], intraoperative complications and early post-operative bleb manipulation [24], which have all been shown to influence post-operative surgical outcomes. Other factors contributing to the heterogeneity of the patient population are also potential confounders of our results, such as ethnicity [25], glaucoma type [26], previous ophthalmic surgery [26], and duration of post-operative follow-up [11], which are known to influence post-operative success and were not controlled for. The focus of our present study was, however, not to analyze the risk factors and outcomes of deep sclerectomy surgery but rather the utility of imaging the microscopic variations in mechanical structural parameters, such as the scleral flap and TDW, which may have a bearing on long-term surgical success. Secondly, the quantification of surgical parameters was performed manually, subjecting these measurements to inaccuracy and bias. We addressed this by keeping a bank of saved images with the measurements superimposed and having a second author verify the measurements. An important limitation of the image analysis was the difficulty in visualizing the parameters of interest in some patients due to poor contrast between these parameters and the surrounding tissues. This was particularly an issue in cases of surgical failure where the flap had likely fibrosed down onto the sclera, making it indistinguishable from the surrounding tissue. The surgical parameters that were indistinguishable had to, therefore, be excluded from the analysis. The amount of data excluded due to poor segmentation represented a minority of the dataset, and we found no systematic difference in outcomes between excluded and included data. Lastly, we used a threshold of 18 mmHg and the criterion of medication use to classify surgical outcomes into complete success, qualified success, and failure in line with recommendations by the World Glaucoma Association consensus on definitions of success. Altering the IOP threshold may, however, change the distribution of our results. We deliberately chose a level of 18 mmHg to reflect the target IOP level we generally aim to achieve in our real-world population of mainly moderate-advanced glaucoma, in line with published outcomes from the Advanced Glaucoma Intervention Study [27]. Considering these limitations, a prospective study using AS-OCT to image trabeculectomy blebs in the post-operative period is required.

5. Conclusions

Swept-source anterior-segment OCT can be used to accurately visualize and quantify the surgical parameters which influence aqueous outflow in deep sclerectomy surgery. Our proposed technique of image capture and processing can help surgeons better understand the influence of these parameters on aqueous outflow, which may help improve surgical outcomes.

Author Contributions: Conceptualization, J.C.K.T. and N.A.V.; methodology, J.C.K.T. and N.A.V.; software, J.C.K.T.; validation, J.C.K.T., H.M., M.B., A.C. and N.A.V.; formal analysis, J.C.K.T.; investigation, J.C.K.T. and N.A.V.; resources, H.M., M.B., A.C. and N.A.V.; data curation, J.C.K.T.; writing—original draft preparation, J.C.K.T.; writing—review and editing, J.C.K.T., H.M., M.B., A.C. and N.A.V.; visualization, J.C.K.T., H.M., M.B., A.C. and N.A.V.; supervision, M.B., A.C. and N.A.V.; project administration, N.A.V.; funding acquisition, N.A.V. All authors have read and agreed to the published version of the manuscript.

Funding: This research received no external funding.

Informed Consent Statement: Informed consent was obtained from all subjects involved in the study.

Data Availability Statement: Samples of the AS-OCT images of blebs used in this study can be shared upon reasonable request.

Acknowledgments: The authors would like to thank Yalin Zheng and Jianyang Xie from the University of Liverpool for providing guidance on the use of ITK Snap to measure surgical parameters in this study.

Conflicts of Interest: The authors declare no conflict of interest.

Abbreviations

Anterior-segment Optical Coherence Tomography (AS-OCT), Deep Sclerectomy (DS), Mitomycin-C (MMC), Complete success (CS), Qualified success (QS), Failure (F), Intraocular pressure (IOP).

References

1. Tham, Y.C.; Li, X.; Wong, T.Y.; Quigley, H.A.; Aung, T.; Cheng, C.Y. Global prevalence of glaucoma and projections of glaucoma burden through 2040: A systematic review and meta-analysis. *Ophthalmology* **2014**, *121*, 2081–2090. [CrossRef] [PubMed]
2. Jayaram, H. Intraocular pressure reduction in glaucoma: Does every mmHg count? *Taiwan J. Ophthalmol.* **2020**, *10*, 255–258. [CrossRef] [PubMed]
3. Wolters, J.E.J.; van Mechelen, R.J.S.; Al Majidi, R.; Pinchuk, L.; Webers, C.A.B.; Beckers, H.J.M.; Gorgels, T. History, presence, and future of mitomycin C in glaucoma filtration surgery. *Curr. Opin. Ophthalmol.* **2021**, *32*, 148–159. [CrossRef]
4. Mendrinos, E.; Mermoud, A.; Shaarawy, T. Nonpenetrating glaucoma surgery. *Surv. Ophthalmol.* **2008**, *53*, 592–630. [CrossRef] [PubMed]
5. Varga, Z.; Shaarawy, T. Deep sclerectomy: Safety and efficacy. *Middle East Afr. J. Ophthalmol.* **2009**, *16*, 123–126. [CrossRef] [PubMed]
6. Seo, J.H.; Kim, Y.A.; Park, K.H.; Lee, Y. Evaluation of Functional Filtering Bleb Using Optical Coherence Tomography Angiography. *Transl. Vis. Sci. Technol.* **2019**, *8*, 14. [CrossRef]
7. Hayek, S.; Labbé, A.; Brasnu, E.; Hamard, P.; Baudouin, C. Optical Coherence Tomography Angiography Evaluation of Conjunctival Vessels During Filtering Surgery. *Transl. Vis. Sci. Technol.* **2019**, *8*, 4. [CrossRef]
8. Oh, L.J.; Wong, E.; Lam, J.; Clement, C.I. Comparison of bleb morphology between trabeculectomy and deep sclerectomy using a clinical grading scale and anterior segment optical coherence tomography. *Clin. Exp. Ophthalmol.* **2017**, *45*, 701–707. [CrossRef]
9. Miura, M.; Kawana, K.; Iwasaki, T.; Kiuchi, T.; Oshika, T.; Mori, H.; Yamanari, M.; Makita, S.; Yatagai, T.; Yasuno, Y. Three-dimensional anterior segment optical coherence tomography of filtering blebs after trabeculectomy. *J. Glaucoma* **2008**, *17*, 193–196. [CrossRef]
10. Heuer, D.; Barton, K.; Grehn, F.; Shaarway, T.; Sherwood, M. *Consensus on Definitions of Success*; Kugler Publications: Amsterdam, The Netherlands, 2018; Volume 2018.
11. Gedde, S.J.; Feuer, W.J.; Lim, K.S.; Barton, K.; Goyal, S.; Ahmed, I.I.; Brandt, J.D. Treatment Outcomes in the Primary Tube Versus Trabeculectomy Study after 5 Years of Follow-up. *Ophthalmology* **2022**, *129*, 1344–1356. [CrossRef]
12. Conlon, R.; Saheb, H.; Ahmed, I.I. Glaucoma treatment trends: A review. *Can. J. Ophthalmol.* **2017**, *52*, 114–124. [CrossRef] [PubMed]
13. Leung, D.Y.; Tham, C.C. Management of bleb complications after trabeculectomy. *Semin. Ophthalmol.* **2013**, *28*, 144–156. [CrossRef] [PubMed]
14. Hoffmann, E.M.; Herzog, D.; Wasielica-Poslednik, J.; Butsch, C.; Schuster, A.K. Bleb grading by photographs versus bleb grading by slit-lamp examination. *Acta Ophthalmol.* **2019**, *98*, e607–e610. [CrossRef]
15. Kudsieh, B.; Fernández-Vigo, J.I.; Canut Jordana, M.I.; Vila-Arteaga, J.; Urcola, J.A.; Ruiz Moreno, J.M.; García-Feijóo, J.; Fernández-Vigo, J. Updates on the utility of anterior segment optical coherence tomography in the assessment of filtration blebs after glaucoma surgery. *Acta Ophthalmol.* **2022**, *100*, e29–e37. [CrossRef]
16. Lenzhofer, M.; Strohmaier, C.; Hohensinn, M.; Hitzl, W.; Sperl, P.; Gerner, M.; Steiner, V.; Moussa, S.; Krall, E.; Reitsamer, H.A. Longitudinal bleb morphology in anterior segment OCT after minimally invasive transscleral ab interno Glaucoma Gel Microstent implantation. *Acta Ophthalmol.* **2019**, *97*, e231–e237. [CrossRef] [PubMed]
17. Konstantopoulos, A.; Yadegarfar, M.E.; Yadegarfar, G.; Stinghe, A.; Macleod, A.; Jacob, A.; Hossain, P. Deep sclerectomy versus trabeculectomy: A morphological study with anterior segment optical coherence tomography. *Br. J. Ophthalmol.* **2013**, *97*, 708–714. [CrossRef] [PubMed]
18. Ibarz Barberá, M.; Morales Fernández, L.; Tañá Rivero, P.; Gómez de Liaño, R.; Teus, M.A. Anterior-segment optical coherence tomography of filtering blebs in the early postoperative period of ab externo SIBS microshunt implantation with mitomycin C: Morphological analysis and correlation with intraocular pressure reduction. *Acta Ophthalmol.* **2022**, *100*, e192–e203. [CrossRef]
19. Gambini, G.; Carlà, M.M.; Giannuzzi, F.; Boselli, F.; Grieco, G.; Caporossi, T.; De Vico, U.; Savastano, A.; Baldascino, A.; Rizzo, C.; et al. Anterior Segment-Optical Coherence Tomography Bleb Morphology Comparison in Minimally Invasive Glaucoma Surgery: XEN Gel Stent vs. PreserFlo MicroShunt. *Diagnostics* **2022**, *12*, 1250. [CrossRef]

20. Anand, N.; Pilling, R. Nd:YAG laser goniopuncture after deep sclerectomy: Outcomes. *Acta Ophthalmol.* **2010**, *88*, 110–115. [CrossRef]
21. Mermoud, A.; Karlen, M.E.; Schnyder, C.C.; Sickenberg, M.; Chiou, A.G.; Hédiguer, S.E.; Sanchez, E. Nd:Yag goniopuncture after deep sclerectomy with collagen implant. *Ophthalmic Surg. Lasers* **1999**, *30*, 120–125. [CrossRef]
22. Wilkins, M.; Indar, A.; Wormald, R. Intra-operative mitomycin C for glaucoma surgery. *Cochrane Database Syst. Rev.* **2005**, *2005*, Cd002897. [CrossRef] [PubMed]
23. Batterbury, M.; Wishart, P.K. Is high initial aqueous outflow of benefit in trabeculectomy? *Eye* **1993**, *7 Pt 1*, 109–112. [CrossRef] [PubMed]
24. Wells, A.P.; Bunce, C.; Khaw, P.T. Flap and suture manipulation after trabeculectomy with adjustable sutures: Titration of flow and intraocular pressure in guarded filtration surgery. *J. Glaucoma* **2004**, *13*, 400–406. [CrossRef] [PubMed]
25. Nguyen, A.H.; Fatehi, N.; Romero, P.; Miraftabi, A.; Kim, E.; Morales, E.; Giaconi, J.; Coleman, A.L.; Law, S.K.; Caprioli, J.; et al. Observational Outcomes of Initial Trabeculectomy With Mitomycin C in Patients of African Descent vs Patients of European Descent: Five-Year Results. *JAMA Ophthalmol.* **2018**, *136*, 1106–1113. [CrossRef] [PubMed]
26. Issa de Fendi, L.; Cena de Oliveira, T.; Bigheti Pereira, C.; Pereira Bigheti, C.; Viani, G.A. Additive Effect of Risk Factors for Trabeculectomy Failure in Glaucoma Patients: A Risk-group From a Cohort Study. *J. Glaucoma* **2016**, *25*, e879–e883. [CrossRef]
27. The Advanced Glaucoma Intervention Study (AGIS): 7. The relationship between control of intraocular pressure and visual field deterioration.The AGIS Investigators. *Am. J. Ophthalmol.* **2000**, *130*, 429–440. [CrossRef]

Disclaimer/Publisher's Note: The statements, opinions and data contained in all publications are solely those of the individual author(s) and contributor(s) and not of MDPI and/or the editor(s). MDPI and/or the editor(s) disclaim responsibility for any injury to people or property resulting from any ideas, methods, instructions or products referred to in the content.

Article

Initial Clinical Experience with Ahmed Valve in Romania: Five-Year Patient Follow-Up and Outcomes

Ramona Ileana Barac [1], Vasile Harghel [1,*], Nicoleta Anton [2], George Baltă [1], Ioana Teodora Tofolean [1], Christiana Dragosloveanu [1,*], Laurențiu Flavius Leuștean [1], Dan George Deleanu [1] and Diana Andreea Barac [1]

[1] Department of Ophthalmology, "Carol Davila" University of Medicine and Pharmacy, 050474 București, Romania; ramona.barac@drd.umfcd.ro (R.I.B.); george.balta@drd.umfcd.ro (G.B.); ioana.tofolean@umfcd.ro (I.T.T.); flavius-laurentiu.leustean@drd.umfcd.ro (L.F.L.); dan-george.deleanu@umfcd.ro (D.G.D.); andreeadiana.e.barac@stud.umfcd.ro (D.A.B.)

[2] Department of Ophthalmology, University of Medicine and Pharmacy "Grigore T. Popa", 700115 Iași, Romania; anton.nicoleta1@umfiasi.ro

* Correspondence: harghel.vasile@gmail.com (V.H.); christiana.dragosloveanu@umfcd.ro (C.D.)

Citation: Barac, R.I.; Harghel, V.; Anton, N.; Baltă, G.; Tofolean, I.T.; Dragosloveanu, C.; Leuștean, L.F.; Deleanu, D.G.; Barac, D.A. Initial Clinical Experience with Ahmed Valve in Romania: Five-Year Patient Follow-Up and Outcomes. *Bioengineering* **2024**, *11*, 820. https://doi.org/10.3390/bioengineering11080820

Academic Editors: Karanjit S. Kooner and Osamah J. Saeedi

Received: 27 June 2024
Revised: 25 July 2024
Accepted: 5 August 2024
Published: 12 August 2024

Copyright: © 2024 by the authors. Licensee MDPI, Basel, Switzerland. This article is an open access article distributed under the terms and conditions of the Creative Commons Attribution (CC BY) license (https://creativecommons.org/licenses/by/4.0/).

Abstract: Background: Glaucoma is a leading cause of irreversible blindness worldwide and is particularly challenging to treat in its refractory forms. The Ahmed valve offers a potential solution for these difficult cases. This research aims to assess the initial clinical experience with Ahmed valve implantation in Romania, evaluating its effectiveness, associated complications, and overall patient outcomes over a five-year period. Methods: We conducted a prospective study on 50 patients who underwent Ahmed valve implantation due to various types of glaucoma. Patients were monitored at several intervals, up to five years post-surgery. Intraocular pressure and visual acuity were the primary measures of success. Results: On average, patients maintained the intraocular pressure within the targeted range, with the mean intraocular pressure being 17 mmHg 5 years post-surgery. Success, defined as maintaining target intraocular pressure without additional surgery, was achieved in 82% at 1 year, 68% at 3 years, and 60% after 5 years postoperative. Conclusion: Ahmed valve implantation is a viable treatment option for refractory glaucoma, demonstrating significant intraocular pressure reduction and manageable complication rates over a five-year follow-up period. Future research should focus on long-term outcomes and optimization of surgical techniques to further reduce complication rates and improve patient quality of life.

Keywords: glaucoma; Ahmed glaucoma valve implant

1. Introduction

The management of glaucoma, a leading cause of irreversible blindness worldwide, continues to present significant challenges to ophthalmologists. Traditional treatment modalities, including pharmacological therapy and conventional surgical procedures, often fall short of adequately controlling intraocular pressure (IOP) in advanced cases. The advent of innovative surgical interventions, such as the Ahmed glaucoma valve (AGV) implant, offers new hope for patients who struggle with refractory glaucoma (resistant to conventional medical, laser, and surgical treatments aimed at reducing intraocular pressure) [1–5].

This article delves into the initial clinical experiences with the AGV implant, highlighting the outcomes and implications of a recent clinical trial involving a cohort of patients who underwent this procedure. By assessing the effectiveness, safety, and patient outcomes associated with the AGV, this study aims to provide a comprehensive evaluation of its role in the modern therapeutic arsenal against glaucoma. This is the first study conducted in Romania on a group of patients with refractory glaucoma, with a follow-up period of 5 years, providing valuable insights into the long-term effectiveness and safety of the

AGV implant in this patient population. Our focus is to thoroughly analyze the sociodemographic characteristics, types of glaucoma, postoperative IOP variations, visual acuity outcomes, and the complications encountered over the extended follow-up period.

Globally, the AGV has become a pivotal tool in the surgical management of refractory glaucoma due to its effectiveness in reducing IOP and its relatively favorable safety profile. Studies from various regions, including long-term analyses, have consistently demonstrated the AGV's capability to achieve significant IOP reduction with manageable complication rates [4–6]. For instance, Kang et al. (2022) reported favorable long-term outcomes, highlighting both the efficacy and the sustained benefits of AGV implantation [4]. Similarly, Christakis et al. (2017) found comparable results when contrasting the AGV with the Baerveldt implant, reinforcing the AGV's standing as an effective option for IOP control in refractory cases [7,8].

The study also places its findings in the context of international research, comparing outcomes with significant studies such as the Tube Versus Trabeculectomy (TVT) study and the Primary Tube Versus Trabeculectomy (PTVT) study, both of which provide extensive data on the efficacy and complications associated with tube shunt surgeries [9–14]. This comparison is crucial in understanding the broader implications of AGV use and its positioning relative to other surgical options. Furthermore, by documenting the sociodemographic characteristics and types of glaucoma treated, this study aims to contribute to a more nuanced understanding of how these factors influence surgical outcomes and complications.

Our findings are expected to contribute significantly to the global body of knowledge on glaucoma management, offering a unique perspective from a Romanian cohort and adding to the diversity of clinical experiences with the AGV. The insights gained from this study will not only enhance clinical practice in Romania but also provide valuable data for ophthalmologists worldwide who are managing patients with refractory glaucoma.

2. Materials and Methods

Data collection for this study was meticulously planned and executed to ensure the accuracy and reliability of the findings. Fifty-eight patients with refractory glaucoma were recruited from the Clinical Hospital for Ophthalmological Emergencies Bucharest between 2014 and 2018. All patients provided written informed consent prior to participation. Patients had different forms of uncontrolled glaucoma, either by medication or surgery.

The choice of AGV over other treatment options was based on several considerations. It has consistently demonstrated significant reductions in IOP across various studies and patient populations [6,7]. Its non-valved mechanism allows for a more controlled release of aqueous humor, reducing the risk of postoperative hypotony [8]. Long-term studies have demonstrated the sustained benefits of AGV implantation in maintaining reduced IOP levels and minimizing complications over extended periods. Also, it has proven to be effective in managing refractory glaucoma, including cases with neovascular glaucoma and other challenging conditions. This makes it a preferred option for patients who have not responded to conventional treatments [9–14].

During the study, three patients dropped out due to death, and five patients missed follow-up visits. For patients who missed visits, the last observation carried forward method was employed to handle missing data, ensuring the robustness of the analysis. The criteria for inclusion are as follows:

- Indication for AGV implant;
- Age ≥ 18;
- Consent of the patient to participate in the study;
- Refractory glaucoma;
- Compliance with follow-ups for a 5-year period.

The criteria for success are as follows:

- IOP less than 22 mmHg and greater than 5 mmHg (with or without medication);
- Decrease in visual acuity by no more than two lines at Snellen optotype;

- Solved postoperative complications (if any occurred);
- had a functional valve on follow-up.

The sample size for this study was calculated based on several critical parameters to ensure sufficient power to detect meaningful clinical differences and to achieve statistically significant results. The following parameters were used for determining the sample size:

- The anticipated difference in the primary outcome measure between baseline and post-intervention.
- The effect size was estimated based on previous studies and clinical expectations of the AGV effectiveness.
- An estimate of the proportion of participants who might drop out of the study or miss follow-up visits.
- Based on similar studies, an anticipated dropout rate of approximately 10% was factored into the sample size calculation to account for potential loss to follow-up.
- The prevalence of refractory glaucoma and the availability of eligible patients within the recruitment period.
- Feasibility and resource constraints were also considered to ensure the study could be practically conducted within the specified timeframe.

Before surgery, a blood sample was taken for laboratory tests. The tests included standard assessments of patients' general health. The approximate amount of blood taken did not exceed 41 mL (3 tablespoons), which corresponds to the total volume of blood taken throughout the study. All information obtained from the samples was kept confidential.

All AGV implantation procedures were performed by the same surgeon and using identical processes, including subconjunctival anesthesia, during the study period. All patients were anesthetized with subconjunctival anesthesia, using Lidocaine 1% mixed with Ropivacaine 1%. A 10 mm conjunctival incision was created along the limbus to construct a conjunctival flap, and a 4–6 mm wide half-layer scleral flap was formed. The body implant was positioned 8 mm from the limbus, outside the limbal healing space. The plate was then sutured to the sclera with a 7.0 non-absorbable suture. The drainage tube was trimmed to permit a 2–3 mm insertion in the AC and was bevel cut to an angle of 30° to facilitate AC entering. An AC paracentesis was performed, and a viscoelastic substance was injected to increase the space. The AC was then entered 1–3 mm posteriorly to the corneoscleral limbus with a 22G needle. The tube was inserted into the AC through the needle tract. An additional 10.0 nylon suture was placed to fixate the tube to the sclera. Afterward, both the scleral flap and conjunctiva were then sutured with 8.0 absorbable suture. We did not use Mitomycin as it is not approved by the National Agency for Medicines and Medical Devices in Romania. We did not use any other antimetabolite at implantation to avoid increasing the risk of denudation and valve expulsion. 5-fluorouracil was used per secundum in the patients who required needling [10,13]. All patients received similar postoperative topical medications: 0.5% chloramphenicol and 0.2% betamethasone four times daily for one month. The model selected for implantation was the FP7 in all patients in the superotemporal quadrant (Figure 1).

The information collected was introduced into a computerized database. Excel was used for data entry and processing, and statistical analyses were conducted using SAS Enterprise Guide 9.4M7 software. We used various statistical analyses, statistical tests, t-student test for hypothesis testing, descriptive analysis, averaging, calculation of absolute and relative frequency in percentages, correlations (using Pearson test), and correlation ratio to interpret the intensity of independent variables as compared to dependent variable for linear regression models, each variable validation, graphics, and histogram frequencies. The results are reported based on the total number of cases remaining in the study.

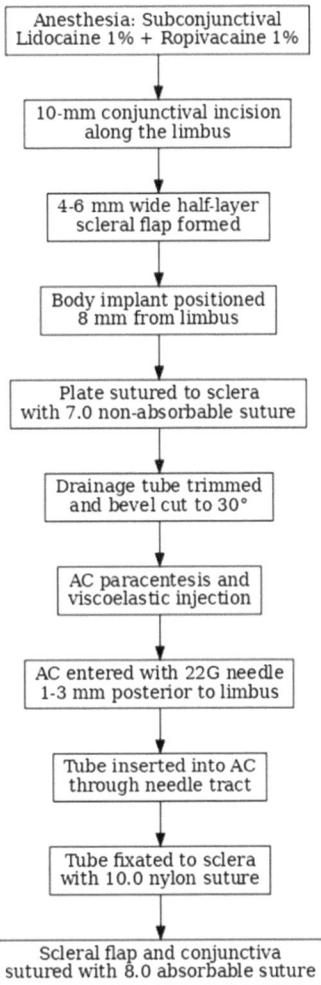

Figure 1. Ahmed glaucoma valve implant: surgery steps.

We measured the visual acuity with the best correction using the Snellen optotype. Intraocular pressure was measured using the Goldmann tonometer with topical anesthesia. We noted the antiglaucoma treatment that patients use with each measurement. Visual field examination results were not tracked as a parameter because the target was to decrease IOP. Moreover, the patients included in the study had mostly significant low acuity and visual field loss at the beginning of the study; thus, substantial visual field changes were not anticipated as primary outcomes. The primary intervention goal was to stabilize and lower IOP to prevent further progression of glaucoma.

We have received ethical approval from the Clinical Emergency Eye Hospital Bucharest Ethic Committee for this study.

3. Results

- Analysis of socio-demographic characteristics

The group comprised 50 patients with an average age of 51.7 years old, with the youngest being 18 and the oldest being 84 years old. We noticed that 40% of the patients had elementary studies, and 60% had at least a professional qualification. There is a discrepancy between the level of professional education (62% of the patients have a qualification) and the job of the adult patients. Only six (12%) patients were employed, and none were employers or workers on their own.

- Distribution of patients by glaucoma type

Most of the patients had an admission diagnostic of glaucoma secondary to vitreoretinal surgery (24%). The cause of this phenomenon is the fact that the clinic where the surgeries were performed has a high number of interventions on the posterior pole (Table 1).

Table 1. Distribution of patients by glaucoma type.

Glaucoma Type Total	Patients n. 50	% 100%
Secondary glaucoma after vitreoretinal surgery	12	24%
Primary open-angle glaucoma	10	20%
Neovascular secondary glaucoma	8	16%
Primary closed-angle glaucoma	7	14%
Aphakic secondary glaucoma	6	12%
Traumatic secondary glaucoma	2	4%
Congenital glaucoma	1	2%
Uveitic secondary glaucoma	1	2%
Juvenile glaucoma	1	2%
Secondary glaucoma after keratoplasty	1	2%
Posterior embryotoxon	1	2%

In second place as ethology in our study, we have primary glaucoma with an open angle. In 10 patients (20%) with primary open-angle glaucoma that was not amendable with medical treatment or other surgical interventions and had a progression to atrophy, we implanted AGV.

The third place stands for neovascular glaucoma, with eight patients (16%). Neovascular glaucoma is only in third place because of the use of anti-VEGF agents in the last few years. This has led to a decrease in the number of patients with neovascular glaucoma to whom an AGV was implanted [15,16].

- Intraocular pressure variation after surgery

The targeted intraocular pressure was below 22 mmHg and above 5 mmHg. Hypotonia is defined as a decrease in the intraocular pressure below 5 mmHg. In the group of patients that we examined, the pressure was measured before surgery, one day, one month, 3 months and 6 months, one year, three years, and five years after surgery and in all patients included in the study when they presented in the hospital for different problems. We calculated the mean values of the intraocular pressure at every step of the follow-up. One day after surgery, only one patient had hypotonia and a small anterior chamber. A visco-elastic substance was injected into the anterior chamber, and the pressure became normal in the following days. During the study, there were no other cases of hypotonia.

The very small number of patients with hypotonia after the implantation of AGV is explained by the fact that the technique used implies covering the tube with a scleral flap. This decreased risk was also described by other authors when using the scleral flap [17,18]. Because of the elastomeric membrane that functions as a valve, the system remains closed when the intraocular pressure decreases below 8 mmHg. There can be losses next to the tube or if the tube is punctured accidentally in the extra cameral portion; in these instances, the valve works, but hypotonia ensues because of the losses outside the valve [9–22].

One patient had an intraocular pressure of 30 mmHg one day after surgery, and this happened because of a blockage of the tube by vitreous. The pressure did not decrease after administration of specific drugs locally or systemically, so we decided to practice anterior vitrectomy.

The IOP a day after surgery had a mean value of 12 mmHg. In patients with an intraocular pressure above 22 mmHg, antiglaucoma medication was administered, or massage of the eye was made, or needling or surgical excision of the fibrous capsule was intended.

In most patients with hypertonia, this was caused by the formation of a capsule of fibrous tissue around the body of the valve with modified collagen because of the presence of the aqueous humor under the tenon, between the tenon and the sclera. This fibrous cyst does not allow the normal drainage of the aqueous humor and creates pressure around the valve's body.

A fibrous cyst appeared in seven patients that were subjected to needling; in four patients, the surgical excision of the cysts was made; in two patients at 1 month, in one at 6 month, and in the other at one year.

The wide range of IOP values (5–30 mmHg) underscores the need for individualized patient monitoring and management. The mean difference in IOP before and after surgery was found to be 7.53 mmHg. This indicates a statistically significant reduction in IOP post-surgery, supporting the effectiveness of the surgical technique employed. While some patients maintained IOP within the target range, others experienced either hypotonic or elevated pressures, highlighting the variability in postoperative outcomes. The skewed distribution towards higher IOP values suggests that a significant portion of measurements were at or above the target upper limit, which may indicate a need for more aggressive IOP management in some cases. The presence of both hypotonic and elevated pressures in the dataset emphasizes the importance of regular follow-up and potentially adjusting treatment strategies to achieve optimal IOP control (Figure 2).

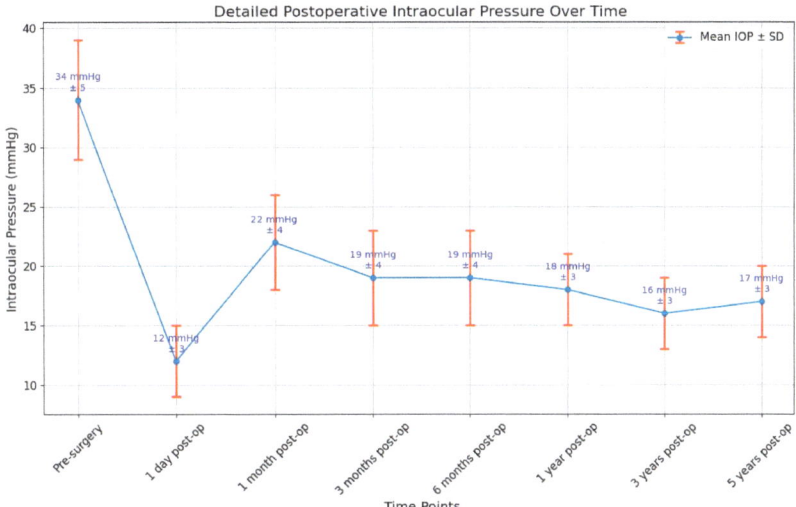

Figure 2. Intraocular pressure variations ± SD before and after surgery.

According to the success and failure criteria defined before, finding an intraocular pressure above 22 mmHg in more than two visits that could not be controlled by medication, massage, needling, or excision of the cysts implied that the patient was in the failure category.

After one year postoperative, failure was declared in six patients (12%). They had elevated IOP during multiple visits due to the cyst formation and valve's body entrapment. At the 3-year mark after surgery, another five patients developed high IOP due to this condition. And 5 years postoperative, valve body entrapment happened in another four patients, making it the most frequent complication (30%) leading to failure.

We concluded that the number of medications necessary for glaucoma control was reduced after surgery, from 3.37 before surgery to 1.21 after surgery. This decrease is noticed in most of the published data [9–14].

- Visual acuity variation after surgery

Visual acuity (VA) is the second criterion used to define successful or unsuccessful intervention. Loss of light perception or a decrease in VA of more than two rows in the optotype was considered a criteria for failure. Our study is a prospective one, allowing a good appreciation of the variation in visual acuity and a good appreciation of the evolution of lens-related problems. Visual acuity was measured using the Snellen optotype with the best correction. During the period of follow-up of a year, 8 patients (16%) out of 50 had a decrease in VA, and 3 of them had a decrease with more than two rows in the optotype, so the intervention was considered unsuccessful in those cases (Table 2).

Table 2. Decrease in sight by glaucoma type at 1 year postoperative.

Type of Glaucoma with a Decrease in Sight	Number of Patients with a Decrease in VA	Number of Patients with a Decrease in VA of More Than 2 Rows	Number of Patients with VA = WLP
Glaucoma secondary to vitreoretinal surgery	3	1	1
Secondary neovascular glaucoma	2	0	1
Primary glaucoma with open angle	2	0	0
Juvenile glaucoma	1	0	0
Total	8	1	2

The first patient, with glaucoma secondary to vitreoretinal surgery, had developed a valve capture with high-pressure values that could not be controlled; we practiced the excision of the fibrous cyst after a year, but the patient's VA changed from LP (light perception) to WLP (without light perception).

The second patient, with neovascular secondary glaucoma, had a suprachoroidal hemorrhage and vitreous hemorrhage with a decrease in VA from LP to WLP.

The third patient with glaucoma secondary to vitreoretinal surgery, with surgically corrected retinal detachment, with surgically treated cataracts, silicone oil extraction, and iridectomy, developed suprachoroidal hemorrhage 3 months after surgery with a decrease in visual acuity from 0.05 to CF (counting fingers).

We noticed that all patients who lost light perception had a very low visual acuity at the beginning of the study, all of them being with only LP. We also noticed that in four patients (8%), visual acuity has increased. The patients had lens opacities before surgery that needed an extracapsular lens extraction with a pseudophakic implant after 6 to 12 months. In all three, the sight became better.

- Considerations regarding surgery complications

An old surgery axiom states that the best treatment for surgery complications is prevention. Each patient must be treated as a unique case.

Most of the complications appeared in the group of patients with glaucoma secondary to vitreoretinal surgery, them being best represented in this study. The second complication frequency was the subgroup of patients with secondary neovascular glaucoma (Figure 2).

Valve's body entrapment was noticed in 15 (31.91%) patients, leading to an increase in intraocular pressure. Leaking of the aqueous humor in the space created between the tenon and sclera leads to an inflammatory reaction in the tenon with fibrous tissue development and modified collagen. A fibrous, hard capsule is formed around the valve's body, and the filtering of the aqueous humor in the subconjunctival space is blocked; this leads to an increase in the pressure inside the valve. The high pressure around the body of the valve by the fibrous capsule leads to restriction on valve opening when intraocular pressure is above 8 mmHg, so the valve will not open when the pressure is even higher. It can be prevented by digital ocular massage, needling using an antimetabolite (5-fluorouracil), and regular check-ups [23,24].

Cyst excision was intended with local subconjunctival anesthesia. An incision at the level of the conjunctiva was made at the highest point of the cyst; the fibrous area behind the conjunctiva was extracted, and the valve stayed in place. After the cyst excision, the conjunctiva was sutured so that the valve was totally covered. The histologic exams of the peri-valve tissues extracted revealed modifications in collagen structure and transformations of fibroblasts to myofibroblast cells.

Hyphema appeared in seven (14.89%) patients a day after surgery. In six cases, hyphema resolved spontaneously in the next few days. In one of the patients, we practiced the washing of the anterior chamber and Avastin (Bevacizumab) injection in the anterior chamber.

Suture dehiscence was the third most frequent complication without exteriorization of the valve. This appeared in five (10.64%) patients. In three of these, after re-suturing, the evolution was good. In two patients at one year postoperative, a reintervention was needed, and the valve coverage with oral mucous tissue was performed. However, it proved to be inefficient as it reappeared, and we had to extract the valve between 1- and 3-years postoperative.

Tube blockage appeared in four patients (8.51%). In two patients, the tube was blocked by blood, and with the massage of the globe, the tube became patent, and there was no need for surgical intervention. In one patient, the iris blocked the tube, and a surgical intervention was needed to reposition the tube. In the fourth patient, the tube was blocked by a vitreous on the day following the surgery, with the patient having a vitreous in the anterior chamber. General and local treatment for glaucoma was intended but with no results. In 4 days after surgery, vitrectomy was intended because the high intraocular pressure could not be controlled.

Rebellious pain was unresponsive to any treatment developed in four (8.51%) patients. In the case of the patient with congenital glaucoma, the valve had to be explanted, while in other cases, the pain was remitted by itself.

Vitreous hemorrhage developed in three (6.38%) patients: two with neovascular glaucoma and one with aphakic glaucoma. In one patient, the hemorrhage resolved spontaneously, and in the other two, vitrectomy was needed.

Suprachoroidal hemorrhage developed in two (4.26%) patients: one with neovascular glaucoma and the other with glaucoma secondary to vitreoretinal surgery.

In the patient with neovascular glaucoma, visual acuity was reduced from LP to WLP. In the patient with secondary glaucoma after vitreoretinal surgery with surgically corrected retinal detachment and cataracts, silicone oil extraction, and iridectomy 3 months after the surgery, suprachoroidal hemorrhage developed, and visual acuity decreased from 0.05 to CF.

Retinal detachment developed in two (4.26%) patients with glaucoma secondary to vitreoretinal surgery in 4 and 6 months from the implant; surgical intervention was intended with re-attachment of the retina.

Diplopia developed in two (4.26%) patients. The first patient had secondary posttraumatic glaucoma. The valve was implanted in the super-nasal quadrant, and no intervention was needed because, at the next visit, the problem was solved. The second patient had glaucoma after vitreoretinal surgery, and the valve was implanted in the superotemporal quadrant. Diplopia developed when the patient was looking up and towards the right; in the primary position, there was no diplopia. The patient was satisfied, and the diplopia resolved spontaneously.

Hypotony occurred in one patient but resolved after injecting viscoelastic into the anterior chamber; there were no patients with athalamy. In the study group, I noticed a very small incidence rate of hypotonia. Unlike other artificial drainage systems, hypotonia is much less present in the AGV because of the valve mechanism. Unlike other studies, we obtained a much lower rate of hypotonia, and we believe that the implantation technique with the scleral flap is responsible for this; introducing the tube through the scleral flap makes a much better seal than covering the tube with preserved sclera.

Cornea-tube contact occurred in one patient; the tube was shortened on the second postoperative day without further damage to the cornea. At one year postoperative, one patient had endothelial damage and required a cornea transplant, after which the AGV tube was repositioned in the posterior chamber.

Iris-tube contact occurred in one patient a year after implant because the valve moved; we had to reposition the tube (Figure 3).

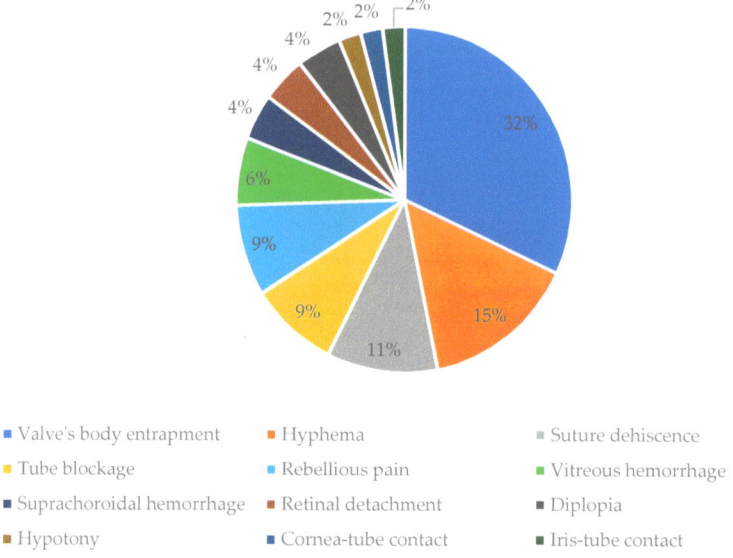

Figure 3. Surgery complications.

According to the success criteria we defined, we achieved a success rate of 82% one year after the implant. After 3 years, it was 68%, and after 5 years, it was 60%.

The predominance of the valve's body entrapment as a cause of failure highlights a potential area for improvement in either the surgical technique or the design and handling of the Ahmed Valve. The variety of causes listed also underscores the complexity of the surgery and the need for comprehensive preoperative assessment and postoperative care to minimize risks. The data suggest that while the Ahmed Valve can be an effective treatment, there are specific complications that need to be addressed to improve overall success rates (Table 3).

Table 3. Patients who experienced failure at 5 years postoperative.

Cause	Nr. of Failed Cases
Valve's body entrapment	15
Suprachoroidal haemorrhage	2
Suture dehiscence	2
Rebellious pain	1

Another objective of the study was to assess whether the rate of occurrence of postoperative complications after the valve implant in patients with previous antiglaucoma surgery is higher than the rate of complications in patients without previous glaucoma surgery. For this, we compared the complications and calculated if there is a correlation between the two variables.

The statistical correlation between the number of surgeries prior to the implant and postoperative complications was 0.876. The correlation is direct and strong, which means that it is statistically significant that patients with multiple ophthalmic surgeries had more frequent postoperative complications. The postoperative complications that occurred more frequently in patients with surgeries prior to the implant were wound dehiscence, valve entrapment, rebellious pain, hyphema, decreased visual acuity, and displaced tubes. The statistical correlation between the number of antiglaucoma surgeries prior to the implant and postoperative complications was 0.925. The correlation is direct and strong. The coefficient of correlation between patient age or area of origin and complications is negative; there is no statistically significant correlation between these two variables.

4. Discussion

The AGV represents a significant advancement in the surgical management of refractory glaucoma. This article examines the initial clinical experience with the AGV implant, focusing on several key areas: socio-demographic characteristics of the patient cohort, distribution of patients by glaucoma type, variations in IOP and visual acuity after surgery, and complications associated with the procedure.

Correlating the incidence of glaucoma with age and gender, we found that our patient cohort predominantly consisted of older adults, with a higher prevalence in males. This demographic trend aligns with the existing literature, which indicates that age and male gender are significant risk factors for glaucoma [2,6,22,25].

In this study, the majority of patients diagnosed with glaucoma had developed the condition secondary to vitreoretinal surgery, specifically due to the use of silicone oil for retinal detachment procedures. These patients experienced elevated IOP as a result of emulsified silicone oil blocking the pupil, causing iris synechia, or obstructing the camerular angle. This secondary glaucoma could not be managed with medication alone.

The second most common type of glaucoma in the study was primary open-angle glaucoma, affecting 20% of patients. These patients had not responded to medical treatments and other surgical interventions and showed progressive atrophy, necessitating the implantation of the AGV.

Neovascular glaucoma was the third most common, affecting 16% of patients. The relatively lower incidence of this type was attributed to the recent use of anti-VEGF agents, which have reduced the number of neovascular glaucoma cases requiring AGV implantation. All of these patients had failed trabeculectomy surgeries. After this study, we changed our therapeutic conduct, and now we implant AGV in secondary neovascular glaucoma per primam.

The targeted IOP post-surgery was set between 5 and 22 mmHg. Hypotonia, defined as an IOP below 5 mmHg, was observed in only one patient the day after AGV implantation, which was successfully corrected with viscoelastic injection. The use of a scleral flap to cover the tube likely contributed to the low incidence of hypotonia, a finding consistent with other studies.

The elastomeric membrane in the AGV helps maintain IOP above 8 mmHg, preventing hypotonia unless there are external losses. Elevated IOP (above 22 mmHg) was managed with medication, massage, needling, or surgical excision of fibrous cysts, which formed around the valve in some patients due to modified collagen from aqueous humor exposure.

Fibrous cysts requiring intervention were observed in seven patients, with varying follow-up times for surgical excision. Overall, the surgery reduced the number of medications needed for glaucoma control from an average of 3.37 to 1.21, aligning with published data.

In this study, the majority of complications occurred in patients with glaucoma secondary to vitreoretinal surgery, followed by those with neovascular glaucoma. The most frequent complications included hyphema, valve entrapment, suture dehiscence, tube blockage, rebellious pain, vitreous hemorrhage, suprachoroidal hemorrhage, retinal detachment, diplopia, hypotony, cornea–tube, and iris–tube contact. The study achieved a success rate of 82% one year after AGV implantation in the group, indicating a generally favorable outcome despite the noted complications. The developed complications were resolved either by re-intervening or spontaneously [26–29].

In comparison to the Tube Versus Trabeculectomy (TVT) study and the Primary Tube Versus Trabeculectomy (PTVT) study, our results show a similar effectiveness in IOP control and a comparable rate of complications. The TVT study, which focused on patients with prior ocular surgery, reported success rates and complication profiles similar to ours, with a notable difference being the type of implant used (Baerveldt versus Ahmed). However, the complication profile in our study showed some differences. Hyphema and valve body entrapment were notable complications in our cohort, particularly among patients with secondary glaucoma due to vitreoretinal surgery. These complications were less frequently highlighted in the TVT study, which may be due to differences in patient populations and surgical techniques. Our use of the AGV, which has a unique flow-restricting mechanism, might account for the lower incidence of hypotony compared to the Baerveldt implant used in the TVT study. The PTVT study, dealing with primary glaucoma cases, also showed effective IOP management but had fewer secondary glaucomas compared to our study. Our study's patient population included a significant number of cases with secondary glaucoma, often post-vitreoretinal surgery, which contrasts with the primary glaucoma cases in the PTVT study. Despite these differences, our findings align with the PTVT study in demonstrating the efficacy of tube shunt surgery (Ahmed valve) in lowering IOP. In terms of postoperative interventions, our study reported several cases requiring additional surgeries, such as excision of fibrous capsules, cataract extractions, and vitrectomies. This is somewhat consistent with the PTVT study, where re-interventions were also necessary, though with different underlying causes and frequencies [9–14].

Recent advancements in glaucoma treatment, such as micro-pulse transscleral cyclophotocoagulation (MP-TSCPC) and continuous wave transscleral cyclophotocoagulation (CW-TSCPC), have shown promising results in managing IOP in neovascular glaucoma. Zemba et al. (2022) compared these two methods, finding that MP-TSCPC offered better IOP control and fewer complications than CW-TSCPC in NVG patients. These findings are relevant to our study as they highlight alternative or adjunctive treatments that could potentially enhance outcomes for patients with refractory glaucoma, particularly those with neovascular glaucoma. Incorporating such advanced treatment options could further improve the efficacy and safety profile of glaucoma management strategies, offering additional hope for patients who do not respond to conventional therapies [30–33].

Another issue that we encountered was patient compliance. Indeed, during the follow-up period, we experienced some cases of missing visits and missing data. Specifically, three patients dropped out of the study due to death, and five patients missed scheduled follow-up visits due to various reasons. Throughout the study, we maintained close contact with our patients, utilizing phone calls, communication with family members, and offering free visits to encourage participation and adherence to the follow-up schedule. Despite these efforts, some data gaps remained due to the reasons mentioned above. However, we believe that the overall impact on our study's findings was minimal and that the strategies we employed helped to maintain the integrity of our data as much as possible. In order to minimize the occurrence of missing visits and data in future studies, several strategies can be implemented, such as electronic reminders, home visits for patients who are unable to attend visits due to mobility issues, offering flexible scheduling options for follow-ups, and enhancing patient engagement through education about the study's importance and their role in it can foster a sense of commitment and responsibility [32–34].

In addition, it is important to note that we did not track the visual field as a parameter in this study. The primary focus was on reducing intraocular pressure, as the included patients already had significantly low visual acuity and visual field at the beginning of the study.

Furthermore, we have not investigated the number of corneal endothelial cells, which is a limitation. We currently plan to address this and investigate the gonioscopic position of the tube in the anterior chamber, assessing whether it is closer or farther from the corneal endothelium.

Analyzing the trends and patterns observed in this study suggests several underlying mechanisms and relationships. The high incidence of secondary glaucoma post-vitreoretinal surgery indicates a strong association between retinal interventions and subsequent glaucoma development. The significant reduction in IOP and medication dependency post-AGV implantation underscores the valve's effectiveness in managing refractory cases. Moreover, the complications profile highlights the need for meticulous surgical techniques and postoperative management to mitigate risks. The correlation between patient demographics and glaucoma incidence emphasizes the importance of tailored treatment approaches based on age and gender, acknowledging these as critical factors in disease progression and management outcomes.

The prognosis of patients after the 5-year postoperative period was generally positive, with sustained IOP control and reduced medication dependency. However, long-term monitoring is crucial to manage potential late-onset complications and to ensure the continued success of the AGV implantation.

The study highlights the complication rate in these groups and outlines various corrective procedures performed to manage these issues. The findings contribute to understanding the effectiveness and challenges of AGV implantation, suggesting future research directions to improve patient outcomes and reduce complication rates.

5. Conclusions

1. The mean number of glaucoma medications used postoperatively is lower; the average in the adult group decreased from 3.37 preoperatively to 1.21 postoperatively.
2. According to the success and failure criteria, a year after the AGV implant, success was recorded in 41 adult patients, with the success rate being 82% at one year, 68% after 3 years, and 60% after 5 years postoperative.
3. AGV is a solution not only for refractory glaucoma after classic surgery but can be used for secondary glaucoma after vitreoretinal surgery or for neovascular glaucoma refractory to drug therapy. It might be the first indication of antiglaucoma surgery in these patients, with a high success rate.
4. Subconjunctival anesthesia is a good option for glaucoma surgery, including AGV insertion, providing good mobility of the eye and no need for a traction suture.
5. AGV can be used as the first option of surgical treatment in secondary neovascular glaucoma, with classic trabeculectomy having a high risk of failure.
6. Even though it is a difficult procedure, with good surgical technique and careful and frequent patient follow-up, in the long run, AGV is a chance for patients with difficult glaucoma to keep their vision.

Author Contributions: Conceptualization, R.I.B.; methodology, N.A.; software, G.B.; validation, R.I.B., N.A. and C.D.; formal analysis, D.A.B.; investigation, D.G.D.; resources, I.T.T.; data curation, G.B. and L.F.L.; writing—original draft preparation, V.H.; writing—review and editing, V.H.; visualization, L.F.L.; supervision, R.I.B.; project administration, I.T.T. All authors have read and agreed to the published version of the manuscript.

Funding: This research received no external funding.

Institutional Review Board Statement: The study was conducted in accordance with the Declaration of Helsinki and approved by the Clinical Emergency Eye Hospital Bucharest Ethics Committee (protocol code 7).

Informed Consent Statement: Informed consent was obtained from all subjects involved in the study.

Data Availability Statement: The original contributions presented in the study are included in the article; further inquiries can be directed to the corresponding author/s.

Conflicts of Interest: The authors declare no conflicts of interest.

References

1. Barac, I.R.; Pop, M.; Balta, F. Refractory secondary glaucoma—Clinical case. *J. Med. Life* **2012**, *5*, 107–109. [PubMed]
2. Ong, S.C.; Aquino, M.C.; Chew, P.; Koh, V. Surgical outcomes of a second ahmed glaucoma valve implantation in Asian eyes with refractory glaucoma. *J. Ophthalmol.* **2020**, *2020*, 8741301. [CrossRef] [PubMed]
3. Jazzaf, A.M.; Netland, P.A.; Charles, S. Incidence and management of elevated intraocular pressure after silicone oil injection. *Eur. J. Gastroenterol. Hepatol.* **2005**, *14*, 40–46. [CrossRef]
4. Kang, Y.K.; Shin, J.P.; Kim, D.W. Long-term surgical outcomes of Ahmed valve implantation in refractory glaucoma according to the type of glaucoma. *BMC Ophthalmol.* **2022**, *22*, 270. [CrossRef] [PubMed]
5. Scott, I.U.; Alexandrakis, G.; Flynn, H.W.; Smiddy, W.E.; Murray, T.G.; Schiffman, J.; Gedde, S.J.; Budenz, D.L.; Fantes, F.; Parrish, R.K. Combined pars plana vitrectomy and glaucoma drainage implant placement for refractory glaucoma. *Arch. Ophthalmol.* **2000**, *129*, 334–341. [CrossRef]
6. Ishida, K.; Netland, P.A. Ahmed Glaucoma Valve Implantation in African American and White Patients. *Arch. Ophthalmol.* **2006**, *124*, 800–806. [CrossRef] [PubMed]
7. Christakis, P.G.; Zhang, D.; Budenz, D.L.; Barton, K.; Tsai, J.C.; Ahmed, I.I.; ABC-AVB Study Groups. Five-Year Pooled Data Analysis of the Ahmed Baerveldt Comparison Study and the Ahmed versus Baerveldt Study. *Arch. Ophthalmol.* **2017**, *176*, 118–126. [CrossRef] [PubMed]
8. Budenz, D.L.; Feuer, W.J.; Barton, K.; Schiffman, J.; Costa, V.P.; Godfrey, D.G.; Buys, Y.M.; Ahmed Baerveldt Comparison Study Group. Postoperative complications in the Ahmed Baerveldt comparison study during five years of follow-up. *Am. J. Ophthalmol.* **2016**, *163*, 75–82.e3. [CrossRef] [PubMed]
9. Gedde, S.J.; Chen, P.P.; Heuer, D.K.; Singh, K.; Wright, M.M.; Feuer, W.J.; Schiffman, J.C.; Shi, W.; Primary Tube Versus Trabeculectomy Study Group. The Primary Tube versus Trabeculectomy (PTVT) Study: Methodology of a multicenter randomized clinical trial comparing tube shunt surgery and trabeculectomy with Mitomycin C. *Ophthalmology* **2018**, *125*, 774–781. [CrossRef]
10. Gedde, S.J.; Feuer, W.J.; Shi, W.; Lim, K.S.; Barton, K.; Goyal, S.; Ahmed, I.I.K.; Brandt, J.; Primary Tube Versus Trabeculectomy Study Group. Treatment outcomes in the Primary Tube Versus Trabeculectomy Study after 1 year of follow-up. *Ophthalmology* **2018**, *125*, 650–663. [CrossRef]
11. Saheb, H.; Gedde, S.J.; Schiffman, J.C.; Feuer, W.J. Outcomes of glaucoma reoperations in the Tube Versus Trabeculectomy (TVT) Study. *Arch. Ophthalmol.* **2014**, *157*, 1179–1189.e2. [CrossRef]
12. Gedde, S.J.; Herndon, L.W.; Brandt, J.D.; Budenz, D.L.; Feuer, W.J.; Schiffman, J.C. Postoperative complications in the Tube Versus Trabeculectomy (TVT) Study during five years of follow-up. *Arch. Ophthalmol.* **2012**, *153*, 804–814.e1. [CrossRef]
13. Gedde, S.J.; Heuer, D.K.; Parrish, R.K., 2nd; Tube Versus Trabeculectomy Study Group. Review of results from the Tube Versus Trabeculectomy Study. *Curr. Opin. Ophthalmol.* **2010**, *21*, 123–128. [CrossRef]
14. Gedde, S.J.; Vinod, K.; Prum, B.E., Jr.; Primary Tube Versus Trabeculectomy Study Group. Results from the Primary Tube Versus Trabeculectomy Study and translation to clinical practice. *Curr. Opin. Ophthalmol.* **2023**, *34*, 129–137. [CrossRef]
15. Xie, Z.; Liu, H.; Du, M.; Zhu, M.; Tighe, S.; Chen, X.; Yuan, Z.; Sun, H. Efficacy of ahmed glaucoma valve implantation on neovascular glaucoma. *Int. J. Med Sci.* **2019**, *16*, 1371–1376. [CrossRef]
16. Ozdamar, A.; Aras, C.; Ustundag, C.; Tamcelik, N.; Ozkan, S. Scleral tunnel for the implantation of glaucoma seton devices. *Ophthalmic Surg. Lasers Imaging Retin.* **2001**, *32*, 432–435. [CrossRef]
17. Lee, E.S.; Kang, S.Y.; Kim, N.R.; Hong, S.; Ma, K.T.; Seong, G.J.; Kim, C.Y. Split-thickness hinged scleral flap in the management of exposed tubing of a glaucoma drainage device. *Eur. J. Gastroenterol. Hepatol.* **2011**, *20*, 319–321. [CrossRef]
18. Ishida, K.; Ahmed, I.I.; Netland, P.A. Ahmed Glaucoma Valve surgical outcomes in eyes with without silicone oil endotamponament. *J. Glaucoma* **2009**, *18*, 325–330. [CrossRef] [PubMed]
19. Minckler, D.S.; Francis, B.A.; Hodapp, E.A.; Jampel, H.D.; Lin, S.C.; Samples, J.R.; Smith, S.D.; Singh, K. Aqueous shunts in glaucoma: A report by the American Academy of Ophthalmology. *Ophthalmology* **2008**, *115*, 1089–1098. [CrossRef] [PubMed]
20. Hong, C.-H.; Arosemena, A.; Zurakowski, D.; Ayyala, R.S. Glaucoma drainage devices: A systematic literature review and current controversies. *Surv. Ophthalmol.* **2005**, *50*, 48–60. [CrossRef]
21. Arora, K.S.; Robin, A.L.; Corcoran, K.J.; Corcoran, S.L.; Ramulu, P.Y. Use of various glaucoma surgeries and procedures in medicare beneficiaries from 1994 to 2012. *Ophthalmology* **2015**, *122*, 1615–1624. [CrossRef] [PubMed]
22. Chen, P.P.; Yamamoto, T.; Sawada, A.; Parrish, R.K.; Kitazawa, Y. Use of antifibrosis agents and glaucoma drainage devices in the American and Japanese Glaucoma Societies. *J. Glaucoma* **2007**, *16*, 14–19. [CrossRef]
23. Siempis, T.; Younus, O.; Makuloluwa, A.; Montgomery, D.; Croghan, C.; Sidiki, S. Long-Term Outcomes of Ahmed Valve Surgery in a Scottish Cohort of Patients with Refractory Glaucoma. *Cureus* **2023**, *15*, e35877. [CrossRef] [PubMed] [PubMed Central]
24. Vinod, K.; Gedde, S.J.; Feuer, W.J.; Panarelli, J.F.; Chang, T.C.; Chen, P.P.; Parrish, R.K.I. Practice Preferences for Glaucoma Surgery: A Survey of the American Glaucoma Society. *Eur. J. Gastroenterol. Hepatol.* **2017**, *26*, 687–693. [CrossRef] [PubMed]

25. Bailey, A.K.; Sarkisian, S.R., Jr. Complications of tube implants and their management. *Curr. Opin. Ophthalmol.* **2014**, *25*, 148–153. [CrossRef] [PubMed]
26. Pakravan, M.; Esfandiari, H.; Yazdani, S.; Doozandeh, A.; Dastborhan, Z.; Gerami, E.; Kheiri, B.; Pakravan, P.; Yaseri, M.; Hassanpour, K. Clinical outcomes of Ahmed glaucoma valve implantation in pediatric glaucoma. *Eur. J. Ophthalmol.* **2019**, *29*, 44–51. [CrossRef] [PubMed]
27. Choo, J.Q.; Chen, Z.; Koh, V.; Liang, S.; Aquino, C.M.; Sng, C.; Chew, P. Outcomes and complications of ahmed tube implantation in Asian Eyes. *Eur. J. Gastroenterol. Hepatol.* **2018**, *27*, 733–738. [CrossRef] [PubMed]
28. Helzner, J. Xen-Augmented Baerveldt vs. Ahmed Glaucoma Valve. *Glaucoma Physician* **2020**, *24*, 44.
29. Baltă, F.; Dinu, V.; Zemba, M.; Baltă, G.; Barac, A.D.; Schmitzer, S.; Dragosloveanu, C.D.M.; Barac, R.I. Choroidal Thickness Increase after Subliminal Transscleral Cyclophotocoagulation. *Diagnostics* **2022**, *12*, 1513. [CrossRef]
30. Zemba, M.; Dumitrescu, O.-M.; Stamate, A.-C.; Barac, I.R. Micropulse Transscleral Cyclophotocoagulation for Glaucoma after Penetrating Keratoplasty. *Diagnostics* **2022**, *12*, 1143. [CrossRef]
31. Zemba, M.; Dumitrescu, O.-M.; Vaida, F.; Dimirache, E.-A.; Pistolea, I.; Stamate, A.C.; Burcea, M.; Branisteanu, D.C.; Balta, F.; Barac, I.R. Micropulse vs. continuous wave transscleral cyclophotocoagulation in neovascular glaucoma. *Exp. Ther. Med.* **2022**, *23*, 278. [CrossRef] [PubMed]
32. Zaharia, A.-C.; Dumitrescu, O.-M.; Radu, M.; Rogoz, R.-E. Adherence to Therapy in Glaucoma Treatment—A Review. *J. Pers. Med.* **2022**, *12*, 514. [CrossRef] [PubMed]
33. Subhan, I.A.; Alosaimy, R.; Alotaibi, N.T.; Mirza, B.; Mirza, G.; Bantan, O. Evaluation of Compliance Issues to Anti-glaucoma Medications before and after a Structured Interventional Program. *Cureus* **2022**, *14*, e25943. [CrossRef] [PubMed] [PubMed Central]
34. Oltramari, L.; Mansberger, S.L.; Souza, J.M.P.; de Souza, L.B.; de Azevedo, S.F.M.; Abe, R.Y. The association between glaucoma treatment adherence with disease progression and loss to follow-up. *Sci. Rep.* **2024**, *14*, 2195. [CrossRef]

Disclaimer/Publisher's Note: The statements, opinions and data contained in all publications are solely those of the individual author(s) and contributor(s) and not of MDPI and/or the editor(s). MDPI and/or the editor(s) disclaim responsibility for any injury to people or property resulting from any ideas, methods, instructions or products referred to in the content.

Systematic Review

Wound Modulations in Glaucoma Surgery: A Systematic Review

Bhoomi Dave [1,2], Monica Patel [1], Sruthi Suresh [1], Mahija Ginjupalli [1], Arvind Surya [1], Mohannad Albdour [3] and Karanjit S. Kooner [1,4,*]

[1] Department of Ophthalmology, University of Texas Southwestern Medical Center, Dallas, TX 75390, USA; bhd27@drexel.edu (B.D.); monica.patel2@utsouthwestern.edu (M.P.); sruthi.prabhasuresh@utsouthwestern.edu (S.S.); mahija.ginjupalli@utsouthwestern.edu (M.G.); arvindsurya52@gmail.com (A.S.)
[2] Drexel University College of Medicine, Philadelphia, PA 19129, USA
[3] Department of Ophthalmology, King Hussein Medical Center Royal Medical Services, Amman 11180, Jordan; drmohannadalbdour@yahoo.com
[4] Department of Ophthalmology, Veteran Affairs North Texas Health Care System Medical Center, Dallas, TX 75216, USA
* Correspondence: karanjit.kooner@utsouthwestern.edu

Abstract: Excessive fibrosis and resultant poor control of intraocular pressure (IOP) reduce the efficacy of glaucoma surgeries. Historically, corticosteroids and anti-fibrotic agents, such as mitomycin C (MMC) and 5-fluorouracil (5-FU), have been used to mitigate post-surgical fibrosis, but these have unpredictable outcomes. Therefore, there is a need to develop novel treatments which provide increased effectiveness and specificity. This review aims to provide insight into the pathophysiology behind wound healing in glaucoma surgery, as well as the current and promising future wound healing agents that are less toxic and may provide better IOP control.

Keywords: glaucoma surgery; wound healing; antifibrotic agents; anti-vascular endothelial growth factors; cytokine inhibitors; anti-LOXL2 monoclonal Ab; integrin inhibitors; growth factor inhibitors

Citation: Dave, B.; Patel, M.; Suresh, S.; Ginjupalli, M.; Surya, A.; Albdour, M.; Kooner, K.S. Wound Modulations in Glaucoma Surgery: A Systematic Review. *Bioengineering* 2024, 11, 446. https://doi.org/10.3390/ bioengineering11050446

Academic Editor: Hiroshi Ohguro

Received: 26 March 2024
Revised: 22 April 2024
Accepted: 26 April 2024
Published: 30 April 2024

Copyright: © 2024 by the authors. Licensee MDPI, Basel, Switzerland. This article is an open access article distributed under the terms and conditions of the Creative Commons Attribution (CC BY) license (https:// creativecommons.org/licenses/by/ 4.0/).

1. Introduction

Glaucoma is the second leading cause of blindness, affecting more than 80 million patients worldwide and over 3 million in the USA [1]. The prevalence of glaucoma is expected to double over the next 30 years, which will pose a major public health challenge [2]. An important modifiable risk factor is elevated intraocular pressure (IOP) due to the blockage of aqueous humor (AH) outflow [3]. Therefore, it is imperative to understand the mechanics and dynamics of AH production and outflow. The drainage of AH occurs mainly through the conventional pathway [trabecular meshwork (TM), Schlemm's canal (SC), collector channels, aqueous veins, and episcleral veins (EVs) 70%], as well as the non-conventional uveoscleral–uveovortex (US-UV) pathway, uveal meshwork, anterior face of the ciliary muscle through the muscle bundles, suprachoroidal space, and out through the sclera (30%), as in Figure 1A,B [4]. Though the dysfunction of the conventional pathway is not well understood, increased TM contractility, changes in extracellular matrix (ECM) composition, decreased pore density of the inner wall of SC, and disruption of local regulatory mediators may contribute to increased AH outflow resistance [5].

The initial conventional treatment options for controlling elevated IOP involve medications and laser procedures. If medications and laser treatment fail to lower IOP, the next step is performing incisional surgery, which includes trabeculectomy, trabeculotomy, glaucoma drainage devices (GDDs), and minimally invasive glaucoma surgeries (MIGS) [6–8]. Trabeculectomy is designed to remove a portion of the TM and SC to allow the flow of AH to the subconjunctival space [9]. Trabeculotomy, on the other hand, is performed either by an ab interno or ab externo approach, to perforate the TM. It is commonly used in children with congenital glaucoma, but rarely in adults. GDDs (Ahmed®, New World Medical Inc., Rancho Cucamonga, CA, USA; Baerveldt®, Advanced Medical Optics Inc., Santa Ana, CA, USA; and Molteno Ophthalmic Limited®, Dunedin, New Zealand), which consist of a tube and a plate, also drain the AH from the AC into the subconjunctival space, but with more controlled AH outflow and somewhat predictable outcomes. Recently, MIGS (Xen Gel Stent® Abbvie/Allergan Co. Dublin, Ireland, PreserFlo® MicroShunt [made of poly(styrene-b-isobutylene-b-styrene), or SIBS], Santen Pharmaceutical Company®, Osaka, Japan, iStent® Glaukos Corp. inject, Hydrus® Alcon microstents, Kahook® New World Medical dual-blade goniotomy, trabectome® MicroSurgical Technology, gonioscopy-assisted transluminal trabeculotomy, Glaucoma Associates of Texas, Trab 360 OMNI® Sight Sciences, Inc., Visco 360 OMNI® Sight Sciences, Inc., Ab interno canaloplasty (ABiC) Ellex Australia, and Streamline® New World Medical Inc. surgical system have gained popularity due to their relatively quick insertion and lesser tissue manipulation [6–8]. A 2017 survey, conducted to assess surgical practice patterns among members of the American Glaucoma Society (AGS), showed a significant increase in the use of GDDs since 1997 [8]. Trabeculectomy remains the procedure of choice, with higher mean percentages of use (59% ± 30%), followed by GDDs (23% ± 23%) and MIGS (14% ± 20%) [8].

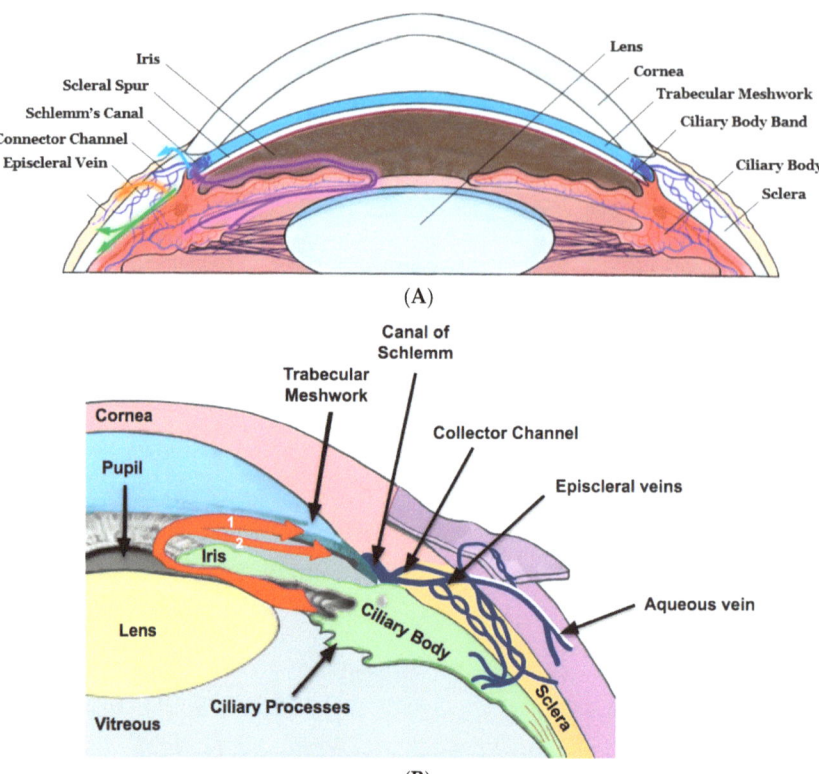

Figure 1. (**A**) Cross-section of an eye illustrating the AH flow dynamics. AH is formed by the ciliary body and flows through the pupil into the anterior chamber (AC). The drainage of AH is mainly via the conventional [TM, SC, and EV] pathway and the non-conventional [US-UV] pathway. (**B**) Higher magnification of (**A**). Red arrow #1 denotes AH flow from the trabecular meshwork through the Schlemm's canal, collector channels, aqueous veins, and into the episcleral veins for drainage into the bloodstream. The uveoscleral pathway (red arrow #2) shows AH flowing directly through the ciliary muscle to the suprachoroidal space, and out through the sclera, eventually reaching general circulation.

A significant postoperative complication of all incisional glaucoma surgeries is a vigorous fibroproliferative response leading to the blockage of AH outflow in the subconjunctival space ("ring-of-steel"), leading to inadequate control of IOP and surgical failure. Therefore, modulating the wound healing process is critical for optimal outcomes in the surgical management of glaucoma [10], hence the reason for this review.

2. Materials and Methods

2.1. Initial Search (Figure 2)

We followed the standards outlined by the Preferred Reporting Items for Systematic Reviews and Meta-Analyses (PRISMA) guidelines during data collection, and the PICOS (Population, Intervention, Comparison, Outcomes and Study) framework was used to create eligibility criteria, Table 1 [11]. We used keywords and MeSH terms, such as "glaucoma" (or "Glaucoma, Angle-Closure" or "Glaucoma, Open-Angle"), "glaucoma wound healing" or "glaucoma filtration surgery" (or "sclerostomy," "trabeculectomy", or "GDDs"), "anti-inflammatory agents" (or "antifibrotic agents").

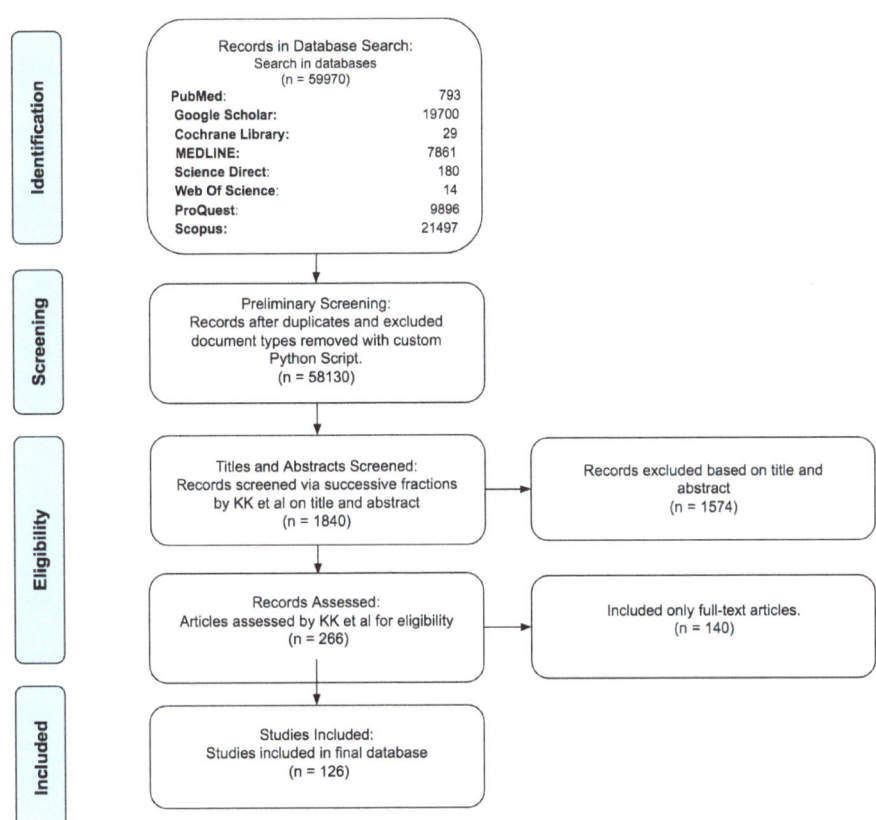

Figure 2. PRISMA Flowchart illustrating the selection process for this systematic review.

Using these terms, we systematically searched the online databases of PubMed (MEDLINE), Cochrane Library, ScienceDirect, Scopus, Google Scholar, ProQuest, and Web of Science up to June 2023. The records from the different databases were compiled in a comma-separated values (CSV) file on Google Sheets.

Table 1. PICOS Criteria for Inclusion of Studies.

Parameter	Description
Population	Patients with glaucoma regardless of the site.
Intervention	Incisional/filtration glaucoma procedures, with or without antifibrotic agents.
Comparison	Results of patients who underwent glaucoma surgery with and without antifibrotic agents.
Outcomes	Quality of IOP control, postoperative complications, visual acuity.
Study Design	Randomized or nonrandomized controlled (or uncontrolled).

2.2. Preliminary Screening

We excluded non-English articles and study types, such as conference abstracts, commentaries, and duplicate papers with the same digital object identifier (DOI) using a script written in the Python programming language (Python Software Foundation, Wilmington, DE, USA, version 3.12.2). The selected manuscripts were then stored in the CSV for eligibility assessment and included information on authors, title, date of publication, journal, and DOI.

2.3. Eligibility Assessment

The four reviewers (KK, SS, MP, and BD) screened every article in the CSV for accuracy and best fit. We included full-text English articles, studies involving animal or human subjects, and clinical trials.

3. Results

After multiple rounds of screening, 126 studies were included in our review. The selected studies discussed the basic principles, development, and applications of wound healing modulation in glaucoma surgery. Figure 2 depicts the eligibility assessment process.

3.1. Overview of the Wound Healing Process

Understanding the conjunctival, episcleral, and scleral wound healing process is critical to evaluate wound healing modulation in glaucoma surgery. The wound repair process can be divided into four key phases: hemostasis, inflammation, proliferation, and remodeling (Figure 3) [12].

3.1.1. Hemostasis

In the first stage, hemostasis, a platelet plug forms to prevent excessive blood loss. This is achieved through activation of the clotting cascades, which begins with vasoconstriction initiated by the release of thromboxane (TXA2) and endothelin-1 from the damaged endothelium [13]. Following this, the interaction of platelet receptors and ECM proteins (collagen, elastin, fibronectin) occurs to promote adherence to the walls of the surrounding blood vessels [14]. Once the platelet receptors adhere to the blood vessels, thrombin is promoted to activate platelets and release granules, which reinforce the coagulation process [15]. Concurrently, platelets produce platelet-derived growth factor (PDGF), which activates endothelial cells to repair damaged vasculature through angiogenesis. Once completed, the hemostasis phase is downregulated by inhibitors, such as activated protein C, prostacyclin, and antithrombin III [15].

3.1.2. Inflammation

The second phase, inflammation, consists of the recruitment of immune cells designed to remove necrotic tissue and pathogens [10]. This phase of wound healing is initiated by the release of damage-associated molecular patterns (DAMP) molecules, pathogen-associated molecular patterns (PAMP) molecules, hydrogen peroxide (H_2O_2), lipid mediators, and chemokines from injured cells [16]. DAMPs are endogenous molecules consisting of DNA, peptides, ECM components, and ATP to activate the innate immune system, while PAMPs work to activate immune cells and release pro-inflammatory cytokines. Both patterns share a common goal of attracting leukocytes to the injured tissue. These modulators lead to the influx of immune cells, specifically neutrophils and leukocytes [17]. Once neutrophils are generated in the bone marrow, they are attracted to the site of injury by the "find me" signals from chemoattractants (molecules that promote movement), including DAMPs, H_2O_2, lipid mediators, and chemokines [17]. After traveling to the wound from damaged vessels, neutrophils remove necrotic tissue, and pathogens trap and kill pathogens with extracellular traps, resulting in wound decontamination [17]. Monocytes work by differentiating into macrophages with variable phenotypes, ultimately initiating the macrophage inflammatory response and further augmenting it by attracting additional monocytes [18].

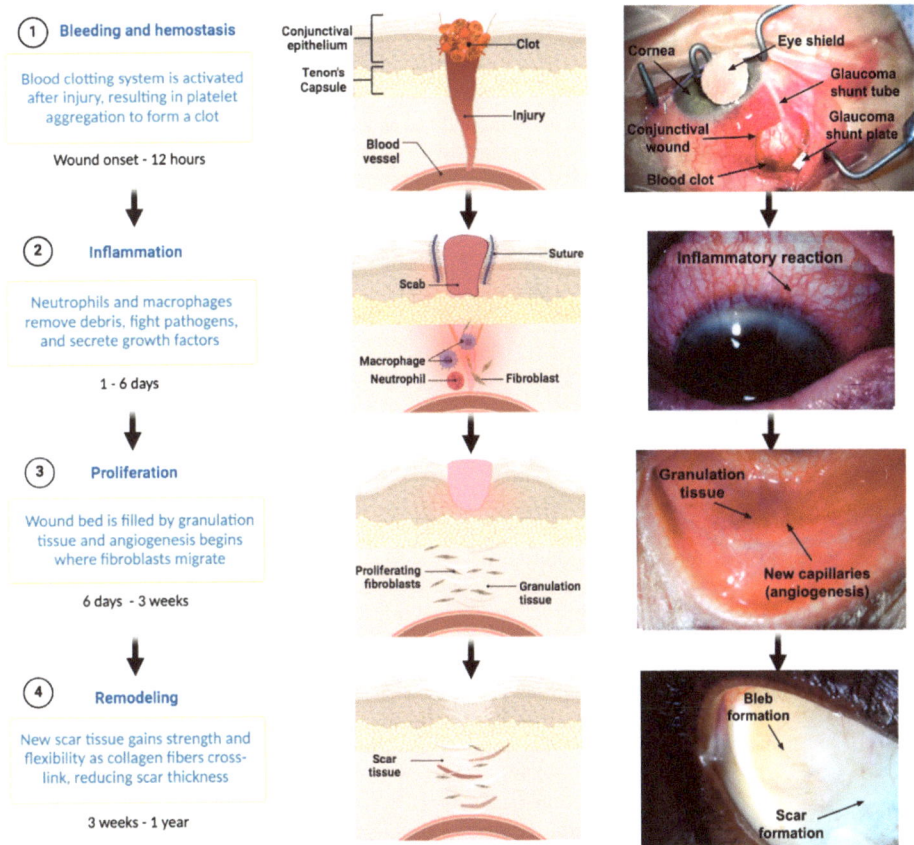

Figure 3. An overview of the chronology seen in the general healthy wound healing process in the eye: From left to right, this figure shows (1) hemostasis, (2) inflammation, (3) proliferation, and (4) remodeling.

3.1.3. Proliferation

The proliferative phase is characterized by wound closure and is essential to wound healing. Proliferation may occur as early as 12 h post-injury, resulting in the formation of highly vascularized granulation tissue. This newly formed tissue allows for ECM synthesis and the activation of fibroblasts. This process occurs simultaneously with neovascularization and immunomodulation, contributing to wound contraction [16]. Wound contraction occurs when myofibroblasts grip the wound edges and pull them together [19]. Microvascular endothelial cells (ECs) lining blood vessels are central to neovascularization. Their activation relies on growth factors (a bioactive molecule released into the environment which affects cell growth) produced by nearby cells, and the production of proteolytic enzymes (matrix metalloproteinases (MMPs), disintegrins, and metalloproteinases) facilitates their navigation through the fibrin/fibronectin clot. ECs initiate angiogenesis by sprouting in response to pro-angiogenic signals (VEGF, FGF, PDGF-β, TGF-β) and angiopoietins, leading to proliferation and migration [15]. The new granulation tissue typically exhibits a red or pink color, attributed to the presence of new blood vessels and other inflammatory agents. The color and condition of the granulation tissue serve as indicators of the progress of wound healing. On the other hand, the dark granulation tissue is an evidence of poor perfusion, ischemia, or infection. This phase of wound healing can span from six days to up to three weeks or longer [14].

3.1.4. Remodeling

In the final phase, remodeling, the granulation tissue is gradually replaced by normal connective tissue. This stage involves a decrease in tissue cellularity due to the massive apoptosis of fibroblasts, myofibroblasts, endothelial cells, and pericytes (cells that are embedded within the vessel wall endothelium) [20]. Integrins play a key role in facilitating cell attachment to the ECM. They have the ability to trigger the activation of latent transforming growth factor beta-1 (TGF-β1), which in turn regulates the processes of wound inflammation and the formation of granulation tissue [21]. The accumulation of ECM molecules, specifically collagen, is a hallmark of the remodeling phase. Type 3 collagen is converted to Type 1, which is a more mature and stiff form. This increases the tensile strength and elasticity of the healed tissue. Although collagen deposition restores most of the strength in the affected tissue, it is estimated that the new scar tissue is 20% weaker and less elastic than pre-injured tissue [22].

3.2. Fibrosis

Fibrosis is the excessive accumulation of connective tissue and ECM components [23]. In the healthy remodeling phase of wound healing, fibrosis is minimal. However, pathological fibrosis can result from an overly aggressive and unchecked healing response secondary to significant tissue injury, poor wound control, predisposed demographics, or an existing immunocompromised patient. In pathologic conditions, such as excessive conjunctival fibrosis, the normally efficient and orderly remodeling phase of wound healing is lost, and the conjunctival epithelium undergoes a state of chronic inflammation characterized by uncontrolled growth factor signaling (Figure 4).

TGF-β, released from macrophages, is the cardinal growth factor involved in the progression of fibrosis during wound healing [24]. Upon TGF-β stimulation, fibroblasts are activated and undergo transition into myofibroblasts, the key effector cells in fibrotic states [24]. Myofibroblasts in the conjunctiva are called Tenon fibroblasts; they augment fibrosis by depositing connective tissue, producing cross linking enzymes, and releasing MMPs during the proliferative stage of wound healing [25,26]. In a normal physiologic state, this process ends with the apoptosis of the myofibroblasts and the cessation of inflammation. However, acceleration into excessive fibrosis is mediated by exaggerated levels of various growth factors and cytokines, including TGF-β, interleukins, such as IL-1, IL-6, IL-10, and PDGF, as illustrated in the graphical abstract and in Figure 3 [27]. These factors ultimately lead to uncontrolled myofibroblast activation and, thus, pathologically excessive deposition of ECM [24].

In regard to glaucoma surgery, uncontrolled postoperative fibrosis is the main cause of procedure failure, resulting in excessive scarring, visual impairment, and subsequent progression of glaucoma. Table 2 (adapted from Yamanka et al., 2015) includes antifibrotic cytokines, growth factors, and signaling pathways relevant to preventing ocular fibrosis [28].

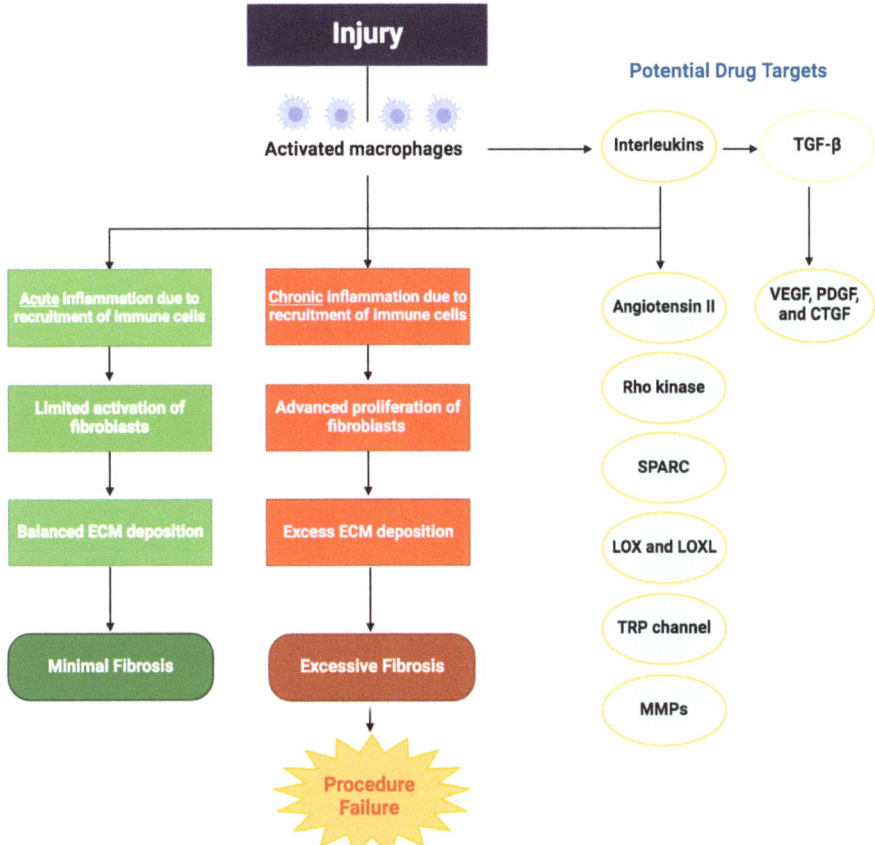

Figure 4. Flow diagram showing two distinct healing outcomes of fibrosis: minimal or excessive fibrosis after glaucoma surgery, as well as potential drug targets.

Table 2. Antifibrotic targets and their mechanism of action, adapted from Yamanka et al. [28].

Antifibrotic Targets	Mechanism of Action	Applications
IL-1 [29]	IL-1 controls integrin expression in leukocytes and endothelial cells.	1-methyl hydrazino analogs are an excellent IL-1 blocker and reduce inflammation.
IL-6 [28,30]	IL-6 stimulates B-cell differentiation, T-cell activation, and immunoglobulin production.	Tocilizumab is an anti- IL-6 receptor antibody, which, in a rheumatoid arthritis clinical study, reduced inflammation and fibrosis.
IL-7 [31,32]	IL-7 is a profibrotic growth factor and activates signaling that suppresses fibroblast-driven ECM expression.	In a septic shock trial, IL-7 application restored CD4+ and CD8 cell count.
IL-10 [33–35]	IL-10 is an anti-inflammatory cytokine which reduces production of inflammatory cytokine mRNA.	In a mice study, IL-10 increased the number of neutrophils and monocytes.

Table 2. Cont.

Antifibrotic Targets	Mechanism of Action	Applications
IL-22 [36–38]	IL-22, a pro-inflammatory cytokine, upregulates acute phase proteins.	In a hepatitis clinical trial, IL-22 protected against epithelial cell injury and reduced inflammation.
Anti-VEGF [39]	VEGF is a potent mediator of angiogenesis, vasculogenesis and vascular endothelial cell permeability.	Anti-VEGF therapies inhibit vascular endothelial growth factor, thus preventing angiogenesis and the disruption of the blood–retinal barrier.
Platelet-derived growth factor (PDGF) [40]	The PDGF family consists of disulphide-linked dimers and induces proliferation of macrophages and fibroblasts migration into a wound site.	ARC126 and ARC127 are PDGFβ inhibitors, and they reduced both epiretinal membrane formation and retinal detachment.
Connective tissue growth factor (CTGF) [41]	CTGF is a fibrogenic cytokine upregulated by TGF-β causes persistent fibrosis through CTGF.	Targeting either CTGF or TGF-β signaling may reduce scar tissue formation.
Matrix metalloproteinases (MMPs) [28,42]	MMPs are a group of proteolytic enzymes which degrade most extracellular matrix proteins during wound remodeling.	Administration of GM6001, an MMP inhibitor, reduced scar formation after glaucoma surgery in rabbits.
Lysyl oxidase (LOX) and lysyl oxidase-like proteins (LOXL) [28,43]	Lysyl oxidase (LOX) and lysyl oxidase-like (LOXL) are ECM enzymes which crosslink collagen and elastin, leading to fibrosis.	Anti LOXL2 monoclonal antibody (GS-607601) reduced inflammation and fibrosis after glaucoma surgery in rabbits.
Rho kinase inhibitors [28,44]	ROCK 1 and 2 are downstream components of Rho-GTPase Rho mediated signaling and play an important role in cytoskeletal organization controlling cellular morphology migration and motility. Rac1 is a low-molecular-weight Rho GTPase.	In a lab experiment, inhibiting Rac1 with NSC23766 or siRNA achieved reduction in conjunctival tissue fibrosis and collagen matrix contraction.
Secreted protein acidic and rich in cysteine (SPARC) inhibitors [28,45]	SPARC is a 43 kDa collagen-binding matricellular glycoprotein that modulates cellular interactions with the surrounding ECM. SPARC contributes to ECM organization and cell migration.	In an in vitro experiment, SPARC knockdown resulted in TGFβ2-driven upregulation of Type I collagen, and fibronectin expression was suppressed. Reducing SPARC expression may suppress subconjunctival fibrosis.
Angiotensin II [28,46]	Angiotensin II is an effector molecule and causes ocular fibrosis. Activation of NF-κB by angiotensin II leads to the survival of corneal myofibroblasts, and, consequently. fibrosis.	In lab experiments, angiotensin-converting enzyme inhibitors (ACE II s) and angiotensin receptor (AT2) antagonists effectively suppressed vascular damage.
Transient receptor potential (TRP) channel antagonists [28,47]	The TRP channels are activated by multiple endogenous and external stimuli and mediate several wound healing functions. Their receptor-induced responses include cell proliferation and migration, along with immune cell activation, tissue infiltration, and fibrosis.	In an alkali-burn mouse wound healing model, treatment with a TRPV1 antagonist effectively suppressed fibrosis. Additionally, in vitro experiments using ocular fibroblasts demonstrated that the TRPV1 antagonist inhibited the transdifferentiation of myofibroblasts.
Transforming growth factor-β (TGF-β) inhibitor [28,48]	TGF-β plays a significant role as an effective mediator in the development of scar tissue in the eye.	In lab experiments, tranilast suppressed TGF-β activation and resulted in the suppression of collagen production. In vitro experiments using siRNA to suppress the TGF-β type II receptor gene demonstrated both suppression of fibronectin production and inhibition of cell migration.

3.3. Wound Healing in Trabeculectomy

A trabeculectomy is a common filtering procedure performed in glaucoma patients [49]. A conjunctival incision is made followed by a partial-thickness scleral flap to expose the trabecular meshwork (Figure 5A). The AC is inserted, and a block of trabecular meshwork and SC are excised. After performing a localized iridectomy (Figure 5B), the scleral flap is reattached using interrupted sutures. Sponge-soaked MMC is applied under the conjunctiva for variable time followed by irrigation with a balanced salt solution. Figure 5C shows an ultrasound biomicroscopy (UBM) of a patient after trabeculectomy.

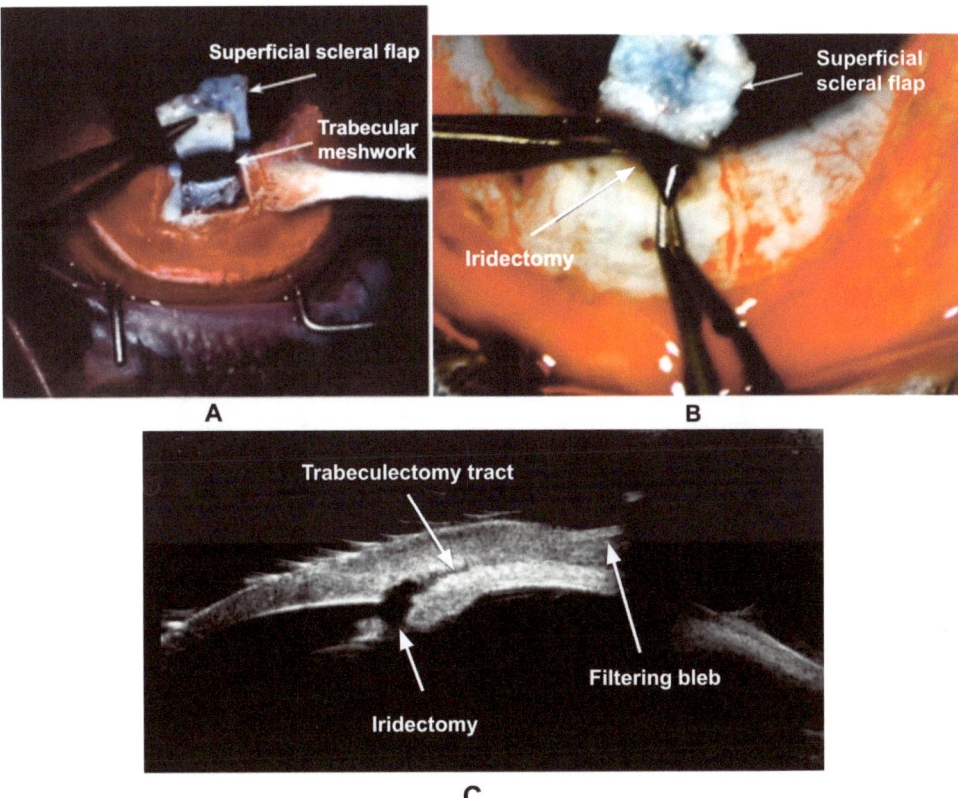

Figure 5. (**A–C**). Steps of trabeculectomy: conjunctival incision, superficial scleral flap, removal of trabecular meshwork/SC block, and iris (iridectomy). (**C**) shows an ultrasound biomicroscopy (UBM) after trabeculectomy.

The AH flows under the scleral flap into the subconjunctival space, forming an aqueous humor reservoir commonly known as a filtering bleb. The formation of a shallow filtering bleb versus a large cystic bleb (which may restrict the flow of AH) is preferable for an optimal reduction in IOP. Unlike other ocular surgeries (cataract, retinal), where complete healing and restoration of the incised tissue is desired, the success of trabeculectomy depends on the optimal flow of AH under the scleral flap [28].

A trabeculectomy bleb undergoes four phases of postoperative wound healing. The first phase consists of an immediate inflammatory response and involves the recruitment of inflammatory cells, such as cytokines and growth factors. These inflammatory cells lay the foundation for the second phase, which involves the formation of highly vascularized granulation tissue, proliferation, and tissue repair. This phase may last through the second or third month after the surgery.

The third phase involves the activation, migration, and proliferation of episcleral fibroblasts, angiogenesis, and formation of collagen bundles. The final remodeling phase is characterized by the contraction of collagen bundles and scar tissue formation. The latter may impede flow of AH and its final absorption in the subconjunctival space. While healing under the scleral flap is important, the fibroblasts in the Tenon's capsule are the main effector cells in the initiation and mediation of trabeculectomy wound healing and fibrotic scar formation [50].

The failure of glaucoma filtration surgery is mainly due to excessive subconjunctival wound fibrosis. Therefore, suppression of wound fibrosis is critical to maintain the smooth flow of AH [50]. Though the use of antifibrotic agents, such as MMC, have increased the success rate, there are still a number of complications, such as cystic blebs, dysaesthesia, wound leaks, blebitis, and endophthalmitis, which present challenges (Figure 6).

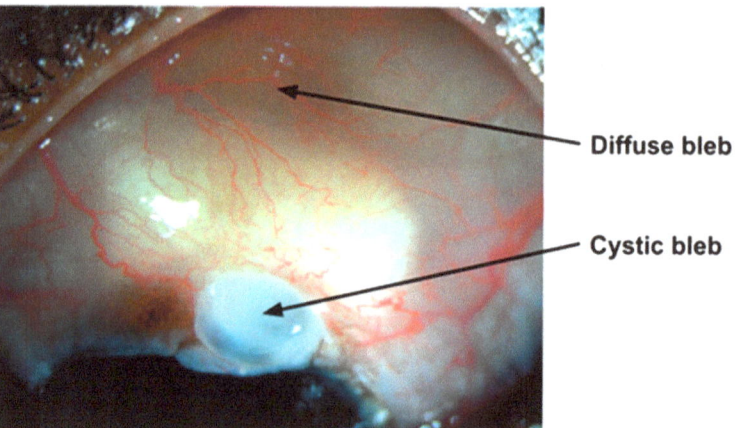

Figure 6. A cystic bleb at the limbus and a diffuse bleb formed after trabeculectomy.

3.4. Wound Healing after Glaucoma Drainage Devices (GDDs) and Bleb-Forming MIGS

In certain high risk patients, such as those who had previously undergone a trabeculectomy, secondary glaucoma, or have African American heritage, the use of glaucoma drainage devices (GDDs) is preferred. The following section will focus on the wound healing process after commonly used GDDs and bleb-forming MIGS. Understanding the complex wound healing process following a GDD or bleb-forming MIGS procedure is crucial, since the healing success depends largely on how the eye responds after surgery [51]. In a GDD procedure, an implant is selected, which shares common features consisting of a biocompatible silicone tube and a plate of varying size that is positioned in the subconjunctival space [51]. Likewise, bleb-forming MIGS channel AH from the AC into the subconjunctival space (Figure 7A,B).

The tissue trauma caused by the aforementioned GDDs and bleb-forming MIGS (peritomy, cauterization, and suturing of the patch grafts) leads to the release of plasma proteins and other inflammatory cells, such as neutrophils, macrophages, and fibroblasts [52]. Additionally, AH has inflammatory properties and is known to contain growth factors (VEGF, FGF, PDGF-β, TGF-β) that can lead to a brisk fibrotic response in the subconjunctival space [53]. A 2013 study verified the presence of TGF-β2 in glaucomatous AH and also identified notably higher levels of chemokine (C-C motif) ligand 2 (CCL2; MCP-1) [54]. Controlling the inflammation caused by these factors is crucial to the success of GDDs and bleb-forming MIGS, since inflammation surrounding the endplate or AH outflow is the leading cause of implant failure [55]. The first-line treatment for decreasing postoperative inflammation is the use of corticosteroids, both topical and oral. Corticosteroids achieve their anti-inflammatory effects primarily by interfering with pro-angiogenic signal transduction pathways [56]. For example, a commonly used synthetic corticosteroid, dexamethasone, is an extremely strong anti-inflammatory agent, with effects up to six times more potent than prednisolone or triamcinolone and twenty-five times more than hydrocortisone [57]. These corticosteroids and broad-spectrum antibiotics are commonly administered subconjunctivally at the conclusion of the procedures [58]. Furthermore, topical application of corticosteroids is continued for 2–3 months following surgery to maintain a decreased inflammatory response.

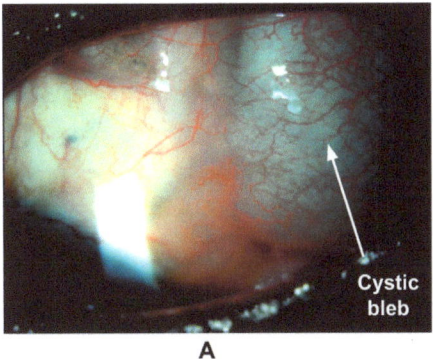

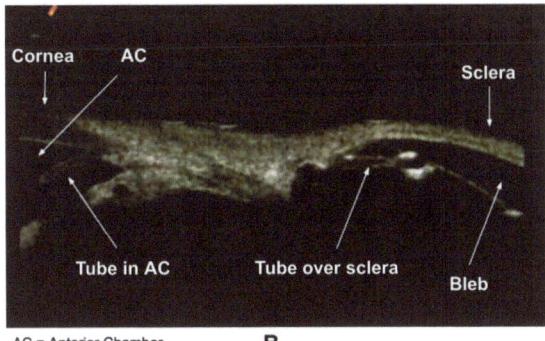

Figure 7. (**A**) A large, encysted bleb superolaterally in the left eye, formed after the insertion of a GDD. (**B**) Ultrasound biomicroscopy of the anterior segment shows the tip of the GDD in the anterior chamber. Posteriorly, the GDD tube is seen laying on the sclera, and a large filtering bleb is clearly visible.

As AH flows into the subconjunctival space, an excessive fibrotic reaction in the filtering bleb may result in bleb failure. The resultant encapsulation of the bleb impedes the AH outflow, resulting in elevated IOP [52]. The use of antimetabolites, namely MMC and 5-FU, have been efficacious in decreasing fibroblast proliferation following trabeculectomy, but their use in GDDs and bleb-forming MIGS is not widely accepted [59]. Some studies highlight the usage of MMC in the success of bleb-forming MIGS, but the benefits of MMC to GDD procedures remains unproven [60,61]. A 1995 study by Perkins et al. showed that while use of MMC with a double-plate Molteno implant showed a one-year success rate of 85% versus 20% in the control eyes, the two-year success rates were comparable for both groups [62,63]. A couple of years later, Lee et al. and Cantor et al. both concluded that adjunct use of MMC with Molteno implants did not offer significantly different outcomes from control groups at one-year post-surgery [64,65]. These studies showed a significantly higher incidence of complications in the MMC groups, including flat ACs and choroidal effusions. Additionally, a 2009 study demonstrating the adjunct use of MMC with the Ahmed glaucoma valve in infants with mostly primary congenital glaucoma (54.8%) or aphakic glaucoma (16.1%) showed that the MMC group had a significantly shorter bleb survival versus the control [66]. Currently, in Xen Gel Stent® or PreserFlo® MicroShunt procedures, surgeons either inject MMC or use MMC-soaked sponges [67]. However, it is still not commonplace to administer MMC during a GDD procedure.

MMC is potentially cytotoxic and may be associated with avascular and cystic blebs that are prone to complications, such as hypotony, blebitis, and endophthalmitis [68]. For this reason, there is a lot of interest in exploring the usage of different antimetabolites during MIGS. For example, in animal studies, valproic acid (VPA) has been used as an adjunct antifibrotic agent during implantation of the PreserFlo® MicroShunt [69]. This study demonstrated that postoperative subconjunctival injections of VPA yielded significantly better outcomes than the control group treated with phosphate buffered saline. After two weeks post-surgery, the control group blebs failed, whereas the VPA group maintained diffused, fluid-filled blebs visible up to 28 days. Histology showed that in the VPA-treated groups, the subconjunctival stromal matrix was made of loosely arranged and thin criss-crossed ECM fibers, compared to the thick, disorganized fibers in the control group. This suggests that VPA improves bleb functionality by facilitating a less dense connective tissue structure. Additionally, VPA was found to suppress collagen and fibronectin gene expression, while enhancing the expression of factors disrupting TGF-β pathways. Another study comparing the concomitant usage of VPA and MMC with varying doses of MMC in a rabbit model of the PreserFlo® MicroShunt found that the combination therapy was less cytotoxic when compared to MMC alone [70]. Moreover, the combination decreased VEGF and collagen gene expression more than MMC alone was able to. Together, these findings suggest that the usage of VPA as an antimetabolite in MIGS may reduce toxicity while more effectively managing the fibrotic response following implantation.

Although the use of steroids and antimetabolites is an integral aspect of managing inflammation and fibrosis in GDD and bleb-forming MIGS procedures, the biocompatibility of materials used in implants also plays a role in modulating wound healing. Most modern glaucoma devices are constructed from polypropylene (PP) and silicones, but their hydrophobic nature can lead to protein buildup and fibrosis [71]. To combat these complications, other materials, like gelatin and SIBS, have been innovatively used in the creation of the Xen Gel Stent® and PreserFlo® MicroShunt, respectively. Gelatin is a protein derived from collagen, and it is crosslinked with glutaraldehyde (GTA) to create the hydrophilic tube used in the Xen Gel Stent® [72]. This combination of materials resulted in a stable implant that showed no signs of hydrolytic degradation. Moreover, implantation of these materials does not cause significant inflammation or a foreign-body tissue reaction [73]. In fact, in an early-stage pilot study, a collagen stent placed into the subconjunctival space without connecting to the AC or allowing AH flow, showed no fibrosis around it after six months [72]. However, a 2010 investigation comparing gelatin hydrogels cross-linked with GTA to those with 1-ethyl-3-(3-dimethyl aminopropyl)carbodiimide (EDC) in rat iris pigment epithelium revealed that the EDC-treated groups exhibited lower levels of cytotoxicity, IL-1β, and TNF-α levels than GTA-treated ones. Furthermore, GTA groups demonstrated significant inflammation, suggesting EDC as a biocompatible alternative for GTA. However, further research is needed for its application in glaucoma implants. In addition, a 2006 study examining the usage of SIBS in a drainage implant instead of silicone demonstrated noncontinuous collagen deposition with no macrophages or myofibroblasts visible around the SIBS tube versus collagen deposition and myofibroblast differentiation induced by silicone [74]. A study conducted in 2022 involving fifteen New Zealand White rabbits that were implanted with PreserFlo® MicroShunts revealed the presence of a wide variety of cells, including polymorphonuclear leukocytes, myofibroblasts, and foreign body giant cells within the bleb and around the microshunt postoperatively [75]. These findings suggest that although the implantation of the SIBS MicroShunt has been efficacious as a bleb-forming MIGS, the presence of certain fibrotic factors may affect long-term outcomes.

Despite the innovation of new postoperative treatments and biocompatible implant materials, fibrosis continues to be a limiting factor in many glaucoma surgeries. Thus, further studies are needed to continue research on novel antifibrotic drugs and materials.

3.5. Current Glaucoma Wound Healing Agents

A common surgical complication after glaucoma surgery is the formation of scarring, which impedes the flow of AH. Therefore, treatment modalities have focused on reducing fibroblast production in order to decrease fibrosis postoperatively [76]. In the early 1990s, MMC and 5-FU were tested, and both showed high effectiveness [77].

MMC is a natural alkaloid synthesized from *Streptomyces caespitosus*, a species of actinobacteria [78]. It reduces fibroblast collagen synthesis by inhibiting DNA-dependent RNA synthesis and inducing DNA crosslinking (Figure 8) [35]. The crosslinked DNA segments block key DNA metabolism steps, including the replication and transcription of fibroblasts, which reduces collagen deposition and ultimately decreases the extent of scar formation at the subconjunctival site [79]. As MMC is most efficiently converted to its active form in Tenon's fibroblasts compared to fibroblasts from other parts of the body, it is widely used as an agent of choice during filtration surgery. In a 1992 study on human Tenon's capsule tissue, MMC administration led to the inhibition of fibroblast proliferation by 31.3% [78]. Additionally, MMC is significantly more potent than 5-FU, and is currently the agent of choice [78].

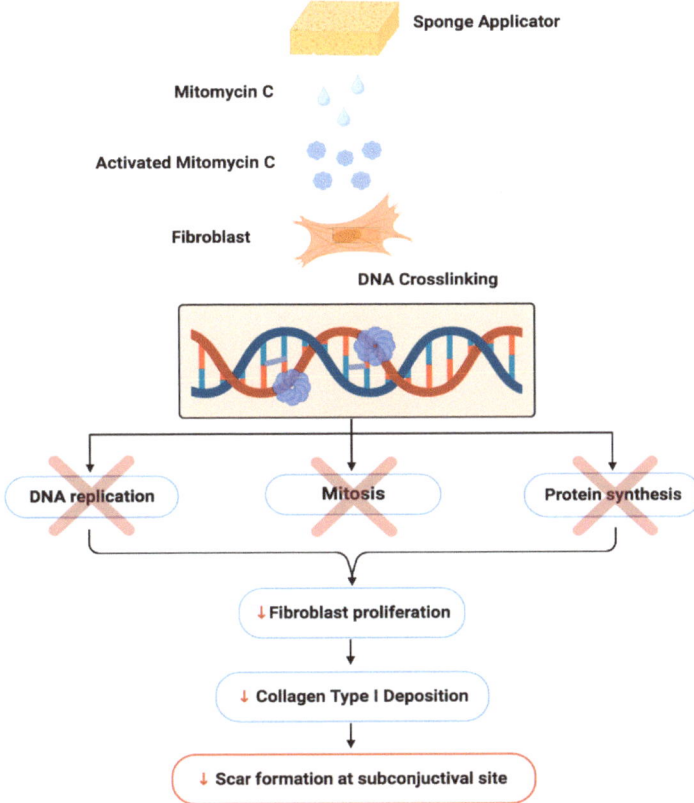

Figure 8. The mechanism of action and effects of mitomycin C.

5-FU is a pyrimidine analog that selectively inhibits both DNA and RNA synthesis, thus halting cellular proliferation and inducing direct cytotoxicity [80]. It is converted to three primary active metabolites: fluorodeoxyuridine monophosphate (FdUMP), fluorodeoxyuridine triphosphate (FdUTP), and fluorouridine triphosphate (FUTP), as shown in Figure 9. Its conversion to FdUMP forms a stable complex with an enzyme called thymidylate synthase, which inhibits DNA replication and repair [80]. In a 2008 study assessing 5-FU's use as an antimetabolite during trabeculectomy, it was shown to significantly reduce the risk of surgical failure in patients undergoing initial trabeculectomy, with a success rate of 81.6% (compared to 20.4% in controls) after 6 months [80].

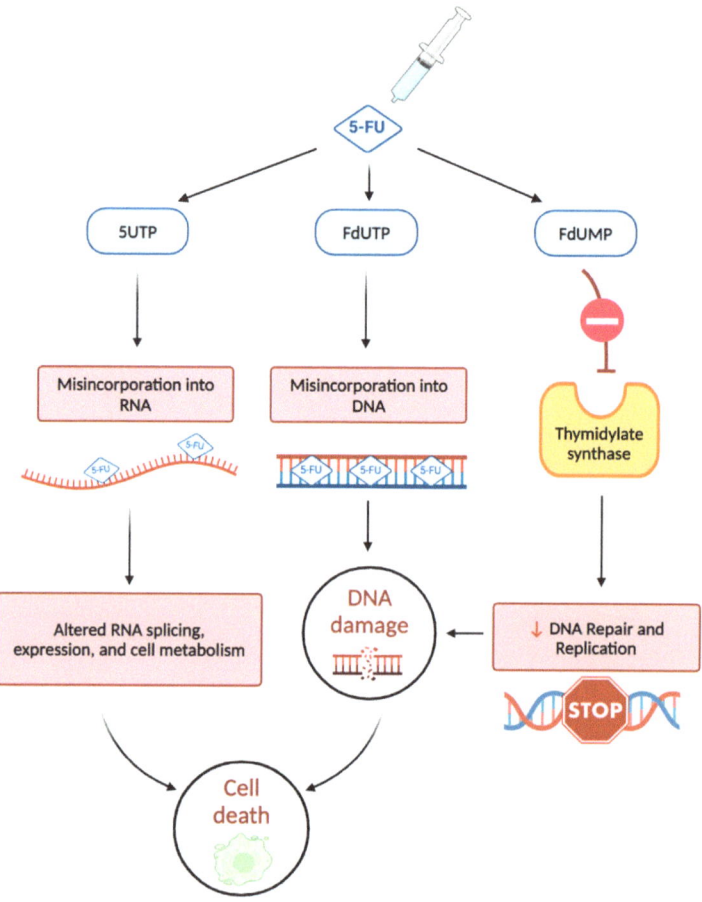

Legend:
5-FU = 5-Fluorouracil;   **5UTP** = 5 Uridine-5'-triphosphate;   **FdUTP** = 5-fluorodeoxyuridine triphosphate;
FdUMP = Fluoro-deoxyuridine monophosphate

Figure 9. The mechanism of action and effects of 5-fluorouracil.

It is well established that the usage of 5-FU and MMC has significantly improved success rates in glaucoma surgery [77]. However, these agents can cause widespread cell death, which increases the risk of several complications, such as prolonged subconjunctival hemorrhage and the formation of thin-walled avascular blebs that are prone to leakage and infection [81]. Therefore, the search for less toxic antifibrotic agents is crucial in reducing postoperative complications.

Secondly, controlling inflammation after glaucoma surgery is also of utmost importance for bleb survival. Topical corticosteroid agents have been used to control inflammation in the postoperative period [82]. They are thought to stimulate a steroid receptor in the nucleus of each cell, resulting in the widespread modification of up to 6000 genes within a few hours of its exposure [83]. Their anti-inflammatory property is largely mediated by the suppression of leukocyte concentration and vascular permeability (characterized by the inflammatory phase of wound healing). Consequently, this leads to decreased local tissue damage, reduced release of pro-fibrotic mediators, and less production of fibrin clots (involved in the hemostasis stage of wound healing) [84]. Broadway et al. were the first to show a significant reversal in macrophages, lymphocytes, and mast cells of conjunctival tissues after one month of preoperative steroids; their surgical success rates were also improved from 50% to 81% [85].

In some patients, steroid response (elevated IOP) is a significant side effect after prolonged topical corticosteroid usage. Its prevalence is approximately 18% to 36%, but it has been reported to be as high as 92% in patients with POAG [86,87]. Thus, clinicians must be watchful for elevated IOP after corticosteroid use and manage it appropriately with anti-glaucoma medications.

Bevacizumab is a recombinant humanized anti-VEGF immunoglobulin, which was initially used in the treatment of metastatic cancers, but which is now widely used in ophthalmology for proliferative diabetic retinopathy, exudative macular degeneration, macular edema, retinal vein occlusions, and neovascular glaucoma [88]. VEGF encourages angiogenesis (proliferative stage of wound healing), which ultimately results in fibrosis [88]. In a study at the Catholic University of Korea, increased amounts of VEGF were found in the vitreous and AH in glaucoma patients undergoing trabeculectomy. This prompted the authors to try anti-VEGF agents to reverse postoperative scarring [89]. Later, in 2012, Ghanem published a study using 55 patients to compare the use of subconjunctival bevacizumab versus a placebo in patients undergoing a primary trabeculectomy with MMC [90,91]. At a one-year follow up, he found a statistically significant reduction in vascularity of the filtering bleb in the bevacizumab + MMC group compared to the placebo group [90]. Table 3 shows a summary of each agent's mechanism of action and administration.

Table 3. Current wound healing agents.

Agent	Mechanism of Action	Administration
Mitomycin C (MMC) [79]	An alkaloid, produced by *Streptomyces caespitosus*; works by inhibiting DNA-dependent RNA synthesis and triggering apoptosis.	Either via MMC-soaked sponge or subconjunctival injection postoperatively.
5-fluorouracil (5-FU) [80]	A pyrimidine analog, interferes with ribosomal RNA synthesis; diminishes episcleral scar formation by inducing apoptosis of fibroblasts in Tenon's capsule.	Similar to MMC.
Corticosteroids [84]	Reduce the expression of cytokines, such as TNF-alpha, IL-1, IL-2, IL-10, and IL-12, which decrease the number of tissue macrophages and blood monocytes during the inflammatory phase of wound healing.	Topical, subconjunctival injection, or oral perioperatively.
Bevacizumab [89]	Selectively binds to and blocks circulating VEGF to reduce micro-angiogenesis, thereby limiting the blood supply to scarred granulation tissue during the proliferative phase of wound healing.	Subconjunctival injection postoperatively.

3.6. Landmark 5-FU and MMC Studies

Author/Year/Country	Results
Kitazawa Y. et al., 1991. Japan [92]	Thirty-two patients undergoing trabeculectomy were assigned to receive either MMC (seventeen eyes) or 5-FU (fifteen eyes). The mean preoperative IOPs (mmHg) were 28.7 ± 7.9 (MMC) and 32.7 ± 10.0 (5-FU). At the final post-op visit, the mean postoperative IOPs were 8.6 ± 3.8 (MMC) and 12.3 ± 4.2 (5-FU). The incidence of corneal complications was lower in the MMC group (12%) compared to the 5-FU group (53%).
Katz GJ et al., 1995. USA [93]	In a high-risk filtration study, 20 patients received MMC and 9 received 5-FU. The mean preoperative IOP's (mmHg, MMC vs. 5-FU) were 32.6 ± 10.5 and 31.5 ± 9.8, respectively ($p = 0.78$). At 32 months, the postoperative IOP's were, similarly, 9.0 ± 4.9 vs. 16.3 ± 4.8 ($p = 0.0003$). The MMC group required fewer medications for IOP control (0.5 vs. 1.6) ($p = 0.01$).
Lamping et al., 1995. USA [94]	A total of 74 pseudophakic patients with glaucoma underwent trabeculectomy, and received either 5-FU (40 eyes) or MMC (40 eyes). Preoperative IOP's (mmHg, MMC vs. 5-FU) were 30.6 vs. 31.5, respectively. At 12 months post-op, the IOP's were, similarly, 12.8 vs. 14.8 mmHg ($p = 0.001$). The MMC-treated eyes required fewer IOP-lowering medications (0.6) compared to 5-FU-treated eyes (1.05) ($p = 0.03$).
Zadok D et al., 1995. Israel [95]	This trabeculectomy study compared postoperative subconjunctival injections of 5-FU (19 eyes) with single intraoperative application of subconjunctival MMC (20 eyes). At 6 months, IOPs averaged 10.9 mmHg (MMC-treated eyes) vs. 14.2 mmHg (5-FU-treated eyes) ($p = 0.14$). The MMC-treated group was on fewer medications (0.3 vs. 1.1, $p < 0.001$).
Cohen et al., 1996. USA [96]	In a combined cataract and trabeculectomy study, 72 eyes were randomized to MMC (0.5 mg/mL) vs. a placebo. At 6 months, significantly fewer medications were required for the MMC group (0.5 vs. 1.2; $p = 0.002$). Similarly, at 12 months, the MMC group had significantly reduced mean IOP (7.65 mmHg vs. 3.84 mmHg; $p = 0.001$). However, the MMC group showed large filtering blebs and more frequent wound leaks.
Costa et al., 1996. Brazil [97]	A total of 28 eyes with advanced POAG were given either MMC (0.2 mg/mL) or saline solution intraoperatively for 3 min. Mean IOPs were significantly lower in the MMC group compared to the controls at the final post-op visit ($p = 0.001$). The IOP (mmHg) was ≤15 in 85.7% (MMC) vs. 28.6% (control, $p = 0.002$). Choroidal effusions (35.7% vs. 14.3%, $p = 0.0065$) and shallow AC (35.7% vs. 7.1%) were more common in the MMC group.
Carlson et al., 1997. USA [98]	In a combined phacoemulsification and trabeculectomy procedure, 29 patients received either MMC [0.5 mg/mL] or a placebo. Pre-op IOPs (mmHg) were 18.4 ± 2.7 (MMC) vs. 19.1 ± 4.0 (placebo). At 8 months, MMC-treated eyes had a lower average IOP (12.3 ± 1.6) compared to the placebo-treated eyes (15.2 ± 1.5). At 12 months, IOPs averaged 12.6 ± 1.0 (MMC) and 16.2 ± 1.5 (placebo). On average, the MMC group had lower post-op IOP levels than the placebo group ($p = 0.04$).
Singh et al., 1997. USA [99]	A total of 101 eyes of black Ghanian patients with POAG were treated with either 5-FU and MMC after trabeculectomy. The 5-FU group (50.0 mg/mL for 5 min) had 57 patients, and the MMC group (0.5 mg/mL for 3.5 min) had 44 patients. Overall mean pre-op IOP (mmHg) was 30.1. Patients receiving MMC (IOP = 14.7) had a lower mean postoperative IOP than those receiving 5-FU (IOP = 16.7; $p = 0.05$).
Singh et al., 1997. USA [100]	In a black West African population, 81 eyes were divided to receive MMC or 5-FU during trabeculectomy. A total of 37 received 5-FU (50 mg/mL for 5 min) and 44 received MMC (0.4 mg/mL for 2 min). Pre-op IOP (mmHg) was 30.7 (MMC) vs. 32 (5-FU). The mean post-op IOP was 13.7 (MMC) vs. 16.3 (5-FU, $p = 0.05$).
Andreanos et al., 1997. Greece [101]	The study assessed MMC in 46 patients (26 M + 20 F) undergoing a repeat trabeculectomy. Patients were randomly assigned to MMC (24) vs. control group (22). Pre-op IOPs (mmHg) ranged from 27 to 38. Post-op complications were higher in the MMC group, including choroidal effusion (8.3% vs. 0%) and shallow AC (29.2% vs. 13.6%). Mean IOP (≤20 mmHg after 18 months) was 83.3% in the MMC group compared to 63.6% in the control group.

Author/Year/Country	Results
Singh et al., 2000. USA [102]	In this trabeculectomy study, 54 eyes received MMC (0.4 mg/mL for 2 min) and 54 eyes received 5-FU (50 mg/mL for 5 min). At 3 years post-op, there was no statistically significant difference between the two groups for mean preoperative IOP, or post-op interventions/complications.
DeBry et al., 2002. USA [68]	In this trabeculectomy study involving 239 eyes, a Kaplan–Meier analysis suggested 5-year probabilities of developing endophthalmitis (7.5%), bleb leaks (17.9%), and blebitis (6.3%). Trabeculectomy with MMC was associated with significant morbidity, and the risk of complications reached 23% at 5 years.
WuDunn et al., 2002. USA [103]	A total of 115 eyes underwent trabeculectomy [57 eyes (5-FU) and 58 eyes (MMC)]. The mean preoperative IOP (mmHg) was 24.3 (5-FU) vs. 21.9 (MMC), with no statistical significance ($p = 0.09$). At 12 months, 94% of 5-FU eyes and 89% of MMC eyes reached the target IOP of 21 mmHg ($p = 0.49$).
Sisto et al., 2007. Italy [104]	A total of 40 eyes with neovascular glaucoma were divided to receive post-op 5-FU (18) vs. intraoperative MMC (22) after filtration surgery. Pre-op IOPs (mmHg) were 40.4 ± 10.3 (5-FU) and 42 ± 11.3 (MMC), respectively. The mean follow-up period was 35.8 (5-FU) and 18.6 (MMC) months. Although the mean IOP significantly decreased in both groups [from 40 to 14.7 (5-FU) group ($p < 0.0001$); vs. 42 to 29.9 (MMC) group ($p = 0.0006$)], the difference between the two groups was not significant.
Mostafaei et al., 2011. Iran [105]	A total of 40 patients with high-risk open angle glaucoma received either MMC or 5-FU. Mean preoperative IOPs (mmHg) were 30.6 (5-FU) and 31.2 (MMC), respectively. At 6 months, the mean IOPs postoperatively for 5-FU (13.6) and MMC (11.4) were similar. The relative success of 5-FU vs. MMC was 0.93 [95% CI: 0.8–1.1].
Fendi et al., 2013. Brazil [106]	A meta-analysis of 5 randomized controlled clinical trials comprising 416 patients comparing MMC against 5-FU was carried out. Pre-op IOP was ≥ 21 mmHg in both groups. Lower IOPs (mean difference 2.17 mmHg) and higher success rates were observed in the MMC arm (92%) than in the 5-FU arm (84.2%, $p = 0.01$).

3.7. Experimental Wound Healing Agents

3.7.1. Nanoparticles

Nanomedicine encompasses the comprehensive regulation, repair, and improvement of human biology at the molecular level [107]. This is achieved by engineered nanodevices and nanostructures that operate in parallel at the single-cell level, with the goal of achieving desired medical benefits [108].

This new technology has prompted the need to develop newer drug delivery systems that allow for the gradual and sustained release of a drug, combined with improving bioavailability and minimizing complications (Figure 10). Many new nanoparticles composed of different structures (hollow, solid, or porous), shapes, and sizes have been developed. They contain or encapsulate certain molecules, such as drugs, DNA, RNA, or antibodies [109].

Common nanodelivery systems include nanoparticles, nanodiamonds (NDs), dendrimers, liposomes, and other devices. Drugs are incorporated into these nanomaterials through encapsulation or surface conjugation. Encapsulated drugs are released as the nanomaterials disassemble at the intended site, while conjugated drugs are released when the bond between the nanomaterial and drug is cleaved at the target site [110]. These nanomaterial-based drug delivery strategies have the potential to overcome limitations of conventional glaucoma treatments. Furthermore, incorporating inorganic nanoparticles into a hydrogel may enhance efficacy at the same or less dosage [109].

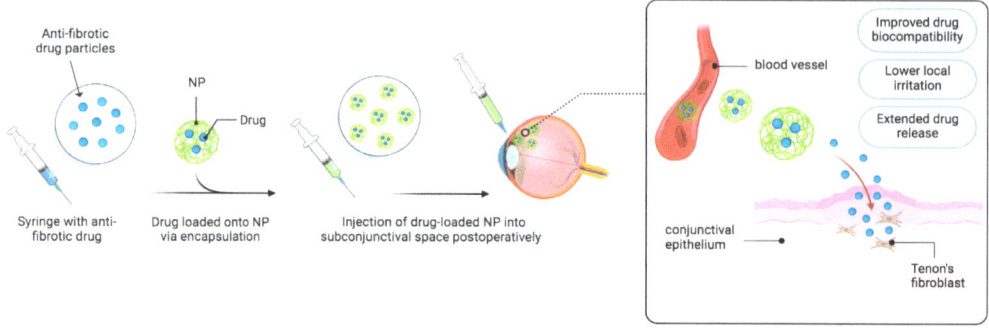

Figure 10. A schematic demonstrating nanotechnology-mediated drug delivery involving an antifibrotic drug encapsulated in a nanoparticle.

3.7.2. Targeting mRNAs

Noncoding RNAs, including long noncoding RNAs (lncRNAs; LINC) and microRNAs (miR; miRNA), are increasingly being studied as key regulators of scarring in bleb formation after glaucoma filtering surgery. Both miRNA-200a and miRNA-200b are believed to promote fibrosis in the glaucoma filtering tract. Studies have shown that the expression of miR-26a in fibrotic bleb tissue varies and is downregulated compared to controls [111]. Enhanced expression of miR-200b has been observed in trabecular meshwork cells treated with TGF-β during post-trabeculectomy scarring [112]. Further investigations by Drewry et al. have shown that miRNA-200b affects the activity of two pathways that regulate cell proliferation, namely p27/kip1 and RND3. They have also shown that inhibition of phosphatase and tensin homolog (PTEN) gene, an inhibitor of the PI3K/Akt pathway (cell growth, proliferation, and migration), results in increased expression of the profibrotic proteins P13K, Akt, α-SMA, and fibronectin [113]. However, the specific genes influenced by miR-200b and their downstream effects remain unclear [113]. Overall, a more in-depth exploration of noncoding RNAs is necessary to comprehend their roles in the development of glaucoma and the identification of potential therapeutic targets [114].

Yu et al. reviewed potential anti/pro-fibrotic noncoding RNA agents that may be used in glaucoma filtering surgery (Table 4) [114].

Table 4. A compiled summary of studies associated with noncoding RNAs, adapted from Yu et al., 2022 [114].

Noncoding RNAs	Authors, Year, Country	Summary	Pro/Anti-Fibrotic Role
miR-26a	Wang et al., 2018, China [115]	miR-26a is significantly downregulated in filtering tract scars and is inversely correlated with connective tissue growth factor (CTGF) mRNA levels.	Anti-fibrotic
miR-29b	Ran et al., 2015, China [116]	TGF-β2 stimulates the proliferation of human tenon fibroblasts (HTF) by suppressing miR-29b expression, which is regulated by Nrf2.	Anti-fibrotic
miR-139	Deng et al., 2019, China [117]	Overexpression of miR-139 effectively counteracted the TGFβ1-induced increase in collagen I and α-smooth muscle actin levels, as well as HTF proliferation.	Anti-fibrotic

Table 4. *Cont.*

Noncoding RNAs	Authors, Year, Country	Summary	Pro/Anti-Fibrotic Role
miR-200a	Peng et al., 2019, China [118]	miR-200a is reduced, while FGF7 is increased in glaucoma. miR-200a has a protective function on the glaucomatous optic nerve injury through its effect by suppressing the MAPK signaling pathway mediated by FGF7.	Anti-fibrotic
miR-200b	Tong et al., 2019, China [119]	The induction of fibrosis in HTFs occurs through TGF-β1-mediated miR-200b by suppressing the PTEN gene signaling pathway.	Pro-fibrotic
miR216b	Xu et al., 2014, China [120]	miR-216b directly targeted and decreased the expression of Beclin 1, a pro-apoptotic molecule. In HTFs treated with hydroxycamptothecin, miR-216b regulates both autophagy and apoptosis by modulating Beclin 1.	Pro-fibrotic
Lnc H19	Zhu et al., 2020, China [121]	TGF-β induced the expression of H19 in HTFs, and suppressing H19 inhibited the effects of TGF-β. The findings suggest that H19 modulates β-catenin expression via miR-200a in TGF-β-treated HTFs. Therefore, suppressing H19 may result in attenuation of scar after glaucoma surgery.	Pro-fibrotic
Lnc NR003923	Zhao et al., 2019, China [122]	Inhibiting NR003923 expression in HTFs resulted in the suppression of cell migration, proliferation, fibrosis, and autophagy induced by TGF-β.	Pro-fibrotic
LINC00028	Sui et al., 2020, China [123]	In HTFs treated with TGFβ1, the decrease in LINC00028 expression inhibits migration, proliferation, invasion, epithelial-mesenchymal transition, fibrosis, and autophagy.	Pro-fibrotic

3.7.3. Infliximab

Infliximab is a chimeric monoclonal antibody that targets tumor necrosis factor (TNF)-α and which is composed of both mouse and human elements (human–murine IgG1). TNF-α acts as a local regulator for leukocytes and endothelial cells, functioning through paracrine and autocrine pathways and influencing immunological and inflammatory cascades [124]. Infliximab works by binding to TNF-α, thereby blocking NF-kB (transcription factor for the inflammatory process) migration, resulting in a decrease in the production of pro-inflammatory cytokines, such as IL-1 and IL-6, and adhesion molecules [125,126]. Therefore, infliximab may be a potential agent in modulating surgical fibrosis.

3.8. Future Directions

To improve the reliability and validity of the findings presented in this review, additional comparative research involving promising new antimetabolite agents is warranted. These future agents include anti-TGFβ agents (lerdelimumab, fresolimnumab, pirfenidone), kinase inhibitors (Nintedanib), anti-TNF-α agents (infliximab), beta-radiation, and nanotechnology-based drug delivery systems. In a study by Shao et. al, researchers concluded that beta radiation during trabeculectomy can reduce fibroblast proliferation and increase the success of glaucoma filtration surgery, but it may also lead to cataract formation [76]. Similarly, nanotechnology-based drug delivery systems have shown great promise in post-surgical wound healing [76]. Sustained-release implants, hydrogels, liposomal systems, and nanoparticles have been explored for targeted delivery and enhanced drug residence time, preventing rapid clearance and improving efficacy of antifibrotic agents [76]. While these new agents show great potential, further studies need to be conducted to optimize the delivery methods and to reduce complications.

4. Conclusions

It is well known that the long-term efficacy of glaucoma surgery is reduced by fibrosis, scar formation, and uncontrolled wound healing. Conventional adjuncts used for mitigating post-surgical fibrosis, such as corticosteroids and anti-fibrotic agents, have unpredictable outcomes and side effects. The ongoing research using promising experimental wound healing agents and new drug targets to prevent fibrosis may improve glaucoma surgery outcomes.

Author Contributions: Conceptualization, K.S.K.; Writing—original draft preparation, K.S.K., B.D., M.P. and S.S.; Writing—review and editing, K.S.K., B.D., M.P., S.S., M.G., M.A. and A.S.; figures, B.D. and S.S. All authors have read and agreed to the published version of the manuscript.

Funding: This research was partly funded by an unrestricted challenge grant from Research to Prevent Blindness, New York, NY, USA.

Data Availability Statement: No new data were created or analyzed in this study. Data sharing is not applicable to this article.

Acknowledgments: We would like to thank Irina Kim Cavder, Priya Mekala, Ibrahim Saleh, Pooja Kumar, Michael Tran, Suyash Jain, and Emily Buchanan for their contributions to this manuscript. BioRender® (Toronto, Ontario, Canada) software was used to make portions of the figures.

Conflicts of Interest: The authors declare no financial disclosures or conflicts of interest in this work.

References

1. Tham, Y.C.; Li, X.; Wong, T.Y.; Quigley, H.A.; Aung, T.; Cheng, C.Y. Global prevalence of glaucoma and projections of glaucoma burden through 2040: A systematic review and meta-analysis. *Ophthalmology* **2014**, *121*, 2081–2090. [CrossRef]
2. Friedman, D.S.; Wolfs, R.C.; O'Colmain, B.J.; Klein, B.E.; Taylor, H.R.; West, S.; Leske, M.C.; Mitchell, P.; Congdon, N.; Kempen, J. Eye Diseases Prevalence Research Group. Prevalence of open-angle glaucoma among adults in the United States. *Arch. Ophthalmol.* **2004**, *122*, 532–538. [CrossRef] [PubMed]
3. Murgoitio-Esandi, J.; Xu, B.X.; Song, B.J.; Zhou, Q.; Oberai, A.A. A Mechanistic Model of Aqueous Humor Flow to Study Effects of Angle Closure on Intraocular Pressure. *Trans. Vis. Sci. Technol.* **2023**, *12*, 16–20. [CrossRef] [PubMed]
4. Johnson, M.; McLaren, J.W.; Overby, D.R. Unconventional aqueous humor outflow: A review. *Exp. Eye Res.* **2017**, *158*, 94–111. [CrossRef] [PubMed]
5. Al-Humimat, G.; Marashdeh, I.; Daradkeh, D.; Kooner, K.S. Investigational Rho Kinase Inhibitors for the Treatment of Glaucoma. *J. Exp. Pharmacol.* **2021**, *13*, 197–212. [CrossRef] [PubMed]
6. SooHoo, J.R.; Seibold, L.K.; Radcliffe, N.M.; Kahook, M.Y. Minimally invasive glaucoma surgery: Current implants and future innovations. *Can. J. Ophthalmol.* **2014**, *49*, 528–533. [CrossRef] [PubMed]
7. Wagner, I.V.; Stewart, M.W.; Dorairaj, S.K. Updates on the Diagnosis and Management of Glaucoma. *Mayo Clin. Proc. Innov. Qual. Outcomes* **2022**, *6*, 618–635. [CrossRef]
8. Vinod, K.; Gedde, S.J.; Feuer, W.J.; Panarelli, J.F.; Chang, T.C.; Chen, P.P.; Parrish, R.K., 2nd. Practice Preferences for Glaucoma Surgery: A Survey of the American Glaucoma Society. *J. Glaucoma* **2017**, *26*, 687–693. [CrossRef] [PubMed]
9. Gedde, S.J.; Schiffman, J.C.; Feuer, W.J.; Herndon, L.W.; Brandt, J.D.; Budenz, D.L. Tube versus Trabeculectomy Study Group. Treatment outcomes in the Tube Versus Trabeculectomy (TVT) study after five years of follow-up. *Am. J. Ophthalmol.* **2012**, *153*, 789–803.e2. [CrossRef]
10. Masoumpour, M.B.; Nowroozzadeh, M.H.; Razeghinejad, M.R. Current and Future Techniques in Wound Healing Modulation after Glaucoma Filtering Surgeries. *Open Ophthalmol. J.* **2016**, *10*, 68–85. [CrossRef]
11. Mehta, A.; De Paola, L.; Pana, T.A.; Carter, B.; Soiza, R.L.; Kafri, M.W.; Potter, J.F.; Mamas, M.A.; Myint, P.K. The relationship between nutritional status at the time of stroke on adverse outcomes: A systematic review and meta-analysis of prospective cohort studies. *Nutr. Rev.* **2022**, *80*, 2275–2287. [CrossRef]
12. Rodrigues, M.; Kosaric, N.; Bonham, C.A.; Gurtner, G.C. Wound Healing: A Cellular Perspective. *Physiol. Rev.* **2019**, *99*, 665–706. [CrossRef] [PubMed]
13. Sangkuhl, K.; Shuldiner, A.R.; Klein, T.E.; Altman, R.B. Platelet aggregation pathway. *Pharmacogenet Genom.* **2011**, *21*, 516–521. [CrossRef]
14. Chaudhary, P.K.; Kim, S.; Kim, S. An Insight into Recent Advances on Platelet Function in Health and Disease. *Int. J. Mol. Sci.* **2022**, *23*, 6022–6031. [CrossRef]
15. Tahery, M.M.; Lee, D.A. Pharmacologic control of wound healing in glaucoma filtration surgery. *J. Ocul. Pharmacol.* **1989**, *5*, 155–179. [CrossRef] [PubMed]

16. Gonzalez, A.C.; Costa, T.F.; Andrade, Z.A.; Medrado, A.R. Wound healing—A literature review. *An. Bras. Dermatol.* **2016**, *91*, 614–620. [CrossRef]
17. Chapple, I.L.C.; Hirschfeld, J.; Kantarci, A.; Wilensky, A.; Shapira, L. The role of the host—Neutrophil biology. *Periodontology 2000* **2023**, 1–47. [CrossRef]
18. Krzyszczyk, P.; Schloss, R.; Palmer, A.; Berthiaume, F. The Role of Macrophages in Acute and Chronic Wound Healing and Interventions to Promote Pro-wound Healing Phenotypes. *Front. Physiol.* **2018**, *9*, 419. [CrossRef] [PubMed]
19. Thiruvoth, F.M.; Mohapatra, D.P.; Kumar, D.; Chittoria, S.R.K.; Nandhagopal, V. Current concepts in the physiology of adult wound healing. *Plast. Aesthetic Res.* **2015**, *2*, 250–256. [CrossRef]
20. Schultz, G.S.; Chin, G.A.; Moldawer, L.; Diegelmann, R.F. *Principles of Wound Healing. Mechanisms of Vascular Disease: A Reference Book for Vascular Specialists*; Fitridge, R., Thompson, M., Eds.; Springer International Publishing: Cham, Switzerland, 2011; pp. 423–450. [CrossRef]
21. Koivisto, L.; Heino, J.; Häkkinen, L.; Larjava, H. Integrins in Wound Healing. *Adv. Wound Care (New Rochelle)* **2014**, *3*, 762–783. [CrossRef]
22. Alhajj, M.; Goyal, A. Physiology, Granulation Tissue. In *StatPearls [Internet]*; StatPearls Publishing: Treasure Island, FL, USA, 2022. Available online: https://www.ncbi.nlm.nih.gov/books/NBK554402/ (accessed on 14 November 2023).
23. Wynn, T.A.; Ramalingam, T.R. Mechanisms of fibrosis: Therapeutic translation for fibrotic disease. *Nat. Med.* **2012**, *18*, 1028–1040. [CrossRef]
24. Henderson, N.C.; Rieder, F.; Wynn, T.A. Fibrosis: From mechanisms to medicines. *Nature* **2020**, *587*, 555–566. [CrossRef] [PubMed]
25. Khaw, P.T.; Bouremel, Y.; Brocchini, S.; Henein, C. The control of conjunctival fibrosis as a paradigm for the prevention of ocular fibrosis-related blindness. "Fibrosis has many friends". *Eye* **2020**, *34*, 2163–2174. [CrossRef] [PubMed]
26. Zhavoronkov, A.; Izumchenko, E.; Kanherkar, R.R.; Teka, M.; Cantor, C.; Manaye, K.; Sidransky, D.; West, M.D.; Makarev, E.; Csoka, A.B. Pro-fibrotic pathway activation in trabecular meshwork and lamina cribrosa is the main driving force of glaucoma. *Cell Cycle* **2016**, *15*, 1643–1652, Erratum in *Cell Cycle* **2016**, *15*, 2087. [CrossRef] [PubMed]
27. Borthwick, L.A.; Wynn, T.A.; Fisher, A.J. Cytokine mediated tissue fibrosis. *Biochim. Biophys. Acta* **2013**, *1832*, 1049–1060. [CrossRef] [PubMed]
28. Yamanaka, O.; Kitano-Izutani, A.; Tomoyose, K.; Reinach, P. Pathobiology of wound healing after glaucoma filtration surgery. *BMC Ophthalmol.* **2015**, *15* (Suppl. S1), 157. [CrossRef] [PubMed]
29. Macleod, T.; Berekmeri, A.; Bridgewood, C.; Stacey, M.; McGonagle, D.; Wittmann, M. The Immunological Impact of IL-1 Family Cytokines on the Epidermal Barrier. *Front. Immunol.* **2021**, *23*, 808012. [CrossRef] [PubMed]
30. Tanaka, T.; Narazaki, M.; Kishimoto, T. IL-6 in inflammation, immunity, and disease. *Cold Spring Harb. Perspect. Biol.* **2014**, *6*, a016295. [CrossRef] [PubMed]
31. Guo, X.; Wang, X.F. Signaling cross-talk between TGF-beta/BMP and other pathways. *Cell Res.* **2009**, *19*, 71–88. [CrossRef]
32. Francois, B.; Jeannet, R.; Daix, T.; Walton, A.H.; Shotwell, M.S.; Unsinger, J.; Monneret, G.; Rimmelé, T.; Blood, T.; Morre, M.; et al. Interleukin-7 restores lymphocytes in septic shock: The IRIS-7 randomized clinical trial. *JCI Insight* **2018**, *3*, e98960. [CrossRef]
33. Huang, Y.H.; Shi, M.N.; Zheng, W.D.; Zhang, L.J.; Chen, Z.X.; Wang, X.Z. Therapeutic effect of interleukin-10 on CCl4-induced hepatic fibrosis in rats. *World J. Gastroenterol.* **2006**, *12*, 1386–1391. [CrossRef] [PubMed]
34. Wang, X.; Wong, K.; Ouyang, W.; Rutz, S. Targeting IL-10 Family Cytokines for the Treatment of Human Diseases. *Cold Spring Harb. Perspect. Biol.* **2019**, *11*, a028548. [CrossRef] [PubMed]
35. Steen, E.H.; Wang, X.; Balaji, S.; Butte, M.J.; Bollyky, P.L.; Keswani, S.G. The Role of the Anti-Inflammatory Cytokine Interleukin-10 in Tissue Fibrosis. *Adv. Wound Care (New Rochelle)* **2020**, *9*, 184–198. [CrossRef] [PubMed]
36. Kong, X.; Feng, D.; Wang, H.; Hong, F.; Bertola, A.; Wang, F.S.; Gao, B. Interleukin-22 induces hepatic stellate cell senescence and restricts liver fibrosis in mice. *Hepatology* **2012**, *56*, 1150–1159. [CrossRef] [PubMed]
37. Arshad, T.; Mansur, F.; Palek, R.; Manzoor, S.; Liska, V. A Double Edged Sword Role of Interleukin-22 in Wound Healing and Tissue Regeneration. *Front. Immunol.* **2020**, *11*, 2148. [CrossRef]
38. Tang, K.Y.; Lickliter, J.; Huang, Z.H.; Xian, Z.S.; Chen, H.Y.; Huang, C.; Xiao, C.; Wang, Y.P.; Tan, Y.; Xu, L.F.; et al. Safety, pharmacokinetics, and biomarkers of F-652, a recombinant human interleukin-22 dimer, in healthy subjects. *Cell. Mol. Immunol.* **2019**, *16*, 473–482. [CrossRef] [PubMed]
39. Keane, P.A.; Sadda, S.R. Development of Anti-VEGF Therapies for Intraocular Use: A Guide for Clinicians. *J. Ophthalmol.* **2012**, *2012*, 483034. [CrossRef] [PubMed]
40. Fredriksson, L.; Li, H.; Eriksson, U. The PDGF family: Four gene products form five dimeric isoforms. *Cytokine Growth Factor Rev.* **2004**, *15*, 197–204. [CrossRef] [PubMed]
41. Wong, C.K.S.; Falkenham, A.; Myers, T.; Légaré, J.F. Connective tissue growth factor expression after angiotensin II exposure is dependent on transforming growth factor-β signaling via the canonical Smad-dependent pathway in hypertensive induced myocardial fibrosis. *J. Renin Angiotensin Aldosterone Syst.* **2018**, *19*, 1470320318759358. [CrossRef]
42. Cabral-Pacheco, G.A.; Garza-Veloz, I.; Castruita-De la Rosa, C.; Ramirez-Acuña, J.M.; Perez-Romero, B.A.; Guerrero-Rodriguez, J.F.; Martinez-Avila, N.; Martinez-Fierro, M.L. The Roles of Matrix Metalloproteinases and Their Inhibitors in Human Diseases. *Int. J. Mol. Sci.* **2020**, *21*, 9739. [CrossRef]

43. Yang, N.; Cao, D.F.; Yin, X.X.; Zhou, H.H.; Mao, X.Y. Lysyl oxidases: Emerging biomarkers and therapeutic targets for various diseases. *Biomed. Pharmacother.* **2020**, *131*, 110791. [CrossRef] [PubMed]
44. Schmandke, A.; Schmandke, A.; Strittmatter, S.M. ROCK and Rho: Biochemistry and neuronal functions of Rho-associated protein kinases. *Neuroscientist* **2007**, *13*, 454–469. [CrossRef] [PubMed]
45. Trombetta-Esilva, J.; Bradshaw, A.D. The Function of SPARC as a Mediator of Fibrosis. *Open Rheumatol. J.* **2012**, *6*, 146–155. [CrossRef] [PubMed]
46. Shi, H.; Zhang, Y.; Fu, S.; Lu, Z.; Ye, W.; Xiao, Y. Angiotensin II as a morphogenic cytokine stimulating fibrogenesis of human tenon's capsule fibroblasts. *Investig. Ophthalmol. Vis. Sci.* **2015**, *56*, 855–864. [CrossRef]
47. Zhong, T.; Zhang, W.; Guo, H.; Pan, X.; Chen, X.; He, Q.; Yang, B.; Ding, L. The regulatory and modulatory roles of TRP family channels in malignant tumors and relevant therapeutic strategies. *Acta Pharm. Sin. B* **2022**, *12*, 1761–1780. [CrossRef] [PubMed]
48. Nakamura, H.; Siddiqui, S.S.; Shen, X.; Malik, A.B.; Pulido, J.S.; Kumar, N.M.; Yue, B.Y. RNA interference targeting transforming growth factor-beta type II receptor suppresses ocular inflammation and fibrosis. *Mol. Vis.* **2004**, *10*, 703–711. [PubMed]
49. Walkden, A.; Au, L.; Fenerty, C. Trabeculectomy Training: Review of Current Teaching Strategies. *Adv. Med. Educ. Pract.* **2020**, *11*, 31–36. [CrossRef] [PubMed]
50. Fan Gaskin, J.C.; Nguyen, D.Q.; Soon Ang, G.; O'Connor, J.; Crowston, J.G. Wound Healing Modulation in Glaucoma Filtration Surgery-Conventional Practices and New Perspectives: The Role of Antifibrotic Agents (Part I). *J. Curr. Glaucoma Pract.* **2014**, *8*, 37–45. [CrossRef] [PubMed]
51. Minckler, D.S.; Francis, B.A.; Hodapp, E.A.; Jampel, H.D.; Lin, S.C.; Samples, J.R.; Smith, S.D.; Singh, K. Aqueous shunts in glaucoma: A report by the American Academy of Ophthalmology. *Ophthalmology* **2008**, *115*, 1089–1098. [CrossRef]
52. Schlunck, G.; Meyer-ter-Vehn, T.; Klink, T.; Grehn, F. Conjunctival fibrosis following filtering glaucoma surgery. *Exp. Eye Res.* **2016**, *142*, 76–82. [CrossRef]
53. Epstein, E. Fibrosing response to aqueous. Its relation to glaucoma. *Br. J. Ophthalmol.* **1959**, *43*, 641–647. [CrossRef] [PubMed]
54. Freedman, J.; Iserovich, P. Pro-inflammatory cytokines in glaucomatous aqueous and encysted Molteno implant blebs and their relationship to pressure. *Investig. Ophthalmol. Vis. Sci.* **2013**, *54*, 4851–4855. [CrossRef] [PubMed]
55. Fuller, J.R.; Bevin, T.H.; Molteno, A.C.; Vote, B.J.; Herbison, P. Anti-inflammatory fibrosis suppression in threatened trabeculectomy bleb failure produces good long term control of intraocular pressure without risk of sight threatening complications. *Br. J. Ophthalmol.* **2002**, *86*, 1352–1354. [CrossRef] [PubMed]
56. Barnes, P.J. Corticosteroid effects on cell signalling. *Eur. Respir. J.* **2006**, *27*, 413–426. [CrossRef] [PubMed]
57. Chang-Lin, J.E.; Attar, M.; Acheampong, A.A.; Robinson, M.R.; Whitcup, S.M.; Kuppermann, B.D.; Welty, D. Pharmacokinetics and pharmacodynamics of a sustained-release dexamethasone intravitreal implant. *Investig. Ophthalmol. Vis. Sci.* **2011**, *52*, 80–86. [CrossRef] [PubMed]
58. Brandt, J.D.; Hammel, N.; Fenerty, C.; Karaconji, T. Glaucoma Drainage Devices. In *Surgical Management of Childhood Glaucoma*; Grajewski, A., Bitrian, E., Papadopoulos, M., Freedman, S., Eds.; Springer: Cham, Switzerland, 2018; pp. 99–127. [CrossRef]
59. Pinchuk, L.; Riss, I.; Batlle, J.F.; Kato, Y.P.; Martin, J.B.; Arrieta, E.; Palmberg, P.; Parrish, R.K., 2nd.; Weber, B.A.; Kwon, Y.; et al. The development of a micro-shunt made from poly(styrene-block-isobutylene-block-styrene) to treat glaucoma. *J. Biomed. Mater. Res. B Appl. Biomater.* **2017**, *105*, 211–221. [CrossRef] [PubMed]
60. Burgos-Blasco, B.; García-Feijóo, J.; Perucho-Gonzalez, L.; Güemes-Villahoz, N.; Morales-Fernandez, L.; Mendez-Hernández, C.D.; Martinez de la Casa, J.M.; Konstas, A.G. Evaluation of a Novel Ab Externo MicroShunt for the Treatment of Glaucoma. *Adv. Ther.* **2022**, *39*, 3916–3932. [CrossRef]
61. Beckers, H.J.M.; Aptel, F.; Webers, C.A.B.; Bluwol, E.; Martínez-de-la-Casa, J.M.; García-Feijoó, J.; Lachkar, Y.; Méndez-Hernández, C.D.; Riss, I.; Shao, H.; et al. Safety and Effectiveness of the PRESERFLO® MicroShunt in Primary Open-Angle Glaucoma: Results from a 2-Year Multicenter Study. *Ophthalmol. Glaucoma* **2022**, *5*, 195–209. [CrossRef]
62. Perkins, T.W.; Cardakli, U.F.; Eisele, J.R.; Kaufman, P.L.; Heatley, G.A. Adjunctive mitomycin C in Molteno implant surgery. *Ophthalmology* **1995**, *102*, 91–97. [CrossRef]
63. Perkins, T.W.; Gangnon, R.; Ladd, W.; Kaufman, P.L.; Libby, C.M. Molteno implant with mitomycin C: Intermediate-term results. *J. Glaucoma* **1998**, *7*, 86–92. [CrossRef]
64. Lee, D.; Shin, D.H.; Birt, C.M.; Kim, C.; Kupin, T.H.; Olivier, M.M.; Khatana, A.K.; Reed, S.Y. The effect of adjunctive mitomycin C in Molteno implant surgery. *Ophthalmology* **1997**, *104*, 2126–2135. [CrossRef] [PubMed]
65. Cantor, L.; Burgoyne, J.; Sanders, S.; Bhavnani, V.; Hoop, J.; Brizendine, E. The effect of mitomycin C on Molteno implant surgery: A 1-year randomized, masked, prospective study. *J. Glaucoma* **1998**, *7*, 240–246. [CrossRef] [PubMed]
66. Al-Mobarak, F.; Khan, A.O. Two-year survival of Ahmed valve implantation in the first 2 years of life with and without intraoperative mitomycin-C. *Ophthalmology* **2009**, *116*, 1862–1865. [CrossRef] [PubMed]
67. Do, A.T.; Parikh, H.; Panarelli, J.F. Subconjunctival microinvasive glaucoma surgeries: An update on the Xen gel stent and the PreserFlo MicroShunt. *Curr. Opin. Ophthalmol.* **2020**, *31*, 132–138. [CrossRef] [PubMed]
68. DeBry, P.W.; Perkins, T.W.; Heatley, G.; Kaufman, P.; Brumback, L.C. Incidence of late-onset bleb-related complications following trabeculectomy with mitomycin. *Arch. Ophthalmol.* **2002**, *120*, 297–300. [CrossRef] [PubMed]
69. Yap, Z.L.; Seet, L.F.; Chu, S.W.; Toh, L.Z.; Ibrahim, F.I.; Wong, T.T. Effect of valproic acid on functional bleb morphology in a rabbit model of minimally invasive surgery. *Br. J. Ophthalmol.* **2022**, *106*, 1028–1036. [CrossRef] [PubMed]

70. Seet, L.F.; Yap, Z.L.; Chu, S.W.L.; Toh, L.Z.; Ibrahim, F.I.; Teng, X.; Wong, T.T. Effects of Valproic Acid and Mitomycin C Combination Therapy in a Rabbit Model of Minimally Invasive Glaucoma Surgery. *Transl. Vis. Sci. Technol.* **2022**, *11*, 30. [CrossRef] [PubMed]
71. Ayyala, R.S.; Duarte, J.L.; Sahiner, N. Glaucoma drainage devices: State of the art. *Expert. Rev. Med. Devices* **2006**, *3*, 509–521. [CrossRef] [PubMed]
72. Lewis, R.A. Ab interno approach to the subconjunctival space using a collagen glaucoma stent. *J. Cataract. Refract. Surg.* **2014**, *40*, 1301–1306. [CrossRef]
73. Shute, T.S.; Dietrich, U.M.; Baker, J.F.; Carmichael, K.P.; Wustenberg, W.; Ahmed, I.I.; Sheybani, A. Biocompatibility of a Novel Microfistula Implant in Nonprimate Mammals for the Surgical Treatment of Glaucoma. *Investig. Ophthalmol. Vis. Sci.* **2016**, *57*, 3594–3600. [CrossRef]
74. Acosta, A.C.; Espana, E.M.; Yamamoto, H.; Davis, S.; Pinchuk, L.; Weber, B.A.; Orozco, M.; Dubovy, S.; Fantes, F.; Parel, J.M. A newly designed glaucoma drainage implant made of poly(styrene-b-isobutylene-b-styrene): Biocompatibility and function in normal rabbit eyes. *Arch. Ophthalmol.* **2006**, *124*, 1742–1749. [CrossRef] [PubMed]
75. van Mechelen, R.; Wolters, J.E.; Herfs, M.; Bertens, C.J.F.; Gijbels, M.; Pinchuk, L.; Gorgels, T.; Beckers, H.J.M. Wound Healing Response After Bleb-Forming Glaucoma Surgery With a SIBS Microshunt in Rabbits. *Trans. Vis. Sci. Technol.* **2022**, *11*, 29–32. [CrossRef] [PubMed]
76. Shao, C.G.; Sinha, N.R.; Mohan, R.R.; Webel, A.D. Novel Therapies for the Prevention of Fibrosis in Glaucoma Filtration Surgery. *Biomedicines* **2023**, *11*, 657. [CrossRef] [PubMed]
77. Cabourne, E.; Clarke, J.C.; Schlottmann, P.G.; Evans, J.R. Mitomycin C versus 5-Fluorouracil for wound healing in glaucoma surgery. *Cochrane Database Syst. Rev.* **2015**, *2015*, CD006259. [CrossRef] [PubMed]
78. Jampel, H.D. Effect of brief exposure to mitomycin C on viability and proliferation of cultured human Tenon's capsule fibroblasts. *Ophthalmology* **1992**, *99*, 1471–1476. [CrossRef] [PubMed]
79. Bass, P.D.; Gubler, D.A.; Judd, T.C.; Williams, R.M. Mitomycinoid alkaloids: Mechanism of action, biosynthesis, total syntheses, and synthetic approaches. *Chem. Rev.* **2013**, *113*, 6816–6863. [CrossRef] [PubMed]
80. Adegbehingbe, B.O.; Oluwatoyin, H.O. Intra-operative 5-FU in Glaucoma Surgery: A Nigerian Teaching Hospital Experience. *Middle East. Afr. J. Ophthalmol.* **2008**, *15*, 57–60. [CrossRef] [PubMed]
81. Horsley, M.B.; Kahook, M.Y. Anti-VEGF therapy for glaucoma. *Curr. Opin. Ophthalmol.* **2010**, *21*, 112–117. [CrossRef] [PubMed]
82. Araujo, S.V.; Spaeth, G.L.; Roth, S.M.; Starita, R.J. A ten-year follow-up on a prospective, randomized trial of postoperative corticosteroids after trabeculectomy. *Ophthalmology* **1995**, *102*, 1753–1759. [CrossRef]
83. Barnes, P.J. How corticosteroids control inflammation: Quintiles Prize Lecture 2005. *Br J Pharmacol.* **2006**, *148*, 245–254. [CrossRef]
84. Lama, P.J.; Fechtner, R.D. Antifibrotics and wound healing in glaucoma surgery. *Surv. Ophthalmol.* **2003**, *48*, 314–346. [CrossRef] [PubMed]
85. Broadway, D.C.; Grierson, I.; Stürmer, J.; Hitchings, R.A. Reversal of topical antiglaucoma medication effects on the conjunctiva. *Arch. Ophthalmol.* **1996**, *114*, 262–267. [CrossRef] [PubMed]
86. Tripathi, R.C.; Parapuram, S.K.; Tripathi, B.J.; Zhong, Y.; Chalam, K.V. Corticosteroids and glaucoma risk. *Drugs Aging* **1999**, *15*, 439–450. [CrossRef] [PubMed]
87. Bernstein, H.N.; Mills, D.W.; Brecker, B. Steroid-induced elevation of intraocular pressure. *Arch. Ophthalmol.* **1963**, *70*, 15–18. [CrossRef] [PubMed]
88. Whitescarver, T.D.; Hobbs, S.D.; Wade, C.I.; Winegar, J.W.; Colyer, M.H.; Reddy, A.; Drayna, P.M.; Justin, G.A. A History of Anti-VEGF Inhibitors in the Ophthalmic Literature: A Bibliographic Review. *J. Vitr. Dis.* **2020**, *5*, 304–312. [CrossRef]
89. Lopilly Park, H.Y.; Kim, J.H.; Ahn, M.D.; Park, C.K. Level of vascular endothelial growth factor in tenon tissue and results of glaucoma surgery. *Arch. Ophthalmol.* **2012**, *130*, 685–689. [CrossRef] [PubMed]
90. Ghanem, A.A. Trabeculectomy with or without Intraoperative Sub-conjunctival Injection of Bevacizumab in Treating Refractory Glaucoma. *J. Clin. Exp. Ophthalmol.* **2011**, *2*, 2. [CrossRef]
91. Muhsen, S.; Compan, J.; Lai, T.; Kranemann, C.; Birt, C. Postoperative adjunctive bevacizumab versus placebo in primary trabeculectomy surgery for glaucoma. *Int. J. Ophthalmol.* **2019**, *12*, 1567–1574. [CrossRef]
92. Kitazawa, Y.; Kawase, K.; Matsushita, H.; Minobe, M. Trabeculectomy with mitomycin. A comparative study with fluorouracil. *Arch. Ophthalmol.* **1991**, *109*, 1693–1698. [CrossRef]
93. Katz, G.J.; Higginbotham, E.J.; Lichter, P.R.; Skuta, G.L.; Musch, D.C.; Bergstrom, T.J. Mitomycin C versus 5-fluorouracil in high-risk glaucoma filtering surgery. Extended follow-up. *Ophthalmology* **1995**, *102*, 1263–1299. [CrossRef]
94. Lamping, K.A.; Belkin, J.K. 5-Fluorouracil and mitomycin C in pseudophakic patients. *Ophthalmology* **1995**, *102*, 70–75. [CrossRef] [PubMed]
95. Zadok, D.; Zadok, J.; Turetz, J.; Krakowski, D.; Nemet, P. Intraoperative mitomycin versus postoperative 5-fluorouracil in primary glaucoma filtering surgery. *Ann. Ophthalmol. Glaucoma* **1995**, *27*, 336–340.
96. Cohen, J.S.; Greff, L.J.; Novack, G.D.; Wind, B.E. A placebo controlled, double-masked evaluation of mitomycin C in combined glaucoma and cataract procedures. *Ophthalmology* **1996**, *103*, 1934–1942. [CrossRef]
97. Costa, V.P.; Comegno, P.E.; Vasconcelos, J.P.; Malta, R.F.; Jose, N.K. Low Dose mitomycin C trabeculectomy in patients with advanced glaucoma. *J. Glaucoma* **1996**, *5*, 193–199. [CrossRef] [PubMed]

98. Carlson, D.W.; Alward, W.L.; Barad, J.P.; Zimmerman, M.B.; Carney, B.L. A randomized study of mitomycin augmentation in combined phacoemulsification and trabeculectomy. *Ophthalmology* **1997**, *104*, 719–724. [CrossRef]
99. Singh, K.; Byrd, S.; Egbert, P.R.; Budenz, D. Risk of hypotony after primary trabeculectomy with antifibrotic agents in a black west African population. *J. Glaucoma* **1998**, *7*, 82–85. [CrossRef]
100. Singh, K.; Egbert, P.R.; Byrd, S.; Budenz, D.L.; Williams, A.S.; Decker, J.H. Trabeculectomy with intraoperative 5-fluorouracil vs mitomycin C. *Am. J. Ophthalmol.* **1997**, *123*, 48–53. [CrossRef] [PubMed]
101. Andreanos, D.; Georgopoulos, G.T.; Vergados, J.; Papaconstantinou, D.; Liokis, N.; Theodossiadis, P. Clinical evaluation of the effect of mitomycin-C in re-operation for primary open angle glaucoma. *Eur. J. Ophthalmol.* **1997**, *7*, 49–54. [CrossRef] [PubMed]
102. Singh, K.; Mehta, K.; Shaikh, N.M.; Tsai, J.C.; Moster, M.R.; Budenz, D.L. Trabeculectomy with intraoperative mitomycin C versus 5-fluorouracil. Prospective randomized clinical trial. *Ophthalmology* **2000**, *107*, 2305–2309. [CrossRef]
103. WuDunn, D.; Cantor, L.B.; Palanca-Capistrano, A.M.; Hoop, J.; Alvi, N.P.; Finley, C. A prospective randomized trial comparing intraoperative 5-fluorouracil vs mitomycin C in primary trabeculectomy. *Am. J. Ophthalmol.* **2002**, *134*, 521–528. [CrossRef]
104. Sisto, D.; Vetrugno, M.; Trabucco, T.; Cantatore, F.; Ruggeri, G.; Sborgia, C. The role of antimetabolites in filtration surgery for neovascular glaucoma: Intermediate-term follow-up. *Acta Ophthalmol. Scand.* **2007**, *85*, 267–271. [CrossRef] [PubMed]
105. Mostafaei, A. Augmenting trabeculectomy in glaucoma with subconjunctival mitomycin C versus subconjunctival 5-fluorouracil: A randomized clinical trial. *Clin. Ophthalmol.* **2011**, *5*, 491–494. [CrossRef] [PubMed]
106. Fendi, L.; Arruda, G.; Scott, I.; Paula, J. Mitomycin C versus 5-fluorouracil as an adjunctive treatment for trabeculectomy: A meta-analysis of randomized clinical trials. *Clin. Exp. Ophthalmol.* **2013**, *41*, 798–806. [CrossRef] [PubMed]
107. Sudhakar, C.K.; Upadhyay, N.; Verma, A.; Jain, A.; Charyulu, R.N.; Jain, S. *Nanomedicine and Tissue Engineering, Nanotechnology Applications for Tissue Engineering*; William Andrew Publishing: Norwich, NY, USA, 2015; pp. 1–19. ISBN 9780323328890. [CrossRef]
108. Zarbin, M.A.; Montemagno, C.; Leary, J.F.; Ritch, R. Nanotechnology in ophthalmology. *Can. J. Ophthalmol.* **2010**, *45*, 457–476. [CrossRef] [PubMed]
109. Kwon, S.; Kim, S.H.; Khang, D.; Lee, J.Y. Potential Therapeutic Usage of Nanomedicine for Glaucoma Treatment. *Int. J. Nanomed.* **2020**, *15*, 5745–5765. [CrossRef] [PubMed]
110. Lu, H.; Wang, J.; Wang, T.; Zhong, J.; Bao, Y.; Hao, H. Recent Progress on nanostructures for drug delivery applications. *J. Nanomater.* **2016**, *2016*, 5762431. [CrossRef]
111. Bao, H.; Jiang, K.; Meng, K.; Liu, W.; Liu, P.; Du, Y.; Wang, D. TGF-β2 induces proliferation and inhibits apoptosis of human Tenon capsule fibroblast by miR-26 and its targeting of CTGF. *Biomed. Pharmacother.* **2018**, *104*, 558–565. [CrossRef]
112. Tong, J.; Fu, Y.; Xu, X.; Fan, S.; Sun, H.; Liang, Y.; Xu, K.; Yuan, Z.; Ge, Y. TGF-B1 Stimulates Human Tenon's Capsule Fibroblast Proliferation by MiR-200b and Its Targeting of P27/Kip1 and RND3. *Investig. Ophthalmol. Vis. Sci.* **2014**, *55*, 2747–2756. [CrossRef]
113. Drewry, M.D.; Challa, P.; Kuchtey, J.G.; Navarro, I.; Helwa, I.; Hu, Y.; Mu, H.; Stamer, W.D.; Kuchtey, R.W.; Liu, Y. Differentially Expressed MicroRNAs in the Aqueous Humor of Patients with Exfoliation Glaucoma or Primary Open-Angle Glaucoma. *Hum. Mol. Genet.* **2018**, *27*, 1263–1275. [CrossRef]
114. Yu, S.; Tam, A.L.C.; Campbell, R.; Renwick, N. Emerging Evidence of Noncoding RNAs in Bleb Scarring after Glaucoma Filtration Surgery. *Cells* **2022**, *11*, 1301. [CrossRef]
115. Wang, W.-H.; Deng, A.-J.; He, S.-G. A Key Role of MicroRNA-26a in the Scar Formation after Glaucoma Filtration Surgery. *Artif. Cells Nanomed. Biotechnol.* **2018**, *46*, 831–837. [CrossRef] [PubMed]
116. Ran, W.; Zhu, D.; Feng, Q. TGF-B2 Stimulates Tenon's Capsule Fibroblast Proliferation in Patients with Glaucoma via Suppression of MiR-29b Expression Regulated by Nrf2. *Int. J. Clin. Exp. Pathol.* **2015**, *8*, 4799–4806. [PubMed]
117. Deng, M.; Hou, S.-Y.; Tong, B.-D.; Yin, J.-Y.; Xiong, W. The Smad2/3/4 Complex Binds MiR-139 Promoter to Modulate TGFβ-Induced Proliferation and Activation of Human Tenon's Capsule Fibroblasts through the Wnt Pathway. *J. Cell. Physiol.* **2019**, *234*, 13342–13352. [CrossRef] [PubMed]
118. Peng, H.; Sun, Y.-B.; Hao, J.-L.; Lu, C.-W.; Bi, M.-C.; Song, E. Neuroprotective Effects of Overexpressed MicroRNA-200a on Activation of Glaucoma-Related Retinal Glial Cells and Apoptosis of Ganglion Cells via Downregulating FGF7-Mediated MAPK Signaling Pathway. *Cell Signal.* **2019**, *54*, 179–190. [CrossRef] [PubMed]
119. Tong, J.; Chen, F.; Du, W.; Zhu, J.; Xie, Z. TGF-B1 Induces Human Tenon's Fibroblasts Fibrosis via MiR-200b and Its Suppression of PTEN Signaling. *Curr. Eye Res.* **2019**, *44*, 360–367. [CrossRef] [PubMed]
120. Xu, X.; Fu, Y.; Tong, J.; Fan, S.; Xu, K.; Sun, H.; Liang, Y.; Yan, C.; Yuan, Z.; Ge, Y. MicroRNA-216b/Beclin 1 Axis Regulates Autophagy and Apoptosis in Human Tenon's Capsule Fibroblasts upon Hydroxycamptothecin Exposure. *Exp. Eye Res.* **2014**, *123*, 43–55. [CrossRef]
121. Zhu, H.; Dai, L.; Li, X.; Zhang, Z.; Liu, Y.; Quan, F.; Zhang, P.; Yu, L. Role of the Long Noncoding RNA H19 in TGF-B1-Induced Tenon's Capsule Fibroblast Proliferation and Extracellular Matrix Deposition. *Exp. Cell Res.* **2020**, *387*, 111802. [CrossRef] [PubMed]
122. Zhao, Y.; Zhang, F.; Pan, Z.; Luo, H.; Liu, K.; Duan, X. LncRNA NR_003923 Promotes Cell Proliferation, Migration, Fibrosis, and Autophagy via the MiR-760/MiR-215-3p/IL22RA1 Axis in Human Tenon's Capsule Fibroblasts. *Cell Death Dis.* **2019**, *10*, 594. [CrossRef]
123. Sui, H.; Fan, S.; Liu, W.; Li, Y.; Zhang, X.; Du, Y.; Bao, H. LINC00028 Regulates the Development of TGFβ1-Treated Human Tenon Capsule Fibroblasts by Targeting MiR-204-5p. *Biochem. Biophys. Res. Commun.* **2020**, *525*, 197–203. [CrossRef]

124. You, K.; Gu, H.; Yuan, Z.; Xu, X. Tumor Necrosis Factor Alpha Signaling and Organogenesis. *Front. Cell Dev. Biol.* **2021**, *9*, 727075. [CrossRef]
125. Collotta, D.; Colletta, S.; Carlucci, V.; Fruttero, C.; Fea, A.M.; Collino, M. Pharmacological Approaches to Modulate the Scarring Process after Glaucoma Surgery. *Pharmaceuticals* **2023**, *16*, 898. [CrossRef] [PubMed]
126. Ebert, E.C. Infliximab and the TNF-alpha system. *Am. J. Physiol. Gastrointest. Liver Physiol.* **2009**, *296*, G612–G620. [CrossRef] [PubMed]

Disclaimer/Publisher's Note: The statements, opinions and data contained in all publications are solely those of the individual author(s) and contributor(s) and not of MDPI and/or the editor(s). MDPI and/or the editor(s) disclaim responsibility for any injury to people or property resulting from any ideas, methods, instructions or products referred to in the content.

Systematic Review

Managing Ocular Surface Disease in Glaucoma Treatment: A Systematic Review

Özlem Evren Kemer [1], Priya Mekala [2], Bhoomi Dave [2,3] and Karanjit Singh Kooner [2,4,*]

1. Department of Ophthalmology, University of Health Sciences, Ankara Bilkent City Hospital, Ankara 06800, Turkey; ozlemvidya@gmail.com
2. Department of Ophthalmology, University of Texas Southwestern Medical Center, Dallas, TX 75390, USA; priya.mekala@utsouthwestern.edu (P.M.); bhd27@drexel.edu (B.D.)
3. Drexel University College of Medicine, Philadelphia, PA 19129, USA
4. Department of Ophthalmology, Veteran Affairs North Texas Health Care System Medical Center, Dallas, TX 75216, USA
* Correspondence: karanjit.kooner@utsouthwestern.edu; Tel.: +1-214-648-4973; Fax: +1-214-648-2469

Abstract: Ocular surface disease (OSD) is a frequent disabling challenge among patients with glaucoma who use benzalkonium chloride (BAK)-containing topical glaucoma medications for prolonged periods. In this comprehensive review, we evaluated the prevalence of OSD and its management, focusing on both current and future alternatives. Preferred Reporting Items for Systematic Reviews and Meta-Analyses (PRISMA) criteria were used to assess a) the impact of active ingredients and preservatives on the ocular surface and b) the efficacy of preservative-free (PF) alternatives and adjunctive therapies. BAK-containing glaucoma medications were found to significantly contribute to OSD by increasing corneal staining, reducing tear film stability, and elevating ocular surface disease index (OSDI) scores. Transitioning to PF formulations or those with less cytotoxic preservatives, such as Polyquad® and SofZia®, demonstrated a marked improvement in OSD symptoms. In particular, the use of adjunct cyclosporine A, through its anti-inflammatory and enhanced tear film stability actions, was shown to be very beneficial to the ocular surface. Therefore, the most effective management of OSD is multi-factorial, consisting of switching to PF or less cytotoxic medications, adjunct use of cyclosporine A, and early incorporation of glaucoma surgical treatments such as laser trabeculoplasty, trabeculectomy, glaucoma drainage devices, or minimally invasive glaucoma surgery (MIGS).

Keywords: ocular surface disease; glaucoma; topical medications; preservatives; benzalkonium chloride

Citation: Kemer, Ö.E.; Mekala, P.; Dave, B.; Kooner, K.S. Managing Ocular Surface Disease in Glaucoma Treatment: A Systematic Review. *Bioengineering* **2024**, *11*, 1010. https://doi.org/10.3390/bioengineering11101010

Academic Editor: Hiroshi Ohguro

Received: 9 September 2024
Revised: 2 October 2024
Accepted: 6 October 2024
Published: 11 October 2024

Copyright: © 2024 by the authors. Licensee MDPI, Basel, Switzerland. This article is an open access article distributed under the terms and conditions of the Creative Commons Attribution (CC BY) license (https://creativecommons.org/licenses/by/4.0/).

1. Introduction

Glaucoma, a global multi-factorial disease, is characterized by progressive degeneration of the optic nerve with or without elevated intraocular pressure (IOP). It is the most common cause of irreversible blindness, and its global prevalence is estimated to be around 4% in patients between the ages of 40 and 80 years [1]. Topical medical therapy has been most commonly used for many years. Research indicates that the average number of medications prescribed is 3.09 and eye drops form the bulk of therapy [2]. This chronic use of multiple topical drugs, combined with other factors such as age and systemic comorbidities and their treatments, profoundly contributes to ocular surface disease (OSD).

OSD is a complex condition that impacts both the tears and the ocular surface (Figure 1), leading to various symptoms such as discomfort, visual disturbances, and tear film instability [3]. It is characterized by increased tear film osmolarity and inflammation, which can manifest clinically as superficial punctate keratitis (SPK), conjunctival hyperemia, and papillary conjunctivitis (Figure 2). The etiology of OSD is diverse and includes environmental and genetic factors, aging, dry eye syndrome, blepharitis, meibomian gland dysfunction

(MGD), and the chronic use of eye drops with preservatives [4] (Figures 3 and 4). Thus, OSD has broader implications than dry eye disease (DED) alone [5,6].

Ocular surface inflammation is thought to play a key role in the pathogenesis of OSD [3]. A 2020 meta-analysis by Roda et al. analyzed 13 articles involving 342 patients with DED and 205 healthy controls. Their systematic review revealed that DED patients had higher levels of tear interleukin (IL)-1β, IL-6, IL-8, IL-10, interferon-γ (IFN-γ), and tumor necrosis factor-α (TNF-α) compared to controls [7]. However, the Dry Eye Assessment and Management (DREAM) study, which analyzed 131 patient tear samples for various tear cytokines levels, including IL-1β, IL-6, IL-8, IL-10, IL-17A, IFNγ, and TNFα, found that only cytokines IL-10, IL-17A, and IFNγ were highly correlated with each other but weakly correlated with some DED signs [8].

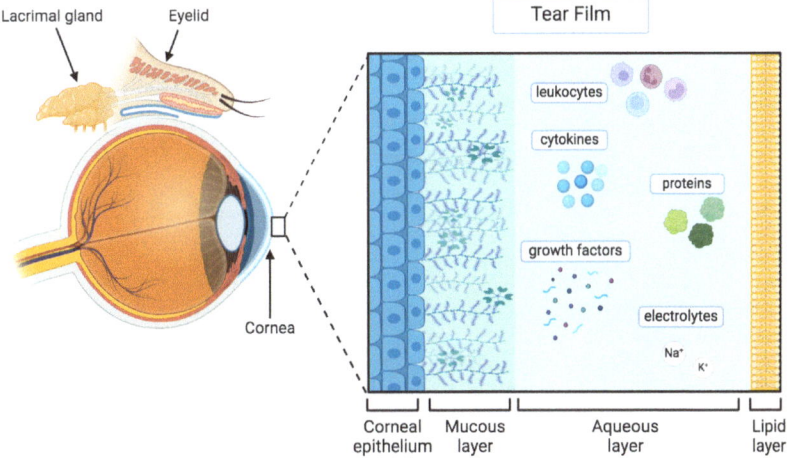

Figure 1. Representation of the ocular surface and tear film composition (corneal epithelium, mucous layer, aqueous layer, and lipid layer). (Figure made using BioRender® software, version 201 and adapted from [9]).

Research indicates that 48–59% of patients with glaucoma experience symptoms of OSD, while 22–78% may exhibit clear clinical signs [10,11]. The long-term use of glaucoma medications, especially those containing benzalkonium chloride (BAK), often exacerbates OSD, leading to decreased quality of life, reduced adherence to treatment, and diminished therapeutic efficacy [12]. Therefore, prompt and effective management of OSD is paramount to maintaining treatment effectiveness, considering the higher prevalence of pre-existing dry eyes in this age group [12].

1.1. Diagnosis of OSD

Various clinical tests, symptom questionnaires (Ocular Surface Disease Index [OSDI]), and imaging modalities are utilized in the diagnosis of OSD. Common clinical tests include Schirmer's test, invasive tear break-up time (TBUT), fluorescein staining, and lissamine green staining [13]. The OSDI questionnaire consists of 12 questions (three for ocular symptoms, six for vision-related functions, and three for environmental triggers). The scores range from 0 to 100, with higher values corresponding to a greater impact on a patient's daily life: 0–12, normal; 13–22, mild; 23–32, moderate; and 33–100, severe. Various corneal imaging devices can provide information regarding the tear meniscus height (TMH), non-invasive tear break-up time (NITBUT), and meibography [14]. Unlike invasive TBUT, NITBUT measurements are performed without fluorescein dye, utilizing videokeratoscopy to detect variations in the placido disks that are reflected on the cornea (Figure 5) [15].

Meibography evaluates the meibomian glands in vivo and a meiboscore can be calculated to quantify loss of meibomian glands (Figure 6).

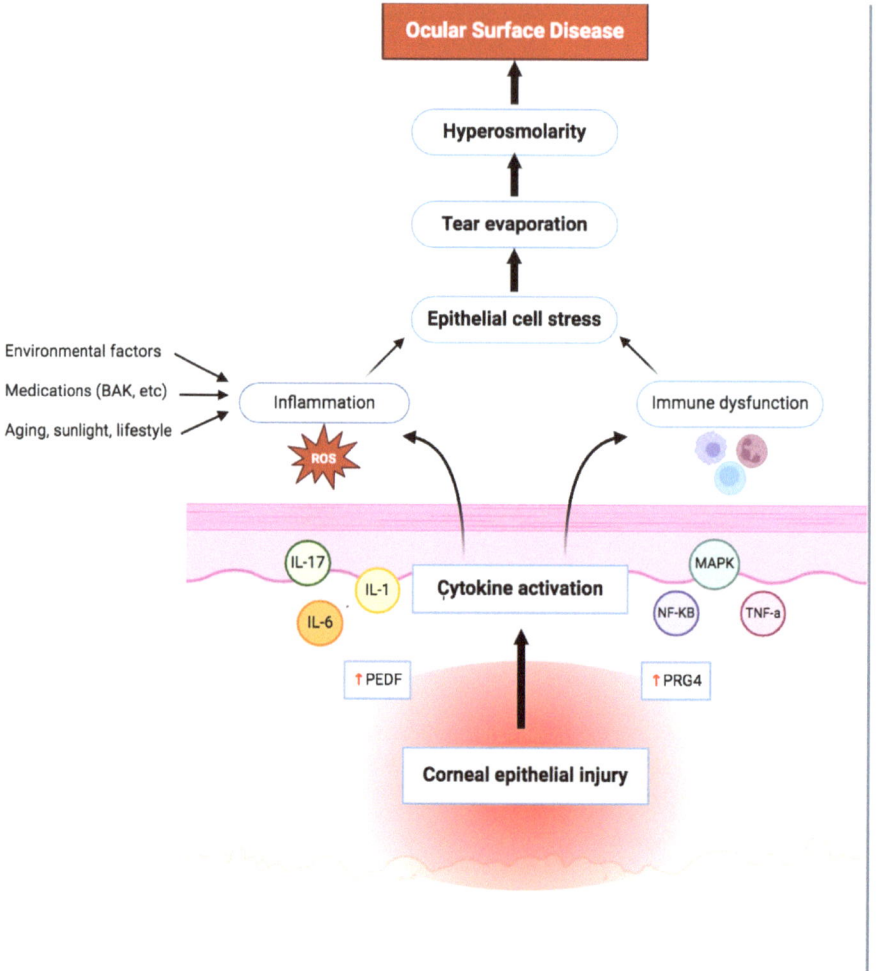

Legend: PEDF = Pigment epithelium-derived factor, PRG4 = proteoglycan 4, IL = interleukin, NF-KB = nuclear factor kappa B, MAPK = mitogen-activated protein kinases, TNF-α = tumor necrosis factor alpha, BAK = benzalkonium chloride, ROS = reactive oxygen species.

Figure 2. A brief overview of the immune-inflammatory mechanisms in the pathogenesis of ocular surface disease (Figure made using BioRender® software, version 201).

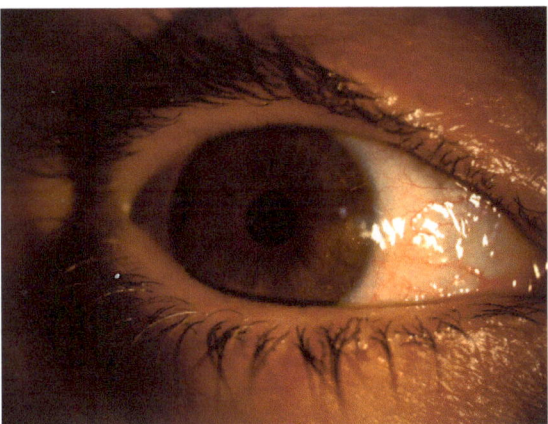

Figure 3. External photograph of an eye with OSD showing MGD, blepharitis, and conjunctival hyperemia. Image courtesy of Karanjit S. Kooner, MD, PhD (University of Texas Southwestern Medical Center, Dallas, TX, USA).

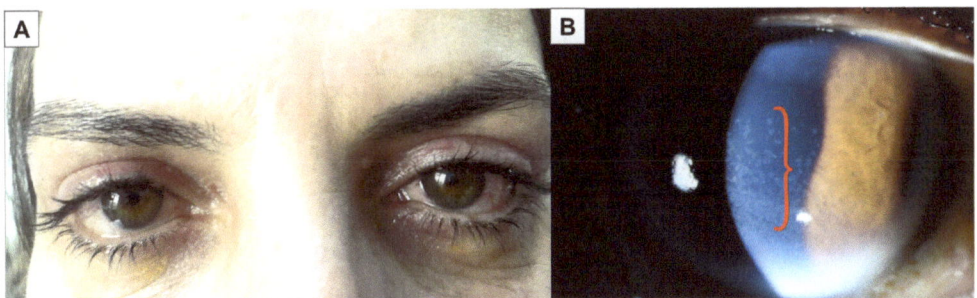

Figure 4. Clinical photographs of ocular surface disease. (**A**) External photograph of a patient with chronic hyperemia and MGD. (**B**) Slit lamp photograph of an eye with superficial punctate keratitis (red curly bracket). Images courtesy of Özlem Evren Kemer, MD (Ankara Bilkent City Hospital, Ankara, Turkey) and Margaret Wang French, MD (University of Texas Southwestern Medical Center, Dallas, TX, USA).

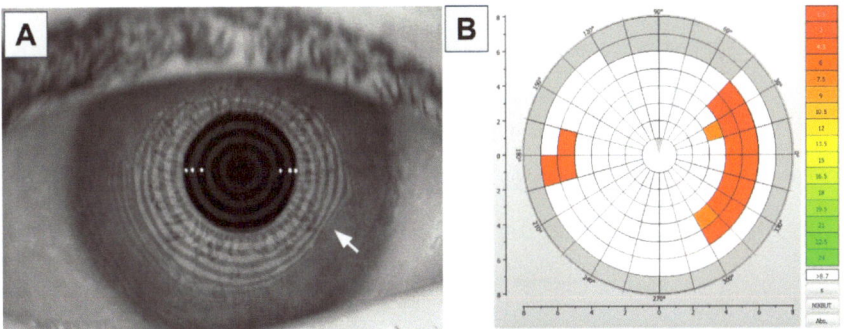

Figure 5. Keratograph of an eye with OSD. (**A**) Keratograph of an eye with areas of dryness (arrow) disrupting the placido disk reflections on the cornea. (**B**) Red-orange areas correspond to faster NITBUT. (OCULUS Keratograph®, OCULUS, Wetzlar, Germany). Images courtesy of Karanjit S. Kooner, MD, PhD (University of Texas Southwestern Medical Center, Dallas, TX, USA).

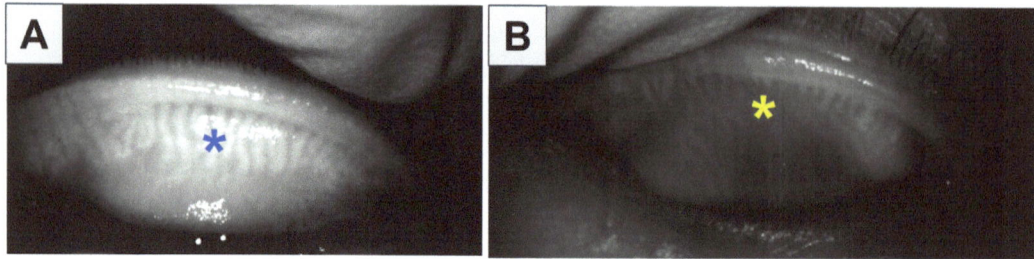

Figure 6. Clinical examples of meibography. (**A**) Meibography in a patient with healthy meibomian glands (asterisk). (**B**) MGD with significant atrophy of meibomian glands with ghosting (pale glands with abnormal meibomian gland architecture, asterisk). Images courtesy of Karanjit S. Kooner, MD, PhD (University of Texas Southwestern Medical Center, Dallas, TX, USA).

1.2. Previous Research

Previous literature reviews describe adverse effects of anti-glaucoma medications on various ocular and periocular structures, mention the effects of some active ingredients and preservatives on the ocular surface, and outline some emerging medication delivery systems [16,17]. However, few studies exist that clearly and thoroughly describe the topical complications of each active ingredient and preservative present in anti-glaucoma treatments and provide a broad overview of the major innovations and future directions.

The purpose of this systematic review is to assess a) the impact of active ingredients and preservatives of anti-glaucoma treatments on the ocular surface and b) the efficacy of preservative-free (PF) alternatives and adjunctive therapies. The article also includes an overview of the future directions and novel therapies in the management of OSD in patients using topical glaucoma medications.

2. Materials and Methods

2.1. Initial Search

Our study was approved by the institutional review board of Ankara Bilkent City Hospital and exempted from full review as no patient information was used. We followed the Preferred Reporting Items for Systematic Reviews and Meta-Analyses (PRISMA) guidelines during data collection and the PICOS (Population, Intervention, Comparison, Outcomes, and Study) framework to create eligibility criteria, Table 1, [18]. The following keywords and MeSH terms were used: "glaucoma" (or "glaucoma, angle-closure", "glaucoma, open-angle"), "dry eye syndromes", "ocular surface disease", "antiglaucoma agents" (or "ophthalmic solutions"), and "preservatives, pharmaceutical" (or "benzalkonium compounds").

Table 1. PICOS criteria for inclusion of studies.

Parameter	Description
Population	Patients with glaucoma regardless of study location
Intervention	Focusing on patients using anti-glaucoma eye drops with or without preservatives
Comparison	Patients using topical eye drops with or without preservatives
Outcomes	OSDI, Schirmer's test, corneal and conjunctival staining (fluorescein, lissamine green), conjunctival hyperemia, meibography, TMH, TBUT, NITBUT
Study Design	Cohort, cross-sectional, case-control, randomized or nonrandomized controlled (or uncontrolled) trials, or reviews

Utilizing these keywords and MeSH terms, we systematically searched the online databases of PubMed (MEDLINE), Cochrane Library (Wiley), ScienceDirect, Scopus, Google

Scholar, ProQuest, and Web of Science up to 20 July 2024. Comma-separated values (CSV) or Microsoft Excel files (Microsoft® Excel, Redmon, WA, USA, version 16.87) were downloaded directly from each database. Considering Google Scholar search results, they were downloaded in CSV format utilizing the Publish or Perish software program (Anne-Wil Harzing, London, England, version 8.12.4612) [19]. All citations were then compiled in a single CSV file. There was a total of 16,119 articles obtained through this preliminary search (Figure 7).

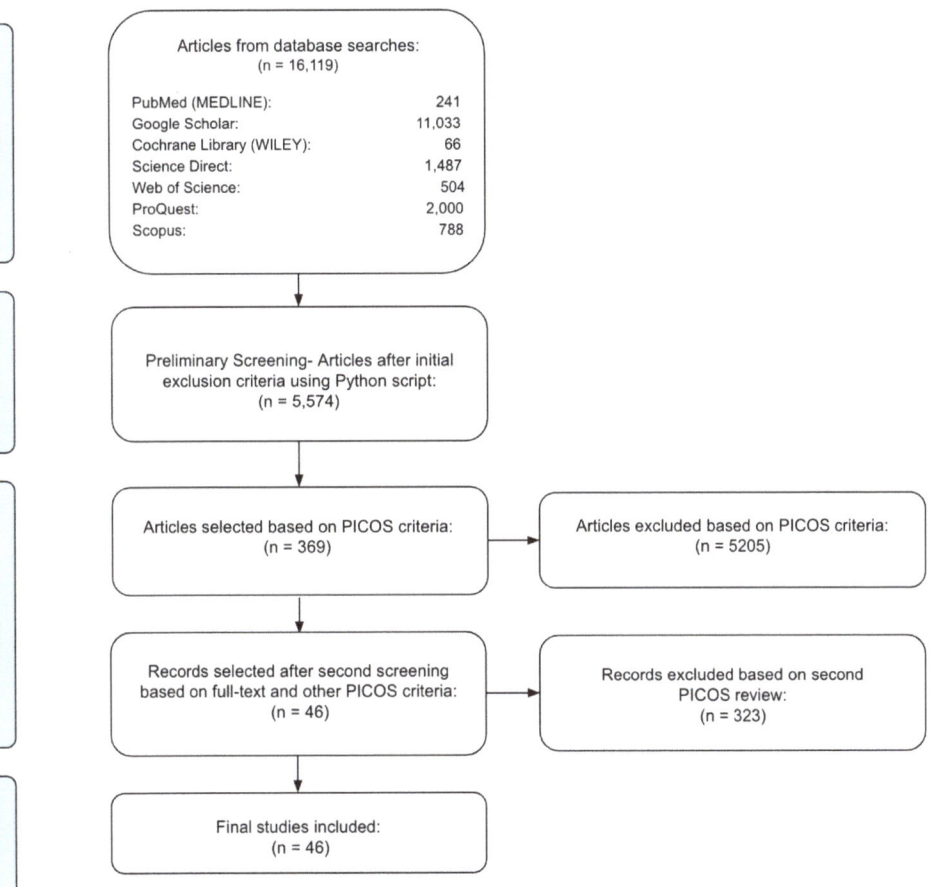

Figure 7. PRISMA flow chart.

2.2. Preliminary Screening

We excluded duplicates, non-English language articles, conference abstracts, and commentaries using a Python script (Python Software Foundation, Wilmington, DE, USA, version 3.12.2). The remaining articles were stored in a single CSV and contained author names, title, date of publication, journal name, and digital object identifier (DOI). A total of 5574 articles remained after preliminary screening.

2.3. Eligibility Assessment

Each article in the CSV was screened utilizing the PICOS criteria mentioned in Table 1, focusing on full-text English articles and studies involving animal or human subjects. After careful screening, an initial 369 articles was finally reduced to 46.

3. Results

Out of 16,119 articles initially identified, only 46 qualified for our final review based on our strict criteria.

3.1. Active Ingredients

There are multiple anti-glaucoma medications available, and they act via different pathways (Table 2). The active ingredients in them may directly irritate and disrupt the ocular surface via several mechanisms, such as toxicity to corneal epithelium leading to cytokine activation, inflammation, immune system dysfunction, epithelial cell stress, tear evaporation, and hyperosmolarity, contributing to the symptoms of OSD (Figure 2). The main clinical studies examining the side effects of glaucoma medications in particular OSD are shown in Table 3.

Table 2. Characteristics of glaucoma medications.

Medications	Mechanism of Action	Dosing & Concentrations	OSD or Other Complications	IOP Reduction
Beta-adrenergic blockers (timolol, levobunolol, betaxolol, metipranolol) [5,20–27]	Decrease aqueous humor (AH) production via blockade of beta-adrenergic receptors on the ciliary epithelium	Once or twice daily; 0.25–0.5%	Conjunctival goblet cell loss, MGD, SPK, and pseudo-pemphigoid cicatrizing conjunctivitis	~20–30%
Prostaglandin analogues (latanoprost, bimatoprost, travoprost, tafluprost) [5,28–30]	Increase uveoscleral outflow by remodeling the ECM and regulating matrix metalloproteinases	Once daily; 0.0015–0.03%	MGD, skin pigmentation, conjunctival hyperemia, pseudo-dendritic keratitis, periorbitopathy, eyelid pigmentation, and hypertrichosis	~25–35%
Alpha-adrenergic agonists (brimonidine, apraclonidine) [31,32]	Selective sympathetic agonists (α2); decrease AH production, and increase uveoscleral and trabecular meshwork (TM) outflow	2–3 times daily; 0.1–0.5%	Allergic follicular conjunctivitis, contact dermatitis, blepharitis, and systemic hypotension	up to 26%
Carbonic anhydrase inhibitors (dorzolamide, brinzolamide), (oral: acetazolamide, methazolamide) [33,34]	Decrease AH production by inhibiting carbonic anhydrase enzyme in the ciliary processes	2–4 times daily; 1–2%	Ocular surface irritation, reduction of basal tear secretion, and blepharitis	~15–20%
Cholinergic agonists (pilocarpine, carbachol) [5,35–37]	Muscarinic receptor agonists; increase TM outflow	4 times daily; 1–4%	MGD, blepharitis, pseudo-pemphigoid cicatrizing conjunctivitis, blurred vision, myopia, miosis, iris cysts, and retinal detachment	~15–25%
Latanoprostene bunod (Vyzulta®) [38]	Induces TM expansion and vasodilation of episcleral veins, thereby increasing AH outflow	Once daily; 0.024%	Hyperemia, hypertrichosis, and eye irritation	~35%
Rho Kinase inhibitors (netarsudil—Rhopressa®) [38,39]	Decrease episcleral venous pressure, increase TM outflow, and decrease AH production via inhibition of rho kinase enzyme	Once daily; 0.02%	Conjunctival hyperemia and hemorrhage, corneal edema, and SPK	~25–30%

Table 2. Cont.

Medications	Mechanism of Action	Dosing & Concentrations	OSD or Other Complications	IOP Reduction
Dorzolamide and timolol maleate solution (combined)	Decrease AH production via a combination of carbonic anhydrase and beta-adrenergic receptor blockade	Twice daily; timolol 0.5%, dorzolamide 2%	Conjunctival goblet cell loss, MGD, SPK, pseudo-pemphigoid cicatrizing conjunctivitis, ocular surface irritation, reduction of basal tear secretion, and blepharitis	~30–35%
Brimonidine tartrate and timolol maleate solution (combined)	Decrease AH production, increase uveoscleral outflow, and increase TM outflow via a combination of alpha and beta-adrenergic receptor blockade	Twice daily; timolol 0.5%, brimonidine 0.2%	Allergic follicular conjunctivitis, contact dermatitis, blepharitis, conjunctival goblet cell loss, MGD, SPK, and pseudo-pemphigoid cicatrizing conjunctivitis	~30–35%
Netarsudil and latanoprost solution (Rocklatan®)	Decrease episcleral venous pressure, increase TM outflow, and decrease AH production via a combination of rho kinase inhibition and prostanoid receptor induction	Once daily; netarsudil 0.02%, latanoprost 0.005%	Hyperemia, conjunctival hemorrhage, MGD, lid pigmentation, pseudo-dendritic keratitis, periorbitopathy, and hypertrichosis	~30–36%
Brimonidine and brinzolamide solution (combined)	Decrease AH production, and increase uveoscleral and TM outflow via inhibition of carbonic anhydrase and alpha-adrenergic receptors	3 times daily; brimonidine 1%, brinzolamide 0.2%	Ocular surface irritation, reduction of basal tear secretion, blepharitis, allergic follicular conjunctivitis, and contact dermatitis	~21–35%

Legend: OSD = ocular surface disease, IOP = intraocular pressure, AH = aqueous humor, MGD = meibomian gland dysfunction, SPK = superficial punctate keratitis, ECM = extracellular matrix, TM = trabecular meshwork.

Table 3. Key studies regarding ocular surface disease in patients on glaucoma medications.

Glaucoma Agents and Patient Characteristics	Study Methods	Study Results	Authors, Country, and Year
Newly diagnosed treatment-naïve POAG patients vs. those on topical anti-glaucoma medications	A prospective cohort study conducted on 120 eyes with POAG (60 on topical anti-glaucoma drops and 60 treatment-naïve eyes).	At 3, 6, and 12 months, the OSDI score, TBUT, Schirmer's test, TMH, and TMD had significantly better values in the treatment-naïve group in comparison to the medicated group ($p < 0.0001$).	Srivastava et al. India, 2024 [40]
Patients with open-angle glaucoma or OHT on topical anti-glaucoma medications vs. healthy subjects	In this cross-sectional study, 75 patients were using topical anti-glaucoma medications and 65 were treatment-naïve subjects. OSDI, Schirmer's test, TBUT, fluorescein staining, and CET were evaluated.	The treatment group had a significantly shorter TBUT, shorter Schirmer's test, and greater fluorescein staining than those of the control group ($p < 0.05$). The mean CET of patients with glaucoma was significantly lower than that of controls in the central, paracentral, mid-peripheral, and peripheral zones (50.6 vs. 53.1 µm; $p < 0.001$). The number of medications and duration of treatment also affected the CET in all zones ($p < 0.05$).	Ye et al. China, 2022 [41]

Table 3. Cont.

Glaucoma Agents and Patient Characteristics	Study Methods	Study Results	Authors, Country, and Year
Glaucoma patients on topical anti-glaucoma medications vs. healthy controls	94 patients with glaucoma on topical medications (study group) and 94 patients in the treatment-naïve control group were assessed using OSDI, TBUT, lissamine green staining, and Schirmer's test.	OSDI scores were significantly higher in the study group (72.4%) vs. controls (44.6%). Similarly, the study group had decreased tear production (84% vs. 53%, respectively), abnormal TBUT (67.1% vs. 47.8%), and positive lissamine green staining (36.2% vs. 31.8%) compared to the control group.	Pai and Reddy India, 2018 [42]
Patients with POAG or OHT on topical anti-glaucoma medications vs. healthy controls	211 eyes of patients with POAG or OHT on topical medication were recruited. Controls consisted of 51 eyes. Outcome measures were fluorescein corneal staining score, TMH, TBUT, and OSDI.	Compared to controls, significantly higher OSDI (10.24 vs. 2.5; $p < 0.001$) and corneal staining (≥ 1: 64.93% vs. 32.61%; $p < 0.001$) scores were recorded in the medication group. No significant differences in TBUT and TMH were observed between groups.	Pérez-Bartolomé et al. Spain, 2017 [43]
Glaucoma patients on topical anti-glaucoma medications vs. OHT patients or relatives of glaucoma patients not on topical medications	In this cross-sectional study, 109 participants (79 on topical medications and 30 controls) were evaluated via OSDI, Schirmer's test, TBUT, and fluorescein staining.	The medication group had significantly shorter TBUT (6.0 vs. 9.5 s; $p < 0.03$), greater fluorescein staining (1.0 vs. 0; $p < 0.001$), and higher impression cytology grade than the control group (1.0 vs. 0.6; $p < 0.001$).	Cvenkel et al. Slovenia, 2015 [44]
Patients with POAG on topical anti-glaucoma medications vs. healthy controls	Age-matched patients were assigned to 2 groups: the glaucoma group (31 patients) and the treatment-naïve control group (30 patients). Each patient was assessed with OSDI, conjunctival/corneal staining, and TBUT.	OSDI scores of the glaucoma group positively correlated to the amount and duration of drops used. The glaucoma group had a higher mean OSDI score than the control group (18.97 vs. 6.25). Abnormal TBUT and staining scores were seen in the glaucoma group compared with the control group (68% vs. 17%).	Saade et al. USA, 2015 [45]
Patients with glaucoma or OHT on 0, 1, or ≥ 2 topical anti-glaucoma medications	39 patients treated for glaucoma or OHT and 9 untreated patients were included in this study. Corneal sensitivity was measured using the Cochet-Bonnet esthesiometer, Schirmer's test, TBUT, corneal and conjunctival fluorescein staining, and OSDI.	Corneal sensitivity of patients treated with IOP-lowering medications was negatively correlated to the number of instillations of P drops ($p < 0.001$) and duration of treatment ($p = 0.001$). There was no significant difference in OSDI or Schirmer's test scores between the groups.	Van Went et al. France, 2011 [46]
Patients with POAG, pseudoexfoliation glaucoma, pigment dispersion glaucoma, or OHT on topical anti-glaucoma medications	This prospective observational study assessed OSDI in 630 patients with POAG, pseudoexfoliation glaucoma, pigment dispersion glaucoma, or OHT who were on topical IOP-lowering medications.	305 patients (48.4%) had an OSDI score indicating either mild, moderate, or severe OSD symptoms. Higher OSDI scores were observed in patients using multiple IOP-lowering medications ($p = 0.0001$).	Fechtner et al. USA, 2010 [47]

Table 3. *Cont.*

Glaucoma Agents and Patient Characteristics	Study Methods	Study Results	Authors, Country, and Year
Patients using P vs. PF topical beta-blocker drops	In a multicenter cross-sectional survey in four European countries, ophthalmologists in private practice enrolled 9658 patients using P or PF beta-blocking eyedrops between 1997 and 2003. Subjective symptoms, conjunctival and palpebral signs, and SPK were assessed before and after a change in therapy.	Palpebral, conjunctival, and corneal signs were significantly more frequent ($p < 0.0001$) in the P-group than in the PF-group, such as pain or discomfort during instillation (48% vs. 19%), foreign body sensation (42% vs. 15%), stinging or burning (48% vs. 20%), and dry eye sensation (35% vs. 16%). A significant decrease ($p < 0.0001$) in all ocular symptoms was observed in patients who switched from P to PF eye drops.	Jaenen et al. Belgium, 2007 [48]
Patients with POAG or OHT using P vs. PF topical anti-glaucoma medications	This prospective epidemiological survey was carried out in 1999 by 249 ophthalmologists on 4107 patients. Ocular symptoms, conjunctiva, and cornea were assessed between P and PF eye drops.	All symptoms were more prevalent with P than with PF drops ($p < 0.001$): discomfort upon instillation (43% vs. 17%), burning-stinging (40% vs. 22%), foreign body sensation (31% vs. 14%), dry eye sensation (23% vs. 14%), and tearing (21% vs. 14%). An increased incidence (>2 times) and duration of ocular signs were seen with P eye drops, which decreased upon switching to PF drops ($p < 0.001$).	Pisella et al. France, 2002 [49]

Legend: POAG = primary open-angle glaucoma, OSDI = ocular surface disease index, TBUT = tear break-up time, TMH = tear meniscus height, TMD = tear meniscus depth, OHT = ocular hypertension, CET = corneal epithelial thickness, IOP = intraocular pressure, P = preserved, PF = preservative-free, SPK = superficial punctate keratitis.

3.1.1. Beta-Adrenergic Blockers

Topical beta-adrenergic blockers reduce aqueous humor (AH) and tear production by blocking beta receptors both on the ciliary epithelium and the main and accessory lacrimal glands [50]. In addition, their sympathomimetic activity may interfere with the epithelial cell viability/homeostasis. Thus, they have several side effects, such as a decrease in tear volume, MGD, conjunctival goblet cell loss, pseudo-pemphigoid cicatrizing conjunctivitis (Figure 8), and nasolacrimal duct obstruction [5,20].

Kuppens et al. reported that the TBUT decreased significantly in patients using both preserved (P) and timolol-PF in comparison to the control group. Thus, timolol-PF and P timolol formulations may both alter the tear film [21]. Other studies have shown that topical beta-blockers may also interfere with the corneal epithelium by inhibiting the sympathetic activity of limbal stem cells, resulting in SPK [22,23]. Laser scanning confocal microscopy and impression cytology have both revealed that beta blockers are toxic to the limbal stem cell microenvironment, thereby delaying corneal epithelial regeneration [23].

In 2003, a Japanese study involving 110 patients with glaucoma (35–88 years with mean age 69.7 ± 10.8) found that SPK was observed in 29.0% of cases [24]. Timolol users had a significantly higher occurrence of SPK (46.2%) compared to those using carteolol (4.2%). Interestingly, the prevalence of SPK was higher in patients using more than two anti-glaucoma eye drops (35.9%) compared to those using no eye drops (19.7%) or only one eye drop (30.9%). Notably, PF timolol still caused tear instability, suggesting that the active ingredient may damage the ocular surface [25]. A cross-sectional study comparing patients on PF timolol maleate (48 eyes) with healthy controls (40 eyes) found that TBUT was significantly higher in controls compared to patients on timolol maleate-PF [26].

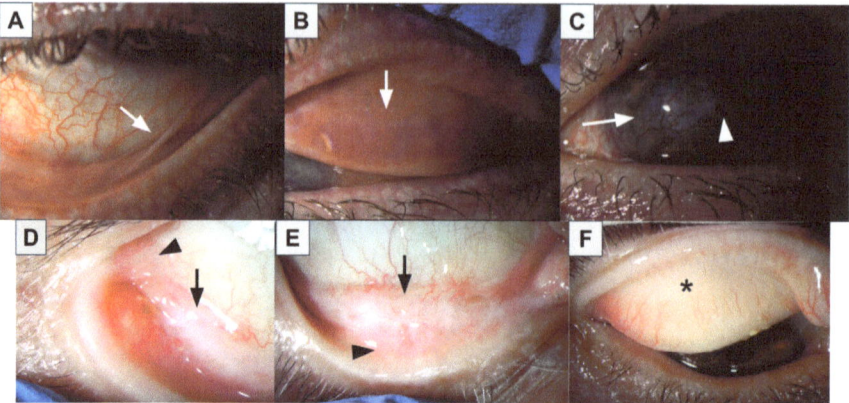

Figure 8. Two patients with ocular cicatricial pemphigoid ((**A**–**C**) patient 1) and ((**D**–**F**) patient 2). (**A**) symblepharon (arrow); (**B**) supratarsal conjunctival scarring (arrow); (**C**) corneal scarring, neovascularization (arrow), and healed descemetocele (arrowhead); (**D**) symblepharon (arrow) and subconjunctival fibrosis (arrowhead); (**E**) symblepharon (arrow), subepithelial fibrosis (arrowhead), inferior forniceal shortening; (**F**) meibomian gland dropout with subepithelial fibrosis (asterisk). Images courtesy of Karanjit S. Kooner, MD, PhD (University of Texas Southwestern Medical Center, Dallas, TX, USA) and Özlem Evren Kemer, MD (Ankara Bilkent City Hospital, Ankara, Turkey).

In an animal study involving New Zealand white rabbits, Russ et al. found that timolol increases subepithelial collagen density and extracellular matrix (ECM) more than prostaglandin analogs (PGAs), thus potentially interfering with glaucoma filtration surgery outcomes [27].

3.1.2. Prostaglandin Analogs

PGAs decrease IOP by remodeling the ECM in the ciliary muscle bundles, iris root, and sclera, thereby increasing uveoscleral outflow. In addition, there may be remodeling of corneal collagen fibers, resulting in decreased central corneal thickness. Other well-documented PGA side effects include skin pigmentation, MGD, conjunctival hyperemia, pseudo-dendritic keratitis, periorbitopathy, eyelid pigmentation, and hypertrichosis [5,28].

In 2016, Yamada et al., using human non-pigmented ciliary epithelial cells, studied bimatoprost, latanoprost, and tafluprost and found elevated matrix metalloproteinase (MMP) levels and reduced levels of tissue inhibitors metalloproteinases (TIMP-1 and TIMP-2) [29].

Similarly, a Turkish study in 2016 involving 70 glaucoma patients found that long-term use of PGAs was significantly associated with a higher prevalence of MGD (92% vs. 58.3% in non-PGA users). These patients also exhibited worse OSDI scores (22.5 ± 24.3 vs. 1.9 ± 3.4), tear film stability, and MGD (95.7%) [30].

3.1.3. Alpha-Adrenergic Agonists

Alpha-adrenergic agonists (brimonidine and apraclonidine) are selective sympathetic agonists of the $\alpha 2$ receptor and thus have multiple effects: (1) decreased AH production, (2) increased uveoscleral outflow, and (3) increased trabecular meshwork (TM) outflow.

Research has shown that the common follicular conjunctivitis may result from alpha-adrenergic agonists' effect on reducing the volume of conjunctival cells, thereby widening intracellular spaces and permitting potential allergens to penetrate subepithelial tissue [31]. The incidence of brimonidine allergy ranges from 4.7% to 25%, with the average time from the start of treatment to the onset of allergic follicular conjunctivitis being six to nine months. However, this interval can vary widely, from as short as 14 days to as long as 12 months and is independent of the presence of BAK [31].

These agents should not be used in children due to the potential for central nervous system depression given that topical alpha-adrenergic agonists are not weight-adjusted [32].

3.1.4. Carbonic Anhydrase Inhibitors

Carbonic anhydrase inhibitors (CAIs) can adversely affect tear film stability, with surface conditions such as hyperemia, blepharitis, dry eyes, and tearing occurring in less than 3% of cases [33,34]. Terai and colleagues discovered that brinzolamide reduced basal tear secretion, although it did not significantly affect TBUT. Specifically, dorzolamide was found to reduce basal tear secretion by 14.3% at 60 min and by 17.3% at 90 min post-application [34]. CAIs are generally avoided in patients who have sulfa allergies or a history of nephrolithiasis.

3.1.5. Cholinergic Agonists

Cholinergic agonists (pilocarpine and carbachol) activate the muscarinic type 3 receptors on ciliary smooth muscle cells, resulting in expansion of the juxtacanalicular portion of the TM and expansion of the Schlemm's canal [35]. It also acts on the iris sphincter muscles, inducing miosis. An in vitro study using immortalized human meibomian gland epithelial cells (IHMGEC) found that pilocarpine led to a dose-dependent decrease in IHMGEC proliferation, leading to cell atrophy and death [36]. Adverse effects of pilocarpine include conjunctival hyperemia, MGD, blepharitis, pseudo-pemphigoid cicatrizing conjunctivitis, burning/stinging, eye pain, blurred vision, increased corneal staining, and headache [5,37].

3.1.6. Latanoprostene Bunod

Latanoprostene bunod (LBN) 0.024%, commercially available as Vyzulta®, is a nitric oxide (NO)-donating prostaglandin F2α analogue which increases the aqueous outflow both by uveoscleral and trabecular pathways. The NO relaxes TM cells and facilitates the trabecular outflow. NO also may regulate ocular blood flow and may promote retinal ganglion cell (RGC) survival in the eye. The latanoprost acid, the second active metabolite, shares the familiar mechanism of action of PGAs by increasing the uveoscleral outflow. The most common ocular adverse effects of LBN were conjunctival hyperemia, hypertrichosis, eye irritation, eye pain, and an increase in iris pigmentation [38].

3.1.7. Netarsudil

Netarsudil 0.02% (Rhopressa®) is a rho-associated kinase (ROCK) inhibitor and a norepinephrine transporter (NET) inhibitor. It has a tri-faceted mechanism of action: it increases the TM outflow, decreases episcleral venous pressure, and decreases AH production [39]. Furthermore, it may decrease RGC loss by improving optic nerve head perfusion by its effect on endothelin 1. Common side effects include conjunctival hyperemia, subconjunctival bleeding, SPK, corneal edema, and whorl or honeycomb keratopathy [38].

3.2. Preservatives

Preservatives in glaucoma medications are crucial for preventing microbial contamination and ensuring their longevity, safety, and efficacy. These preservatives can be broadly categorized as detergents, oxidative agents, and ionic tamponade agents (Table 4).

Table 4. Common preservatives in ocular formulations [51].

Category	Examples
Detergents	benzalkonium chloride (BAK) polidronium chloride (polyquaternium-1, Polyquad®)
Oxidative agents	stabilized oxychloro complex (SOC, Purite®) sodium perborate (GenAqua®)
Ionic tamponade agents	borate, sorbitol, propylene glycol, and zinc (SofZia®)

3.2.1. Detergents

Detergents act by disrupting the cell membranes of microbials, thus preventing contamination. BAK and Polyquad® (polidronium chloride) are among the most well-known in this category. BAK, a cationic detergent, is commonly used in approximately 70% of multi-dose glaucoma drops at concentrations ranging from 0.003% to 0.02% [52]. BAK is highly cytotoxic to the corneal and conjunctival epithelial cells, including the limbal stem cells. Its mode of action is by damaging the deoxyribonucleic acid (DNA), disrupting tight junctions, and inducing cell death through apoptosis or necrosis [53]. Although its lipophilic nature allows easier penetration of topical drugs through the corneal epithelium, it may also cause ocular surface irritation and inflammation [54]. Polyquad®, a quaternary ammonium compound, has a polymeric structure that limits its penetration into cell membranes, making it less cytotoxic compared to BAK.

Several researchers found wide variation in the prevalence of OSD among patients using topical glaucoma medications (37–91%) [55]. Ramli et al. found higher rates of corneal staining (63% vs. 36%), abnormal Schirmer's tests (39% vs. 25%), and moderate OSDI symptoms (17% vs. 7%) in patients using BAK-containing medications compared to the control group [55]. They also found a strong association between the number of eye drops, the presence of preservatives, and the severity of OSD.

Another multicenter cross-sectional study in 9658 patients with open-angle glaucoma assessed the prevalence of toxicity when using beta-blocker eye drops with or without preservatives [48]. The researchers found that patients using P drops reported significantly more symptoms, such as pain during instillation (48% vs. 19%), foreign body sensation (42% vs. 15%), stinging or burning (48% vs. 20%), and dry eye sensation (35% vs. 16%), compared to those on PF drops. When patients were switched from P to PF drops, there was a significant reduction in ocular symptoms and signs, highlighting the benefits of PF formulations.

Chronic use of topical anti-glaucoma medications may interfere with wound healing after glaucoma filtering procedures [56]. Histological specimens from patients undergoing filtering surgery have shown reductions in goblet cells and increased inflammatory cells, such as macrophages, lymphocytes, fibroblasts, and mast cells.

In comparison, Polyquad® (polyquaternium-1)-containing drops have fewer adverse effects than BAK. For instance, OSDI scores were significantly lower in patients using Polyquad®-preserved travoprost compared to BAK-preserved travoprost [57]. Additionally, in vitro studies with human TM cells showed higher cell viability with Polyquad®-preserved formulations versus BAK-preserved ones [58]. Interestingly, compared to Polyquad®, BAK has been shown to be associated with dose-dependent reductions in TM cell viability and increased levels of MMP-9, a factor in glaucoma pathogenesis [58].

3.2.2. Oxidative Agents

The most common oxidative agent used in ocular pharmacology is stabilized oxychloro complex (SOC, Purite®). It disrupts microbial protein synthesis through the production of chlorine dioxide. SOC is most suitable for chronic use because of its unique ability to break down into components already found in the tears (Na^+, Cl^-, O_2, and H_2O). This property enhances its tolerability, reduces toxicity, and improves patient compliance [51].

In a 12-month, randomized, multicenter, double-masked study, brimonidine-Purite® 0.15% and 0.2% were compared to brimonidine-BAK 0.2% in patients with glaucoma or ocular hypertension. The results showed that brimonidine-Purite® 0.15% provided comparable IOP reduction to brimonidine 0.2% with significantly lower incidence of allergic conjunctivitis and hyperemia, higher patient satisfaction, and comfort ratings [59].

3.2.3. Ionic Tamponade Agents

These agents, such as SofZia®, are buffers that maintain the pH and osmolarity of the solution and enhance its comfort and stability. SofZia® contains borate, sorbitol, propylene glycol, and zinc and has both antibacterial and antifungal properties. It degrades quickly upon contact with cations on the ocular surface, resulting in less cytotoxicity compared to BAK [60].

Kanamoto's group, in 2015, studied the ocular surface tolerability of tafluprost with 0.001% BAK versus travoprost preserved with SofZia® in 195 patients with glaucoma. They found that SPK and conjunctival hyperemia scores were lower in the tafluprost group compared to the travoprost group ($p = 0.038$) [61].

3.3. Penetration Enhancers

Penetration enhancers are used in topical ocular medications to enable active ingredients to penetrate the ocular surface through the transcellular or paracellular routes. One of the most common penetration enhancers used in anti-glaucoma medications is BAK. Other examples include chelating agents, cyclodextrins, crown ethers, bile acids, salts, cell-penetrating peptides, saponin, ethylenediaminetetraacetic acid (EDTA), paraben, and Transcutol®. Each penetration enhancer, however, has its own specific side effects [62].

4. Discussion

4.1. Management of OSD Caused by Glaucoma Medications

Managing glaucoma patients with OSD requires a multi-faceted approach focused on reducing ocular surface toxicity, improving tear film stability, and controlling inflammation. Switching to PF medications, using supportive treatments for the ocular surface, and regular monitoring are key components of this strategy. In addition, advanced therapies and surgical options can be considered for patients with severe or refractory OSD.

4.1.1. Step 1: Modify Glaucoma Therapy

Transitioning to a PF version of the identical medication enhances OSD outcomes while maintaining the same hypotensive effect. A 2010 study from Finland found that replacing a P prostaglandin analog with a PF variant resulted in a significant reduction in OSD symptoms such as itching (46.8% to 26.5%), irritation/burning/stinging (56.3% to 28.4%), dry eye sensation (64.6% to 39.4%), abnormal fluorescein staining of the cornea (81.6% to 40.6%), conjunctival hyperemia (84.2% to 60%). Furthermore, TBUT increased from 4.5 ± 2.5 to 7.8 ± 4.9 s [63].

Similarly, an Italian study also found that switching BAK-containing beta-blocker formulations to PF versions resulted in a notable reduction of OSD symptoms, specifically, burning and stinging (40% to 20%), foreign body sensation (31% to 14%), dryness sensation (23% to 14%), and tearing (21% to 14%) [64].

Similar findings have also been reported even for combined PF brimonidine tartrate medications and resulted in improved patient comfort, satisfaction, and adherence to the treatment. When PF options are not available, one may try formulations containing Polyquad® or SofZia® [65].

4.1.2. Step 2: Ocular Surface Lubrication, Anti-Inflammatory Treatment, and Other Supplemental Therapies

The Dry Eye Workshop (DEWS) II Subcommittee's recommendations have simplified diagnosing DED. The diagnosis can be made if a patient exhibits a NITBUT (less than 10 s), high tear osmolarity (>308 mOsm/L), ocular surface staining (more than five spots on the cornea), accompanied by symptomatic evaluation using validated scoring systems like the OSDI [66]. In the early stages of OSD, the elimination of P medications combined with the use of PF artificial tears may be sufficient [67]. Proper eyelid hygiene with a frequent cleansing routine and warm compresses may help alleviate associated blepharitis.

Anti-Inflammatory Treatment (Cyclosporine A and Topical Steroids)

In addition, clinicians may consider starting anti-inflammatory treatment using topical steroids or cyclosporine A (CsA) drops. A 2023 South Korean randomized clinical trial demonstrated that 0.05% topical CsA significantly improved OSD parameters, increased Schirmer's test scores, TBUT, and TMH, and decreased ocular staining and MMP-9 positivity in treated eyes [68].

In some patients not responding to conventional treatment, the use of topical steroids may be essential, though, physicians must be aware of their role in elevating IOP and inducing cataracts. Carbon-20 ester steroids (loteprednol) are often preferred over carbon-20 ketone steroids (prednisolone, dexamethasone, and fluorometholone) [69].

Omega-3 Fatty Acid Supplementation

According to a 2019 meta-analysis consisting of 17 randomized clinical trials, omega-3 fatty acid supplementation has been associated with decreased dry eye symptoms and corneal staining along with increased TBUT and Schirmer's test values [70].

Vitamin A Eye Gel

Vitamin A may offer a promising therapeutic option for individuals with dry eye syndrome. The use of vitamin A palmitate as an eye gel has demonstrated beneficial effects on the morphology of the conjunctival epithelium and density of goblet cells in patients undergoing long-term treatment with topical PGAs [71].

Autologous Serum Eye Drops

Autologous serum eye drops, prepared by centrifuging a patient's own blood to separate the liquid and cellular components, may be used for moderate to severe OSD. Studies have shown that autologous serum eye drops contain cytokines and biochemical factors that are important for ocular surface health, including epithelial growth factor, TGF-β, and fibronectin [72].

Cryopreserved Amniotic Membranes

For patients with refractory OSD, a self-retained sutureless cryopreserved amniotic membrane (cAM) with a poly-carbonate ring frame can be placed under topical anesthetic in the clinic for an average duration of five days. cAM exhibits anti-inflammatory properties and promotes corneal healing through mechanical protection of the epithelial surface. In a multi-center, retrospective study involving 89 eyes from 77 patients with moderate-to-severe OSD, placement of a cAM for two days improved DEWS scores, corneal staining, visual symptoms, and ocular discomfort at one-week, one-month, and three-month follow-up [73]. Reported side effects include foreign body sensation and temporary blurred vision.

4.1.3. Step 3: Surgical Treatment

Regarding patients who are intolerant to topical medications and show no improvement with PF medications or oral CAIs, surgical procedures may be considered, such as selective laser trabeculoplasty, trabeculectomy, glaucoma drainage devices (GDDs), and minimally invasive glaucoma surgery (MIGS) [74]. However, each surgical intervention is fraught with its own complications and must be considered carefully.

4.2. Future Directions in the Management of Ocular Surface Diseases

There are exciting new therapies and technologies in the pipeline for managing OSD in patients using topical glaucoma medications. This study could be improved further by conducting a detailed systematic literature review regarding the efficacy of these novel treatments. A brief overview of some of these innovative therapies is described below, including sustained-release drug delivery systems (extraocular and intraocular) [75], intense pulsed light therapy, thermal pulsation devices, photobiomodulation, nanoparticles, gene alteration, stem cell applications, umbilical cord blood serum eye drops, and acupuncture (Table 5).

Table 5. Future directions in the management of ocular surface disease in glaucoma.

Product	Product Status	Mechanism of Action
Extraocular Drug Delivery Systems		
Gel-forming drops A. SoliDrop® gel solution (Otero Therapeutics) [76]	Preclinical	The higher viscosity gel-containing drops stay on the surface of the eyes for a longer period of time, thereby providing greater surface protection.
Ocular inserts A. Bimatoprost Ocular Ring® (AbbVie) [77] B. Topical Ophthalmic Drug Delivery Device® (TODDD®, Amorphex Therapeutics) [78]	Bimatoprost Ocular Ring® is in Phase 2, and TODDD® is in Phase 1.	Ocular rings containing anti-glaucoma medications may be inserted in the upper and lower fornices for slow release, thickening the precorneal tear film and protecting the eye.
Passive Diffusion Contact Lenses (PDCLs) A. Vitamin-E CLs loaded with timolol (University of Florida, USA) [79] B. Methafilcon lenses loaded with latanoprost (Harvard Medical School, USA) [80]	Preclinical	Anti-glaucoma drug impregnated CLs release active ingredients through passive diffusion.
Molecular Imprinted Contact Lenses (MICLs) A. Timolol maleate loaded MICL (University of Kerala, India) [81]	Preclinical	During the fabrication of MICLs, molecular sites akin to drug receptor sites are embedded in the polymer, increasing loading and sustained release of anti-glaucoma drugs.
Punctal Plugs (PPs) A. Evolute® (travoprost-loaded, Mati Therapeutics) [82] B. OTX-TP® (travoprost-loaded, Ocular Therapeutix) [83]	Evolute® is in Phase 2, and OTX-TP® is in Phase 3.	PPs block tear drainage and increase tear film contact time with the ocular surface.
Intraocular Drug Delivery Systems		
Anterior Chamber (AC) Intracameral Implants (II) A. DURYSTA® (bimatoprost, AbbVie) [84] B. ENV515® (travoprost, Envisia Therapeutics) [85] C. OTX-TIC® (travoprost, Ocular Therapeutix) [86] D. iDose® (travoprost, Glaukos Corporation) [87]	Phase 2 or 3	II are injected in the AC or anchored in the trabecular meshwork (TM) and slowly release medications over months. They are either biodegradable hydrogel or titanium implants.
Subconjunctival Implants (SI) A. Eye-D VS-101® (latanoprost, Biolight Life Sciences) [78,88]	Phase 1 or 2a	SI impregnated with glaucoma drugs are injected subconjunctivally to provide slow drug release.
Innovative Technological Devices		
Intense Pulsed Light (IPL) Therapy A. OptiLight® (Lumenis) [89,90]	Phase 4	High intensity light pulses are directed around the eyes, which may destroy abnormal blood vessels and alter meibomian gland architecture and function.

Table 5. Cont.

Product	Product Status	Mechanism of Action
Thermal Pulsation Devices (TPD) A. LipiFlow® (Johnson & Johnson Vision) [91]	Phase 4	TPDs consist of disposable eyepieces which direct heat and pressure over the eyelids to liquefy and express meibomian gland secretions.
Photobiomodulation A. Low-level light therapy with near-infrared light-emitting diodes (Dankook University, South Korea) [92] B. Photobiomodulation With REd vs. BluE Light (REBEL) Study (Aston University, United Kingdom) [93]	Phase 2	Photobiomodulation uses a mask to emit light over the face and eyelids. Blue light inhibits microbial growth while red light generates heat, promotes tissue repair, and decreases inflammation.
Other Emerging Therapies		
Nanoparticles A. Timolol-loaded gold nanoparticles (Uka Tarsadia University, India) [94–96]	Preclinical	Nanoparticles consisting of certain polymers, lipids, or metals may improve drug bioavailability, enabling slow release and reducing adverse effects.
Gene Therapy A. Recombinant adeno-associated virus (AAV) vector-mediated gene therapy targeting prostaglandin F2α synthesis in the AC [97] B. Intravitreal injections of AAV-F-iTrkB (AAV farnesylation of the intracellular domain of TrkB) [98]	Preclinical	Ocular gene therapy can target the TM to increase AH outflow and offer neuroprotection by limiting retinal ganglion cell (RGC) loss.
Stem Cell Applications A. Bone marrow-derived Mesenchymal Stem Cells (MSCs) injected into the AC [99] B. Human-induced pluripotent stem cells-derived RGCs [100] C. Human adipose-derived MSCs conditioned-medium ocular instillation [101]	Preclinical	Stem cells can be used to improve TM structure and function, promote RGC survival, and improve corneal barrier dysfunction.
Umbilical Cord Blood Serum (CBS) Eye Drops A. Singapore Cord Blood Bank CBS eye drops (Singapore National Eye Center) [102]	Phase 2	CBS drops contain high levels of growth factors and anti-inflammatory cytokines.
Acupuncture A. Niemtzow Acupuncture Protocol (University of Pittsburgh, USA) [103]	Phase 3	Acupuncture may downregulate proinflammatory cytokines and increase the release of acetylcholine in the lacrimal glands, promoting tear secretion.

Legend: TODDD® = Topical Ophthalmic Drug Delivery Device®, PDCLs = passive diffusion contact lenses, CLs = contact lenses, MICLs = molecular imprinted contact lenses, PPs = punctual plugs, AC = anterior chamber, II = intracameral implants, TM = trabecular meshwork, SI = subconjunctival implants, IPL = intense pulsed light, TPD = thermal pulsation devices, AAV = adeno-associated virus, AH = aqueous humor, RGC = retinal ganglion cell, CBS = cord blood serum.

4.2.1. Sustained-Release Drug Delivery Systems

Sustained-release systems are broadly categorized into extraocular or intraocular delivery platforms that offer a consistent drug concentration at the target site over a longer duration [75]. They offer promising alternatives to current challenges of ocular surface toxicity, inadequate IOP control, and non-compliance.

Extraocular Drug Delivery Platforms

Extraocular systems include gel-forming eye drops, ocular inserts, contact lenses, and punctal plugs [75]. Gel-forming formulations, such as SoliDrop® (Otero Therapeutics, University of Pittsburgh, Pittsburgh, PA, USA) transform into a semi-solid gel upon contact with tears, extending drug residence time [76].

Ocular inserts, such as Bimatoprost Ocular Ring® (AbbVie, Chicago, IL, USA) and Topical Ophthalmic Drug Delivery Device® (TODDD) (Amorphex Therapeutics, Andover, MA, USA), are placed in the conjunctival fornix, releasing the drug through diffusion and bioerosion [77,78].

Contact lenses are well-tolerated and are used in various forms to deliver ocular medications with the added benefits of minimal interference with vision and prolonged drug residence time. For example, contact lenses impregnated with timolol-vitamin E complex and Methafilcon lenses loaded with latanoprost utilize passive diffusion [79,80]. Another contact lens option uses molecular imprinting during the fabrication and polymerization process to create drug receptor sites, enhancing drug retention and release [81].

Punctal plugs, such as Evolute® (Mati Therapeutics, Austin, TX, USA) and OTX-TP® (Ocular Therapeutix, Bedford, MA, USA), are inserted in the lid puncta, delivering travoprost through diffusion while maintaining tear film integrity [82,83].

Intraocular Drug Delivery Systems

Intraocular drug delivery devices are designed to be inserted in the eye for prolonged drug release. They, however, require surgical intervention and may carry risks, such as damage to the eye, hypotony, IOP spikes, retinal detachment, and endophthalmitis [75].

Among the intracameral implants, DURYSTA® (AbbVie) is an FDA-approved biodegradable bimatoprost implant that lowers IOP for 4–6 months [84]. Similarly, ENV515® (Envisia Therapeutics, Durham, NC, USA) and OTX-TIC® (Ocular Therapeutix) release travoprost over a similar period [85,86]. Glaukos' product iDose Travoprost® (Aliso Viejo, CA, USA) is inserted within the TM, offering a year-long IOP control, but requires a more prolonged and invasive procedure [87].

Another alternative, such as subconjunctival implant Eye-D VS-101® (Biolight Life Sciences, Tel Aviv, Israel), offers a less invasive option and ease of injection, and still provides sustained release of latanoprost over several months [78,88].

4.2.2. Innovative Technological Devices

New technological innovations in OSD management include intense pulsed light (IPL) therapy, thermal pulsation devices, and photobiomodulation.

Intense Pulsed Light Therapy

During IPL treatments (OptiLight®, Lumenis, Yokneam, Isreal), protective eyepieces cover the eyes and high intensity light pulses are directed above the eyebrows, lower eyelids, zygomatic region, and nose, leading to destruction of abnormal blood vessels while maintaining meibomian gland architecture and function [89]. IPL therapy consists of four 20 minute sessions at three-week intervals followed by maintenance therapy every three to six months. A 2022 meta-analysis of 15 randomized controlled clinical trials found that compared to controls, patients who received IPL treatments had improved OSDI scores, standard patient evaluation of eye dryness (SPEED) scores, artificial tear usage, tear film lipid layer, meibomian gland quality, meibomian gland expression, corneal fluorescein staining, TBUT, and NITBUT [90].

Thermal Pulsation Devices

Thermal pulsation devices such as LipiFlow® (Johnson & Johnson Vision, Jacksonville, FL, USA) consist of disposable eyepieces with attached lenses that protect the cornea while direct heat and pressure are applied over the eyelids to liquefy and express meibomian gland secretions. Although IPL and thermal pulsation devices are FDA-approved, they have not yet solidified their role in daily clinical practice. A 2022 meta-analysis found that compared to controls, patients who received LipiFlow® treatments had improvements in OSDI scores, SPEED scores, and meibomian glands yielding secretion scores [91].

Photobiomodulation

Photobiomodulation, also known as low-level light therapy, uses a mask that covers the face and eyelids and emits light in the red (633 nm) or blue (428 nm) wavelength for 15–30 min. Blue light has been shown to inhibit microbial growth while red light generates heat, promotes tissue repair, and decreases inflammation. A US prospective pilot study with 30 patients who received three 15 minute sessions at one-week intervals of photobiomodulation with red light found a statistically significant improvement in NITBUT, TMH, tear film lipid layer thickness, and Schirmer's test [92]. A triple-masked, randomized controlled trial, Photobiomodulation With REd vs. BluE Light (REBEL), is currently being conducted at Aston University, United Kingdom [93].

4.2.3. Other Emerging Therapies

A number of emerging therapies that have shown promise in improving aqueous outflow and neuroprotection include nanoparticles, gene therapy, and stem cell applications [94]. Umbilical cord blood serum eye drops and acupuncture may improve OSD as well.

Nanoparticles

Nanoparticles consisting of certain polymers, lipids, or metals may improve drug bioavailability, enabling slow release while reducing adverse effects. In a lab study using New Zealand white rabbits, timolol-loaded gold nanoparticles embedded in contact lenses led to increased timolol concentrations in tear fluid, conjunctiva, and iris-ciliary muscles [95,96].

Gene Therapy

Preclinical ocular gene therapy alters gene expression via nanoparticles or viral vectors. One arm targets the TM to increase AH outflow. A US study with 30 Brown Norway rats utilized recombinant adeno-associated virus (AAV) vector-mediated gene therapy to target de novo prostaglandin F2α synthesis in the AC and found a reduction in IOP over 12 months [97].

The second arm of gene research focuses on neuroprotection by targeting RGC cell loss by increasing the expression of neurotrophins, such as brain-derived neurotrophic factor (BDNF) and ciliary neurotrophic factor (CNTF), antioxidant genes, anti-inflammatory genes, cell cycle regulators, and protease inhibitors. To prove this hypothesis, Japanese investigators used optic nerve crush (ONC) glaucoma mouse models and injected them with intravitreal injections of AAV-F-iTrkB (AAV farnesylation of the intracellular domain of TrkB) and found increased axon regeneration [98].

Stem Cell Applications

Stem cells can be used to improve TM structure and function, promote RGC survival, and improve corneal barrier dysfunction [94]. An animal study using a Long-Evans rat model of ocular hypertension found that when bone-marrow derived mesenchymal stem cells (MSCs) were tagged and injected into the AC, there was a significant decrease in IOP and MSCs were located in the ciliary processes and TM [99]. In a study conducted at the University of Pennsylvania, human-induced pluripotent stem cells (hiPSCs) were differentiated to mature RGCs in vitro and then injected intravitreally into mice, and it was found that hiPSCs integrated into the RGC layer for a successful transplantation rate of 94% at five-months follow up [100]. In a 2023 Japanese study, a conditioned-medium containing factors secreted from human adipose-derived MSCs was shown to decrease BAK-induced inflammation of human corneal epithelial cells in an in vitro model [101]. Next, the researchers, using a DED rat model, found decreased corneal fluorescein staining and improved tear production [101].

Umbilical Cord Blood Serum Eye Drops

Umbilical cord blood serum (CBS), readily available from blood banks, can be used as eye drops. These drops contain high levels of growth factors (epithelial growth factor and TGF-β1) and anti-inflammatory cytokines. In an observational, longitudinal, interventional study conducted in Singapore, 40 patients with refractory OSD were started on CBS eye drops. On average, the patients used the CBS drops 2.23 times per day with an average of 5.5-months follow-up and were found to show improvement in kerato-epitheliopathy staining score, TBUT, and SPEED score [102].

Acupuncture

Acupuncture may be beneficial in OSD based on its ability to downregulate proinflammatory cytokines and increase the release of acetylcholine in the lacrimal glands promoting tear secretion. A 2022 meta-analysis with 394 patients who underwent acupuncture showed significant improvement in OSDI scores and Schirmer's test scores, including TBUT, compared to controls [103].

5. Conclusions

Clinicians must be aware of the close association and high prevalence between OSD and long-term glaucoma therapy. If left unchecked, OSD may affect quality of life and treatment adherence, thus negatively impacting glaucoma care. Initially, transitioning to PF glaucoma medications is a crucial step, and combining with CsA or topical steroids may be beneficial. For patients with refractory OSD or uncontrolled IOP, surgical interventions may offer some benefits. Overall, a comprehensive and multifaceted management approach is essential to optimize both ocular surface health and effective glaucoma treatment.

Author Contributions: Conceptualization, Ö.E.K., P.M., B.D. and K.S.K.; methodology, Ö.E.K., P.M., B.D. and K.S.K.; investigation, Ö.E.K., P.M., B.D. and K.S.K.; resources, Ö.E.K., P.M., B.D. and K.S.K.; data curation, Ö.E.K., P.M., B.D. and K.S.K.; writing—original draft preparation, Ö.E.K., P.M., B.D. and K.S.K.; writing—review and editing, Ö.E.K., P.M., B.D. and K.S.K.; visualization, Ö.E.K., P.M., B.D. and K.S.K.; supervision, K.S.K.; project administration, K.S.K.; and funding acquisition, K.S.K. All authors have read and agreed to the published version of the manuscript.

Funding: This research was partly funded by NIH/NEI Core Grant for Vision Research (P30 EY030413), a Challenge Grant from Research to Prevent Blindness, New York, NY, USA, and the National Center for Advancing Translational Sciences of the National Institutes of Health under award number UL1TR001105.

Institutional Review Board Statement: Not applicable.

Informed Consent Statement: Not applicable.

Data Availability Statement: No new data were created or analyzed in this study. Data sharing is not applicable to this article.

Acknowledgments: BioRender® (Toronto, Ontario, Canada) software was used to make Figures 1 and 2, and the graphical abstract. We would like to thank Margaret Wang French (University of Texas Southwestern Medical Center, Dallas, TX, USA) for providing clinical photographs used in Figure 4, and Sruthi P. Suresh, Alexandra Pawlowicz for their contributions.

Conflicts of Interest: The authors declare no conflicts of interest.

References

1. Tham, Y.C.; Li, X.; Wong, T.Y.; Quigley, H.A.; Aung, T.; Cheng, C.Y. Global prevalence of glaucoma and projections of glaucoma burden through 2040: A systematic review and meta-analysis. *Ophthalmology* **2014**, *121*, 2081–2090. [CrossRef] [PubMed]
2. Pooja Prajwal, M.R.; Gopalakrishna, H.N.; Kateel, R. An exploratory study on the drug utilization pattern in glaucoma patients at a tertiary care hospital. *J. App. Pharm. Sci.* **2013**, *3*, 151–155. [CrossRef]
3. Craig, J.P.; Nichols, K.K.; Akpek, E.K.; Caffery, B.; Dua, H.S.; Joo, C.K.; Liu, Z.; Nelson, J.D.; Nichols, J.J.; Tsubota, K.; et al. TFOS DEWS II definition and classification report. *Ocul. Surf.* **2017**, *15*, 276–283. [CrossRef]
4. Gomes, J.A.P.; Azar, D.T.; Baudouin, C.; Efron, N.; Hirayama, M.; Horwath-Winter, J.; Kim, T.; Mehta, J.S.; Messmer, E.M.; Pepose, J.S.; et al. TFOS DEWS II iatrogenic report. *Ocul. Surf.* **2017**, *15*, 511–538. [CrossRef]
5. Ruiz-Lozano, R.E.; Azar, N.S.; Mousa, H.M.; Quiroga-Garza, M.E.; Komai, S.; Wheelock-Gutierrez, L.; Cartes, C.; Perez, V.L. Ocular surface disease: A known yet overlooked side effect of topical glaucoma therapy. *Front. Toxicol.* **2023**, *5*, 1067942. [CrossRef]
6. Baudouin, C.; Liang, H.; Hamard, P.; Riancho, L.; Creuzot-Garcher, C.; Warnet, J.M.; Brignole-Baudouin, F. The ocular surface of glaucoma patients treated over the long term expresses inflammatory markers related to both T-helper 1 and T-helper 2 pathways. *Ophthalmology* **2008**, *115*, 109–115. [CrossRef]
7. Roda, M.; Corazza, I.; Bacchi Reggiani, M.L.; Pellegrini, M.; Taroni, L.; Giannaccare, G.; Versura, P. Dry eye disease and tear cytokine levels-a meta-analysis. *Int. J. Mol. Sci.* **2020**, *21*, 3111. [CrossRef] [PubMed]
8. Roy, N.S.; Wei, Y.; Ying, G.S.; Maguire, M.G.; Asbell, P.A. Association of tear cytokine concentrations with symptoms and signs of dry eye disease: Baseline data from the Dry Eye Assessment and Management (DREAM) study. *Curr. Eye Res.* **2023**, *48*, 339–347. [CrossRef]
9. Scarpellini, C.; Ramos Llorca, A.; Lanthier, C.; Klejborowska, G.; Augustyns, K. The potential role of regulated cell death in dry eye diseases and ocular surface dysfunction. *Int. J. Mol. Sci.* **2023**, *24*, 731. [CrossRef]
10. Fineide, F.; Magnø, M.; Dahlø, K.; Kolko, M.; Heegaard, S.; Vehof, J.; Utheim, T.P. Topical glaucoma medications-possible implications on the meibomian glands. *Acta Ophthalmol.* **2024**, *102*, 1–14. [CrossRef]
11. Kolko, M.; Gazzard, G.; Baudouin, C.; Beier, S.; Brignole-Baudouin, F.; Cvenkel, B.; Fineide, F.; Hedengran, A.; Hommer, A.; Jespersen, E.; et al. Impact of glaucoma medications on the ocular surface and how ocular surface disease can influence glaucoma treatment. *Ocul. Surf.* **2023**, *29*, 456–468. [CrossRef] [PubMed]
12. Li, G.; Akpek, E.K.; Ahmad, S. Glaucoma and ocular surface disease: More than meets the eye. *Clin. Ophthalmol.* **2022**, *16*, 3641–3649. [CrossRef] [PubMed]
13. Zhang, X.; Vadoothker, S.; Munir, W.M.; Saeedi, O. Ocular surface disease and glaucoma medications: A clinical approach. *Eye Contact Lens* **2019**, *45*, 11–18. [CrossRef]
14. Garcia-Terraza, A.L.; Jimenez-Collado, D.; Sanchez-Sanoja, F.; Arteaga-Rivera, J.Y.; Morales Flores, N.; Pérez-Solórzano, S.; Garfias, Y.; Graue-Hernández, E.O.; Navas, A. Reliability, repeatability, and accordance between three different corneal diagnostic imaging devices for evaluating the ocular surface. *Front. Med.* **2022**, *9*, 893688. [CrossRef]
15. Schmidl, D.; Schlatter, A.; Chua, J.; Tan, B.; Garhöfer, G.; Schmetterer, L. Novel approaches for imaging-based diagnosis of ocular surface disease. *Diagnostics* **2020**, *10*, 589. [CrossRef] [PubMed]
16. Andole, S.; Senthil, S. Ocular surface disease and anti-glaucoma medications: Various features, diagnosis, and management guidelines. *Semin. Ophthalmol.* **2023**, *38*, 158–166. [CrossRef]
17. Scelfo, C.; ElSheikh, R.H.; Shamim, M.M.; Abbasian, J.; Ghaffarieh, A.; Elhusseiny, A.M. Ocular surface disease in glaucoma patients. *Curr. Eye Res.* **2023**, *48*, 219–230. [CrossRef]
18. Cochrane Library. What Is PICO? Available online: https://www.cochranelibrary.com/about-pico (accessed on 1 October 2024).
19. Harzing, A.W. Publish or Perish. Available online: https://harzing.com/resources/publish-or-perish (accessed on 22 August 2024).
20. Seider, N.; Miller, B.; Beiran, I. Topical glaucoma therapy as a risk factor for nasolacrimal duct obstruction. *Am. J. Ophthalmol.* **2008**, *145*, 120–123. [CrossRef]
21. Kuppens, E.V.; de Jong, C.A.; Stolwijk, T.R.; de Keizer, R.J.; van Best, J.A. Effect of timolol with and without preservative on the basal tear turnover in glaucoma. *Br. J. Ophthalmol.* **1995**, *79*, 339–342. [CrossRef]
22. Yuan, X.; Ma, X.; Yang, L.; Zhou, Q.; Li, Y. β-blocker eye drops affect ocular surface through β2 adrenoceptor of corneal limbal stem cells. *BMC Ophthalmol.* **2021**, *21*, 419. [CrossRef]
23. Mastropasqua, R.; Agnifili, L.; Fasanella, V.; Curcio, C.; Brescia, L.; Lanzini, M.; Fresina, M.; Mastropasqua, L.; Marchini, G. Corneoscleral limbus in glaucoma patients: In vivo confocal microscopy and immunocytological study. *Investig. Ophthalmol. Vis. Sci.* **2015**, *56*, 2050–2058. [CrossRef]
24. Inoue, K.; Okugawa, K.; Kato, S.; Inoue, Y.; Tomita, G.; Oshika, T.; Amano, S. Ocular factors relevant to anti-glaucomatous eyedrop-related keratoepitheliopathy. *J. Glaucoma* **2003**, *12*, 480–485. [CrossRef]
25. Zhou, X.; Zhang, X.; Zhou, D.; Zhao, Y.; Duan, X. A narrative review of ocular surface disease related to anti-glaucomatous medications. *Ophthalmol. Ther.* **2022**, *11*, 1681–1704. [CrossRef]
26. Rolle, T.; Spinetta, R.; Nuzzi, R. Long term safety and tolerability of Tafluprost 0.0015% vs. Timolol 0.1% preservative-free in ocular hypertensive and in primary open-angle glaucoma patients: A cross sectional study. *BMC Ophthalmol.* **2017**, *17*, 136. [CrossRef]

27. Russ, H.H.; Costa, V.P.; Ferreira, F.M.; Valgas, S.R.; Correa Neto, M.A.; Strobel, E.; Truppel, J.H. Conjunctival changes induced by prostaglandin analogues and timolol maleate: A histomorphometric study. *Arq. Bras. Oftalmol.* **2007**, *70*, 910–916. [CrossRef]
28. Yoshino, T.; Fukuchi, T.; Togano, T.; Seki, M.; Ikegaki, H.; Abe, H. Eyelid and eyelash changes due to prostaglandin analog therapy in unilateral treatment cases. *Jpn. J. Ophthalmol.* **2013**, *57*, 172–178. [CrossRef]
29. Yamada, H.; Yoneda, M.; Gosho, M.; Kato, T.; Zako, M. Bimatoprost, latanoprost, and tafluprost induce differential expression of matrix metalloproteinases and tissue inhibitor of metalloproteinases. *BMC Ophthalmol.* **2016**, *16*, 26. [CrossRef]
30. Mocan, M.C.; Uzunosmanoglu, E.; Kocabeyoglu, S.; Karakaya, J.; Irkec, M. The association of chronic topical prostaglandin analog use with meibomian gland dysfunction. *J. Glaucoma* **2016**, *25*, 770–774. [CrossRef]
31. Yeh, P.H.; Cheng, Y.C.; Shie, S.S.; Lee, Y.S.; Shen, S.C.; Chen, H.S.; Wu, W.C.; Su, W.W. Brimonidine related acute follicular conjunctivitis: Onset time and clinical presentations, a long-term follow-up. *Medicine* **2021**, *100*, e26724. [CrossRef]
32. Trotta, D.; Zucchelli, M.; Salladini, C.; Ballerini, P.; Rossi, C.; Aricò, M. Brimonidine eye drops within the reach of children: A possible foe. *Children* **2024**, *11*, 317. [CrossRef]
33. Rohrschneider, K.; Koch, H.-R. [Effects of acetazolamide (Diamox®, Glaupax®) on tear production]. *Klin. Monatsblätter für Augenheilkd. [Clin. Mon. Newslett. Ophthalmol.]* **1991**, *199*, 79–83. [CrossRef] [PubMed]
34. Terai, N.; Müller-Holz, M.; Spoerl, E.; Pillunat, L.E. Short-term effect of topical antiglaucoma medication on tear-film stability, tear secretion, and corneal sensitivity in healthy subjects. *Clin. Ophthalmol.* **2011**, *5*, 517–525. [CrossRef] [PubMed]
35. Skaat, A.; Rosman, M.S.; Chien, J.L.; Mogil, R.S.; Ren, R.; Liebmann, J.M.; Ritch, R.; Park, S.C. Effect of pilocarpine hydrochloride on the schlemm canal in healthy eyes and eyes with open-angle glaucoma. *JAMA Ophthalmol.* **2016**, *134*, 976–981. [CrossRef]
36. Zhang, Y.; Kam, W.R.; Liu, Y.; Chen, X.; Sullivan, D.A. Influence of pilocarpine and timolol on human meibomian gland epithelial cells. *Cornea* **2017**, *36*, 719–724. [CrossRef]
37. Hartenbaum, D.; Maloney, J.; Vaccarelli, L.; Liss, C.; Wilson, H.; Gormley, G.J. Comparison of dorzolamide and pilocarpine as adjunctive therapy in patients with open-angle glaucoma and ocular hypertension. *Clin. Ther.* **1999**, *21*, 1533–1538. [CrossRef]
38. Mehran, N.A.; Sinha, S.; Razeghinejad, R. New glaucoma medications: Latanoprostene bunod, netarsudil, and fixed combination netarsudil-latanoprost. *Eye* **2020**, *34*, 72–88. [CrossRef]
39. Patton, G.N.; Lee, H.J. Chemical insights into topical agents in intraocular pressure management: From glaucoma etiopathology to therapeutic approaches. *Pharmaceutics* **2024**, *16*, 274. [CrossRef] [PubMed]
40. Srivastava, K.; Bhatnagar, K.R.; Shakrawal, J.; Tandon, M.; Jaisingh, K.; Pandey, L.; Roy, F. Ocular surface changes in primary open-angle glaucoma on anti-glaucoma medications versus treatment-naïve patients. *Indian J. Ophthalmol.* **2024**, *72*, 374–380. [CrossRef]
41. Ye, Y.; Xu, Y.; Yang, Y.; Fan, Y.; Liu, P.; Yu, K.; Yu, M. Wide corneal epithelial thickness mapping in eyes with topical antiglaucoma therapy using optical coherence tomography. *Transl. Vis. Sci. Technol.* **2022**, *11*, 4. [CrossRef]
42. Pai, V.; Reddy, L.S.H. Prevalence of ocular surface disease in patients with glaucoma on topical medications. *Asian J. Ophthalmol.* **2018**, *16*, 101–109. [CrossRef]
43. Pérez-Bartolomé, F.; Martínez-de-la-Casa, J.M.; Arriola-Villalobos, P.; Fernández-Pérez, C.; Polo, V.; García-Feijoó, J. Ocular surface disease in patients under topical treatment for glaucoma. *Eur. J. Ophthalmol.* **2017**, *27*, 694–704. [CrossRef] [PubMed]
44. Cvenkel, B.; Štunf, Š.; Srebotnik Kirbiš, I.; Strojan Fležar, M. Symptoms and signs of ocular surface disease related to topical medication in patients with glaucoma. *Clin. Ophthalmol.* **2015**, *9*, 625–631. [CrossRef] [PubMed]
45. Saade, C.E.; Lari, H.B.; Berezina, T.L.; Fechtner, R.D.; Khouri, A.S. Topical glaucoma therapy and ocular surface disease: A prospective, controlled cohort study. *Can. J. Ophthalmol.* **2015**, *50*, 132–136. [CrossRef] [PubMed]
46. Van Went, C.; Alalwani, H.; Brasnu, E.; Pham, J.; Hamard, P.; Baudouin, C.; Labbé, A. [Corneal sensitivity in patients treated medically for glaucoma or ocular hypertension]. *J. Fr. Ophtalmol.* **2011**, *34*, 684–690. [CrossRef] [PubMed]
47. Fechtner, R.D.; Godfrey, D.G.; Budenz, D.; Stewart, J.A.; Stewart, W.C.; Jasek, M.C. Prevalence of ocular surface complaints in patients with glaucoma using topical intraocular pressure-lowering medications. *Cornea* **2010**, *29*, 618–621. [CrossRef]
48. Jaenen, N.; Baudouin, C.; Pouliquen, P.; Manni, G.; Figueiredo, A.; Zeyen, T. Ocular symptoms and signs with preserved and preservative-free glaucoma medications. *Eur. J. Ophthalmol.* **2007**, *17*, 341–349. [CrossRef]
49. Pisella, P.J.; Pouliquen, P.; Baudouin, C. Prevalence of ocular symptoms and signs with preserved and preservative free glaucoma medication. *Br. J. Ophthalmol.* **2002**, *86*, 418–423. [CrossRef]
50. Petounis, A.D.; Akritopoulos, P. Influence of topical and systemic beta-blockers on tear production. *Int. Ophthalmol.* **1989**, *13*, 75–80. [CrossRef]
51. Kaur, I.P.; Lal, S.; Rana, C.; Kakkar, S.; Singh, H. Ocular preservatives: Associated risks and newer options. *Cutan. Ocul. Toxicol.* **2009**, *28*, 93–103. [CrossRef]
52. Goldstein, M.H.; Silva, F.Q.; Blender, N.; Tran, T.; Vantipalli, S. Ocular benzalkonium chloride exposure: Problems and solutions. *Eye* **2022**, *36*, 361–368. [CrossRef]
53. Debbasch, C.; Brignole, F.; Pisella, P.J.; Warnet, J.M.; Rat, P.; Baudouin, C. Quaternary ammoniums and other preservatives' contribution in oxidative stress and apoptosis on Chang conjunctival cells. *Investig. Ophthalmol. Vis. Sci.* **2001**, *42*, 642–652.
54. Zhang, R.; Park, M.; Richardson, A.; Tedla, N.; Pandzic, E.; de Paiva, C.S.; Watson, S.; Wakefield, D.; Di Girolamo, N. Dose-dependent benzalkonium chloride toxicity imparts ocular surface epithelial changes with features of dry eye disease. *Ocul. Surf.* **2020**, *18*, 158–169. [CrossRef] [PubMed]

55. Ramli, N.; Supramaniam, G.; Samsudin, A.; Juana, A.; Zahari, M.; Choo, M.M. Ocular surface disease in glaucoma: Effect of polypharmacy and preservatives. *Optom. Vis. Sci.* **2015**, *92*, e222–e226. [CrossRef] [PubMed]
56. Sherwood, M.B.; Grierson, I.; Millar, L.; Hitchings, R.A. Long-term morphologic effects of antiglaucoma drugs on the conjunctiva and Tenon's capsule in glaucomatous patients. *Ophthalmology* **1989**, *96*, 327–335. [CrossRef] [PubMed]
57. Kumar, S.; Singh, T.; Ichhpujani, P.; Vohra, S. Ocular surface disease with BAK preserved travoprost and polyquaternium 1(Polyquad) preserved travoprost. *Rom. J. Ophthalmol.* **2019**, *63*, 249–256. [CrossRef]
58. Ammar, D.A.; Kahook, M.Y. Effects of benzalkonium chloride- or polyquad-preserved fixed combination glaucoma medications on human trabecular meshwork cells. *Mol. Vis.* **2011**, *17*, 1806–1813.
59. Katz, L.J. Twelve-month evaluation of brimonidine-purite versus brimonidine in patients with glaucoma or ocular hypertension. *J. Glaucoma* **2002**, *11*, 119–126. [CrossRef] [PubMed]
60. Ryan, G., Jr.; Fain, J.M.; Lovelace, C.; Gelotte, K.M. Effectiveness of ophthalmic solution preservatives: A comparison of latanoprost with 0.02% benzalkonium chloride and travoprost with the sofZia preservative system. *BMC Ophthalmol.* **2011**, *11*, 8. [CrossRef]
61. Kanamoto, T.; Kiuchi, Y.; Tanito, M.; Mizoue, S.; Naito, T.; Teranishi, S.; Hirooka, K.; Rimayanti, U. Comparison of the toxicity profile of benzalkonium chloride-preserved tafluprost and SofZia-preserved travoprost applied to the ocular surface. *J. Ocul. Pharmacol. Ther.* **2015**, *31*, 156–164. [CrossRef]
62. Moiseev, R.V.; Morrison, P.W.J.; Steele, F.; Khutoryanskiy, V.V. Penetration enhancers in ocular drug delivery. *Pharmaceutics* **2019**, *11*, 321. [CrossRef]
63. Uusitalo, H.; Chen, E.; Pfeiffer, N.; Brignole-Baudouin, F.; Kaarniranta, K.; Leino, M.; Puska, P.; Palmgren, E.; Hamacher, T.; Hofmann, G.; et al. Switching from a preserved to a preservative-free prostaglandin preparation in topical glaucoma medication. *Acta Ophthalmol.* **2010**, *88*, 329–336. [CrossRef] [PubMed]
64. Iester, M.; Telani, S.; Frezzotti, P.; Motolese, I.; Figus, M.; Fogagnolo, P.; Perdicchi, A. Ocular surface changes in glaucomatous patients treated with and without preservatives beta-blockers. *J. Ocul. Pharmacol. Ther.* **2014**, *30*, 476–481. [CrossRef]
65. Jandroković, S.; Vidas Pauk, S.; Lešin Gaćina, D.; Skegro, I.; Tomić, M.; Masnec, S.; Kuzman, T.; Kalauz, M. Tolerability in glaucoma patients switched from preserved to preservative-free prostaglandin-timolol combination: A prospective real-life study. *Clin. Ophthalmol.* **2022**, *16*, 3181–3192. [CrossRef] [PubMed]
66. Wolffsohn, J.S.; Arita, R.; Chalmers, R.; Djalilian, A.; Dogru, M.; Dumbleton, K.; Gupta, P.K.; Karpecki, P.; Lazreg, S.; Pult, H.; et al. TFOS DEWS II diagnostic methodology report. *Ocul. Surf.* **2017**, *15*, 539–574. [CrossRef]
67. Jones, L.; Downie, L.E.; Korb, D.; Benitez-Del-Castillo, J.M.; Dana, R.; Deng, S.X.; Dong, P.N.; Geerling, G.; Hida, R.Y.; Liu, Y.; et al. TFOS DEWS II management and therapy report. *Ocul. Surf.* **2017**, *15*, 575–628. [CrossRef]
68. Kim, J.G.; An, J.H.; Cho, S.Y.; Lee, C.E.; Shim, K.Y.; Jun, J.H. Efficacy of topical 0.05% cyclosporine A for ocular surface disease related to topical anti-glaucoma medications. *J. Ocul. Pharmacol. Ther.* **2023**, *39*, 389–397. [CrossRef]
69. Pleyer, U.; Ursell, P.G.; Rama, P. Intraocular pressure effects of common topical steroids for post-cataract inflammation: Are they all the same? *Ophthalmol. Ther.* **2013**, *2*, 55–72. [CrossRef] [PubMed]
70. Giannaccare, G.; Pellegrini, M.; Sebastiani, S.; Bernabei, F.; Roda, M.; Taroni, L.; Versura, P.; Campos, E.C. Efficacy of omega-3 fatty acid supplementation for treatment of dry eye disease: A meta-analysis of randomized clinical trials. *Cornea* **2019**, *38*, 565–573. [CrossRef]
71. Cui, X.; Xiang, J.; Zhu, W.; Wei, A.; Le, Q.; Xu, J.; Zhou, X. Vitamin A palmitate and carbomer gel protects the conjunctiva of patients with long-term prostaglandin analogs application. *J. Glaucoma* **2016**, *25*, 487–492. [CrossRef]
72. Vazirani, J.; Sridhar, U.; Gokhale, N.; Doddigarla, V.R.; Sharma, S.; Basu, S. Autologous serum eye drops in dry eye disease: Preferred practice pattern guidelines. *Indian J. Ophthalmol.* **2023**, *71*, 1357–1363. [CrossRef]
73. McDonald, M.; Janik, S.B.; Bowden, F.W.; Chokshi, A.; Singer, M.A.; Tighe, S.; Mead, O.G.; Nanda, S.; Qazi, M.A.; Dierker, D.; et al. Association of treatment duration and clinical outcomes in dry eye treatment with sutureless cryopreserved amniotic membrane. *Clin. Ophthalmol.* **2023**, *17*, 2697–2703. [CrossRef]
74. Conlon, R.; Saheb, H.; Ahmed, I.I.K. Glaucoma treatment trends: A review. *Can. J. Ophthalmol.* **2017**, *52*, 114–124. [CrossRef]
75. Al-Qaysi, Z.K.; Beadham, I.G.; Schwikkard, S.L.; Bear, J.C.; Al-Kinani, A.A.; Alany, R.G. Sustained release ocular drug delivery systems for glaucoma therapy. *Expert Opin. Drug Deliv.* **2023**, *20*, 905–919. [CrossRef]
76. M Grover, L.; Moakes, R.; Rauz, S. Innovations in fluid-gel eye drops for treating disease of the eye: Prospects for enhancing drug retention and reducing corneal scarring. *Expert Rev. Ophthalmol.* **2022**, *17*, 175–181. [CrossRef]
77. Brandt, J.D.; Sall, K.; DuBiner, H.; Benza, R.; Alster, Y.; Walker, G.; Semba, C.P. Six-month intraocular pressure reduction with a topical bimatoprost ocular insert: Results of a phase II randomized controlled study. *Ophthalmology* **2016**, *123*, 1685–1694. [CrossRef]
78. Kesav, N.P.; Young, C.E.C.; Ertel, M.K.; Seibold, L.K.; Kahook, M.Y. Sustained-release drug delivery systems for the treatment of glaucoma. *Int. J. Ophthalmol.* **2021**, *14*, 148–159. [CrossRef]
79. Hsu, K.H.; Carbia, B.E.; Plummer, C.; Chauhan, A. Dual drug delivery from vitamin E loaded contact lenses for glaucoma therapy. *Eur. J. Pharm. Biopharm.* **2015**, *94*, 312–321. [CrossRef]
80. Ciolino, J.B.; Ross, A.E.; Tulsan, R.; Watts, A.C.; Wang, R.F.; Zurakowski, D.; Serle, J.B.; Kohane, D.S. Latanoprost-eluting contact lenses in glaucomatous monkeys. *Ophthalmology* **2016**, *123*, 2085–2092. [CrossRef]

81. Anirudhan, T.S.; Nair, A.S.; Parvathy, J. Extended wear therapeutic contact lens fabricated from timolol imprinted carboxymethyl chitosan-g-hydroxy ethyl methacrylate-g-poly acrylamide as a onetime medication for glaucoma. *Eur. J. Pharm. Biopharm.* **2016**, *109*, 61–71. [CrossRef]
82. Clinical Trials. Safety and Intraocular Lowering Effect of Delivery of Travoprost Evolute® in Subjects with Elevated Intraocular Pressure. Available online: https://clinicaltrials.gov/study/NCT04962009?term=EVOLUTE&rank=2 (accessed on 1 September 2024).
83. Perera, S.A.; Ting, D.S.; Nongpiur, M.E.; Chew, P.T.; Aquino, M.C.; Sng, C.C.; Ho, S.W.; Aung, T. Feasibility study of sustained-release travoprost punctum plug for intraocular pressure reduction in an Asian population. *Clin. Ophthalmol.* **2016**, *10*, 757–764. [CrossRef]
84. Weinreb, R.N.; Bacharach, J.; Brubaker, J.W.; Medeiros, F.A.; Bejanian, M.; Bernstein, P.; Robinson, M.R. Bimatoprost implant biodegradation in the Phase 3, randomized, 20-month ARTEMIS studies. *J. Ocul. Pharmacol. Ther.* **2023**, *39*, 55–62. [CrossRef]
85. Clinical Trials. Safety and Efficacy of ENV515 Travoprost Extended Release (XR) in Patients with Bilateral Ocular Hypertension or Primary Open Angle Glaucoma. Available online: https://clinicaltrials.gov/study/NCT02371746?term=ENV515&rank=1 (accessed on 1 September 2024).
86. Clinical Trials. Safety, and Efficacy of OTX-TIC in Participants with Open Angle Glaucoma or Ocular Hypertension. Available online: https://clinicaltrials.gov/study/NCT04360174?term=OTX-TIC&rank=1 (accessed on 1 September 2024).
87. Berdahl, J.P.; Sarkisian, S.R., Jr.; Ang, R.E.; Doan, L.V.; Kothe, A.C.; Usner, D.W.; Katz, L.J.; Navratil, T. Efficacy and safety of the travoprost intraocular implant in reducing topical iop-lowering medication burden in patients with open-angle glaucoma or ocular hypertension. *Drugs* **2024**, *84*, 83–97. [CrossRef]
88. Rafiei, F.; Tabesh, H.; Farzad, F. Sustained subconjunctival drug delivery systems: Current trends and future perspectives. *Int. Ophthalmol.* **2020**, *40*, 2385–2401. [CrossRef]
89. Safir, M.; Twig, G.; Mimouni, M. Dry eye disease management. *BMJ* **2024**, *384*, e077344. [CrossRef]
90. Miao, S.; Yan, R.; Jia, Y.; Pan, Z. Effect of intense pulsed light therapy in dry eye disease caused by meibomian gland dysfunction: A systematic review and meta-analysis. *Eye Contact Lens* **2022**, *48*, 424–429. [CrossRef]
91. Hu, J.; Zhu, S.; Liu, X. Efficacy and safety of a vectored thermal pulsation system (Lipiflow®) in the treatment of meibomian gland dysfunction: A systematic review and meta-analysis. *Graefes. Arch. Clin. Exp. Ophthalmol.* **2022**, *260*, 25–39. [CrossRef]
92. Antwi, A.; Schill, A.W.; Redfern, R.; Ritchey, E.R. Effect of low-level light therapy in individuals with dry eye disease. *Ophthalmic Physiol. Opt.* **2024**, 1–8. [CrossRef]
93. Clinical Trials. Photobiomodulation with REd vs. BluE Light (REBEL). Available online: https://clinicaltrials.gov/study/NCT06371300 (accessed on 3 September 2024).
94. Ciociola, E.C.; Fernandez, E.; Kaufmann, M.; Klifto, M.R. Future directions of glaucoma treatment: Emerging gene, neuroprotection, nanomedicine, stem cell, and vascular therapies. *Curr. Opin. Ophthalmol.* **2024**, *35*, 89–96. [CrossRef]
95. Occhiutto, M.L.; Maranhão, R.C.; Costa, V.P.; Konstas, A.G. Nanotechnology for medical and surgical glaucoma therapy-a review. *Adv. Ther.* **2020**, *37*, 155–199. [CrossRef]
96. Maulvi, F.A.; Patil, R.J.; Desai, A.R.; Shukla, M.R.; Vaidya, R.J.; Ranch, K.M.; Vyas, B.A.; Shah, S.A.; Shah, D.O. Effect of gold nanoparticles on timolol uptake and its release kinetics from contact lenses: In vitro and in vivo evaluation. *Acta Biomater.* **2019**, *86*, 350–362. [CrossRef]
97. Chern, K.J.; Nettesheim, E.R.; Reid, C.A.; Li, N.W.; Marcoe, G.J.; Lipinski, D.M. Prostaglandin-based rAAV-mediated glaucoma gene therapy in Brown Norway rats. *Commun. Biol.* **2022**, *5*, 1169. [CrossRef] [PubMed]
98. Nishijima, E.; Honda, S.; Kitamura, Y.; Namekata, K.; Kimura, A.; Guo, X.; Azuchi, Y.; Harada, C.; Murakami, A.; Matsuda, A.; et al. Vision protection and robust axon regeneration in glaucoma models by membrane-associated Trk receptors. *Mol. Ther.* **2023**, *31*, 810–824. [CrossRef] [PubMed]
99. Roubeix, C.; Godefroy, D.; Mias, C.; Sapienza, A.; Riancho, L.; Degardin, J.; Fradot, V.; Ivkovic, I.; Picaud, S.; Sennlaub, F.; et al. Intraocular pressure reduction and neuroprotection conferred by bone marrow-derived mesenchymal stem cells in an animal model of glaucoma. *Stem Cell Res. Ther.* **2015**, *6*, 177. [CrossRef] [PubMed]
100. Vrathasha, V.; Nikonov, S.; Bell, B.A.; He, J.; Bungatavula, Y.; Uyhazi, K.E.; Murthy Chavali, V.R. Transplanted human induced pluripotent stem cells- derived retinal ganglion cells embed within mouse retinas and are electrophysiologically functional. *iScience* **2022**, *25*, 105308. [CrossRef] [PubMed]
101. Imaizumi, T.; Hayashi, R.; Kudo, Y.; Li, X.; Yamaguchi, K.; Shibata, S.; Okubo, T.; Ishii, T.; Honma, Y.; Nishida, K. Ocular instillation of conditioned medium from mesenchymal stem cells is effective for dry eye syndrome by improving corneal barrier function. *Sci. Rep.* **2023**, *13*, 13100. [CrossRef]
102. Wong, J.; Govindasamy, G.; Prasath, A.; Hwang, W.; Ho, A.; Yeo, S.; Tong, L. Allogeneic umbilical cord plasma eyedrops for the treatment of recalcitrant dry eye disease patients. *J. Clin. Med.* **2023**, *12*, 6750. [CrossRef]
103. Prinz, J.; Maffulli, N.; Fuest, M.; Walter, P.; Hildebrand, F.; Migliorini, F. Acupuncture for the management of dry eye disease. *Front. Med.* **2022**, *16*, 975–983. [CrossRef]

Disclaimer/Publisher's Note: The statements, opinions and data contained in all publications are solely those of the individual author(s) and contributor(s) and not of MDPI and/or the editor(s). MDPI and/or the editor(s) disclaim responsibility for any injury to people or property resulting from any ideas, methods, instructions or products referred to in the content.

Article

Exploring New Therapeutic Avenues for Ophthalmic Disorders: Glaucoma-Related Molecular Docking Evaluation and Bibliometric Analysis for Improved Management of Ocular Diseases

Flaviu Bodea [1], Simona Gabriela Bungau [1,2,*], Andrei Paul Negru [1,3], Ada Radu [4], Alexandra Georgiana Tarce [5], Delia Mirela Tit [1,2], Alexa Florina Bungau [1,3], Cristian Bustea [6,*], Tapan Behl [7] and Andrei-Flavius Radu [1,3]

[1] Doctoral School of Biomedical Sciences, University of Oradea, 410087 Oradea, Romania; dr.flaviu.bodea@gmail.com (F.B.); negrupaul59@gmail.com (A.P.N.); dtit@uoradea.ro (D.M.T.); pradaalexaflorina@gmail.com (A.F.B.); andreiflavius.radu@uoradea.ro (A.-F.R.)
[2] Department of Pharmacy, Faculty of Medicine and Pharmacy, University of Oradea, 410028 Oradea, Romania
[3] Department of Preclinical Disciplines, Faculty of Medicine and Pharmacy, University of Oradea, 410073 Oradea, Romania
[4] Ducfarm Pharmacy, 410514 Oradea, Romania; adaroman96@gmail.com
[5] Medicine Program of Study, Faculty of Medicine and Pharmacy, University of Oradea, 410073 Oradea, Romania; tarce_alexandra@yahoo.com
[6] Department of Surgery, Oradea County Emergency Clinical Hospital, 410169 Oradea, Romania
[7] School of Health Sciences &Technology, University of Petroleum and Energy Studies, Dehradun 248007, India; tapanbehl31@gmail.com
* Correspondence: sbungau@uoradea.ro (S.G.B.); cristibustea@yahoo.com (C.B.)

Citation: Bodea, F.; Bungau, S.G.; Negru, A.P.; Radu, A.; Tarce, A.G.; Tit, D.M.; Bungau, A.F.; Bustea, C.; Behl, T.; Radu, A.-F. Exploring New Therapeutic Avenues for Ophthalmic Disorders: Glaucoma-Related Molecular Docking Evaluation and Bibliometric Analysis for Improved Management of Ocular Diseases. *Bioengineering* 2023, 10, 983. https://doi.org/10.3390/bioengineering10080983

Academic Editors: Karanjit S. Kooner and Osamah J. Saeedi

Received: 5 July 2023
Revised: 14 August 2023
Accepted: 17 August 2023
Published: 20 August 2023

Copyright: © 2023 by the authors. Licensee MDPI, Basel, Switzerland. This article is an open access article distributed under the terms and conditions of the Creative Commons Attribution (CC BY) license (https://creativecommons.org/licenses/by/4.0/).

Abstract: Ophthalmic disorders consist of a broad spectrum of ailments that impact the structures and functions of the eye. Due to the crucial function of the retina in the vision process, the management of eye ailments is of the utmost importance, but several unmet needs have been identified in terms of the outcome measures in clinical trials, more proven minimally invasive glaucoma surgery, and a lack of comprehensive bibliometric assessments, among others. The current evaluation seeks to fulfill several of these unmet needs via a dual approach consisting of a molecular docking analysis based on the potential of ripasudil and fasudil to inhibit Rho-associated protein kinases (ROCKs), virtual screening of ligands, and pharmacokinetic predictions, emphasizing the identification of new compounds potentially active in the management of glaucoma, and a comprehensive bibliometric analysis of the most recent publications indexed in the Web of Science evaluating the management of several of the most common eye conditions. This method resulted in the finding of ligands (i.e., ZINC000000022706 with the most elevated binding potential for ROCK1 and ZINC000034800307 in the case of ROCK2) that are not presently utilized in any therapeutic regimen but may represent a future option to be successfully applied in the therapeutic scheme of glaucoma following further comprehensive testing validations. In addition, this research also analyzed multiple papers listed in the Web of Science collection of databases via the VOSviewer application to deliver, through descriptive analysis of the results, an in-depth overview of publications contributing to the present level of comprehension in therapeutic approaches to ocular diseases in terms of scientific impact, citation analyses, most productive authors, journals, and countries, as well as collaborative networks. Based on the molecular docking study's preliminary findings, the most promising candidates must be thoroughly studied to determine their efficacy and risk profiles. Bibliometric analysis may also help researchers set targets to improve ocular disease outcomes.

Keywords: glaucoma; molecular docking; retinal diseases; bibliometric analysis; ocular disease therapy; micropulse laser therapy

1. Introduction

The anterior section of the human eye is solely responsible for focusing a sharp image of the visual world onto the retina. The primary symptom of retinal disorders is visual impairment, which is not localized and may result from an abnormality at any point along the visual pathway, such as optical aberrations, functional visual impairment, or cortical disorders [1].

Central serous chorioretinopathy, age-related macular degeneration, diabetic macular edema, diabetic retinopathy, glaucoma, and retinal vein occlusion are among the most prevalent eye disorders in which inflammatory processes and oxidative stress are essential causative factors [2,3]. Retinal and choroidal vascular disorders constitute the most common causes of visual impairment. The approximate worldwide prevalence of age-related macular degeneration was 196 million in 2020, and due to the aging of the global population, 288 million are anticipated for 2040 [4]. In 2012, the worldwide prevalence of diabetic macular edema was 21 million and of diabetic retinopathy was 93 million. In addition, retinal vein occlusion was the second most common retinal vascular condition, with a worldwide estimated prevalence of 16.4 million individuals in 2008, and an important percentage of those suffering from it developed macular edema [5]. According to studies, central serous chorioretinopathy develops approximately six times more often in men than in women, with an incidence rate of 10 per 100,000 men per year [6]. Moreover, 57.5 million individuals in the world have primary open-angle glaucoma. Individuals over the age of 60, persons with a family history of glaucoma, patients under treatment with glucocorticoids, diabetics, those with high myopia, elevated blood pressure, and a central corneal thickness of less than 5 mm, as well as those who have sustained an eye injury, have an elevated risk of developing glaucoma. Glaucoma is anticipated to impact around 112 million individuals by 2040 [7].

Each of the aforementioned diseases has particular signs and mechanisms of disease development, but they all share the same effect on the retina's structure and functionality. Given the essential role of the retina in the vision process, the treatment of eye conditions is of the utmost importance. Successful approaches to improving the management of these conditions seek to maintain or recover visual function, ameliorate symptoms, limit the condition's progression, and improve sufferers' standard of living [8,9].

Due to its complexity, glaucoma is one of the most studied pathologies in the ophthalmologic field. Glaucoma represents an ensemble of degenerative eye conditions characterized by damage to the optic nerve that, if neglected, can lead to permanent visual deterioration. A disturbance in the production and outflow of aqueous humor causes high intraocular pressure (IOP), which is the primary risk indicator for glaucoma [10].

To preserve vision, glaucoma treatments seek to reduce IOP and slow the progression of the disease. The most common treatment is the application of topical eye medications that either decrease or enhance aqueous humor discharge. Beta-blockers, inhibitors of carbonic anhydrase, and alpha-agonists are among the compounds with beneficial pharmacological action in glaucoma. In certain instances, oral drugs may be recommended for IOP reduction [11].

In conjunction with medication, laser therapy can be utilized to increase or decrease the production of aqueous humor. Argon laser trabeculoplasty (ALT) and selective laser trabeculoplasty (SLT) are widespread procedures that enhance fluid drainage by targeting the trabecular meshwork [12]. Furthermore, micropulse laser therapy (MLT) has emerged as a promising treatment approach for glaucoma. This novel strategy offers several benefits for the treatment of glaucoma. Reducing thermal damage to adjacent tissues is one of the primary benefits. By delivering laser energy in micropulses, MLT enables precise targeting of the trabecular meshwork and additional structures associated with aqueous outflow without causing substantial thermal damage. This not only reduces the possibility of complications but also improves the patient's comfort throughout the process. MLT's capacity to accomplish a more uniform distribution of energy is an additional advantage [13].

Surgical procedures may be required when drugs and laser therapy are inadequate to control IOP. Among the surgical techniques accessible are trabeculectomy, a procedure that constructs a new drainage channel, and minimally invasive glaucoma surgery [11].

Rho kinases (ROCKs) have been identified as essential proteins in glaucoma pathogenesis. ROCK1 and ROCK2 are involved in multiple cellular processes, such as the modulation of cell contractility, cytoskeletal structure, and cell adhesion. In glaucoma, their improper functioning has been linked to the development of elevated IOP and consequent optic nerve injury [14].

The ROCK signaling pathway is triggered in the trabecular meshwork, an essential tissue that influences the discharge of aqueous humor. Furthermore, elevated ROCK activity disrupts the actin cytoskeleton, leading to reduced aqueous humor drainage and increased IOP. In addition, it has been demonstrated that ROCKs induce inflammation, fibrotic alterations, and oxidative stress in the trabecular meshwork, thus leading to the progression of the disease [15].

Ripasudil and fasudil, both significant ROCK inhibitors, have emerged as prospective therapeutic options for the treatment of glaucoma. These pharmacological compounds demonstrated their beneficial impact by blocking the activity of ROCKs, resulting in trabecular meshwork rest, enhanced discharge of aqueous humor, and a reduction in IOP [16,17].

Given the implications of the ROCK pathway in glaucoma pathophysiology and the identification of unmet needs in current glaucoma management, in particular the need for better outcome measures in clinical trials [18], the present study was designed to take a dual approach. Starting from two compounds with known and proven activity (i.e., IOP lowering, antioxidant activity, wound-healing activity, increasing drainage from the eye) [16,17], an in silico examination was designed to evaluate the binding potential of compounds without current medical applications to ROCK1 and 2, with the objective of discovering potential uses for new compounds that will require further validation and endorsement studies. The second section of the research is devoted to a thorough and distinct bibliometric assessment of therapy management for ocular diseases, with emphasis on certain more prevalent ocular pathologies, including glaucoma, but also on laser therapies, in particular micropulse laser therapy, as can be seen in the search algorithm used in Web of Science. By reviewing the recent scientific literature in a distinct manner, publications addressing treatment strategies and therapeutic advances in ocular diseases were assessed.

The current study provides a two-part investigation aimed at incorporating computational approaches (i.e., molecular coupling analysis, virtual screening of ligands, and estimations of some pharmacokinetic parameters) to identify novel compounds not currently used in medical practice but with potential for future applications due to their high binding affinities and at performing a comprehensive bibliometric analysis for the evaluated field. The molecular docking study was based on ligands with demonstrated biological action in the scientific literature (i.e., ripasudil and fasudil) and targeting key proteins (i.e., ROCK1 and ROCK2) identified as being involved in the modulation of glaucoma pathophysiological mechanisms.

The improvement brought to existing scientific information relies on the demand to develop novel pharmacological strategies to enhance the care of glaucoma patients who do not respond to the current standard of treatment. Furthermore, the contributions to the field of study generated by the present bibliometric analysis are based on the distinct and comprehensive design with emphasis on therapy, including micropulse laser therapy, a topic less addressed in the literature. The results generated by the software and the interpretation of the data contribute to providing an overview of the most relevant and prolific journals, countries, authors, and organizations. The scientific information provided can be an instrument that saves time for researchers in the pre-publication period, facilitating the identification of significant journals, unmet needs, the status of contemporary comprehension in this domain, and the possibility of opening new inter- and multidisciplinary as well as international collaborative networks.

2. Materials and Methods

2.1. Ligand Preparation

Ripasudil and fasudil, known Rho kinase inhibitors, were selected for the molecular docking investigations. The structures of ripasudil and fasudil were downloaded in the Simulation Description Format (.SDF) format from the PubChem online collection (https://pubchem.ncbi.nlm.nih.gov/, accessed on 20 June 2023). An essential first step in the ligand preparation process consisted of converting the ligands from the SDF format, which is not compatible with AutoDockTools 1.5.7 (https://vina.scripps.edu/, accessed on 20 June 2023), into the native AutoDockTools format, which is the Protein Data Bank, Partial Charge and Atom Type (PDBQT) format, utilizing the Open Babel GUI 2.3.1 program (https://openbabel.org/docs/current/GUI/GUI.html/, accessed on 20 June 2023) [19,20]. The second step involved importing the ligands (i.e., in PDBQT format) into AutoDockTools, where the necessary charges and rotatable bonds were added.

2.2. Protein Preparation

The structures of proteins were retrieved from the Research Collaboratory for Structural Bioinformatics Protein Data Bank (PDB) (https://www.rcsb.org/, accessed on 23 June 2023). Specifically, the following protein molecules with their corresponding PDB identifiers were selected: 2ESM (ROCK1 bound to fasudil) [21] and 7JNT (ROCK2 complexed with a selective inhibitor) [22]. The first step in the protein preparation process for molecular docking consisted of removing the co-crystalized water and the co-crystalized ligands using Molegro Molecular Viewer (http://molexus.io/molegro-molecular-viewer/, accessed on 23 June 2023). In the second stage, the protein was imported into the software AutoDockTools, during which hydrogens with polarity as well as Gasteiger charges were inserted. Finally, we saved the protein in PDBQT format.

2.3. Molecular Coupling Assessments

The computational studies dealing with the assessment of molecular couplings were conducted via AutoDockTools, a commonly utilized software application. Discovery Studio Visualizer 4.5 (https://www.3ds.com/products-services/biovia/products/molecular-modeling-simulation/biovia-discovery-studio/visualization/, accessed on 26 June 2023) was utilized to produce the 2D and 3D representations, which offer an in-depth representation of the evaluation outcomes (i.e., the bonds between protein and ligand and the spatial arrangement of docked structures). The grid box size for all molecular docking simulations was preset to $60 \times 60 \times 60$ Ångstroms (Å). This box defines the region where the docking instruments will examine potential arrangements. A verification process involving the re-docking of the native ligand of the researched molecule (i.e., the compound that crystallizes with the molecule) along with the comparison of the docked configuration to the native position was conducted prior to docking the targeted ligands. Moreover, by comparing the native molecule to the re-docked compound, the magnitude of similarities between the atomic coordinates of the two positions is used to calculate the root mean square deviation (RMSD). In general, docking methods that produce docking configurations with RMSD values lower than 2 Å (i.e., lower values imply a better match) are efficient at anticipating ligand poses. As calculated by the AutoDock-Tools 1.5.7 program, the RMSD for all docking configurations of the re-docked native compound was lower than 2 Å in the current study.

An important criterion to evaluate the possible affinities and interaction strengths in this context is the emphasis on intra-ligand and protein binding energies. Although direct comparisons between several compounds may not always be possible using binding energies' absolute values, a given compound's relative differences in binding sites to a protein are likely to produce results with greater accuracy. This distinction makes identifying more advantageous binding conformations easier and boosts confidence when choosing good candidates.

To calculate the binding free energy of the ligand-receptor complex, this scoring function integrates several energy factors, such as van der Waals interactions, hydrogen bonds, electrostatic interactions, and torsional strain. The produced poses are sorted according to how well they are projected to bind the chosen protein. Lower rating values in this ranking algorithm denote more reliable and favorable interactions. While the molecular docking process and scoring functions provide valuable insights into the protein–ligand interaction and identify the most probable binding pose, it is essential to complement these computational results with in vivo and in vitro studies. The results from these experimental approaches help validate the accuracy of the predicted binding poses and provide a comprehensive understanding of the ligand's interaction with the target protein [23].

2.4. In Silico Screening

The primary goal of the virtual screening investigation was to find novel compounds that have potential Rho kinase inhibitor properties. Compounds with chemical structures that are comparable were identified using the SwissSimilarity (http://www.swisssimilarity.ch/, accessed on 27 June 2023) web-based instrument. The newly discovered compounds were extracted from the ZINC online database in .SDF format and reformatted with OpenBabelGUI to a form appropriate for AutoDock Vina assessments (https://vina.scripps.edu/, accessed on 27 June 2023), in which charges have been included and rotational bonds were identified. Furthermore, the following stage of investigation consisted of docking them to the targeted molecules in an attempt to identify compounds with significant binding capacity and with prospects for future integration into thorough in silico, in vitro, and in vivo investigations.

2.5. Estimates of Pharmacokinetic Data

The SwissADME (http://www.swissadme.ch/, accessed on 28 June 2023) online tool was utilized for predicting some relevant pharmacokinetic properties of molecules considered to have the greatest potential based on the results of the molecular coupling and virtual screening of ligands examinations.

3. Results

3.1. Molecular Docking of Compounds against ROCK1

The preliminary step in the docking procedure is the reattachment of the molecule with which the molecule of protein is co-crystallized. The grid-box dimensions and coordinates are $60 \times 60 \times 60$ Å, and X = 52.29, Y = 99.79, and Z = 28.53, respectively. The RMSD of the re-docked compound compared to the native ligand is 0.88, which fits into the general values accepted by the literature, indicating a good molecular docking algorithm. In the case of 2ESM, the native ligand is represented by fasudil, a potent Rho kinase inhibitor. The best re-docked position of the native ligand has a binding potential in the form of an affinity of -8.3 kcal/mol for the evaluated protein. Figure 1 displays the values of affinity for the protein of the top 9 docked poses and the docked arrangement of fasudil overlaid on the co-crystallized framework.

The docking results in the case of ripasudil concluded that the best docking pose of this compound had a higher affinity (-8.6 kcal/mol) for the protein compared to the native ligand, fasudil (-8.3 kcal/mol). Figure 2 depicts the bonds between specific aminoacids from the protein and the ligands. The illustration contains 2D representations of the interactions and 3D diagrams of the ligands within the protein's binding pocket.

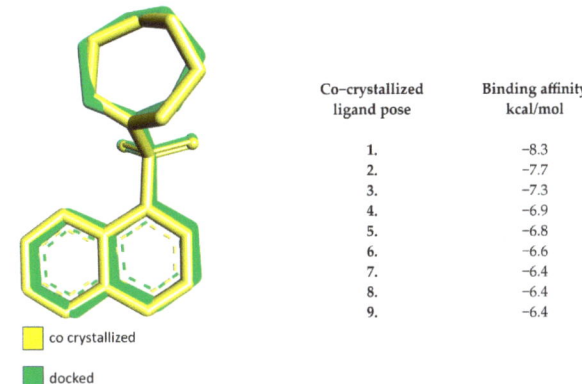

Figure 1. The binding capacity of the re-docked native compound for ROCK1 and the structure of the native ligand overlaid with the re-docked compound.

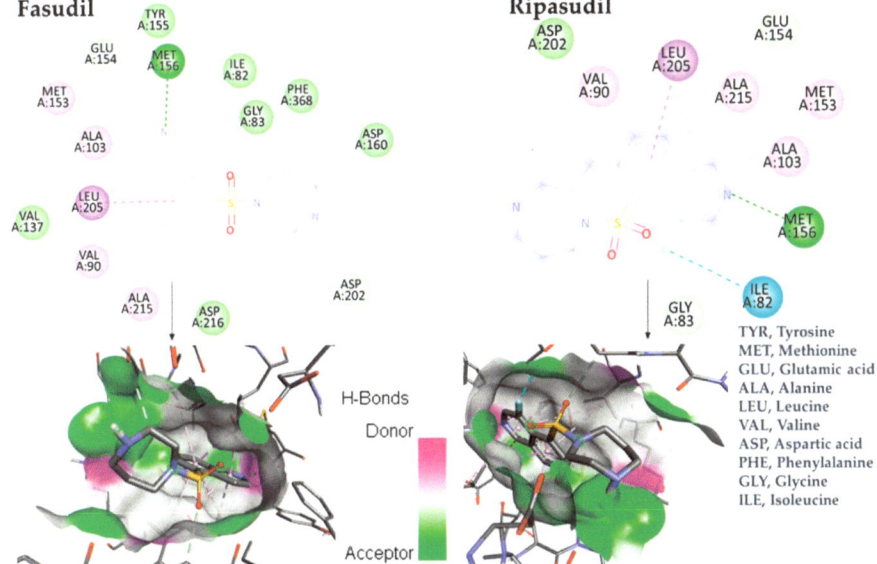

Figure 2. Comprehensive two-dimensional and three-dimensional interactions of ligand–ROCK1.

The two-dimensional representations show that fasudil is involved in interactions with MET156, GLU154, MET153, ALA103, LEU205, VAL90, ALA215, and ASP202. Ripasudil interacts with VAL90, LEU205, ALA215, GLU154, MET153, ALA103, MET156, ILE82, and GLY83. The common amino acids that interact with fasudil and ripasudil include MET156, MET153, ALA103, LEU205, VAL90, ALA215, and GLU154. Additionally, ripasudil interacts with ILE82 through the fluorine atom, which could be responsible for the stronger binding affinity of this compound for ROCK1.

In the context of our docking study, a compelling interaction in which both compounds form essential hydrogen bonds with MET156, an amino acid residue located within the binding region of the protein. It is essential to acknowledge from the stability of the protein–ligand complex that the creation of these hydrogen bonds plays a crucial part in coordinating the binding mechanism. It is vital that these hydrogen bonds with MET156 form because they increase the stability of the protein–ligand complex. This interaction will likely improve how strongly the compounds bind and how specifically they interact. This emphasizes how vital MET156 is in ensuring the protein complex is stable, which could potentially positively impact its therapeutic benefits.

3.2. Molecular Docking of Compounds against ROCK2

As in the case of ROCK1, the first step was to re-dock the protein's (ROCK2) co-crystallized ligand, N-[(3-methoxyphenyl) methyl]-5H-[1] benzopyrano[3,4-c]pyridine-8-carboxamide. The grid-box dimensions and coordinates were set at 60 × 60 × 60 Å and X = 47.05, Y = 68.16, and Z = 36.18, respectively. The RMSD value of the re-docked native ligand compared to the co-crystallized was 1.065, a value within the acceptable range stated in the literature. The best position of the re-docked compound presents a binding affinity of −10.7 kcal/mol. Moreover, the affinity values of the top nine docked positions and an overlay representation of the docked and co-crystallized structures are displayed in Figure 3.

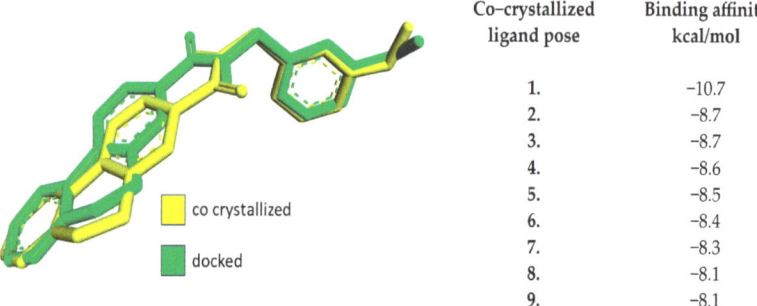

Co–crystallized ligand pose	Binding affinity kcal/mol
1.	−10.7
2.	−8.7
3.	−8.7
4.	−8.6
5.	−8.5
6.	−8.4
7.	−8.3
8.	−8.1
9.	−8.1

Figure 3. The representation of the most elevated binding affinities of the re-docked native compound for ROCK2 and the structural representation of the native ligand superimposed with the re-docked ligand.

The docking results in the case of ripasudil concluded that the best docking position of this compound presented a binding potential of −9.1 kcal/mol for the evaluated protein. fasudil had a binding affinity of −8.7 kcal/mol. Ripasudil and fasudil showed lower affinities for the protein than the native ligand. Figure 4 depicts the interactions between the ligands and the protein. The illustration contains two-dimensional representations of the interactions and three-dimensional models of the compounds within the protein's binding site.

According to the two-dimensional model, the native ligand interrelates to the following amino acids: ASP232, PHE136, LEU123, PHE103, GLY101, ARG100, LYS121, ALA231, LEU221, TYR171, MET172, ALA119, and VAL106. Fasudil interacts with MET172, ALA119, LEU221, ALA231, VAL106, MET169, ASN219, ASP218, and ARG100. Ripasudil interacts with MET172, GLU170, ALA119, LEU221, MET169, VAL106, ALA231, ASP218, ASN219, and ASP232. The common amino acids interacting with fasudil and ripasudil are represented by: MET172, ALA119, LEU221, ALA231, VAL106, and ARG100. As for the native ligand, it has two common amino acids with fasudil and ripasudil, namely, ALA119 and MET172.

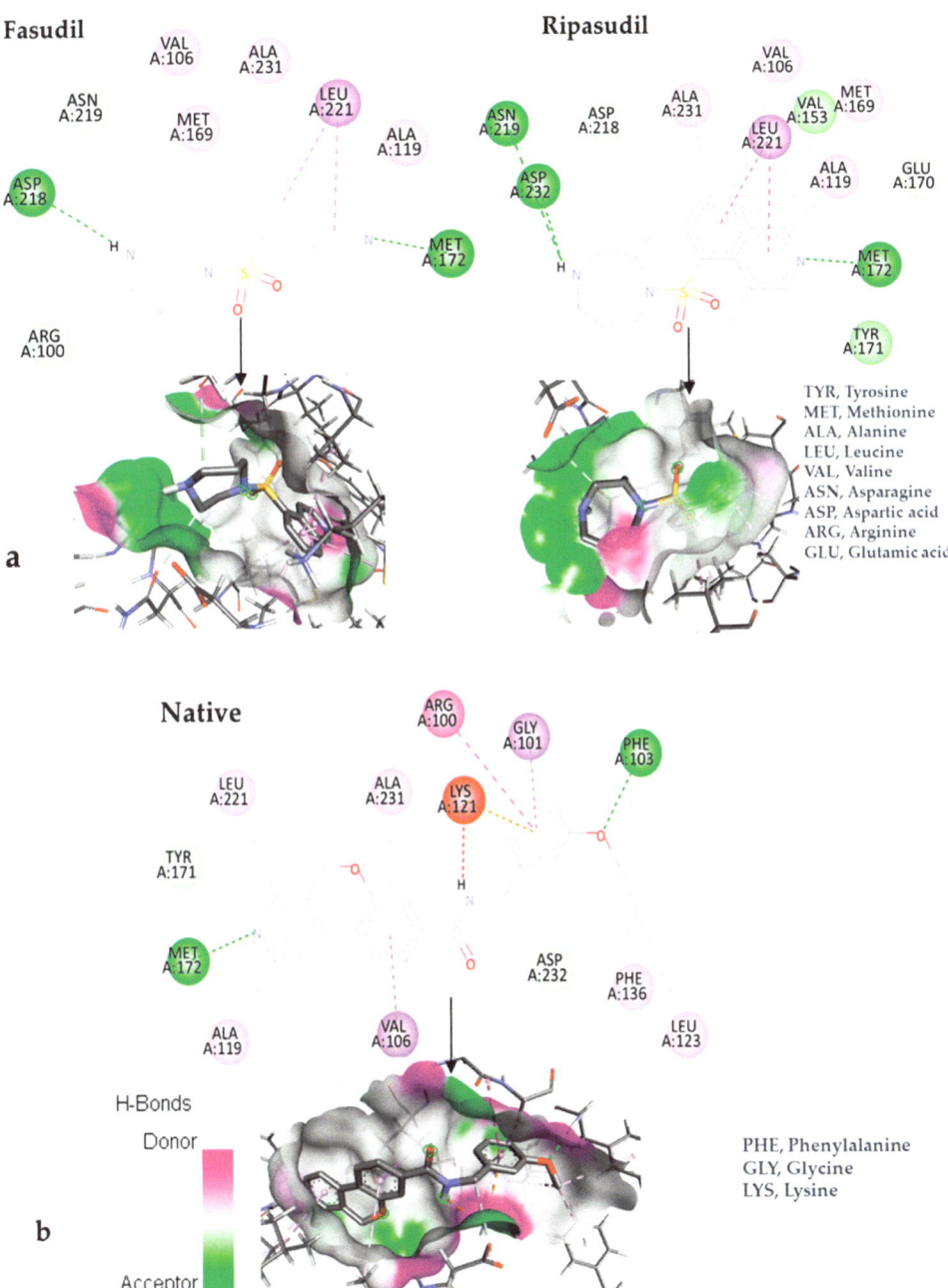

Figure 4. In-depth two-dimensional and three-dimensional interplays of ligand–ROCK1. (**a**) Interactions between different structures of fasudil and ripasudil with the substructures of the target protein; (**b**) Interactions between different structures of the native ligand with the substructures of the target protein.

During the in-depth analysis of protein–ligand interactions, a notable observation emerges. Although the three compounds' chemical structures are different from those of the native ligand, they share contact with MET172 via a hydrogen bond. This finding underscores the pivotal role of this amino acid in establishing a stable complex between the protein and the ligand. One of the amino acids that could play an essential role in stabilizing the ligand–protein complex is represented by ASP218, which forms a hydrogen bond in the case of fasudil. Additionally, in the case of ripasudil, interactions with ASN219 and ASP232 are evident. For the native ligand, interaction with PHE103 might also have a crucial role in complex formation.

3.3. In Silico Screening of the Candidate with the Most Potential

Among the investigated molecules, ripasudil showed the highest affinity towards both proteins (ROCK1 and ROCK2). Starting from the chemical structure of this compound, a simulated screening of ligands via the SiwssSimilarity online tool has been conducted. The objective of this specific investigation was to discover compounds with comparable chemical structures to ripasudil and determine their affinity for ROCK1 and ROCK2 via molecular docking studies.

SwissSimilarity's digital platform analyzes and identifies similarities between compounds using a range of chemical fingerprinting methodologies. In these systems, two- and three-dimensional strategies of structural comparison are utilized. Furthermore, in order to seek compounds with similar physicochemical properties, SwissSimilarity provides data regarding the physical and chemical properties of the compounds [24].

To display and distinguish molecular structures, extended-connectivity chemical fingerprints (ECFPs) are commonly used. As circular fingerprints, ECFPs contain information about a molecule's parts. The ECFP algorithm generates these fingerprints using a graph-based methodology by analyzing all possible paths across atoms in a molecular structure of a particular dimension. It has been demonstrated that ECFPs are widely utilized in a variety of applications, such as virtual screening, compound library clustering, and similarity research [25]. The screening method was set as ECFP, and the search was performed on the ZINC (Lead-like) database (https://zinc.docking.org/, accessed on 20 April 2023). From this search, twenty molecules presenting similarity values ranging between 0.722 and 0.475 were identified. Furthermore, a higher similarity score indicates a more significant similarity to the parent substance.

Table 1 provides the scores of similarities, chemical structures, and affinities of the five most prospective molecules, as determined by molecular docking evaluation, with regard to their affinity to ROCK1.

ZINC000000022706 showed the most elevated affinity (−9.0 kcal/mol) for ROCK1, surpassing the parent molecule ripasudil (−8.6 kcal/mol) and the native ligand fasudil (−8.3 kcal/mol).

Ripasudil was kept as the parent compound in the screening process for ROCK2 because it has a higher affinity for the protein compared to fasudil. Table 2 provides the scores of similarities, chemical structures, and affinities of the five most prospective molecules, as determined by molecular docking evaluation, with regard to their affinity to ROCK2.

ZINC000034800307 showed a higher affinity (−8.8 kcal/mol) for the protein than fasudil (−8.7 kcal/mol) but a lower affinity than ripasudil (−9.1 kcal/mol) and the native ligand (−10.7 kcal/mol). The observed values are also consistent with the average values found in published research evaluating the binding potential of various compounds to ROCK2 (i.e., −7.39 to −9.07 kcal/mol) [26].

Figure 5 depicts the interactions between the identified ligands and the target protein. The illustration contains two-dimensional representations of the ligands and three-dimensional models of the compounds within the protein's binding site.

Table 1. Five compounds with the greatest affinity for ROCK1.

Compound	Chemical Structure	Binding Affinity (kcal/mol)	Score of Similarity
ZINC000000022706		−9.0	0.772
ZINC000193357696		−7.3	0.508
ZINC000193358334		−7.3	0.492
ZINC000034800306		−7.1	0.590
ZINC000034800307		−7.1	0.590

In accordance with the two-dimensional representation, ZINC000000022706 interacts with the following ROCK1 amino acids: GLY83, MET156, ALA103, GLU154, VAL90, MET153, ALA215, and LEU205. Both ripasudil (parent compound) and ZINC000000022706 interact with VAL90, LEU205, ALA215, GLU154, MET153, ALA103, and MET156, indicating a similar binding process to ROCK1. Even though the parent compound additionally interacts with ILE82, the binding affinity of ZINC000000022706 is higher, indicating a more stable ligand–protein complex. A similar binding mechanism was revealed when we examined the ZINC000000022706 protein's interaction with fasudil in greater detail. Specifically, most interactions occur through the isoquinoline ring, displaying a high resemblance. An exception to this pattern is observed with GLY83, which establishes its binding to ZINC000000022706 via a sulfonyl group. Consequently, considering the comparable binding energies, the significant number of shared amino acids, and the chemically akin structure of ZINC000000022706 compared to fasudil, this substance emerges as a plausible candidate with therapeutic potential. Given the elevated number of shared amino acids, it is reasonable to assume that ZINC000000022706 could participate in comparable molecular recognition processes, perhaps targeting similar biological pathways as fasudil. The alignment of binding energies additionally offers a theoretical framework for assessing the potency of these interactions.

Table 2. Five compounds with the greatest affinity for ROCK2.

Compound	Chemical Structure	Binding Affinity (kcal/mol)	Score of Similarity
ZINC000034800307		−8.8	0.590
ZINC000000022706		−8.6	0.772
ZINC000054371104		−7.7	0.492
ZINC000193357696		−7.6	0.590
ZINC000534634918		−7.6	0.590

Therefore, evidence of similar binding patterns to the protein as the parent compound is apparent in the case of ZINC000000022706. Notably, the recurrence of hydrogen bond formation with MET156 is a noteworthy observation. This recurrence substantiates the hypothesis that ZINC000000022706 can potentially emerge as a therapeutic candidate characterized by similar attributes to the parent compound.

ZINC000034800307 interacts with the following ROCK2 amino acids: ASP218, ALA231, VAL106, LEU221, GLU170, MET172, MET169, and ALA119. Comparing the ligand–protein interactions for both ZINC000034800307 and the parent compound, ripasudil, we can see that both ZINC000034800307 and ripasudil interact with the following amino acids: ASP218, ALA231, VAL106, LEU221, GLU170, MET172, MET169, and ALA119, indicating a similar binding method to the protein. Ripasudil's or fasudil's interactions with ROCK2 and those with ROCK1 are similar, suggesting a possible inference even if the parent chemical of ROCK2 is not included among the known medications with established effects on the pathophysiology under study. A similar binding pattern is seen when the interaction mode between the parent compound–protein and ZINC000034800307–protein is examined. This resemblance encompasses both the chemical groups and the interacting amino acids. Notably, the isoquinoline ring mediates the majority of interactions with the protein. There are a few exceptions, though: ripasudil interacts with ASP232, ASN219, and ASP218 through the 1,4-diazepane ring in this scenario.

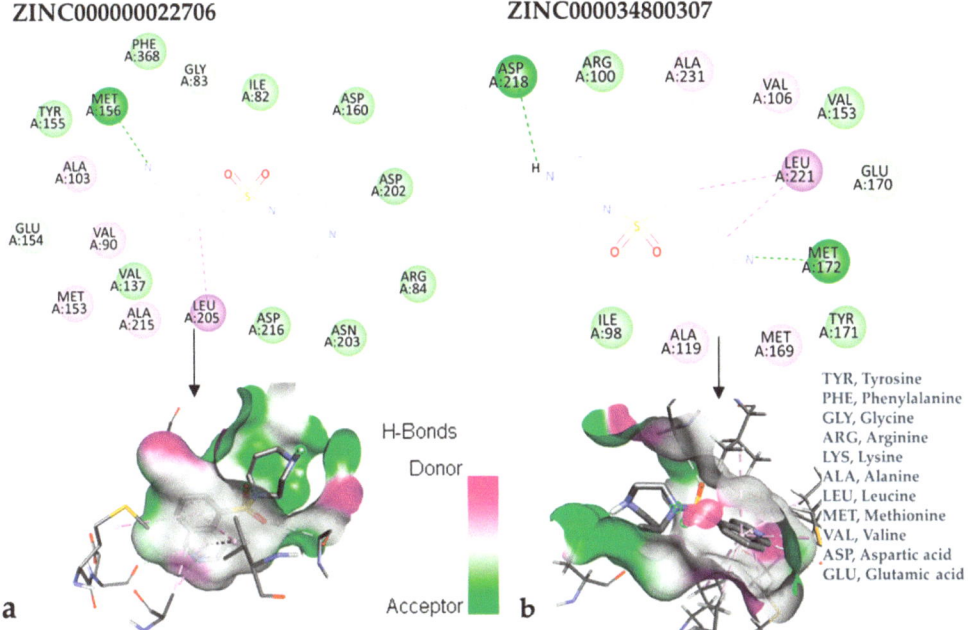

Figure 5. Comprehensive two-dimensional and three-dimensional interactions of ligand–target protein. (**a**) Interactions with ROCK1; (**b**) interactions with ROCK2.

In contrast, contact occurs through a piperazine ring in the instance of ZINC000034800307 instead of a 1,4-diazepane ring. The isoquinoline-mediated interactions have a recurring motif, highlighting a consistent and possibly meaningful protein binding method. The distinctive interactions with particular amino acids, such as ASP232, ASN219, and ASP218 in the case of ripasudil, and the piperazine ring in the case of ZINC000034800307, point to structural modifications that may be responsible for the selectivity and affinity of these interactions. Moreover, the comparable binding patterns and shared structural elements between the interactions of ZINC000034800307–protein and the parent compound–protein highlight the potential of ZINC000034800307 as a promising candidate for further investigation. Its interaction similarities, especially the mode of interaction with protein residues and functional groups, underscore the potential for targeted bioactivity, warranting deeper exploration for potential therapeutic implications.

A parallel trend becomes apparent with ZINC000034800307, mirroring the findings elucidated by ZINC000000022706. Precisely, a reminiscent mode of interaction with MET172 emerges. A hydrogen bond ensues between the ligand and this specific amino acid, substantiating its instrumental function in bolstering the stability of the ligand–protein complex.

3.4. Computational Assessments of Relevant Pharmacokinetic Data for Newly Found Molecules

By utilizing SwissADME, important pharmacokinetic data for two possible Rho kinase inhibitors has been evaluated. The findings of the computational ADME assessment are outlined in Table 3.

Table 3. The findings of the computational ADME evaluation of possible Rho kinase inhibitors.

Characteristics	ZINC000000022706 (ROCK1)	ZINC000034800307 (ROCK2)
Formula	$C_{16}H_{21}N_3O_2S$	$C_{14}H_{17}N_3O_2S$
Molecular weight	319.42 g/mol	291.37 g/mol
Num. rotatable bonds	2	2
Num. H-bond acceptors	5	5
Num. H-bond donors	1	1
Log P	2.50	2.10
Gastrointestinal absorption	High	High
CYP2C19 inhibitor	No	No
CYP2C9 inhibitor	No	No
CYP2D6 inhibitor	Yes	No
CYP3A4 inhibitor	Yes	No
Lipinski	Yes	Yes

Ripasudil has a molecular weight of 359.8 g/mol and fasudil has a molecular weight of 291.37 g/mol, while the investigated compounds have a molecular weight of 319.42 g/mol and 291.37 g/mol. Furthermore, the investigated compounds also have under five H-bond donors (one for both), under ten H-bond acceptors (five for both), and Log P values of 2.50 (ZINC000000022706) and 2.10 (ZINC000034800307), which fall under Lipinski's list of five concepts [27], suggesting the fact that the identified molecules have good oral bioavailability and drug-like characteristics. ZINC000000022706 is a CYP2D6 and CYP3A4 inhibitor, suggesting that it has the potential to alter drug metabolism and interactions with other compounds processed by these enzymes. ZINC000034800307 does not affect the investigated enzymes.

Utilizing digital models and algorithms, SwissADME can assess permeability, solubility, pharmacokinetics, and lipophilicity. Computational approaches to assessing the ADME characteristics of a compound are beneficial to preliminary drug design, but they cannot replace in vivo assays. In contrast, they seek to develop a rapid and cost-effective approach to assessing the pharmacokinetic characteristics of small molecules [28].

The integration of advanced computational methodologies, including molecular docking, ligand-based virtual screening, and ADME evaluation, within the realm of glaucoma research offers promising avenues for revolutionizing patient care and augmenting therapeutic options in clinical practice. These computational tools hold the potential to yield profound insights into the intricate landscape of drug discovery and development, thereby fostering the emergence of more efficacious and tailored treatment strategies for individuals afflicted by glaucoma.

Molecular docking, for instance, empowers researchers to predict the binding affinity and conformational preferences of potential therapeutic agents directed toward specific protein targets, as exemplified by the case of ROCK1 and ROCK2 in glaucoma. Such predictive capabilities facilitate the identification of novel molecular entities that exhibit a heightened propensity to engage with these targets and elicit favorable modulatory effects upon their biological functions.

Ligand-based virtual screening, tailored to the context of glaucoma, serves to substantially broaden the array of candidate compounds warranting consideration. Through an analytical lens rooted in structural and chemical similarity to established active agents, this methodology effectively enlarges the repertoire of potential drug candidates, thus augmenting the likelihood of discovering efficacious treatment regimens.

The judicious inclusion of ADME evaluations at the outset of the drug discovery journey endows researchers with the capacity to discern and prioritize compounds with enhanced potential for favorable outcomes within clinical trials and eventual real-world application. This anticipatory consideration of pharmacokinetic characteristics guides the selection of candidates that hold promise for efficacious therapeutic interventions. Furthermore, the integration of computational techniques, such as molecular docking and virtual screening, offers the advantageous prospect of diminishing the reliance on early stage animal testing.

The subsequent characteristics reflect the research's strengths: estimates of the binding capacity and the mechanism of interaction between a protein of interest and the ligand with documented biological impacts, which can be used to guide the design of novel ligands; evaluation of a large database of molecules for possible lead compounds; and a comprehensive analysis of the physicochemical and pharmacokinetic properties of newly discovered molecules.

4. Bibliometric Analysis

Bibliometric studies are essential in gaining a comprehensive understanding of publication trends, assessing the impact of the research, and understanding the research landscape in a specific field. While the selection of the search algorithm is of utmost importance, the choice of the source database is also crucial. The Web of Science (W.o.S) collection of databases has been used for the present investigation due to its comprehensive accumulation of documents across multiple fields, citation indexing capabilities, and global coverage. The following search algorithm was used to identify relevant articles related to therapy and micropulse in the field of ophthalmology: ALL = (ophthalmology OR retinal diseases OR Diabetic macular edema* OR Retinal vein occlusion OR Glaucoma OR Central serous chorioretinopathy OR Age related macular degeneration) AND ALL = (therapy OR micropulse). A total of 43,450 documents were identified, of which 31,301 (70.04%) were articles, 5927 (13.64%) were review articles, 3643 (8.38%) were meeting abstracts, and 2524 (5.81%) were proceeding papers. The remaining document types had fewer than 1000 classified documents each.

Regarding the language distribution of the identified documents, we found that English is the most commonly used language, accounting for 95.94% (41,685) of the total papers. German was the second most common language, accounting for 2.64% (1149) of the complete paper count. The percentages for French and Portuguese were 0.757% (329) and 0.168% (73). The other languages had fewer than 50 documents.

The identified documents were classified into 165 W.o.S categories, with the critical caveat that a single manuscript might be allocated to more than one category. The following categories had the highest number of assigned manuscripts: "Ophthalmology" 30,317 documents; "Medicine Research Experimental" 2494; "Pharmacology Pharmacy" 2281; "Medicine General Internal" 1704; "Biochemistry Molecular Biology" 1349; "Genetics Heredity" 1108; and "Cell Biology" 1037; other categories had under 1000 manuscripts assigned to them. Figure 6 represents the tree map of the top 10 most populated categories according to W.o.S.

Only English-language articles were considered for the present study, limiting the number of papers assessed to 29,882. The analysis was performed using VOSviewer version 1.6.19 [29,30] and the built-in analysis tools available within the W.o.S system. Furthermore, the necessary information has been extracted from W.o.S as tab-delimited files encompassing the complete record and cited documents' references using the Export function.

The first paper indexed in W.o.S collection of databases matching the search algorithm was published in 1945. For a more comprehensive approach targeting novel elements in the management of ocular diseases (i.e., micropulse laser therapy), the years 2011–2023 were chosen as the period of bibliometric evaluation and science mapping research. For the evaluated time period, we determined the most prolific nations, journals, authors, articles, and organizations in the field under consideration.

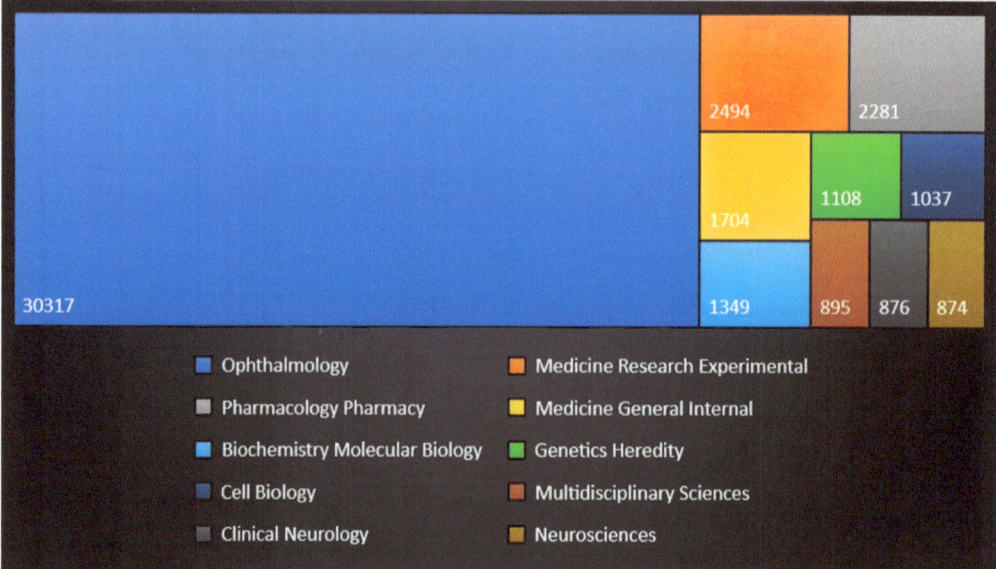

Figure 6. Treemap visualization of the Top 10 Categories.

The country collaboration system diagram was generated using VOSviewer to determine the collaborative relationships between countries. The size of each individual bubble is determined by the number of published articles. The width of the band connecting two nations is directly related to the collaboration between those countries, while the color of each bubble is determined by the cluster in which the country was categorized. Countries that often publish articles together are usually classified in the same cluster.

The median publication year and citation mapping of the journals examined were determined as well. In the journal's median publication year diagram, the color of each sphere reveals the average publication year, and the size of the sphere is directly related to the overall number of papers published in that journal. The bubbles in the map are color-coded from dark blue to yellow. Darker colors represent an early publication year, while lighter colors, specifically yellow, indicate a later publication year. This color gradient visually represents the time distribution of the articles on the map. To improve precision, the years are represented fractionally (e.g., 2005.50 indicates the midpoint of 2005). In the context of the citation system diagram, the size of the sphere is directly correlated with the total number of published papers, the sphere color shows the cluster, and journals that frequently cite one another are typically grouped together. The width of the band connecting two journals is proportional to the frequency of citations between them.

Lastly, the keyword co-occurrence network and the keyword bubble maps were generated for each period. The hue of the keyword sphere diagram shows the mean number of citations that an item containing the keyword has obtained. Moreover, darker hues indicate fewer citations, whereas lighter hues, particularly yellow, indicate more citations. The magnitude of each sphere indicates its frequency of occurrence. In the keyword co-occurrence system diagram, the dimension of the spheres indicates the number of occurrences, the width of the band linking two words is directly related to the number of co-occurrences, and the color reveals the cluster, with frequently occurring keywords typically clustered together.

4.1. Period 2011–2023

4.1.1. Assessment of the Most Prolific Nations

During the period under examination, the overall number of nations contributing to scientific output rose from 103 to 151, indicating the growing interest of more countries in this field. The United States remains the most significant contributor, with 6736 (35.02%) published papers. The average citation/article published by the United States is 26.66, indicating that these articles had a significant impact on the field. China occupies second place in regard to the number of papers that have been published (2503, 13.01%), and it has an average citation/article of 12.71. Ranked third is England, with 1555 (8.09%) published documents and an average citation/article of 30.28, indicating the high impact of these articles. Out of the top-ranked countries, France stands out with the highest average citation/article (32.62). Table 4 lists the top ten nations in terms of publication prolificacy in the discipline assessed from 2011 to 2023.

Table 4. Ten nations with the highest level of output and productivity.

Country	Papers	Citations	Average Citation/Article	Total Link Strength (TLS)
United States	6736	179,553	26.66	4451
China	2503	31,820	12.71	1151
England	1555	47,090	30.28	2482
Japan	1444	30,626	21.21	745
Germany	1432	39,223	27.39	2088
Italy	1188	25,012	21.05	1559
India	973	13,759	14.14	923
France	750	24,467	32.62	1572
Australia	739	22,583	30.56	1378
South Korea	713	12,330	17.29	354

4.1.2. Evaluation of the Most Productive Journals

A total of 1952 journals published documents that fit the search parameters between 2011 and 2023. The most productive journal of this period is *Investigative Ophthalmology & Visual Science*, with a total of 896 (4.66%) published documents. Ranked second is *Retina—The Journal of Retinal and Vitreous Diseases*, with a total of 799 (4.15%) published documents, and ranked third is *Graefe's Archive for Clinical and Experimental Ophthalmology*, with 615 (3.20%) published documents. The journal *Ophthalmology* stands out with the highest total citations received (compared to the other journals included in the top 10), 39,344, and also with the highest average citation/article (66.24). Table 5 shows some of the most prolific journals that published papers between 2011 and 2023.

4.1.3. Assessment of the Most Prolific Authors

During the evaluated period, 60,211 authors significantly supported scientific advancement. Bandello F., affiliated with Vita-Salute San Raffaele University, Italy, is the most productive author of this period, with 112 published documents. Ranked second is Hauswirth WW, with 92 published documents, affiliated with the University of Florida, United States. Holz, F.G., has the highest average citation per document out of the authors listed among the first ten, receiving a total of 4809 citations for the 73 published documents, resulting in an average citation per document of 65.88. The ten most prolific authors of the evaluated period are listed in Table 6.

Table 5. Top 10 prolific journals and their metrics.

Journals	No. of Papers	No. of Citations	Average No. of Citations per Article	IF	IF without Self-Citations	Publishing Entity
Investigative Ophthalmology & Visual Science	896	24,621	27.48	4.925	4.589	Assoc Research Vision Ophthalmology Inc., Rockville, MD, USA
Retina—The Journal of Retinal and Vitreous Diseases	799	17,913	22.42	3.975	3.617	Lippincott Williams & Wilkins, Philadelphia, PA, USA
Graefe's Archive for Clinical and Experimental Ophthalmology	615	8208	13.35	3.535	3.372	Springer, Berlin/Heidelberg, Germany
British Journal of Ophthalmology	610	13,254	21.73	5.907	5.565	BMJ Publishing Group, London, UK
American Journal of Ophthalmology	601	19,127	31.83	5.488	5.128	Elsevier Science Inc., Amsterdam, The Netherlands
Ophthalmology	594	39,344	66.24	14.277	13.741	Elsevier Science Inc., Amsterdam, The Netherlands
European Journal of Ophthalmology	466	3257	6.99	1.922	1.743	Sage Publications Ltd., New York, NY, USA
BMC Ophthalmology	452	3712	8.21	2.086	1.992	BMC, London, UK
Ophthalmology and Therapy	415	1594	3.84	4.927	4.759	Springer Int Publ Ag, Cham, Switzerland
PLoS ONE	322	6701	20.81	3.752	3.608	Public Library Science, San Francisco, CA, USA

IF, impact factor.

Table 6. The most productive authors in the field between 2011 and 2023.

Authors' Name	Latest Affiliation	Nation	No.	No. of Citations	Average Citations per Document
Bandello, F.	Vita-Salute San Raffaele University	Italy	112	2034	18.16
Hauswirth, W.W.	University of Florida	United States	92	4404	47.87
Maclaren, R.E.	University of Oxford	England	91	2916	32.04
Liu, Y.	-	-	85	1043	12.27
Zhao, M.W.	-	-	83	613	7.39
Chhablani, J.	University of Pittsburgh	United States	81	1172	14.47
Sahel, J.A.	National Institute of Health and Medical Research (Inserm)	France	76	3819	50.25
Shields, C.L.	Jefferson University	United States	75	2441	32.55
Freund, K.B.	Vitreous Retina Macula Consultants of New York	United States	73	3476	47.62
Holz, F.G.	University of Bonn	Germany	73	4809	65.88

4.1.4. Citation Analysis

The number of published articles increased from 10,649 (1945–2010) to 19,233 (2011–2023). The article that had the most citations during this period was published by Sawcer, S., in 2011 and titled "Genetic risk and a primary role for cell-mediated immune mechanisms in multiple sclerosis" in the journal *Nature*, which has an impressive IF of 69.504. Ranked second in terms of citations is the article titled "Intravitreal Aflibercept (VEGF Trap-Eye) in Wet

Age-related Macular Degeneration", published by Heier, J.S., in the journal *Ophthalmology* in 2012. Table 7 presents the most cited articles of this period.

Table 7. Top 10 most cited articles in the period 2011–2023.

Main Author (Year)	Title of the Paper	Scientific Periodical	IF	C	Ref.
Sawcer, S. (2011)	Genetic risk and a primary role for cell-mediated immune mechanisms in multiple sclerosis	Nature	69.504	1942	[31]
Heier, J.S. (2012)	Intravitreal Aflibercept (VEGF Trap-Eye) in Wet Age-related Macular Degeneration	Ophthalmology	14.277	1562	[32]
Martin, D.F. (2012)	Ranibizumab and Bevacizumab for Treatment of Neovascular Age-Related Macular Degeneration	Ophthalmology	14.277	1315	[33]
Okita, K. (2011)	A more efficient method to generate integration-free human iPS cells	Nature Methods	47.99	1298	[34]
Lim, L.S. (2012)	Age-related macular degeneration	Lancet	202.731	1216	[35]
Mintz-Hittner (2011)	Efficacy of Intravitreal Bevacizumab for Stage 3+Retinopathy of Prematurity.	New England Journal of Medicine	176.082	909	[36]
Quigley, H.A. (2011)	Glaucoma	Lancet	202.731	848	[37]
Mandai, M. (2017)	Autologous Induced Stem-Cell-Derived Retinal Cells for Macular Degeneration	New England Journal of Medicine	176.082	833	[38]
Tang, J. (2011)	Inflammation in diabetic retinopathy	Progress In Retinal and Eye Research	19.704	763	[39]
Rofagha, S. (2013)	Seven-Year Outcomes in Ranibizumab-Treated Patients in ANCHOR, MARINA, and HORIZON	Ophthalmology	14.277	701	[40]

C, number of citations; Ref, references.

4.1.5. The Topic's Most Involved Organizations

The count of active organizations rose from 4474 in 2011 to 11,369 during the current period, indicating the increase in attention that this subject receives. The University of California System remains the most active organization, with a total of 885 (4.60%) publications. Ranked second is the University of London with 812 published documents, which is closely followed by University College London with 749 publications. Table 8 presents the most productive organizations during the 2011–2023 period.

Table 8. The most active organizations during the 2011–2023 period.

Affiliations	Record Count	% of 19,233
University of California System	885	4.60
University of London	812	4.22
University College London	749	3.89
Harvard University	608	3.16
Moorfields Eye Hospital NHS Foundation Trust	553	2.88
Johns Hopkins University	521	2.71
Harvard Medical School	515	2.68
Udice French Research Universities	456	2.37
University of Pennsylvania	389	2.02
Johns Hopkins Medicine	381	1.98

4.2. Science Mapping

4.2.1. Networks of Collaboration between Nations

Figure 7 is an interconnected diagram displaying collaboration paths between nations. For a country to be represented in the network map, a minimum requirement of 50 published papers was imposed, thus guaranteeing a robust dataset, resulting in the inclusion of 45 nations that met this condition. These nations were separated into four distinct groups. The red cluster consists of 19 nations, led by Germany, based on published papers. Furthermore, this cluster mainly contains countries from the European continent, indicating a strong collaborative relationship between countries located in Europe. The green cluster consists of 12 countries, and the United States leads based on the papers that have been published. The blue cluster consists of eight nations, led by England. Furthermore, the yellow cluster includes six countries, and China leads in terms of published documents. Strong collaborative relationships form between the United States and the following countries: China (458), England (445), Germany (281), Canada (228), and Italy (214). A strong collaborative relationship was also formed between England and the following countries: Germany (214), Australia (145), Italy (144), and France (143).

4.2.2. Resource Average Publication Year and Citation System Diagram

Figure 8 depicts a node diagram showcasing the mean year of publication in the case of journals that have published a minimum of 50 articles. *Investigative Ophthalmology & Visual Science*, the most prolific scientific periodical of this time period, possesses an average publication year of 2015.53, indicating that during the 2011–2023 period, documents were published steadily without a substantial increase near the end of the time frame. *Retina—The Journal of Retinal and Vitreous Diseases* ranked second based on the papers that have been published and possesses a mean publication year of 2016.41, whereas *Graefe's Archive for Clinical and Experimental Ophthalmology*, which is the third most prolific scientific periodical of this time frame, encounters an average publication year of 2017.47, suggesting that more papers have been released at the termination of the period. The majority of the period's papers were published towards the completion of the period in subsequent journals: *Ophthalmology and Therapy* (2021.25), *Clinical Ophthalmology* (2020.27), *Journal of Clinical Medicine* (2021.10), *Frontiers in Medicine* (2021.71), *Ophthalmology Retina* (2020.25), and *International Journal of Molecular Sciences* (2020.90).

Figure 9 depicts a diagram of source citations. The journal requirements for inclusion remained unchanged from the previous figure. The scientific periodicals are organized into three separate clusters, with the red cluster containing 26 journals and being led on the basis of published documents by the *British Journal of Ophthalmology*. *Investigative Ophthalmology & Visual Science*, the most influential journal of this time frame, leads the group of 17 journals included in the green cluster. Furthermore, the blue cluster consists of 15 scientific journals and is led by the period's second-highest-producing journal, *Retina—The Journal of Retinal and Vitreous Diseases*. The red cluster and the blue cluster are closely intertwined, indicating that the subjects found in journals included in these clusters are closely related, and they frequently cite each other. To further validate this fact, on a closer analysis of the network map, we can notice that articles from *Retina—The Journal of Retinal and Vitreous Diseases* were often cited by articles from *Ophthalmology* (link strength: 994), the *American Journal of Ophthalmology* (847), and *Graefe's Archive for Clinical and Experimental Ophthalmology*.

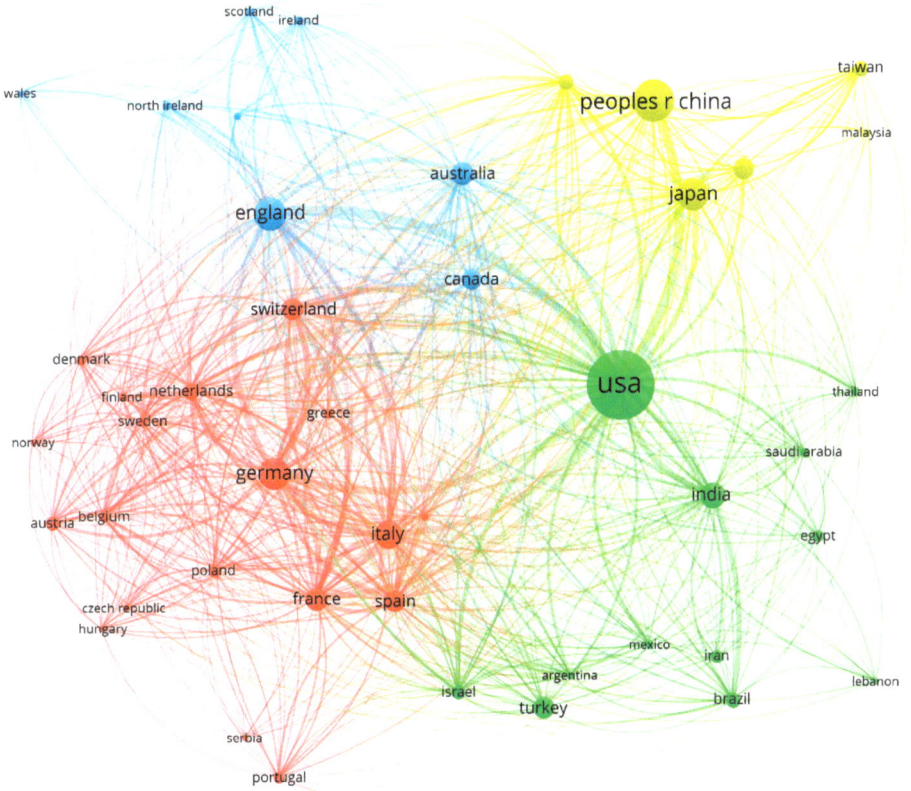

Figure 7. Collaborative writing networks between countries for the period under review, revealed by VOSviewer.

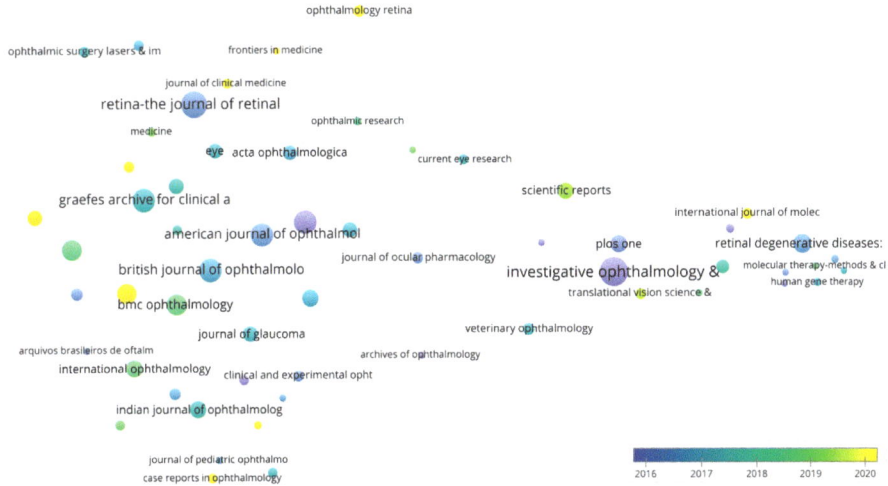

Figure 8. Average journal publication year 2011–2023 (VOSviewer).

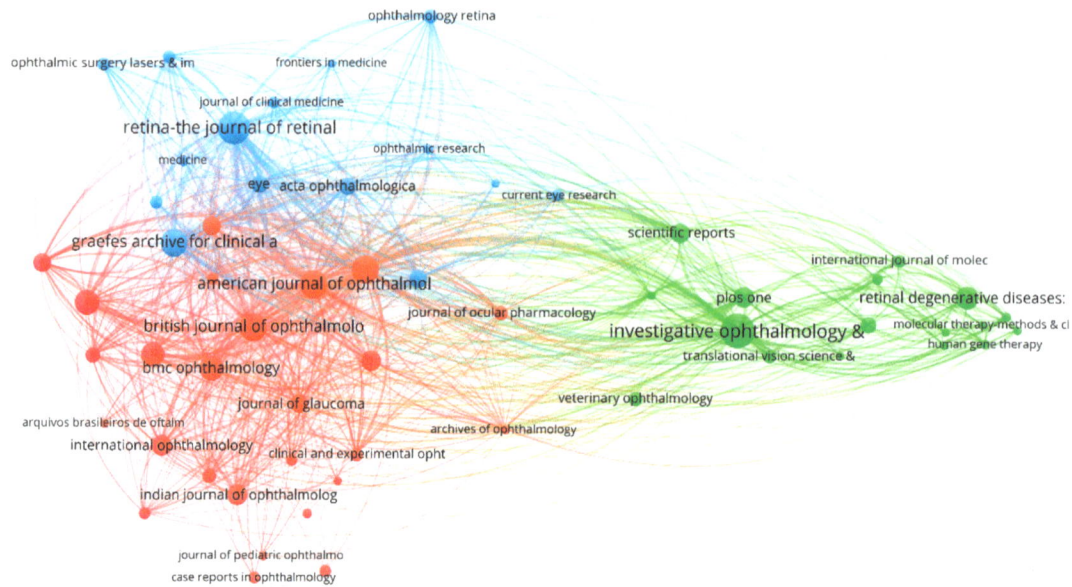

Figure 9. 2011–2023 resource citation system map utilizing VOSviewer.

4.2.3. Keyword System Mapping and Terms Co-Occurrence System Diagram

Figure 10 depicts the bubble map of extremely regular terms utilized for searches in the field in the period 2011–2023. Only words with a minimum occurrence of 200 are represented in the figure. The following words had a high occurrence: "therapy (3279 occurrences, 17.92 average citations/document)", "ranibizumab (2327, 17.27)", "macular degeneration (1651, 21.79)", "bevacizumab (1503, 17.75)", "glaucoma (1478, 12.72)", and "optical coherence therapy (1477, 20.21)". Terms that had high average citations/document are represented by: "indocyanine green angiography (209, 34.35)", "mouse model (457, 31.56)", "geographic atrophy (236, 30.22)", "in vivo (308, 29.99)", "differentiation (309, 29.78)", "Avastin (211, 28.20)", and "intravitreal ranibizumab (250, 25.31)".

Figure 11 depicts the network map of keyword co-occurrence. The inclusion criteria for the terms were kept unmodified from the previous figure. A total of four clusters are formed. The red cluster includes 41 terms that are mainly focused on treatment approaches, retinal health, and various diseases. The green sphere contains 32 terms associated with glaucoma medical management and therapy. Moreover, 28 keywords are contained in the blue cluster. This cluster appears to focus on the topic of treatment approaches and conditions related to the retina, particularly anti-vascular endothelial growth factor treatments. The yellow cluster includes 17 terms that are mainly focused on various aspects related to retinal disorders, particularly age-related macular degeneration and associated conditions.

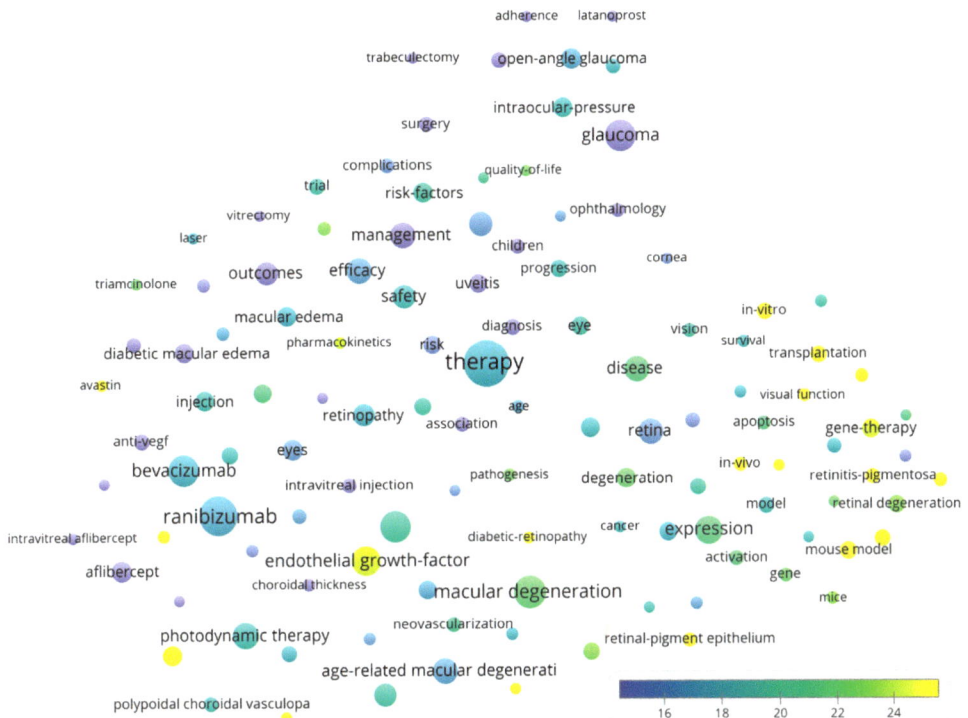

Figure 10. Map of items used highly frequently in the field.

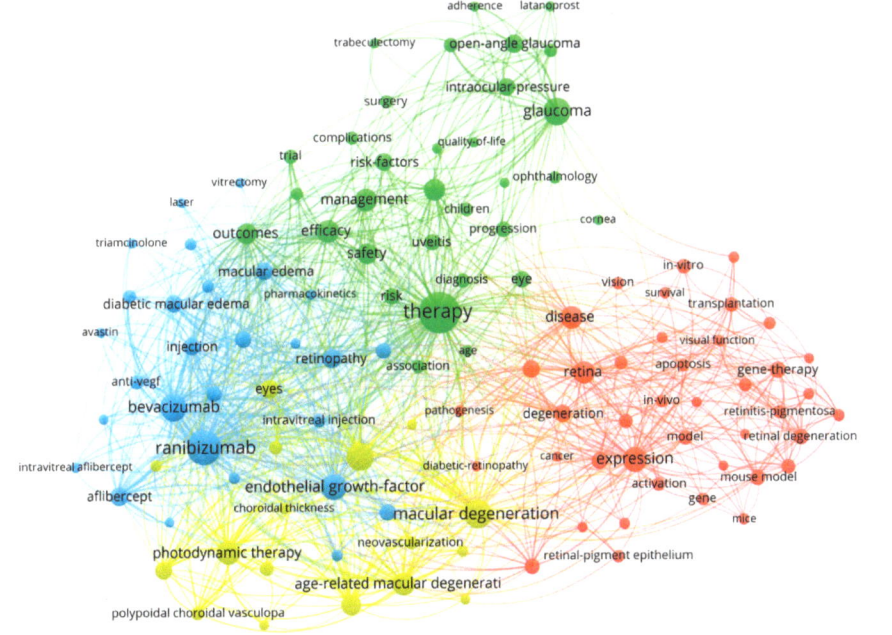

Figure 11. High-frequency term co-occurrence network map.

4.3. Discussions

The United States was the most prolific nation during the assessed time period, indicating that authors from this country are highly interested in the evaluated topic. In 2010, 50% of the countries in the top 10 were from Europe, while in the period 2011–2023, the number of European countries reduced to 40%. China is worth mentioning as it published only 218 documents until 2010 but experienced a rapid increase in interest in the subject by publishing 2503 articles in the period 2011–present.

A noteworthy observation emerges through an analysis of the co-authorship network map. Starting in 2011, a distinct separation between the clusters became evident, indicating that collaboration networks had solidified during the last decade.

Figure 12 shows the total number of papers published annually to illustrate the increasing trend with regard to the number of published papers and researchers' keenness on this topic. Because the total number of papers released before 1990 was less than 40 per year, the chart covers data from 1990 to 2023, thus providing better clarity. The number of papers published has steadily increased throughout the years, reaching a record in 2021 with 2234 articles published. However, the number of published papers fell slightly in 2022 compared to the previous year, with a total of 2208. This could be the result of a temporary decrease in research productivity or a shift in publication trends. More research would be required to uncover the underlying variables that contributed to this slight decline.

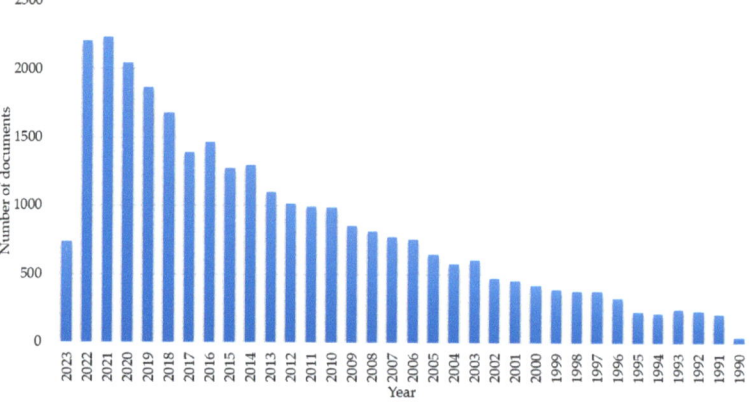

Figure 12. Number of published documents (1990–2023).

Although authors from countries grouped in the same cluster are more likely to collaborate, we cannot ignore the fact that collaborative relationships are also formed with authors from countries not located in the exact same group. Identifying the most prolific nations in this scientific area of interest and the countries that are most likely to collaborate should provide a solid foundation of knowledge and insights for authors interested in this topic and those looking for potential collaborators.

Overall, the most productive journal is the *American Journal of Ophthalmology*, followed by the journals *Ophthalmology* and *Investigative Ophthalmology Visual Science*. Although the journals primarily focused on ophthalmology have published numerous influential and highly regarded articles, the journals that focus on the broader medical field, like Nature and Lancet, for example, should not be dismissed. Despite having fewer publications in this topic area, these journals have a significant influence on the field of ophthalmology. The metric we provided should be useful for future authors who are interested in publishing articles in this field. Also, the bubble maps we provided for the last period should be useful in identifying the journals that actively published during this period.

The present bibliometric analysis provides a powerful set of tools and measures that provide insightful information about research output, impact, trends, and collaboration prospects, thus helping academics better understand this explored subject. One significant advantage of this investigation is its analytical approach. The subjectivity of the authors is not influenced by the use of quantitative tools, resulting in a more objective judgment. Furthermore, bibliometric analysis is cost-effective and reproducible, making it a practical and efficient tool. Another advantage of bibliometric analysis is the ability to study a substantial number of documents. This enables a thorough analysis of the topic area and a larger view of the research landscape. Moreover, the information systematized and classified according to certain parameters constitutes an important time-saving tool for authors during the pre-publication and research topic setting period, facilitating the selection of journals publishing on the desired topic, the most prolific articles as a starting point in identifying the current state of knowledge and unmet needs, as well as author collectives and nations publishing more often in the field for the creation or improvement of collaborative networks between authors interested in research in this field.

Although this field employs programs with a modern interface and efficient algorithms, it is crucial to recognize that there is still room for progress. The most remarkable advancement in this subject arises from the constant improvement of databases. As a result, better and more complete article indexing would result in significant advances in the field of bibliometric analysis. Another important feature contributing to its progress is the precise and accurate indexing of author names. The accurate indexing of authors would provide an accurate representation of their contribution to any given field and improve the identification of significant collaboration networks, leading to more valuable tools for future researchers.

5. Limitations

5.1. Molecular Docking Approach

The present in silico study has a number of limitations, such as its inability to reliably estimate the biological effects of a compound and its focus on a single static interaction between the ligand and the protein of interest, despite the fact that protein molecules are extremely flexible and adaptable in the biological milieu and may undergo significant conformational changes during compound binding.

Furthermore, the inability to assess the solvent effect may also have an impact on ligand binding. These limitations are noticeable at this initial phase of research investigation, yet they may be addressed through further studies in the future, such as the use of more advanced computational methods (e.g., molecular dynamics simulation, network pharmacology, etc.). Still, all findings obtained will require additional experimental validation prior to undergoing all the necessary steps for therapy approval.

Even though the most promising compounds selected and proposed for experimental endorsement based on binding affinity to ROCK1 or ROCK2 do not contain fluorine atoms, it is important to mention that among the first five results offered by SwissSimilarity's digital platform are compounds containing fluorine atoms. The incorporation of fluorine atoms into compounds identified through ligand-based virtual screening poses challenges due to their high electronegativity and potential for forming strong protein interactions. Therefore, rigorous validation and optimization of the parent ligand and subsequent experimental investigations are imperative. While fluorine can enhance compound stability, its introduction can significantly influence metabolic behavior and toxicity, potentially altering a compound's viability as a drug candidate. Some fluorinated substances may exhibit prolonged existence or distinct metabolic pathways, impacting safety profiles. Additionally, the presence of fluorine may intricately affect synthesis, demanding specialized reagents and conditions for fluorination reactions.

The present analysis proposes two candidates following the molecular docking study based on the highest potential binding to the target proteins impacting glaucoma, ROCK1 and ROCK2, representing a preliminary step in drug design studies, a step that needs to be confirmed by extensive computational studies (i.e., molecular dynamics, network pharmacology), in vivo in animal models, and clinical studies in different phases where efficacy and safety profiles are evaluated.

5.2. Bibliometric Analysis

The present bibliometric analysis, which emphasizes solely English articles evaluating treatments for ocular diseases, reveals a number of notable limitations. Primarily, the accuracy of the analysis may be limited by language bias, as the exclusion of papers written in languages other than English may exclude valuable perspectives and insights from various research communities.

In addition, by focusing exclusively on article-type publications, the analysis may have overlooked other valuable sources of data, such as case reports, books, and book chapters, which might offer supplementary approaches and knowledge on ocular disease therapeutics.

The focus on ocular disease therapies in the articles selected could additionally result in a low representation of broader studies investigating associated aspects, such as disease etiology, pathogenesis, and diagnostics. In the setting of ocular diseases, this may impede a comprehensive comprehension of the therapeutic landscape. Moreover, in the present bibliometric research, a large number of documents are included, and as a result, there may be some false positives among the initial results.

Lastly, bibliometric measures, such as citation counts, might not offer a comprehensive evaluation of the quality and influence of the articles that were selected. Variations in citation practices and the prominence of particular journals within the field may affect the perceived value of individual articles, thereby influencing the results of the bibliometric analysis.

6. Conclusions

The dual approach in this article proposes two compounds with possible application in the control of glaucoma conditions, but it is necessary that the results of this study be coupled with extensive computational assessments, in vitro experiments, in vivo validation investigations, and a bibliometric analysis focused on therapy, including laser-based therapy, for some of the most prevalent ocular pathologies, including glaucoma. These findings suggest that the parent compound ripasudil and the compounds identified via ligand-based virtual screening, ZINC000000022706 and ZINC000034800307, have similar binding patterns and affinity for ROCK1 and ROCK2. The compounds also showed a higher affinity for the proteins than fasudil, a known Rho kinase inhibitor. Due to the similarity in ligand–protein interactions and favorable pharmacokinetic properties, ZINC000000022706 and ZINC000034800307 should be further researched as potential ROCK1 and ROCK2 inhibitors. Additional research is necessary to evaluate their efficacy and safety profiles for possible future use as pharmacological agents, including experimental validations and a thorough study of their pharmacokinetic profile.

Examining publishing patterns found a constantly growing trend in published publications, indicating increased researcher interest and active involvement. Moreover, the United States has developed into the most prolific nation in this field, highlighting the region's substantial investment in authors. China has shown a tremendous increase in interest over time, emphasizing its expanding prominence in the sector. The field of ophthalmology is developing more structured and defined collaboration networks. These findings provide useful insights into the dynamics of collaborative research and emphasize the necessity of encouraging interdisciplinary collaboration.

Author Contributions: Conceptualization, F.B., S.G.B., A.P.N. and A.-F.R.; methodology, A.P.N. and A.-F.R.; software, A.P.N. and A.-F.R.; validation, A.R., A.G.T. and A.F.B.; formal analysis, A.P.N., C.B. and A.-F.R.; investigation, F.B., A.P.N. and A.-F.R.; resources, T.B. and A.-F.R.; data curation, A.P.N. and A.-F.R.; writing—original draft preparation, F.B., A.P.N., A.R., A.G.T., A.F.B. and A.-F.R.; writing—review and editing, S.G.B., C.B. and A.-F.R.; visualization, D.M.T. and T.B.; supervision, S.G.B., D.M.T. and A.-F.R.; project administration, A.-F.R.; funding acquisition, A.-F.R. All authors have read and agreed to the published version of the manuscript and have equal contribution to the first author.

Funding: The University of Oradea, Oradea, Romania, supported the APC through an internal project.

Institutional Review Board Statement: Not applicable.

Informed Consent Statement: Not applicable.

Data Availability Statement: All the information in the manuscript is supported by the mentioned references.

Acknowledgments: The authors would like to express their gratitude to the University of Oradea, Oradea, Romania, who supported the APC.

Conflicts of Interest: The authors declare that they have no known competing financial interests or personal relationships that could have appeared to influence the work reported in this paper.

References

1. Landau, K.; Kurz-Levin, M. Retinal Disorders. In *Neuro-Ophthalmology*; Kennard, C., Leigh, R.J.r., Eds.; Elsevier: Amsterdam, The Netherlands, 2011; Volume 102, pp. 97–116, ISBN 0072-9752.
2. Bungau, S.; Abdel-Daim, M.M.; Tit, D.M.; Ghanem, E.; Sato, S.; Maruyama-Inoue, M.; Yamane, S.; Kadonosono, K. Health Benefits of Polyphenols and Carotenoids in Age-Related Eye Diseases. *Oxid. Med. Cell Longev.* **2019**, *2019*, 9783429. [CrossRef] [PubMed]
3. Bodea, F.; Bungau, S.G.; Bogdan, M.A.; Vesa, C.M.; Radu, A.; Tarce, A.G.; Purza, A.L.; Tit, D.M.; Bustea, C.; Radu, A.-F. Micropulse Laser Therapy as an Integral Part of Eye Disease Management. *Medicina* **2023**, *59*, 1388. [CrossRef]
4. Wong, W.L.; Su, X.; Li, X.; Cheung, C.M.G.; Klein, R.; Cheng, C.-Y.; Wong, T.Y. Global Prevalence of Age-Related Macular Degeneration and Disease Burden Projection for 2020 and 2040: A Systematic Review and Meta-Analysis. *Lancet. Glob. Health* **2014**, *2*, e106-16. [CrossRef] [PubMed]
5. Campochiaro, P.A. Retinal and Choroidal Vascular Diseases: Past, Present, and Future: The 2021 Proctor Lecture. *Investig. Ophthalmol. Vis. Sci.* **2021**, *62*, 26. [CrossRef]
6. Liew, G.; Quin, G.; Gillies, M.; Fraser-Bell, S. Central Serous Chorioretinopathy: A Review of Epidemiology and Pathophysiology. *Clin. Exp. Ophthalmol.* **2013**, *41*, 201–214. [CrossRef]
7. Allison, K.; Patel, D.; Alabi, O. Epidemiology of Glaucoma: The Past, Present, and Predictions for the Future. *Cureus* **2020**, *12*, e11686. [CrossRef]
8. Hanumunthadu, D.; Tan, A.C.S.; Singh, S.R.; Sahu, N.K.; Chhablani, J. Management of Chronic Central Serous Chorioretinopathy. *Indian J. Ophthalmol.* **2018**, *66*, 1704–1714.
9. Brand, C.S. Management of Retinal Vascular Diseases: A Patient-Centric Approach. *Eye* **2012**, *26*, S1–S16. [CrossRef]
10. Weinreb, R.N.; Aung, T.; Medeiros, F.A. The Pathophysiology and Treatment of Glaucoma: A Review. *JAMA* **2014**, *311*, 1901–1911. [CrossRef]
11. Schuster, A.K.; Erb, C.; Hoffmann, E.M.; Dietlein, T.; Pfeiffer, N. The Diagnosis and Treatment of Glaucoma. *Dtsch. Arztebl. Int.* **2020**, *117*, 225–234. [CrossRef]
12. Hutnik, C.; Crichton, A.; Ford, B.; Nicolela, M.; Shuba, L.; Birt, C.; Sogbesan, E.; Damji, K.F.; Dorey, M.; Saheb, H.; et al. Selective Laser Trabeculoplasty versus Argon Laser Trabeculoplasty in Glaucoma Patients Treated Previously with 360° Selective Laser Trabeculoplasty: A Randomized, Single-Blind, Equivalence Clinical Trial. *Ophthalmology* **2019**, *126*, 223–232. [CrossRef] [PubMed]
13. Ma, A.; Yu, S.W.Y.; Wong, J.K.W. Micropulse Laser for the Treatment of Glaucoma: A Literature Review. *Surv. Ophthalmol.* **2019**, *64*, 486–497. [CrossRef] [PubMed]
14. Abbhi, V.; Piplani, P. Rho-Kinase (ROCK) Inhibitors—A Neuroprotective Therapeutic Paradigm with a Focus on Ocular Utility. *Curr. Med. Chem.* **2020**, *27*, 2222–2256. [CrossRef] [PubMed]
15. Wang, J.; Liu, X.; Zhong, Y. Rho/Rho-Associated Kinase Pathway in Glaucoma. *Int. J. Oncol.* **2013**, *43*, 1357–1367. [CrossRef]
16. Testa, V.; Ferro Desideri, L.; Della Giustina, P.; Traverso, C.E.; Iester, M. An Update on Ripasudil for the Treatment of Glaucoma and Ocular Hypertension. *Drugs Today* **2020**, *56*, 599–608. [CrossRef]
17. Khallaf, A.M.; El-Moslemany, R.M.; Ahmed, M.F.; Morsi, M.H.; Khalafallah, N.M. Exploring a Novel Fasudil-Phospholipid Complex Formulated as Liposomal Thermosensitive in Situ Gel for Glaucoma. *Int. J. Nanomed.* **2022**, *17*, 163–181. [CrossRef]
18. Cursiefen, C.; Cordeiro, F.; Cunha-Vaz, J.; Wheeler-Schilling, T.; Scholl, H.P.N. Unmet Needs in Ophthalmology: A European Vision Institute-Consensus Roadmap 2019–2025. *Ophthalmic Res.* **2019**, *62*, 123–133. [CrossRef]
19. Sanner, M.F. Python: A Programming Language for Software Integration and Development. *J. Mol. Graph. Model* **1999**, *17*, 57–61.
20. O'Boyle, N.M.; Banck, M.; James, C.A.; Morley, C.; Vandermeersch, T.; Hutchison, G.R. Open Babel: An Open Chemical Toolbox. *J. Cheminform.* **2011**, *3*, 33. [CrossRef]
21. Crystal Structure of ROCK1 Bound to Fasudil. Available online: https://www.wwpdb.org/pdb?id=pdb_00002esm (accessed on 23 June 2023).
22. Crystal Structure of Rho-Associated Protein Kinase 2 (ROCK2) in Complex with a Potent and Selective Dual ROCK Inhibitor. Available online: https://www.wwpdb.org/pdb?id=pdb_00007jnt (accessed on 23 June 2023).
23. Morris, G.M.; Ruth, H.; Lindstrom, W.; Sanner, M.F.; Belew, R.K.; Goodsell, D.S.; Olson, A.J. AutoDock4 and AutoDockTools4: Automated Docking with Selective Receptor Flexibility. *J. Comput. Chem.* **2009**, *30*, 2785–2791. [CrossRef]

24. Bragina, M.E.; Daina, A.; Perez, M.A.S.; Michielin, O.; Zoete, V. The SwissSimilarity 2021 Web Tool: Novel Chemical Libraries and Additional Methods for an Enhanced Ligand-Based Virtual Screening Experience. *Int. J. Mol. Sci.* **2022**, *23*, 811. [CrossRef] [PubMed]
25. Rogers, D.; Hahn, M. Extended-Connectivity Fingerprints. *J. Chem. Inf. Model.* **2010**, *50*, 742–754. [CrossRef] [PubMed]
26. Appunni, S.; Gupta, D.; Rubens, M.; Singh, A.K.; Swarup, V.; Himanshu, N. Targeting ROCK2 Isoform with Its Widely Used Inhibitors for Faster Post-Stroke Recovery. *Indian J. Biochem. Biophys.* **2021**, *58*, 27–34.
27. Lipinski, C.A.; Lombardo, F.; Dominy, B.W.; Feeney, P.J. Experimental and Computational Approaches to Estimate Solubility and Permeability in Drug Discovery and Development Settings. *Adv. Drug Deliv. Rev.* **1997**, *23*, 3–25. [CrossRef]
28. Daina, A.; Michielin, O.; Zoete, V. SwissADME: A Free Web Tool to Evaluate Pharmacokinetics, Drug-Likeness and Medicinal Chemistry Friendliness of Small Molecules. *Sci. Rep.* **2017**, *7*, 42717. [CrossRef]
29. van Eck, N.J.; Waltman, L. Citation-Based Clustering of Publications Using CitNetExplorer and VOSviewer. *Scientometrics* **2017**, *111*, 1053–1070. [CrossRef]
30. van Eck, N.J.; Waltman, L. Software Survey: VOSviewer, a Computer Program for Bibliometric Mapping. *Scientometrics* **2010**, *84*, 523–538. [CrossRef]
31. Sawcer, S.; Hellenthal, G.; Pirinen, M.; Spencer, C.C.A.; Patsopoulos, N.A.; Moutsianas, L.; Dilthey, A.; Su, Z.; Freeman, C.; Hunt, S.E.; et al. Genetic Risk and a Primary Role for Cell-Mediated Immune Mechanisms in Multiple Sclerosis. *Nature* **2011**, *476*, 214–219.
32. Heier, J.S.; Brown, D.M.; Chong, V.; Korobelnik, J.F.; Kaiser, P.K.; Nguyen, Q.D.; Kirchhof, B.; Ho, A.; Ogura, Y.; Yancopoulos, G.D.; et al. Intravitreal Aflibercept (VEGF Trap-Eye) in Wet Age-Related Macular Degeneration. *Ophthalmology* **2012**, *119*, 2537–2548. [CrossRef]
33. Martin, D.F.; Maguire, M.G.; Fine, S.L.; Ying, G.S.; Jaffe, G.J.; Grunwald, J.E.; Toth, C.; Redford, M.; Ferris, F.L. Ranibizumab and Bevacizumab for Treatment of Neovascular Age-Related Macular Degeneration: Two-Year Results. *Ophthalmology* **2012**, *119*, 1388–1398. [CrossRef]
34. Okita, K.; Matsumura, Y.; Sato, Y.; Okada, A.; Morizane, A.; Okamoto, S.; Hong, H.; Nakagawa, M.; Tanabe, K.; Tezuka, K.; et al. A More Efficient Method to Generate Integration-Free Human IPS Cells. *Nat. Methods* **2011**, *8*, 409–412. [CrossRef] [PubMed]
35. Lim, L.S.; Mitchell, P.; Seddon, J.M.; Holz, F.G.; Wong, T.Y. Age-Related Macular Degeneration. *Lancet* **2012**, *379*, 1728–1738. [CrossRef] [PubMed]
36. Mintz-Hittner, H.A.; Kennedy, K.A.; Chuang, A.Z. Efficacy of Intravitreal Bevacizumab for Stage 3+ Retinopathy of Prematurity. *N. Engl. J. Med.* **2011**, *364*, 603–615. [CrossRef] [PubMed]
37. Quigley, H.A. Glaucoma. *Lancet* **2011**, *377*, 1367–1377. [CrossRef] [PubMed]
38. Mandai, M.; Watanabe, A.; Kurimoto, Y.; Hirami, Y.; Morinaga, C.; Daimon, T.; Fujihara, M.; Akimaru, H.; Sakai, N.; Shibata, Y.; et al. Autologous Induced Stem-Cell-Derived Retinal Cells for Macular Degeneration. *N. Engl. J. Med.* **2017**, *376*, 1038–1046. [CrossRef]
39. Tang, J.; Kern, T.S. Inflammation in Diabetic Retinopathy. *Prog. Retin. Eye Res.* **2011**, *30*, 343–358. [CrossRef]
40. Rofagha, S.; Bhisitkul, R.B.; Boyer, D.S.; Sadda, S.R.; Zhang, K. Seven-Year Outcomes in Ranibizumab-Treated Patients in ANCHOR, MARINA, and HORIZON: A Multicenter Cohort Study (SEVEN-UP). *Ophthalmology* **2013**, *120*, 2292–2299. [CrossRef]

Disclaimer/Publisher's Note: The statements, opinions and data contained in all publications are solely those of the individual author(s) and contributor(s) and not of MDPI and/or the editor(s). MDPI and/or the editor(s) disclaim responsibility for any injury to people or property resulting from any ideas, methods, instructions or products referred to in the content.

Article

Development and Verification of a Novel Three-Dimensional Aqueous Outflow Model for High-Throughput Drug Screening

Matthew Fung [1,*], James J. Armstrong [1,2], Richard Zhang [1], Anastasiya Vinokurtseva [1,2], Hong Liu [2] and Cindy Hutnik [2,3]

1. Schulich School of Medicine & Dentistry, Western University, London, ON N6A 3K7, Canada
2. Department of Ophthalmology, Schulich School of Medicine & Dentistry, Western University, London, ON N6A 3K7, Canada
3. Department of Ophthalmology, Ivey Eye Institute, St. Joseph's Health Center, London, ON N6A 4V2, Canada
* Correspondence: matthew.fung1@ucalgary.ca

Citation: Fung, M.; Armstrong, J.J.; Zhang, R.; Vinokurtseva, A.; Liu, H.; Hutnik, C. Development and Verification of a Novel Three-Dimensional Aqueous Outflow Model for High-Throughput Drug Screening. *Bioengineering* **2024**, *11*, 142. https://doi.org/10.3390/bioengineering11020142

Academic Editor: Hiroshi Ohguro

Received: 30 December 2023
Revised: 20 January 2024
Accepted: 25 January 2024
Published: 31 January 2024

Copyright: © 2024 by the authors. Licensee MDPI, Basel, Switzerland. This article is an open access article distributed under the terms and conditions of the Creative Commons Attribution (CC BY) license (https://creativecommons.org/licenses/by/4.0/).

Abstract: Distal outflow bleb-forming procedures in ophthalmic surgery expose subconjunctival tissue to inflammatory cytokines present in the aqueous humor, resulting in impaired outflow and, consequently, increased intraocular pressure. Clinically, this manifests as an increased risk of surgical failure often necessitating revision. This study (1) introduces a novel high-throughput screening platform for testing potential anti-fibrotic compounds and (2) assesses the clinical viability of modulating the transforming growth factor beta-SMAD2/3 pathway as a key contributor to post-operative outflow reduction, using the signal transduction inhibitor verteporfin. Human Tenon's capsule fibroblasts (HTCFs) were cultured within a 3D collagen matrix in a microfluidic system modelling aqueous humor drainage. The perfusate was augmented with transforming growth factor beta 1 (TGFβ1), and afferent pressure to the tissue-mimetic was continuously monitored to detect treatment-related pressure elevations. Co-treatment with verteporfin was employed to evaluate its capacity to counteract TGFβ1 induced pressure changes. Immunofluorescent studies were conducted on the tissue-mimetic to corroborate the pressure data with cellular changes. Introduction of TGFβ1 induced treatment-related afferent pressure increase in the tissue-mimetic. HTCFs treated with TGFβ1 displayed visibly enlarged cytoskeletons and stress fiber formation, consistent with myofibroblast transformation. Importantly, verteporfin effectively mitigated these changes, reducing both afferent pressure increases and cytoskeletal alterations. In summary, this study models the pathological filtration bleb response to TGFβ1, while demonstrating verteporfin's effectiveness in ameliorating both functional and cellular changes caused by TGFβ1. These demonstrate modulation of the aforementioned pathway as a potential avenue for addressing post-operative changes and reductions in filtration bleb outflow capacity. Furthermore, the establishment of a high-throughput screening platform offers a valuable pre-animal testing tool for investigating potential compounds to facilitate surgical wound healing.

Keywords: glaucoma; intraocular pressure; microfluidics; inflammation; fibrosis; tissue-mimetic; drug screening; organ modelling; three-dimensional aqueous outflow

1. Introduction

The efficacy of glaucoma surgery hinges on sustained improvements in aqueous humor (AH) outflow to control intraocular pressure (IOP), and, consequently, further glaucoma progression. AH outflow is a critical factor influenced by iatrogenic tissue damage during surgery. Following the surgical insult, damaged tissue initiates a wound healing response, exposing the subconjunctiva to acute surgical inflammation and endogenous pro-inflammatory and pro-fibrotic cytokines in the AH [1]. This pro-inflammatory cascade contributes to excessive wound healing and post-operative subconjunctival fibrosis, ultimately impairing outflow capacity.

A key player in this fibrotic cascade is the transforming growth factor-beta (TGFβ) family of AH cytokines. Present at elevated levels in AH of glaucoma patients [2,3], TGFβ induces alterations in the extracellular matrix (ECM) of the trabecular meshwork, resulting in heightened resistance to aqueous outflow and subsequently elevated intraocular pressure (IOP) [4–6]. Beyond the deleterious microenvironmental effects of TGFβ, similar pathophysiological changes can be observed post-operatively in the filtration bleb [7,8]. Thus, specific targeting of cellular processes downstream of TGFβ may be an avenue to mitigate fibrotic processes contributing to surgical failure while preserving processes essential for normal wound healing and homeostasis.

Myofibroblasts, characterized by the increased expression of α-smooth muscle actin (α-SMA), which is stimulated via the SMAD 2/3 pathway downstream of TGFβ, play a crucial role in wound healing [9,10]. This fibroblast-to-myofibroblast transdifferentiation leads to the formation of actin–myosin stress fiber bundles akin to those found in muscle cells, earning them the moniker myofibroblasts. These cells exhibit increased ECM secretion and contractile activity, critical in normal wound-healing processes [11]. After healthy wound healing, myofibroblasts are deactivated or removed via apoptosis. However, in bleb-forming procedures, excessive myofibroblast activity can be harmful, contributing to scar tissue formation and impairing outflow tract patency, both of which contribute to surgical failure [12–14]. Thus, drugs targeting TGFβ-mediated myofibroblast activity may help maintain outflow patency throughout the post-operative period.

Despite substantial advancements in surgical glaucoma intervention, progress in developing compounds to address post-operative fibrosis has been slow [15,16]. Current strategies rely on local anti-metabolite treatments, such as mitomycin C (MMC) or 5-fluorouracil (5FU). However, their short half-lives may yield insufficient anti-fibrotic efficacy and the non-specific targeting of all cell types often leads to undesirable off-target effects, including tissue thinning, leaks, and endophthalmitis [17,18]. Additionally, ocular steroid therapy, used to control post-surgical inflammation, poses a major risk factor for cataracts and paradoxically increases the risk of steroid-induced glaucoma [19]. Overall, these interventions carry significant risks, necessitating the exploration of novel avenues to address inflammatory changes after surgery.

The current drug development process begins with 2D cell culture and progresses to animal models. However, cells in 2D cell culture are separated from physiological factors affecting cellular behavior. In the eye, fibroblasts are typically associated with an ECM that modulates cell growth and activity [20]. Additionally, fibroblasts are sensitive to the "stiffness" of the surface on which they are grown, where solid plastic culture plates have been known to inherently induce myofibroblast transdifferentiation, potentially confounding experimental results [21–23]. Conversely, animal models do account for the physiological context of fibroblasts within a perfused organ system. However, these experiments tend to have significant ethical and financial cost associated with them, are significantly more time-consuming than in vitro work, and, as such, have challenges that prevent them from being applicable in the early drug screening process. In addition, interspecies variability cannot fully mimic the conditions observed in humans [24,25].

Thus, the purpose of this study is to elucidate the effects of TGFβ on AH outflow capacity in distal outflow bleb-forming procedures. We aimed to adapt previous methods of in vitro blood vessel modeling [26] to develop a high-throughput drug testing platform that better mirrors the subconjunctival environment to screen potential post-surgical anti-fibrotic adjuvants. The current study introduces a novel perfused 3D collagen-based subconjunctival tissue mimetic to assess the effects of TGFβ and fibrosis-modulating treatments on AH outflow. In doing so, the model aims to better replicate the physiological milieu afforded by animal models while providing a cost-effective and high-throughput in vitro testing platform. This study hypothesizes that exogenous TGFβ1 will increase outflow resistance within this novel in vitro model and using a small molecule to inhibit a secondary messenger downstream of the TGFβ1 pathway will mitigate this effect.

2. Methods

2.1. Primary Human Tenon's Capsule Fibroblast (HTCF) Procurement and Culture

Primary HTCFs were obtained from 2–4 mm^3 resections of Tenon's capsule during ocular surgeries at the Ivey Eye Institute, London, ON, Canada, before anti-metabolite treatment [27]. This study adhered to the tenets of the Declaration of Helsinki. Ethical approval (HSREB: 106783) was obtained for patient demographic data collection associated with tissue explants. Informed consent was secured before tissue procurement, with no alteration to patient surgeries. Samples were cultured in Dulbecco Modified Eagle's Minimum Essential Medium (DMEM, Gibco, Waltham, MA, USA, Cat# 12430054), supplemented with 10% Fetal Bovine Serum (FBS, Gibco, Waltham, MA, USA, Cat# A38403), 2mM L-glutamine (Gibco, Waltham, MA, USA, Cat# A2916801), 1% amphotericin (Gibco, Waltham, MA, USA, Cat# 15290018), and 1% penicillin/streptomycin (P/S, Gibco, Waltham, MA, USA, Cat# 15240096) from Sigma-Aldrich (Oakville, ON, Canada). Fibronectin (Sigma-Aldrich) was used to coat 6-well culture plates (9.5 cm^2) for seven days. Migrating fibroblasts were trypsinized and stored in liquid nitrogen. HTCFs were incubated in DMEM with 10% FBS at 37 °C in a 5% CO_2 atmosphere until experimental use or up to the 5th cell passage (Figure 1).

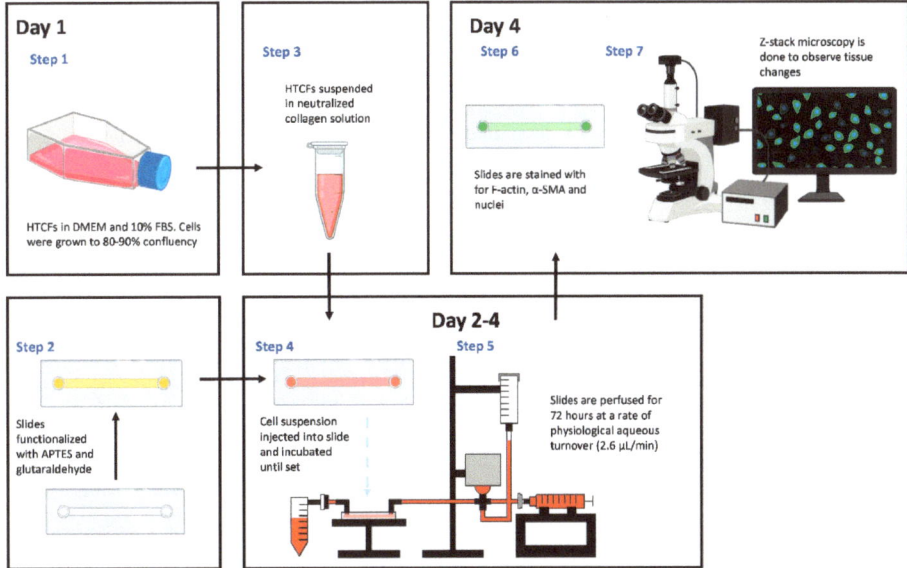

Figure 1. Microfluidic workflow: Pre-clinical applicability is assessed through pressure changes throughout the experiment within the microfluidic model. Fluorescent microscopy was used to link morphological changes in hydrogel to functional changes in the model.

2.2. Microfluidic Slide Chamber Preparation

To covalently link collagen to the slide chamber, a μ-slide I microfluidic slide (Ibidi #80176 Fitchburg, MA, USA) was salinized with vaporized 3-Aminopropyl-triethoxysilane (APTES) (Sigma-Aldrich, Oakville, ON, Canada). APTES was heated to 150 °C, and the fumes were pushed through the slide with a syringe pump (Harvard apparatus) for one hour at a rate of 60 mL/h. Unfunctionalized APTES was purged with 10 mL PBS (pH 7.4). After salinization, glutaraldehyde was bonded to the APTES by incubating a 6% glutaraldehyde (Sigma-Aldrich, Oakville, ON, Canada) solution for 30 min at ambient temperature, followed by washing with 10mL PBS (Figure 2). Finally, the chamber was sterilized by perfusing 70% ethanol followed by PBS before manipulation in a sterile hood [28,29].

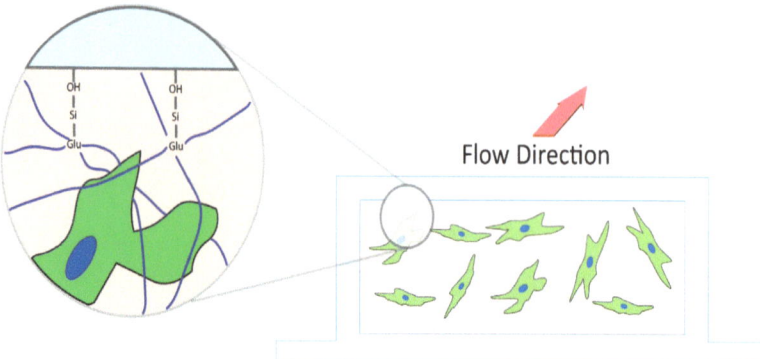

Figure 2. Cross section schematic of microfluidics chamber. APTES salinizes the hydroxide groups on the slide, which then allows glutaraldehyde to bind with the silicon group. Glutaraldehyde crosslinks with the loose unincorporated collagen (navy) anchoring it into the microfluidics chamber to prevent extrusion of matrix and Tenon's capsule fibroblast upon perfusion.

2.3. HTCF Preparation for Hydrogel Formation

HTCFs were grown until 90–100% confluent in a 25 mL Falcon flask, dissociated using trypsin (Fisher Scientific, Waltham, UK), and quantified. The pellet was resuspended in DMEM supplemented with 2% FBS to a concentration of 6.25×10^4 cells/µL. To create the hydrogel, 80 µL of rat tail collagen solution (1.8 mg/mL), 8 µL Waymouth media (Gibco, Cat# 15290018), and 8 µL sterile NaOH solution (0.275M) were mixed, and 4 µL of the cell suspension was added [27]. A total of 50 µL of the collagen–HTCF liquid hydrogel was cast into the functionalized microfluidics slide and incubated at 37 °C in 5% CO_2 for 30 min until solidified. The final HTCF concentration within the hydrogel was 2.5×10^6 cells/mL.

2.4. Perfusion Track Construction and Experimental Setup

The model utilized Sanipure BDP 1/8′ ID × 1/4′ OD tubing (Cole-pharma, Quebec, QC, Canada) and polypropylene connectors (Harvard Apparatus, Saint-Laurent, QC, Canada) to construct the track (Figure 3). The track connected the syringe pump to the blood pressure transducer and the tissue slide. Connectors and tubing were autoclaved before assembly. The perfusate in the syringe pump comprised DMEM with 2% FBS, 1% P/S, and 0.25 mmol/L ascorbic acid 2-phosphate (AA2P). Experimental compounds were added to the perfusate according to the treatment groups. The track was primed with media, and slides were connected bead-to-bead on the influent and effluent side to prevent air bubbles.

Two runs were set up in parallel to account for pump error, allowing for paired comparison. For each run, equal lengths of tubing were measured to account for interference from tubing friction. To test the tissue mimetic's ability to modulate outflow resistance, exogenous TGFβ1 at a concentration of 2 ng/mL was added to the perfusate to simulate post-surgical aqueous humor conditions. This was compared to the tissue-mimetic perfused with vehicle perfusate. To inhibit TGFβ1 mediated SMAD2/3 activity, TGFβ1 perfusate was co-supplemented with 10 µM verteporfin (Sigma Aldrich, Oakville, ON, Canada), and this was compared to the TGFβ1 only group.

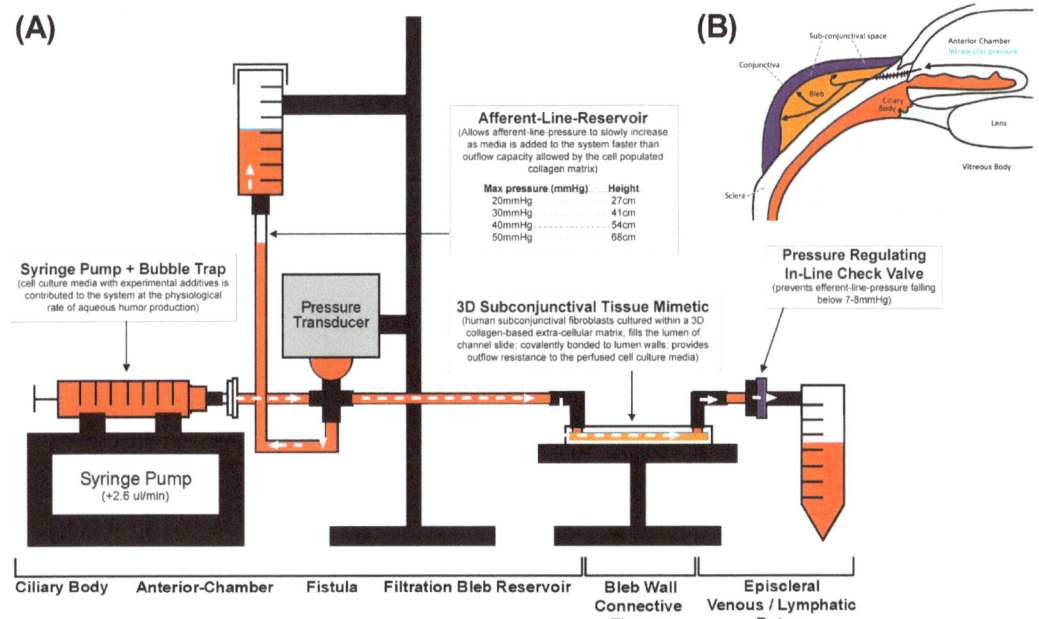

Figure 3. Microfluidics model (A) mimics anatomical qualities of surgical filtration bleb (B). Perfusate from the syringe pump mimics aqueous humor production in the ciliary bodies (red). The surgical filtration bleb (orange) is modelled by the collagen hydrogel in the slide chamber. Finally, a pressure gated one-way valve is down located downstream of the slide chamber to mimic episcleral back pressure (purple). As the cells within the hydrogel proliferate, hydrodynamic changes lead to impaired flow. Consequently, perfusate accumulates in the influent pressure column, which can be read manually or measured through the pressure transducer upstream representing IOP.

2.5. Afferent Line Pressure Analysis

Afferent line pressure, serving as an analogue for intraocular pressure (IOP), was measured in real-time with a blood pressure transducer (Harvard Apparatus 72-4498). The pressure was measured at a rate of 10 readings per second, recorded on Labchart 8 (AD instruments). Raw data were down-sampled by 600 times to provide average pressure readings every minute. Baseline afferent pressure was established by averaging the readings in the first hour to account for system pressure equilibration (P_{equil}). Every subsequent pressure value was normalized to P_{equil} to give pressure changes relative to baseline (ΔP). ΔP for vehicle control (ΔP_v) and TGFβ1 (ΔP_t) were compared to test whether TGFβ1mediated changes to outflow resistance occurred. Differences in ΔP for TGFβ1+ verteporfin treatment (ΔP_{vert}) and ΔP_t were used to examine verteporfin's ability to attenuate TGFβ1mediated pressure changes. Afferent line pressure changes served as the primary outcome to test whether a compound may have potential clinical applicability.

2.6. Tissue-Mimetic Fixation and Histological Staining

At the end of perfusion experiments, the tissue-mimetic was washed with PBS at 2.6 µL/min. The matrix was fixed by perfusing a 4% paraformaldehyde (Sigma, Oakville, ON, Canada) solution [26] and blocked with a 2% bovine serum albumin (Sigma, Oakville, ON, Canada) solution by perfusing tissue-mimetics for 1 h at 2.6 µL/min. Subsequently, 50 µL of anti-actin, α-smooth muscle—Cy3 mouse monoclonal antibody (5 µg/mL, C6198, Sigma-Aldrich) was pushed into the slide chamber and allowed to incubate at 4 °C overnight. Finally, the slide was stained with 50 µL of F-actin cytopainter (Abcam, Waltham,

UK) and 50 µL Hoechst (Sigma-Aldrich, Oakville, ON, Canada) for 30 min and 10 min, respectively, at ambient temperature.

2.7. Confocal Immuno-Fluorescent Microscopy

Z-stack microscopy was performed from the upper to the lower tissue margin of the slide to assess three-dimensional structure. Z-stack steps were separated at intervals of 2.54 µm for a total height of 200 µm. Analysis of Z-stack was performed on ImageJ. The number of cells was quantified by projecting the Hoechst (nuclear) channel into various 2D projections and using the particle measurement tool to identify shapes measuring at least 10×10 pixels to quantify the number of nuclei visible. For FITC (F-actin) channels, each slide within the stack was thresholded, and the area of staining was measured. The average cellular content of F-actin was calculated by summing the total area of staining for each slide in the FITC channels and dividing by the number of nuclei detected from the DAPI channel. This allowed the comparison of average F-actin expression on a per-nucleus basis to estimate the degree of myofibroblast transdifferentiation.

2.8. Western Blot

HTCFs were cultivated to confluency in 6-well plates, subjected to a 24-h serum starvation period, and treated with varied concentrations of verteporfin (Sigma Aldrich) up to 20 µM. Cells underwent a 48-h treatment period and were subsequently lysed in 250 µL of lysis buffer (PhosphoSafe Extraction Reagent, Novagen, Oakville, ON, Canada) supplemented with a protease inhibitor cocktail (P2714, Sigma-Aldrich, Oakville, ON, Canada).

The raw lysate underwent homogenization through a 27 G needle, followed by centrifugation at $13,000 \times g$ for 10 min to extract the supernatant. Protein content normalization was achieved using a Pierce BCA protein assay kit (ThermoFisher, Rockford, IL, USA), to enable loading of 10 µg of protein per well in a 10% polyacrylamide gel. Electrophoresis ensued for 1 h, and an iBlot Gel Transfer Device (Invitrogen, Rockford, IL, USA) facilitated protein transfer onto a nitrocellulose membrane.

The membrane was subsequently blocked with 5% (w/v) bovine serum albumin (Sigma-Aldrich, Oakville, ON, Canada) solubilized in Tris-buffered saline (TBST) at ambient temperature for 1 h. Afterwards, the membranes were incubated overnight at 4 °C involving primary antibodies (0.5 µg/mL) against αSMA (ab5694, Abcam, Waltham, UK) and GAPDH (Santa Cruz Biotechnology, Inc., Dallas, TX, USA), diluted in TBST containing 5% BSA (w/vol). Washed membranes were then incubated with 1:3000 (v/v) dilutions of goat anti-rabbit.t IgG conjugated with horseradish peroxidase (Santa Cruz Biotechnology, Inc., Dallas, UK).

Protein loading control utilized WesternBright Quantum chemiluminescent reagent (Advansta, Inc., San Jose, UK) molecular markers in conjunction with GAPDH. Densitometric analysis and imaging were conducted using the ChemiDoc MP System (Bio-Rad Laboratories, Inc., Hercules, UK) connected to Image Lab (Version 6, Bio-Rad).

3. Results

3.1. Analysis of Three-Dimensional Tissue Mimetic Structure

To assess cellular morphology and integration into the collagen matrix, slides were imaged with light microscopy daily at pre-set regions of interest to observe cellular morphology within the hydrogel. HTCFs within the chamber were observed for at least 5 days within the chamber with maximum cell density achieved approximately at days 3 to 4 (Figure 4A). Z-stack images confirmed hydrogel integrity after the 5-day run and the presence of three-dimensional proliferation of HTCFs throughout the collagen matrix (Figure 4B).

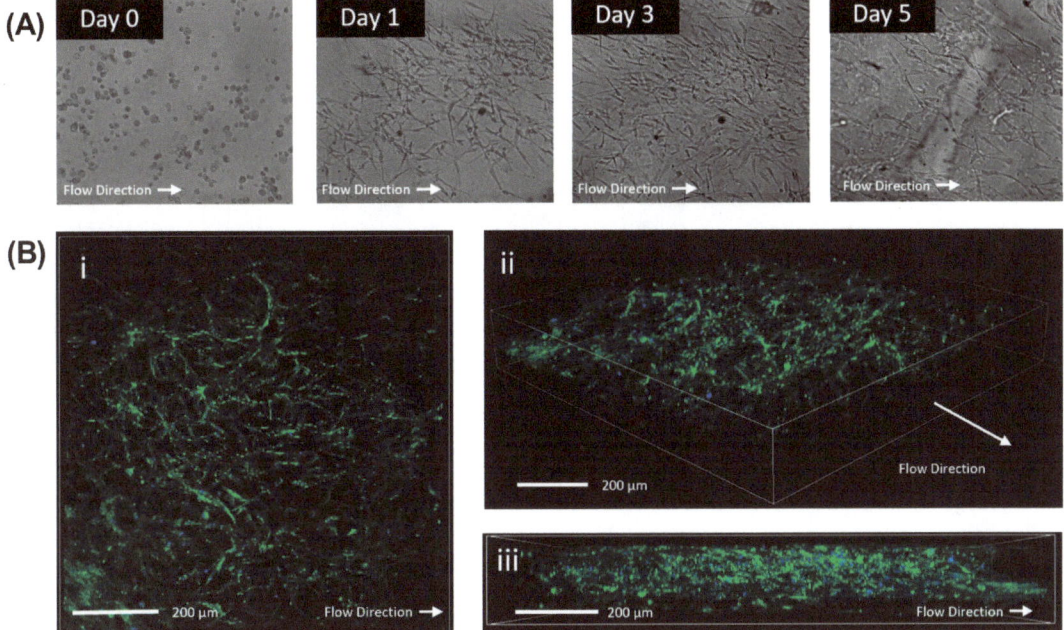

Figure 4. Verification of three dimensional proliferation of cells within the hydrogel. Light microscopy of HTCFs within the collagen matrix perfused with perfusion media. Cells demonstrated growth and proliferation within the slides (**A**). Confocal microscopy of slide after being perfused in the model for 72 h. Cells were stained for F-actin stress fibres (FITC) and Hoest (Nuclear). HTCFs demonstrate proliferation within the length (**B**) (i) width (ii), and height (iii) of the flow chamber demonstrating growth three dimensionally.

3.2. Morphological Changes Due to TGFβ1 and Verteporfin Co-Treatment

HTCFs treated with TGFβ1 demonstrated increased proliferation and elevated presence of actin stress fibers compared to vehicle control (Figure 5). Interestingly, TGFβ1 treated HTCFs arrange themselves more uniformly with neighboring cells, suggesting organized ECM contraction. HTCFs cotreated with verteporfin demonstrated decreased area of fluorescence for F-actin and α-SMA resembling vehicle control.

3.3. Afferent Line Pressure Changes

Initially, afferent line pressure was measured with an acellular collagen matrix, which demonstrated no pressure changes after initial system equilibration, and afferent line pressure remained constant for up to 72 h. Treatment of HTCF-containing collagen matrices with TGFβ1 demonstrated continual increases in afferent line pressure throughout the experimental duration (Figure 6). The difference in afferent line pressure between TGFβ1 only and VC was the greatest towards the end of the experiment, where the mean difference in afferent line pressure between ΔP_v and ΔP_t was 18.74 mmHg (SE = 5.726 mmHg). Significant pressure differences were achieved between VC and TGFβ1 after 26 h of perfusion.

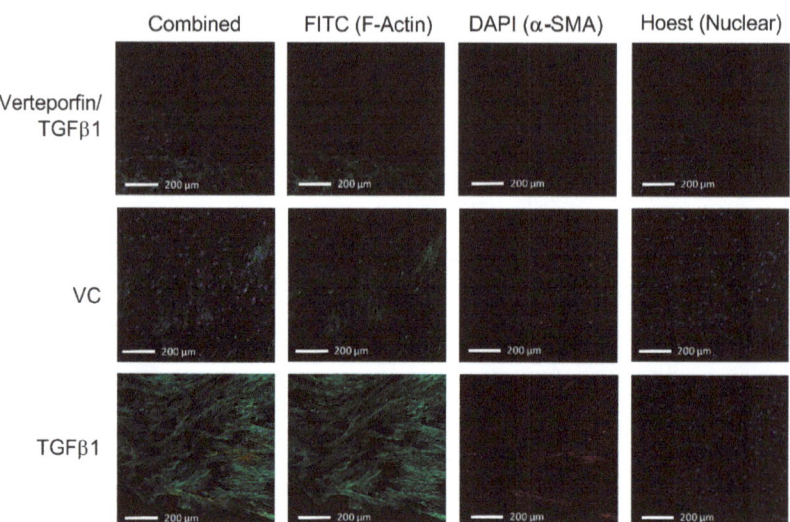

Figure 5. TGFβ1 leads to increased expression of F-actin stress fibres and α-SMA. Represented is an overlay of all 2.54 μm images within the Z-stack overlayed into a two-dimensional image to depict differences in fluorescent intensity between groups following the 72-h treatment period. Addition of exogenous TGFβ1 in the perfusate leads to the increased fluorescent intensity from both FITC (F-actin) and DAPI (α-SMA) channels. Verteporfin and TGFβ1 co-treatment attenuates the intensity of both FITC and DAPI channels.

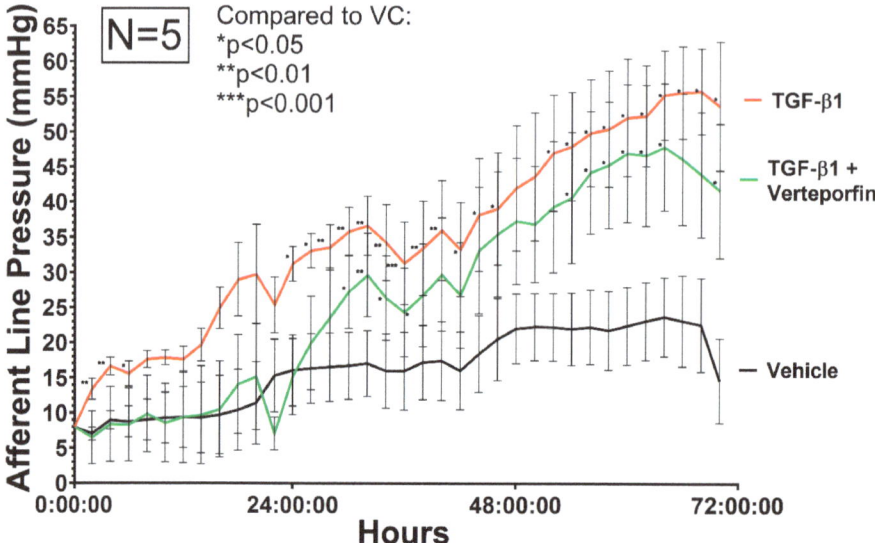

Figure 6. Relative afferent line pressure is increased in HTCF subconjunctival tissue-mimetic treated with TGF-β1 (red). Verteporfin (green) co-treatment decreased afferent line pressure. Pressure is measured in real-time with a blood pressure transducer representing model IOP. Relative pressure changes are normalized to average pressure values in the first hour of the experiment. Data shown are the means ± standard error (SEM) from N = 5 primary HTCF patient samples. (*) indicate significant differences between samples (* $p \leq 0.05$; ** $p \leq 0.01$; *** $p \leq 0.001$) by 2 way ANOVA.

HTCFs co-treated with TGFβ1 and verteporfin demonstrated a significant delay in the generation of outflow resistance. The addition of verteporfin maintained afferent line pressure at levels similar to vehicle treated replicates up until 26 h after the start of the experiment. ΔP_{vert} demonstrated a mean afferent line pressure reduction of 8.51 mmHg (SE = 7.21 mmHg) compared to ΔP_t. A two-way ANOVA revealed that there was a statistically significant difference between the effects of TGFβ1 and vehicle control (F(1, 8), p = 0.011). However, two-way ANOVA revealed that there was overall not a statistically significant difference between the effects of TGFβ1 and TGFβ1/verteporfin cotreatment (F(1, 8), p = 0.272) over the entire experimental duration.

3.4. Semi-Quantification of TGFβ1 Mediated Phenotypic Changes

Three-dimensional Z-stack confocal microscopy was performed and quantification of cytoskeletal fluorescence to the number of nuclei was compared between cells treated with TGFβ1, TGFβ1 and verteporfin, and vehicle control. Vehicle control cells expressed 49.4% of adjusted F-actin expression relative to TGFβ1 treated cells.

Cells treated with TGFβ1 and verteporfin demonstrated 54.3% of adjusted F-actin expression relative to TGFβ1 alone (Figure 7D). Concurrent Western blot revealed decreased levels of αSMA expression and demonstrated dose-response starting at 10 μM of verteporfin (Figure 7E).

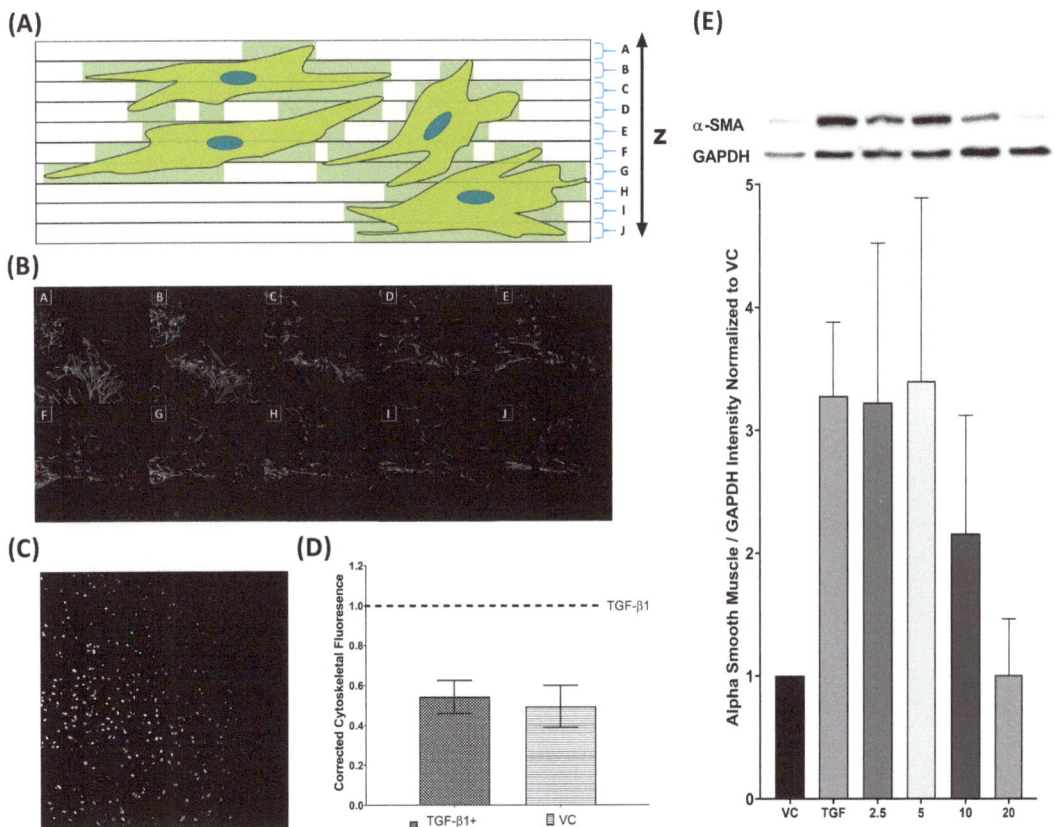

Figure 7. TGFβ1 induced cytoskeletal proliferation is inhibited by verteporfin cotreatment. Collagen slides are imaged as a Z-stack and divided in regular intervals from the bottom to the top of

the slide (**A**). Cumulative cytoskeletal fluorescence (FITC) is summed to give total volume of fluorescence to account for the three dimensionalities of the cytoskeleton (**B**). Nuclei (DAPI) channels are projected onto a flat image and counted to quantify number of cells on each slide (**C**). Cytoskeletal fluorescence was normalized to the number of nuclei on the slide (**D**). Three technical replicates are performed for parallel runs of VC/TGFβ1 and TGFβ1/TGFβ1–verteporfin cotreatment. Verteporfin co-treatment also inhibits TGFβ1 mediated α-SMA (42 kDA) upregulation showing dose effect starting at 10 μM. α-SMA was normalized to GAPDH (36kDA) (**E**).

4. Discussion

4.1. Development of a Three-Dimensional Flow Model

Three-dimensional tissue models more closely mimic the physiological milieu of living tissues, thereby circumventing many limitations inherent to 2D cell culture. Knowing that interstitial flow also affects tissue orientation and fibroblast transformation [26], the primary aim of the study was to develop a novel three-dimensional subconjunctival tissue mimetic and microfluidics platform in which potential novel antifibrotic compounds could be rapidly screened. One benefit of developing a model with interstitial flow is that as fibroblasts begin to transform, they begin to alter the properties of the hydrogel in which they were seeded in. Pressure changes can be appreciated with the pressure transducer upstream or by the height of media in the afferent line reservoir (Figure 3). In doing so, the functional implications of these drugs can be appreciated, as HTCF-mediated changes in overall outflow facility within the collagen matrix will directly affect afferent line pressure.

The model was composed of loose ECM similar to what HTCFs would reside in vivo. This cellular hydrogel was then chemically locked to the inner surfaces of the flow chamber as the APTES and glutaraldehyde irreversibly crosslink the collagen onto the slide (Figure 2). Without this process, the collagen hydrogel would extrude through the system, and perfusate would flow around, not through, the HTCF–collagen suspension. Over the course of the experiment, fibroblasts within the hydrogel were integrated into the ECM. TGFβ1 naïve fibroblasts are capable of generating a moderate amount of outflow resistance within a collagen matrix as signified by afferent line pressure increases with the vehicle control group. This effect was dramatically increased with the addition of exogenous TGFβ1. The timeline over which afferent line pressure increased suggested that fibroblast transdifferentiation into myofibroblasts slowly changed the characteristics of the ECM, progressively increasing its resistance to outflow. As additional ECM proteins were synthesized and remodeled through myofibroblast activity, microscopic flow channels inside the hydrogel likely stenosed and overall tissue patency decreased, similar to the failing filtration bleb (Figure 8) [30,31].

4.2. Phenotypic Effects of TGF β1 Are Reflected within the Model

TGFβs have a major role in fibroproliferative disease [32,33]. Previous research has established that TGFβ2, another isotype, is found in high concentrations within the aqueous humor of patients with primary open-angle glaucoma and is implicated in instability in the blood–aqueous barrier [2]. Meanwhile, TGFβ1 is upregulated as a consequence of general wound healing as seen in surgical insult to tissue as well as neovascular glaucoma [34–36]. Both isotypes have been implicated in overexpression of genes responsible for ECM construction and stress fibre formation [37,38]. Canonically, TGFβ1 mediated SMAD2/3 activity contributes to the contraction of fibrillar collagen and inhibits metalloproteinases expression responsible for extracellular matrix (ECM) degradation [39]. Clinically, dysregulation of the aforementioned pathway has been implicated in fibrotic diseases in multiple organ systems [40,41].

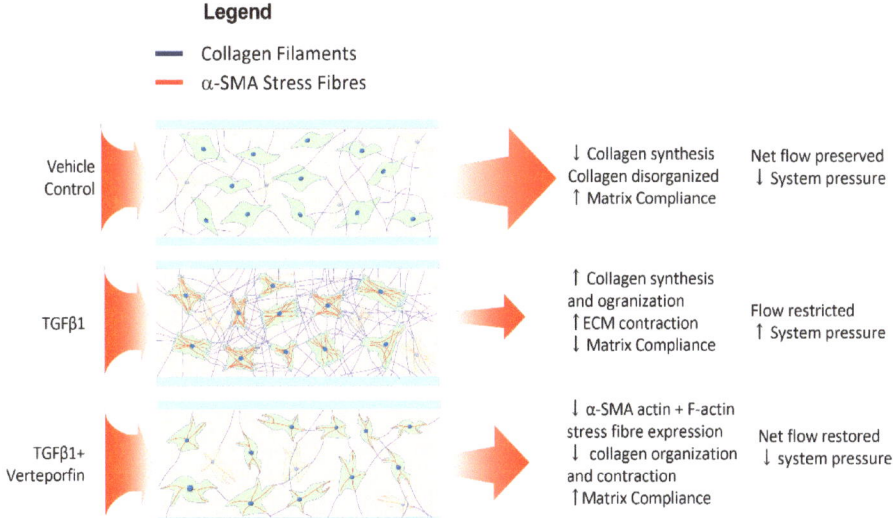

Figure 8. Verteporfin attenuates TGF-β1-mediated changes within HTCFs in the hydrogel. The net effect of TGF-β1 leads to myofibroblastic transdifferentiation indicated by increased cell bulk and α-SMA production. Myofibroblasts organize and contract ECM, lowering hydrogel permeability to perfusate and increasing system pressure. Verteporfin attenuates TGF-β1 mediated myofibroblastic transformation thereby lessening pressure increases.

In the presented model, exogenous TGFβ1 leads to increased myofibroblasts transdifferentiation consistent with previous work [27]. This was evident as TGFβ1 treatment alone was able to induce a-SMA upregulation (Figure 7E). Additionally, F-actin stress fibres, which allow tension generation and ECM remodeling, are elevated in HTCFs treated with exogenous TGFβ1 [42]. These biochemical findings correlate with the corresponding pressure data, as remodeling the hydrogel reduced the patency of the matrix to perfusate and reduce flow. As hydrodynamic resistance increased, the afferent pre-chamber pressure rose as perfusate began to accumulate. Overall, these results reaffirm that TGFβ1 alone can induce fibroblast transdifferentiation into its pathological myofibroblastic phenotype. More importantly, the clinical implications of these pathological changes were able to be captured by the microfluidics model through afferent line pressure increases.

4.3. TGFβ1 Mediated SMAD2/3 Inhibition Leads to Attenuation of Fibroblast Transdifferentiation

Verteporfin is a benzoporphyrin derivative and first made its pharmacological debut as a photosensitizer in photodynamic therapy [43]. Recent work has shown that verteporfin is also a potent inhibitor of the YAP/TAZ signaling pathway [44,45]. Inhibition of YAP/TAZ augments inhibitory SMAD7 activity levels and prevents the SMAD2/3 complex from translocating into the nucleus to induce transcription of myofibroblast-related genes [46,47]. Furthermore, increased SMAD7 leads to matrix metalloproteases upregulation that is responsible for collagen degradation [48,49]. As discussed, the de novo deposition and contraction of the ECM are major contributors to surgical failure. It was demonstrated that YAP inhibition through verteporfin decreases contractility of trabecular meshwork cells [50], and previous reports have shown that verteporfin mitigates kidney myofibroblast transdifferentiation [51]. Likewise, this study revealed that verteporfin leads to decreased TGFβ1 mediated a-SMA expression (Figure 7E) and F-actin stress fibre formation (Figure 5) in HTCFs, suggesting that TGFβ1 mediated myofibroblast transformation is similarly inhibited in HTCFs. Furthermore, it was demonstrated that inhibiting myofibroblast transdifferentiation may have a tangible impact on outflow capacity as afferent line pressure increases were mitigated albeit only over the first 32 h of perfusion (Figure 6). Verteporfin photodegrades readily and has an estimated half-life of 5–6 h; in turn, its inhibitory effect could have been blunted as the experiment continued [52]. This is supported by the fact that the greatest mean difference in pressure was observed in the first 32 h of the experiment between TGFβ1 treatment and TGFβ1/verteporfin co-treatment.

This model does have several limitations relating to incomplete recreation of the physiologic milieu within the filtration bleb. Firstly, inter-patient and cell line variability leads to differing degrees of response to verteporfin leading to large standard deviation between runs. Despite this, the trend suggests that afferent line pressure is attenuated with verteporfin. Additionally, this fibroblast-only model is devoid of other immunomodulating cells. These cells help with cellular debris clearance and alter the activity of signaling molecules such as TGFβ1 in the subconjunctival tissues of the eye. Thus, future work incorporating THP-1 monocytes into this model as a co-culture with the fibroblasts within the collagen matrix is underway as macrophages are known to play a key role in wound healing. Finally, the length of each experiment was limited by how much volume was in the syringe pump as disconnecting to refill it with media would lead to system depressurization and potential contamination. Despite these limitations, proof of principle was established for this model. Future iterations of the model should incorporate a pneumatic pump system that allows for the continuous addition of new perfusate media without depressurization to provide uninterrupted perfusion of the system.

5. Conclusions

In this study, a three-dimensional HTCF collagen culture model was developed that permitted the perfusion of culture media and the ability to measure changes in afferent line pressure as an analogue for IOP. HTCFs in a collagen matrix were successfully used as an in vitro model to demonstrate the potential clinical implications of deranged TGFβ activity following a surgical-induced insult of the subconjunctival milieu. By incorporating a three-dimensional tissue mimetic, the model was able to reflect the behaviour of a filtration bleb while being able to monitor surrogate intraocular pressure when exposed to exogenous TGFβ1 as seen in post-surgical AH. Exogenous TGFβ1 increased afferent line pressure analogous to the clinical changes associated with impending surgical failure. The effects of TGFβ1 activity were able to be transiently attenuated following treatment with verteporfin. This model can serve as an in vitro platform for high-throughput screening of drug candidates that may potentially attenuate post-operative reductions to subconjunctival outflow capacity and pave the way for clinical translation and the benefit of future glaucoma surgery patients.

Author Contributions: Conceptualization, M.F., J.J.A. and C.H.; methodology, M.F. and J.J.A.; investigation, M.F., J.J.A., R.Z., A.V. and H.L.; formal analysis, M.F., J.J.A. and R.Z.; software, R.Z.; writing—original draft preparation, M.F.; writing—review and editing, M.F., J.J.A., R.Z., A.V. and C.H. Supervision, funding Acquisition, and project administration, C.H. All authors have read and agreed to the published version of the manuscript.

Funding: This study was supported from a grant by the Glaucoma Research Society of Canada (889178695 RR0001).

Institutional Review Board Statement: Research ethics approval was acquired prior to experimentation (HSREB: 106783).

Informed Consent Statement: Informed consent was obtained from all subjects involved in the study.

Data Availability Statement: All the information in the manuscript is supported by the mentioned references.

Acknowledgments: We acknowledge the students and staff of the Hutnik laboratory, David O'Gorman, and David Tingey of the Ivey Eye Institute for their support, communication, mentorship, and material that were critical in making the project possible.

Conflicts of Interest: The authors report no conflicts of interest. The authors report funding from the Canadian Glaucoma Society and Schulich School of Medicine and Dentistry, Western University. The content of this submission has not been published or submitted for publication elsewhere. All authors have contributed significantly and agree with the content of this manuscript.

References

1. Huang, W.; Chen, S.; Gao, X.; Yang, M.; Zhang, J.; Li, X.; Wang, W.; Zhou, M.; Zhang, X.; Zhang, X. Inflammation-Related Cytokines of Aqueous Humor in Acute Primary Angle-Closure Eyes. *Investig. Ophthalmol. Vis. Sci.* **2014**, *55*, 1088–1094. [CrossRef] [PubMed]
2. Tripathi, R.C.; Li, J.; Chan, W.F.A.; Tripathi, B.J. Aqueous Humor in Glaucomatous Eyes Contains an Increased Level of TGF-B2. *Exp. Eye Res.* **1994**, *59*, 723–728. [CrossRef] [PubMed]
3. Tran, M.N.; Medveczki, T.; Besztercei, B.; Torok, G.; Szabo, A.J.; Gasull, X.; Kovacs, I.; Fekete, A.; Hodrea, J. Sigma-1 Receptor Activation Is Protective against TGFβ2-Induced Extracellular Matrix Changes in Human Trabecular Meshwork Cells. *Life* **2023**, *13*, 1581. [CrossRef]
4. Prendes, M.A.; Harris, A.; Wirostko, B.M.; Gerber, A.L.; Siesky, B. The Role of Transforming Growth Factor β in Glaucoma and the Therapeutic Implications. *Br. J. Ophthalmol.* **2013**, *97*, 680–686. [CrossRef] [PubMed]
5. Shepard, A.R.; Cameron Millar, J.; Pang, I.H.; Jacobson, N.; Wang, W.H.; Clark, A.F. Adenoviral Gene Transfer of Active Human Transforming Growth Factor-B2 Elevates Intraocular Pressure and Reduces Outflow Facility in Rodent Eyes. *Investig. Ophthalmol. Vis. Sci.* **2010**, *51*, 2067–2076. [CrossRef] [PubMed]
6. Zhao, S.; Fang, L.; Yan, C.; Wei, J.; Song, D.; Xu, C.; Luo, Y.; Fan, Y.; Guo, L.; Sun, H.; et al. MicroRNA-210-3p Mediates Trabecular Meshwork Extracellular Matrix Accumulation and Ocular Hypertension—Implication for Novel Glaucoma Therapy. *Exp. Eye Res.* **2023**, *227*, 109350. [CrossRef]
7. Robertson, J.V.; Siwakoti, A.; West-Mays, J.A. Altered Expression of Transforming Growth Factor Beta 1 and Matrix Metalloproteinase-9 Results in Elevated Intraocular Pressure in Mice. *Mol. Vis.* **2013**, *19*, 684–695. [PubMed]
8. Picht, G.; Welge-Luessen, U.; Grehn, F.; Lütjen-Drecoll, E. Transforming Growth Factor Beta 2 Levels in the Aqueous Humor in Different Types of Glaucoma and the Relation to Filtering Bleb Development. *Graefes Arch. Clin. Exp. Ophthalmol.* **2001**, *239*, 199–207. [CrossRef] [PubMed]
9. Desmouliere, A.; Geinoz, A.; Gabbiani, F.; Gabbiani, G. Transforming Growth Factor-B1 Induces α-Smooth Muscle Actin Expression in Granulation Tissue Myofibroblasts and in Quiescent and Growing Cultured Fibroblasts. *J. Cell Biol.* **1993**, *122*, 103–111. [CrossRef]
10. Lee, S.Y.; Chae, M.K.; Yoon, J.S.; Kim, C.Y. The Effect of CHIR 99021, a Glycogen Synthase Kinase-3β Inhibitor, on Transforming Growth Factor β-Induced Tenon Fibrosis. *Investig. Ophthalmol. Vis. Sci.* **2021**, *62*, 25. [CrossRef]
11. Chen, J.; Li, H.; SundarRaj, N.; Wang, J.H.-C. Alpha-Smooth Muscle Actin Expression Enhances Cell Traction Force. *Cell Motil. Cytoskelet.* **2007**, *64*, 248–257. [CrossRef] [PubMed]
12. Hecker, L.; Jagirdar, R.; Jin, T.; Thannickal, V.J. Reversible Differentiation of Myofibroblasts by MyoD. *Exp. Cell Res.* **2011**, *317*, 1914–1921. [CrossRef] [PubMed]
13. Talele, N.P.; Fradette, J.; Davies, J.E.; Kapus, A.; Hinz, B. Expression of α-Smooth Muscle Actin Determines the Fate of Mesenchymal Stromal Cells. *Stem Cell Rep.* **2015**, *4*, 1016–1030. [CrossRef]
14. Desmoulière, A.; Badid, C.; Bochaton-Piallat, M.L.; Gabbiani, G. Apoptosis during Wound Healing, Fibrocontractive Diseases and Vascular Wall Injury. *Int. J. Biochem. Cell Biol.* **1997**, *29*, 19–30. [CrossRef] [PubMed]
15. Conlon, R.; Saheb, H.; Ahmed, I.I.K. Glaucoma Treatment Trends: A Review. *Can. J. Ophthalmol.* **2017**, *52*, 114–124. [CrossRef]
16. Pillunat, L.E.; Erb, C.; Jünemann, A.G.; Kimmich, F. Micro-Invasive Glaucoma Surgery (MIGS): A Review of Surgical Procedures Using Stents. *Clin. Ophthalmol.* **2017**, *11*, 1583–1600. [CrossRef]
17. Kessing, S.V.; Flesner, P.; Jensen, P.K. Determinants of Bleb Morphology in Minimally Invasive, Clear-Cornea Micropenetrating Glaucoma Surgery with Mitomycin C. *J. Glaucoma* **2006**, *15*, 84–90. [CrossRef]
18. Kankainen, T.; Harju, M. Endophthalmitis and Blebitis Following Deep Sclerectomy and Trabeculectomy with Routine Use of Mitomycin C. *Acta Ophthalmol.* **2023**, *101*, 285–292. [CrossRef]
19. Phulke, S.; Kaushik, S.; Kaur, S.; Pandav, S. Steroid-Induced Glaucoma: An Avoidable Irreversible Blindness. *J. Curr. Glaucoma Pract.* **2017**, *11*, 67.
20. Kanta, J. Collagen Matrix as a Tool in Studying Fibroblastic Cell Behavior. *Cell Adhes. Migr.* **2015**, *9*, 308–316. [CrossRef] [PubMed]
21. Hinz, B. The Myofibroblast: Paradigm for a Mechanically Active Cell. *J. Biomech.* **2010**, *43*, 146–155. [CrossRef] [PubMed]
22. Olsen, A.L.; Bloomer, S.A.; Chan, E.P.; Gaça, M.D.A.; Georges, P.C.; Sackey, B.; Uemura, M.; Janmey, P.A.; Wells, R.G. Hepatic Stellate Cells Require a Stiff Environment for Myofibroblastic Differentiation. *Am. J. Physiol. Gastrointest. Liver Physiol.* **2011**, *301*, G110–G118. [CrossRef] [PubMed]
23. Burgstaller, G.; Gerckens, M.; Eickelberg, O.; Königshoff, M. Decellularized Human Lung Scaffolds as Complex Three-Dimensional Tissue Culture Models to Study Functional Behavior of Fibroblasts. *Methods Mol. Biol.* **2021**, *2299*, 447–456. [CrossRef]
24. Bracken, M.B. Why Animal Studies Are Often Poor Predictors of Human Reactions to Exposure. *J. R. Soc. Med.* **2009**, *102*, 120–122. [CrossRef]

25. von Scheidt, M.; Zhao, Y.; Kurt, Z.; Pan, C.; Zeng, L.; Yang, X.; Schunkert, H.; Lusis, A.J. Applications and Limitations of Mouse Models for Understanding Human Atherosclerosis. *Cell Metab.* **2017**, *25*, 248–261. [CrossRef]
26. Ng, C.P.; Hinz, B.; Swartz, M.A. Interstitial Fluid Flow Induces Myofibroblast Differentiation and Collagen Alignment in Vitro. *J. Cell Sci.* **2005**, *118*, 4731–4739. [CrossRef]
27. Armstrong, J.; Denstedt, J.; Trelford, C.; Li, E.; Hutnik, C. Differential Effects of Dexamethasone and Indomethacin on Tenon's Capsule Fibroblasts: Implications for Glaucoma Surgery. *Exp. Eye Res.* **2019**, *182*, 65–73. [CrossRef] [PubMed]
28. Vandenberg, E.; Elwing, H.; Askendal, A.; Lundström, I. Protein Immobilization of 3-Aminopropyl Triethoxy Silaneglutaraldehyde Surfaces: Characterization by Detergent Washing. *J. Colloid. Interface Sci.* **1991**, *143*, 327–335. [CrossRef]
29. Gunda, N.S.K.; Singh, M.; Norman, L.; Kaur, K.; Mitra, S.K. Optimization and Characterization of Biomolecule Immobilization on Silicon Substrates Using (3-Aminopropyl)Triethoxysilane (APTES) and Glutaraldehyde Linker. *Appl. Surf. Sci.* **2014**, *305*, 522–530. [CrossRef]
30. Leung, D.Y.L.; Tham, C.C.Y. Management of Bleb Complications after Trabeculectomy. *Semin. Ophthalmol.* **2013**, *28*, 144–156. [CrossRef]
31. Yoshida, M.; Kokubun, T.; Sato, K.; Tsuda, S.; Yokoyama, Y.; Himori, N.; Nakazawa, T. DPP-4 Inhibitors Attenuate Fibrosis after Glaucoma Filtering Surgery by Suppressing the TGF-β/Smad Signaling Pathway. *Investig. Ophthalmol. Vis. Sci.* **2023**, *64*, 2. [CrossRef]
32. Saika, S. TGFbeta Pathobiology in the Eye. *Lab. Investig.* **2006**, *86*, 106–115. [CrossRef] [PubMed]
33. McDowell, C.M.; Tebow, H.E.; Wordinger, R.J.; Clark, A.F. Smad3 Is Necessary for Transforming Growth Factor-Beta2 Induced Ocular Hypertension in Mice. *Exp. Eye Res.* **2013**, *116*, 419–423. [CrossRef] [PubMed]
34. Chong, D.L.W.; Trinder, S.; Labelle, M.; Rodriguez-Justo, M.; Hughes, S.; Holmes, A.M.; Scotton, C.J.; Porter, J.C. Platelet-Derived Transforming Growth Factor-B1 Promotes Keratinocyte Proliferation in Cutaneous Wound Healing. *J. Tissue Eng. Regen. Med.* **2020**, *14*, 645–649. [CrossRef] [PubMed]
35. Kane, C.J.M.; Hebda, P.A.; Mansbridge, J.N.; Hanawalt, P.C. Direct Evidence for Spatial and Temporal Regulation of Transforming Growth Factor B1 Expression during Cutaneous Wound Healing. *J. Cell Physiol.* **1991**, *148*, 157–173. [CrossRef] [PubMed]
36. Yu, X.B.; Sun, X.H.; Dahan, E.; Guo, W.Y.; Qian, S.H.; Meng, F.R.; Song, Y.L.; Simon, G.J.B. Increased Levels of Transforming Growth Factor-Beta1 and -Beta2 in the Aqueous Humor of Patients with Neovascular Glaucoma. *Ophthalmic Surg. Lasers Imaging* **2007**, *38*, 6–14. [CrossRef] [PubMed]
37. Li, J.; Tripathi, B.J.; Tripathi, R.C. Modulation of Pre-MRNA Splicing and Protein Production of Fibronectin by TGF-2 in Porcine Trabecular Cells. *Investig. Ophthalmol. Vis. Sci.* **2000**, *41*, 3437–3443.
38. Zhao, X.; Ramsey, K.E.; Stephan, D.A.; Russell, P. Gene and Protein Expression Changes in Human Trabecular Meshwork Cells Treated with Transforming Growth Factor-Beta. *Investig. Ophthalmol. Vis. Sci.* **2004**, *45*, 4023–4034. [CrossRef]
39. Flanders, K.C. Smad3 as a Mediator of the Fibrotic Response. *Int. J. Exp. Pathol.* **2004**, *85*, 47–64. [CrossRef]
40. Meng, X.M.; Chung, A.C.K.; Lan, H.Y. Role of the TGF-β/BMP-7/Smad Pathways in Renal Diseases. *Clin. Sci.* **2013**, *124*, 243–254. [CrossRef]
41. Chen, J.; Xia, Y.; Lin, X.; Feng, X.H.; Wang, Y. Smad3 Signaling Activates Bone Marrow-Derived Fibroblasts in Renal Fibrosis. *Lab. Investig.* **2014**, *94*, 545–556. [CrossRef]
42. Tojkander, S.; Gateva, G.; Lappalainen, P. Actin Stress Fibers—Assembly, Dynamics and Biological Roles. *J. Cell Sci.* **2012**, *125*, 1855–1864. [CrossRef]
43. Bressler, N.M. Photodynamic Therapy of Subfoveal Choroidal Neovascularization in Age- Related Macular Degeneration with Verteporfin: One-Year Results of 2 Randomized Clinical Trials—TAP Report 1. *Arch. Ophthalmol.* **1999**, *117*, 1329–1345. [CrossRef]
44. Szeto, S.G.; Narimatsu, M.; Lu, M.; He, X.; Sidiqi, A.M.; Tolosa, M.F.; Chan, L.; De Freitas, K.; Bialik, J.F.; Majumder, S.; et al. YAP/TAZ Are Mechanoregulators of TGF-b-Smad Signaling and Renal Fibrogenesis. *J. Am. Soc. Nephrol.* **2016**, *27*, 3117–3128. [CrossRef] [PubMed]
45. Liu-Chittenden, Y.; Huang, B.; Shim, J.S.; Chen, Q.; Lee, S.J.; Anders, R.A.; Liu, J.O.; Pan, D. Genetic and Pharmacological Disruption of the TEAD-YAP Complex Suppresses the Oncogenic Activity of YAP. *Genes. Dev.* **2012**, *26*, 1300–1305. [CrossRef]
46. Piersma, B.; Bank, R.A.; Boersema, M. Signaling in Fibrosis: TGF-β, WNT, and YAP/TAZ Converge. *Front. Med.* **2015**, *2*, 59. [CrossRef]
47. Miyazawa, K.; Miyazono, K. Regulation of TGF-β Family Signaling by Inhibitory Smads. *Cold Spring Harb. Perspect. Biol.* **2017**, *9*, a022095. [CrossRef]
48. Wang, B.; Omar, A.; Angelovska, T.; Drobic, V.; Rattan, S.G.; Jones, S.C.; Dixon, I.M.C. Regulation of Collagen Synthesis by Inhibitory Smad7 in Cardiac Myofibroblasts. *Am. J. Physiol. Heart Circ. Physiol.* **2007**, *293*, H1282–H1290. [CrossRef] [PubMed]
49. Zheng, H.; Huang, N.; Lin, J.-Q.; Yan, L.-Y.; Jiang, Q.-G.; Yang, W.-Z. Effect and Mechanism of Pirfenidone Combined with 2-Methoxy-Estradiol Perfusion through Portal Vein on Hepatic Artery Hypoxia-Induced Hepatic Fibrosis. *Adv. Med. Sci.* **2023**, *68*, 46–53. [CrossRef]
50. Chen, W.S.; Cao, Z.; Krishnan, C.; Panjwani, N. Verteporfin without Light Stimulation Inhibits YAP Activation in Trabecular Meshwork Cells: Implications for Glaucoma Treatment. *Biochem. Biophys. Res. Commun.* **2015**, *466*, 221–225. [CrossRef]

51. Jin, J.; Wang, T.; Park, W.; Li, W.; Kim, W.; Park, S.K.; Kang, K.P. Inhibition of Yes-Associated Protein by Verteporfin Ameliorates Unilateral Ureteral Obstruction-Induced Renal Tubulointerstitial Inflammation and Fibrosis. *Int. J. Mol. Sci.* **2020**, *21*, 8184. [CrossRef] [PubMed]
52. Houle, J.M.; Strong, A. Clinical Pharmacokinetics of Verteporfin. *J. Clin. Pharmacol.* **2002**, *42*, 547–557. [CrossRef] [PubMed]

Disclaimer/Publisher's Note: The statements, opinions and data contained in all publications are solely those of the individual author(s) and contributor(s) and not of MDPI and/or the editor(s). MDPI and/or the editor(s) disclaim responsibility for any injury to people or property resulting from any ideas, methods, instructions or products referred to in the content.

Review

Addressing Glaucoma in Myopic Eyes: Diagnostic and Surgical Challenges

Kateki Vinod [1] and Sarwat Salim [2,*]

[1] Department of Ophthalmology, Icahn School of Medicine at Mount Sinai, New York Eye and Ear Infirmary of Mount Sinai, New York, NY 10003, USA

[2] Department of Ophthalmology, Tufts University School of Medicine, Boston, MA 02116, USA

* Correspondence: ssalim@tuftsmedicalcenter.org; Tel.: +617-636-4900; Fax: +617-636-4866

Abstract: Epidemiological and genetic studies provide strong evidence supporting an association between myopia and glaucoma. The accurate detection of glaucoma in myopic eyes, especially those with high myopia, remains clinically challenging due to characteristic morphologic features of the myopic optic nerve in addition to limitations of current optic nerve imaging modalities. Distinguishing glaucoma from myopia is further complicated by overlapping perimetric findings. Therefore, longitudinal follow-up is essential to differentiate progressive structural and functional abnormalities indicative of glaucoma from defects that may result from myopia alone. Highly myopic eyes are at increased risk of complications from traditional incisional glaucoma surgery and may benefit from newer microinvasive glaucoma surgeries in select cases.

Keywords: glaucoma; high myopia; myopia; optic disc; optical coherence tomography; perimetry; cataract surgery

Citation: Vinod, K.; Salim, S. Addressing Glaucoma in Myopic Eyes: Diagnostic and Surgical Challenges. *Bioengineering* **2023**, *10*, 1260. https://doi.org/10.3390/bioengineering10111260

Academic Editors: Karanjit S. Kooner and Osamah J. Saeedi

Received: 16 October 2023
Revised: 23 October 2023
Accepted: 27 October 2023
Published: 29 October 2023

Copyright: © 2023 by the authors. Licensee MDPI, Basel, Switzerland. This article is an open access article distributed under the terms and conditions of the Creative Commons Attribution (CC BY) license (https://creativecommons.org/licenses/by/4.0/).

1. Introduction

The global prevalence of myopia is rapidly increasing, with particularly high rates in Asia [1]. Approximately half the world population is predicted to have myopia by 2050 [2]. This increased prevalence raises additional concerns for vision-threatening ocular sequelae related to myopia, including myopic degeneration, retinal detachment, cataract, amblyopia, and glaucoma. A recent meta-analysis of 24 studies from 11 countries found a 20% increased risk of glaucoma for every one diopter increase in myopia [3].

Epidemiological studies provide robust data and support for myopia as a risk factor for open-angle glaucoma. The Blue Mountains Eye Study was a population-based study of Australian adults aged 49 or older in which glaucoma was observed more commonly in myopes (4.2% of eyes with spherical equivalent −1.0 to −3.0 diopters and 4.4% of eyes with spherical equivalent >−3.0 diopters) than those without myopia (1.5%) [4]. High myopia, commonly defined as a spherical equivalent >−6.0 diopters and/or an axial length > 26.5 mm, appears to confer an even greater risk of glaucoma. In the Beijing Eye Study, a population-based study of Chinese adults aged 40 or older, glaucoma was observed more frequently in eyes with marked (−6.0 to −8.0 diopters) and high myopia (>−8.0 diopters) versus those with moderate myopia (−3.0 to −6.0 diopters; $p = 0.075$), low myopia (−0.5 to −3.0 diopters; $p = 0.001$), emmetropia (−0.5 to +2.0 diopters; $p < 0.001$), and hyperopia (>+2.0 diopters; $p = 0.005$) [5]. In a 10-year follow-up study, low, moderate, and high myopia conferred a 3.2-, 4.2-, and 7.3-fold increased risk of glaucoma, respectively, as compared with emmetropia ($p < 0.01$) [6].

In the Beijing Eye Study, glaucoma was diagnosed based upon morphologic assessments of optic disc photographs. In general, myopic eyes often exhibit optic nerve features, optical coherence tomography of the retinal nerve fiber layer (OCT RNFL) and ganglion cell complex (GCC) thinning, and visual field defects that resemble, but are not always indicative of, glaucoma, thereby complicating its detection. This review focuses on recent advances and research on myopia as it relates to the pathophysiology, diagnosis, and surgical management of glaucoma.

2. Pathophysiology of Glaucoma in Myopic Eyes

The pathophysiology underlying the increased susceptibility of optic nerves in myopic eyes to glaucomatous damage is not fully understood. Optic nerves are thought to be more vulnerable to damage due to poor structural support for retinal nerve fibers in myopic eyes, particularly those with high myopia. Axial elongation likely results in thinning of the sclera and produces greater shearing forces across the stretched lamina cribrosa, leaving the optic nerve more susceptible to injury [7].

Intraocular pressure (IOP) is well established as the main and only modifiable risk factor for glaucoma and likely plays an important role in the pathophysiology of glaucoma in myopic eyes. Recent Mendelian randomization studies suggest that myopia and primary open-angle glaucoma (POAG) share a genetic basis, and that IOP is the primary mediator underlying their bidirectional causal relationship [8,9]. In the Singapore Epidemiology of Eye Diseases Study, eyes with myopia >-3.0 diopters and IOP ≥ 20 mm Hg were over four times more likely to have POAG versus eyes without myopia and normal IOP. Eyes with an axial length >25.5 mm and IOP ≥ 20 mm Hg were over 16 times more likely to have POAG versus eyes with an axial length <23.5 mm and normal IOP [10]. Axial length may therefore be a clinically relevant metric for risk stratification of myopic glaucoma suspects. Importantly, myopic eyes, especially those with high myopia, may develop glaucoma within the normal range of IOP due to underlying structural abnormalities in the sclera and lamina cribrosa [7].

Recent studies offer mixed results regarding a possible genetic correlation between POAG and myopia. Iglesias and associates analyzed data from the Australian & New Zealand Registry of Advanced Glaucoma study (N = 798 POAG patients and 1992 control patients) and the Rotterdam Study (N = 11,097) and did not find evidence supporting an association between polygenic risk scores of POAG and myopia [11]. Using data from the UK Biobank and Genetic Epidemiology Research on Adult Health and Aging (GERA) cohort (N = 154,018), Choquet and colleagues performed a genetic correlation analysis and found that patients with POAG had decreased mean spherical equivalents and a greater probability of having myopia or high myopia when compared with patients without POAG [8]. Additional studies in more diverse patient populations are needed to better elucidate a potential genetic basis for the increased glaucoma risk in myopic individuals that has been found in observational studies.

3. Diagnostic Structural and Functional Testing of Glaucoma in Myopic Eyes

Diagnosing glaucoma in a myopic eye presents numerous challenges, especially when the IOP is within the normal range. Inherent differences in corneal biomechanics may exist in eyes with high myopia as compared with emmetropic eyes and eyes with low myopia, and should be taken into consideration when interpreting measured IOP values. Importantly, rebound tonometry is likely influenced by corneal biomechanics to a greater extent than Goldmann applanation tonometry, and may underestimate IOP in eyes that develop glaucoma within the normal range of IOP [12]. Corneal biomechanical parameters include corneal hysteresis, a biomarker that quantifies the viscous dampening of the cornea, and corneal resistance factor, which is a largely IOP-independent measure of corneal resistance. A meta-analysis of eleven studies using the Ocular Response Analyzer (ORA) to evaluate corneal biomechanical indices observed significantly lower corneal hysteresis and corneal resistance factor among eyes with high (≥ -6.0 diopters; N = 1027 eyes) versus low myopia (≤ -3.0 diopters; N = 835 eyes), and found no difference in central corneal thickness [13]. These findings suggest that high myopia may be associated not only with decreased rigidity of the sclera, but also of the cornea. Clinicians may therefore consider obtaining corneal hysteresis measurements in highly myopic patients, particularly in those who exhibit glaucomatous progression despite seemingly "normal" IOP. Patients with myopia should also be asked about any history of corneal refractive surgery, which is known to further influence the accuracy of IOP measurements due to its effects on corneal biomechanical properties.

Clinically, myopic optic nerves may be difficult to differentiate from glaucomatous optic nerves. Unique anatomical features of myopic optic discs can confound quantitative structural and functional assessments, which can lead to overtreatment or undertreatment in many cases. Myopic eyes often exhibit such characteristics as large-diameter optic discs, optic disc tilting or torsion, and beta-zone peripapillary atrophy. Such features can result in erroneous OCT measurements of the circumpapillary RNFL (Figure 1). In general, retinal thinning is common in myopic eyes and may be misinterpreted as glaucomatous thinning in the setting of a globally reduced circumpapillary RNFL measurement [14]. Kang and colleagues found that average RNFL thickness decreases by approximately 1.3 microns for every additional diopter of myopia [15]. Leung and associates observed an increasingly temporal orientation of the superotemporal and inferotemporal RNFL bundles with higher degrees of myopia, resulting in greater RNFL areas deemed "abnormal" in comparison with a normative database [16]. As a result of this deviation, temporal sectors may appear more robust while nasal sectors may appear thinner on circumpapillary RNFL measurements in myopic eyes, and RNFL probability maps may demonstrate arcuate artifacts mimicking glaucoma. Additionally, misalignment of the scan circle or the presence of extensive peripapillary atrophy crossing its margin may result in erroneous measurements. Retinal nerve fiber layer segmentation artifacts may also result from vitreopapillary traction or peripapillary retinoschisis. Therefore, clinicians must exercise vigilance in identifying "red disease", whereby OCT RNFL may misclassify a healthy myopic nerve as "abnormal" based upon limited normative databases in commercially available OCT machines that exclude highly myopic eyes [14,17]. Specific attention must be paid not only to the global and sectoral circumpapillary RNFL thickness, but also to b-scans, circumpapillary RNFL plots, and RNFL and macular thickness maps [18].

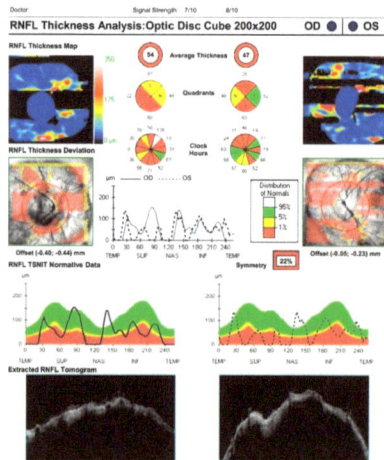

Figure 1. OCT RNFL demonstrating segmentation artifacts in a patient with high myopia, characteristic of "red disease".

While some studies suggest improved diagnostic accuracy of macular thickness over circumpapillary RNFL in identifying glaucoma in highly myopic eyes [19,20], others do not [21,22]. Like circumpapillary RNFL thickness, macular thickness is often globally reduced in high myopes. Segmentation errors also occur with macular imaging in myopic eyes. Hwang and colleagues observed a segmentation error rate of 9.7% in macular thickness scans from 538 eyes with and without glaucoma while excluding those with macular pathology, and these errors were reproducible in 46% of eyes. A higher degree of myopia was found to be significantly associated with the occurrence of segmentation errors ($p < 0.001$) [23].

Advances in OCT RNFL imaging have the potential to improve glaucoma detection in highly myopic eyes. Biswas and colleagues evaluated a new myopic normative OCT RNFL database in 180 eyes with high myopia (average spherical equivalent -8.0 ± 1.8 diopters), which showed improved specificity versus the standard normative database while maintaining sensitivity for detecting glaucoma [24]. Baek and associates incorporated topographic RNFL and ganglion cell–inner plexiform layer (GCIPL) parameters to develop a scoring system to identify glaucomatous damage in the setting of myopia. This combined approach topographic scoring system outperformed individual RNFL and GCIPL thickness parameters in both myopic and highly myopic eyes [25]. Kim and colleagues found that swept-source OCT wide-field maps displayed better accuracy for diagnosing glaucomatous defects in myopic eyes than spectral domain OCT, likely due to the former modality's wider area of measurement [26].

While OCT RNFL and macular imaging have largely supplanted the use of stereoscopic disc photographs in the longitudinal assessment of glaucoma, the latter remain useful when following patients with myopia, particularly as OCT technology continues to evolve (Figure 2). Future directions to improve glaucoma detection in myopic and highly myopic eyes may include OCT angiography as well as artificial intelligence and deep learning strategies.

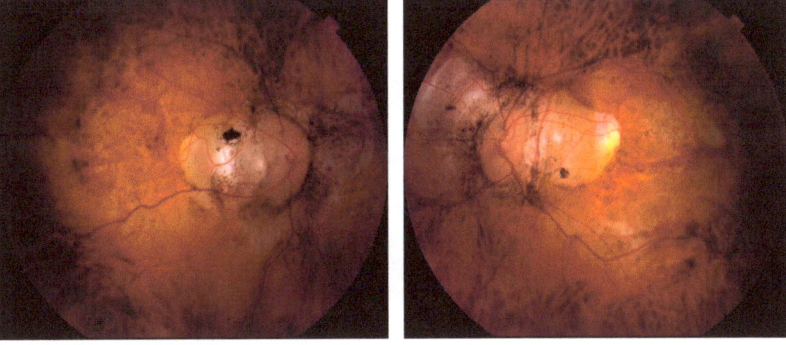

Figure 2. Optic disc photographs from a patient with high myopia demonstrating optic disc tilting and extensive peripapillary atrophy, making optic nerve assessment difficult.

Myopia is known to cause various visual field abnormalities that mimic glaucoma but typically do not progress over time as would glaucomatous defects in untreated eyes. A key component of glaucoma evaluation in myopic eyes, therefore, is longitudinal follow-up. Enlarged blind spots are common in high myopes (Figure 3), and tilted discs often produce superotemporal visual field defects [27]. A retrospective study of myopic glaucoma suspects aged 50 or younger found that myopia was associated with visual field defects resembling glaucoma, such as nasal steps, arcuate defects, and paracentral scotomas [28]. Doshi and colleagues demonstrated nonprogressive visual field defects mirroring glaucoma over a seven-year period in 16 myopic Chinese men (age range 25 to 66 years; mean 38.9), only approximately half of whom were using IOP-lowering medications. The authors postulated that visual field defects in myopic eyes may develop and progress during axial elongation earlier in life and later stabilize in adulthood in this specific subpopulation [29]. Han and associates performed a retrospective longitudinal study in Korea with a minimum follow-up of seven years and found that inferiorly tilted discs were more likely than temporally tilted discs to produce localized, single-hemifield visual field defects and to demonstrate visual field progression. Given that the large majority (92.5%) of visual field defects remained limited to the superior hemifield in eyes with inferiorly tilted optic discs over the study period, the authors proposed that there may be a finite extent to which progression can occur related to the focal vulnerability of a myopic optic nerve in the direction of optic disc tilt [30]. The authors also suggested that a new visual field defect in the opposite hemifield correlating to a previously healthy area of neuroretinal rim might be more suggestive of true glaucoma.

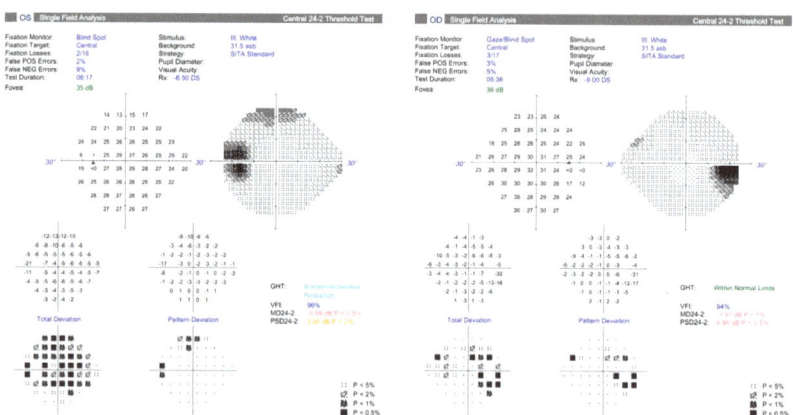

Figure 3. Visual fields from a patient with high myopia demonstrating enlarged blind spots.

Interestingly, Lin and colleagues developed a classification system for visual field defects observed in highly myopic eyes without maculopathy. The authors used a database of 1893 visual fields from Chinese patients (mean age ± SD, 30.94 ± 9.75 years) enrolled in a prospective longitudinal registry study of high myopia, defined as a spherical equivalent ≥ -6.00 diopters or axial length ≥ 26.5 mm. Visual fields were classified as being normal or as having high myopia-related abnormalities (i.e., enlarged blind spot, vertical step, partial peripheral rim, or nonspecific), glaucoma-related abnormalities (i.e., nasal step, partial or full arcuate, or paracentral), or combined defects (i.e., nasal step plus enlarged blind spot). Two trained graders twice analyzed a common set of 1000 visual fields representing all four categories. The intraobserver and interobserver agreements were found to be over 89%. The authors observed that 10.8% of highly myopic eyes showed glaucoma-like visual field defects, and the prevalence of such defects was associated with increased axial length [31]. Additional prospective studies are needed to better understand how morphologic features of myopic optic discs, including tilt, torsion, and peripapillary atrophy, predispose certain myopic eyes to the development and progression of visual field defects.

4. Surgical Considerations in Myopic Eyes

Numerous surgical options exist to treat glaucoma in myopic eyes, and each has its advantages and disadvantages (Table 1). An inevitable trade-off between safety and efficacy exists with regard to glaucoma surgery and is particularly relevant in highly myopic eyes. Decreased scleral rigidity in high myopes increases the likelihood of hypotony-related complications following glaucoma filtration surgery [32–34]. Hypotony maculopathy has been reported to occur with a delayed onset of 14 years following trabeculectomy in the absence of a bleb leak [35]. Given these risks, selective laser trabeculoplasty may be attempted earlier in myopic and highly myopic patients, prior to incisional surgery. Caution should be exercised when performing laser trabeculoplasty in myopic eyes with pigment dispersion syndrome or pigmentary glaucoma, as severe IOP elevation may occur postoperatively due to greater absorption of laser energy by the heavily pigmented trabecular meshwork characteristic of these conditions. The use of lower energy settings and/or a limited extent of treatment (i.e., 90 to 180 degrees, rather than 360 degrees) often mitigate this risk. If laser trabeculoplasty fails to control the IOP adequately or is not feasible, consideration may be given to microinvasive glaucoma surgery (MIGS) as a first surgical intervention in highly myopic eyes with early to moderate glaucoma. These newer procedures generally offer an improved safety profile over traditional incisional glaucoma surgery [36]. Furthermore, myopic eyes typically have wide open angles that facilitate trabecular and Schlemm's canal-based procedures. However, subconjunctival MIGS, such as the XEN gel stent, may be associated with an increased incidence of hypotony and hypotony-related sequelae in highly myopic eyes, akin to those observed with trabeculectomy [37,38].

Table 1. Advantages and disadvantages of glaucoma surgical procedures in myopic eyes.

Glaucoma Surgical Procedure	Advantages	Disadvantages	Special Considerations in Myopic Eyes
Microinvasive glaucoma surgery (MIGS)	Improved safety profile	Angle-based MIGS offer lower efficacy and may not be appropriate for patients with advanced glaucoma	Perform preoperative gonioscopy to assess angle anatomy in candidates for angle-based MIGS
		Hypotony may occur with bleb-based MIGS	Use antimetabolites judiciously if performing bleb-based MIGS to avoid hypotony
Tube shunt surgery	Improved efficacy	Risk of hypotony and hypotony-related complications	Exercise caution while securing endplate to thin sclera
			Consider using valved or smaller surface area nonvalved implant
			Consider performing planned laser tube ligature release if using nonvalved implant
Trabeculectomy	Improved efficacy	Risk of hypotony and hypotony-related complications	Exercise caution during scleral flap dissection as sclera may be thin
			Use an increased number of, and tighter, scleral flap sutures to avoid hypotony
			Use antimetabolites judiciously to avoid hypotony
Trans-scleral cyclophotocoagulation	Nonincisional	Risk of hypotony and hypotony-related complications	Consider alternative to retrobulbar or peribulbar block to avoid risk of scleral perforation
			Consider scleral transillumination to confirm ciliary body location

Angle-based MIGS may not offer sufficient IOP lowering in eyes with severe or rapidly progressive glaucomatous damage, and trabeculectomy may be required in certain cases. Intraoperative modifications to the standard trabeculectomy technique in highly myopic eyes may include an increased number of, and tighter, scleral flap sutures and cautious use of antimetabolites to mitigate the risk of hypotony in the early postoperative period. Alternatively, the use of valved glaucoma drainage implants or smaller-surface-area nonvalved implants may reduce the risk of hypotony and its sequelae. Additionally, scleral flap dissection during trabeculectomy and endplate placement during glaucoma drainage implant surgery must be performed carefully in the presence of thin sclera to avoid scleral perforation. Consideration should be given to performing a planned laser tube ligature release for nonvalved glaucoma drainage implants rather than allowing spontaneous tube ligature as the latter strategy may be associated with a higher likelihood of hypotony, anterior chamber shallowing, and other hypotony-related complications. During the early postoperative course following traditional incisional glaucoma surgery, surgeons should maintain a lower threshold to perform anterior chamber injection of cohesive viscoelastic in highly myopic eyes with hypotony to minimize complications.

Trans-scleral cyclophotocoagulation is a reasonable alternative to incisional glaucoma surgery in highly myopic eyes. Cycloablation is generally associated with intraoperative pain, typically requiring the use of a peribulbar or retrobulbar block to achieve adequate anesthesia. However, peribulbar or retrobulbar blocks carry a risk of globe perforation in highly myopic eyes, and a sub-Tenon's block may be a safer alternative. Intraoperative transillumination of the globe may be advisable during cyclophotocoagulation to confirm the location of the ciliary body, which may vary in highly myopic eyes [39]. Cyclophotocoagulation is a nontitrable procedure that is associated with numerous risks, including macular edema and hypotony. Surgeons should be aware of any history of previous cyclophotocoagulation when performing subsequent glaucoma surgery in patients with high myopia. Late hypotony with maculopathy has been described following uneventful viscocanalostomy in a highly myopic eye with a history of prior cycloablation [40].

Special considerations also exist when planning and performing phacoemulsification cataract surgery in myopic eyes with and without glaucoma. Patients should be counseled about the increased risk of intraoperative and postoperative complications as well as postoperative refractive surprises. In a study of 115 eyes with axial length ≥ 27 mm, Kora and colleagues found that erroneous axial length measurement was the main contributor to postoperative refractive errors [41]. Axial length may be overestimated by ultrasound biometry in highly myopic eyes with posterior staphylomas, resulting in hyperopic surprises. In patients with adequate fixation and without dense lens opacities, optical biometry is more likely to provide accurate axial length measurements than ultrasound [42]. Newer intraocular lens (IOL) calculation formulas, such as the fourth-generation Barrett Universal II and modified Wang–Koch adjustment to the third-generation Holladay I, may further reduce the risk of refractive surprises. Although published studies offer conflicting results, the majority suggest more optimal refractive outcomes when using the Haigis or Barrett Universal II formulas for IOL calculations in eyes with axial length <30 mm and the Barrett Universal II formula in eyes with axial length >30 mm [43]. Novel IOL calculation formulas that use artificial intelligence, including Kane and Hill-RBG3.0, appear to offer even greater accuracy in eyes with high myopia [44]. A history of prior refractive surgery in myopic eyes necessitates adjustment to standard formulas for intraocular lens (IOL) selection, and intraoperative aberrometry may help provide more predictable refractive outcomes.

In general, high myopes are at increased risk for retinal detachment following cataract surgery and should undergo preoperative dilated examination of the retinal periphery to assess for lattice degeneration and/or breaks [45]. Intraoperatively, maintaining the anterior chamber via careful wound construction to avoid leaks and using viscoelastic and irrigation to avoid sudden decompression of the eye are essential to mitigate the risk of vitreous movement. Additionally, a lower bottle height and shorter clear corneal wound facilitate operating in a deeper anterior chamber. Use of a chopping technique for nuclear disassembly may be preferable over a divide-and-conquer technique, both to enable operating in a deep chamber and to reduce stress on the zonules, which are known to be weaker in highly myopic eyes [43]. Surgeons should be alert to sudden posterior iris bowing, anterior chamber deepening, and pupillary dilation, which are typical of lens–iris diaphragm retropulsion syndrome [46]. This phenomenon, common in highly myopic and/or vitrectomized eyes, results from reverse pupillary block and can be resolved by gently separating the iris from the anterior capsule using a second instrument to permit equilibration of pressure in the anterior and posterior chambers. Thorough irrigation and aspiration to remove viscoelastic and residual lens particles from a deeper ciliary sulcus space in a myopic eye is important to prevent IOP elevation in the immediate postoperative period. Anterior capsular polishing is advisable to lower the risk of anterior capsular contraction syndrome, which is more common in myopic eyes and can lead to IOL subluxation [47,48].

Laser keratorefractive surgery is commonly performed in myopic eyes, and patients should be counseled preoperatively regarding the association between myopia and increased risk of glaucoma. Obtaining preoperative baseline glaucoma testing, including perimetry, OCT RNFL, and disc photography, is important in refractive surgery candidates who are deemed glaucoma suspects based on optic disc morphology and/or family history. Flap construction during laser-assisted in situ keratomileusis (LASIK), whether using a femtosecond laser or microkeratome, is associated with acute, severe IOP elevation that can further compromise an optic nerve with existing glaucomatous damage [49]. The newer small incision lenticule extraction (SMILE) procedure may offer an improved safety profile over LASIK in myopic glaucoma suspects given its flapless technique and generally limited postoperative steroid course. While photorefractive keratectomy (PRK) also avoids creation of a flap and concomitant IOP spikes, prolonged postoperative treatment with topical steroids required after PRK may result in IOP elevation in susceptible eyes. Interface fluid syndrome (IFS) is a rare postoperative complication that may develop after LASIK, and less commonly SMILE, in which high IOP (often, but not always, resulting from steroid use) causes an accumulation of fluid within the interface between the flap and the stroma [50–52]. Patients typically present with decreased vision and corneal haze, with or without visible fluid layering within the interface. A high degree of vigilance is required to detect IFS as IOP readings may be artificially low, especially when measured using a Goldmann applanation tonometer. Management of IFS consists of discontinuing steroids and initiating IOP-lowering medications, and incisional glaucoma surgery may be required in refractory cases [51]. The presence of an interface cyst or loose flap after LASIK or SMILE can also result in falsely low IOP readings. In general, measuring the IOP using several different methods, such as dynamic contour tonometry or pneumotonometry, is recommended in eyes with prior keratorefractive surgery given its effects on corneal biomechanics.

5. Conclusions

As myopia prevalence increases worldwide, our understanding of its complex pathophysiologic relationship to glaucoma must improve to lessen the global burden of functional glaucomatous vision loss in myopic eyes. Currently available imaging modalities of the peripapillary RNFL and macula are limited in their glaucoma diagnostic utility in myopic eyes due to nonrepresentative normative databases and artifacts resulting from characteristic anatomic features of myopic eyes. Visual field defects that are typical in glaucoma may also be observed in myopic eyes without glaucoma, further complicating its diagnosis. Given these diagnostic challenges, longitudinal follow-up of myopic eyes using perimetry, OCT RNFL, macular imaging, and disc photography is imperative to identify conversion to, or progression of, glaucoma. Surgical options for myopic eyes with glaucoma have expanded to include newer angle-based MIGS procedures, which may mitigate some of the vision-threatening complications more commonly observed with traditional glaucoma surgery, albeit with less efficacy. Myopic eyes are at increased risk of specific intraoperative and postoperative complications associated with cataract surgery, requiring careful surgical planning and modifications to standard phacoemulsification techniques. Laser keratorefractive surgery is associated with a risk of intraoperative and postoperative IOP elevation as well as unique complications that require a high degree of vigilance to detect. Additional research and developments in ocular imaging and perimetry will enhance our ability to more accurately detect glaucoma in myopic eyes, while advances in surgical innovation will improve the safety profile of glaucoma operations and help preserve vision in myopic patients.

6. Future Directions

Myopia represents a significant public health burden with the potential for far-reaching clinical, economic, and societal impacts. Curbing the myopia epidemic will depend upon the identification of risk factors and the development and implementation of strategies to prevent myopia onset or detect its presence early and slow its progression. Various risk factors for myopia development and progression have been studied in recent years, including environmental contributors such as urbanization [53] and decreased outdoor time [54]. Targeted interventions include increasing outdoor time during childhood. While robust evidence exists supporting the favorable effect of outdoor time on delaying or preventing myopia onset, published studies offer mixed results regarding its impact on reducing myopia progression [55]. Orthokeratology appears to correct low-to-moderate myopia and slow its progression via corneal reshaping, but concerns exist regarding its associated risk of infectious keratitis [56]. Low-dose atropine offers improved tolerability and less rebound myopic progression following its discontinuation compared to higher doses, but at the expense of decreased treatment effect [57]. Additional studies of myopia prevention and control strategies in diverse patient populations and geographic regions with long-term follow-up are needed to identify safe and effective interventions that can be implemented on a global scale.

New strategies will also be necessary to improve early diagnosis of myopia as well as of glaucoma in myopic eyes, given the constraints of existing diagnostic techniques. Artificial intelligence, machine learning, and deep learning have the potential to revolutionize the future of myopia detection and prediction of disease progression. In addition, these strategies may allow for early identification of potentially vision-threatening sequelae of myopia, including glaucoma. Early work shows promise in using fundus photographs to train artificial intelligence platforms to detect glaucoma in populations with a high prevalence of myopia [58]. Using other existing imaging modalities, including OCT and visual fields, to train and validate artificial intelligence algorithms would transform our ability to diagnose and predict glaucomatous progression, which is especially important in complex clinical scenarios such as co-existing myopia and glaucoma. However, several limitations to the widespread implementation of artificial intelligence currently exist. These include reliance on a large number of high-quality images, the expense of imaging equipment, and medicolegal considerations. Moreover, validation of artificial intelligence algorithms would require data sharing among institutions, which may be limited by concerns regarding patient privacy and Health Insurance Portability and Accountability Act (HIPAA) compliance. Surmounting these and other as-yet-unforeseen challenges relating to artificial intelligence will require collaboration among clinicians, scientists, and regulatory agencies.

Author Contributions: Conceptualization, K.V. and S.S.; methodology, K.V. and S.S.; data curation, K.V.; writing—original draft preparation, K.V.; writing: K.V., review and editing: K.V. and S.S.; supervision, S.S. All authors have read and agreed to the published version of the manuscript.

Funding: This research received no external funding.

Institutional Review Board Statement: Not applicable.

Informed Consent Statement: Not applicable.

Data Availability Statement: The data used in this review article are derived from published studies and are publicly available.

Conflicts of Interest: The authors declare no conflict of interest.

References

1. Morgan, I.G.; Ohno-Matsui, K.; Saw, S.M. Myopia. *Lancet* **2012**, *379*, 1739–1748. [CrossRef] [PubMed]
2. Holden, B.A.; Fricke, T.R.; Wilson, D.A.; Jong, M.; Naidoo, K.S.; Sankaidurg, P.; Wong, T.Y.; Naduvilath, T.J.; Resnikoff, S. Global Prevalence of Myopia and High Myopia and Temporal Trends from 2000 through 2050. *Ophthalmology* **2016**, *123*, 1036–1042. [CrossRef] [PubMed]
3. Ha, A.; Kim, C.Y.; Shim, S.R.; Chang, I.B.; Kim, Y.K. Degree of Myopia and Glaucoma Risk: A Dose-Response Meta-analysis. *Am. J. Ophthalmol.* **2022**, *236*, 107–119. [CrossRef] [PubMed]
4. Mitchell, P.; Hourihan, F.; Sandbach, J.; Wang, J.J. The relationship between glaucoma and myopia: The Blue Mountains Eye Study. *Ophthalmology* **1999**, *106*, 2010–2015. [CrossRef] [PubMed]
5. Xu, L.; Wang, Y.; Wang, S.; Wang, Y.; Jonas, J.B. High myopia and glaucoma susceptibility the Beijing Eye Study. *Ophthalmology* **2007**, *114*, 216–220. [CrossRef]

6. Wang, Y.X.; Yang, H.; Wei, C.C.; Xu, L.; Wei, W.B.; Jonas, J.B. High myopia as risk factor for the 10-year incidence of open-angle glaucoma in the Beijing Eye Study. *Br. J. Ophthalmol.* **2022**, *107*, 935–940. [CrossRef]
7. Cahane, M.; Bartov, E. Axial length and scleral thickness effect on susceptibility to glaucomatous damage: A theoretical model implementing Laplace's law. *Ophthalmic Res.* **1992**, *24*, 280–284. [CrossRef]
8. Choquet, H.; Khawaja, A.P.; Jiang, C.; Yin, J.; Melles, R.B.; Glymour, M.M.; Hysi, P.G.; Jorgenson, E. Association Between Myopic Refractive Error and Primary Open-Angle Glaucoma: A 2-Sample Mendelian Randomization Study. *JAMA Ophthalmol.* **2022**, *140*, 864–871. [CrossRef]
9. Chong, R.S.; Li, H.; Cheong, A.J.Y.; Fan, Q.; Koh, V.; Raghavan, L.; Nongpiur, M.E.; Cheng, C.Y. Mendelian Randomization Implicates Bidirectional Association between Myopia and Primary Open-Angle Glaucoma or Intraocular Pressure. *Ophthalmology* **2023**, *130*, 394–403. [CrossRef]
10. Tham, Y.C.; Aung, T.; Fan, Q.; Saw, S.M.; Siantar, R.G.; Wong, T.Y.; Cheng, C.Y. Joint Effects of Intraocular Pressure and Myopia on Risk of Primary Open-Angle Glaucoma: The Singapore Epidemiology of Eye Diseases Study. *Sci. Rep.* **2016**, *6*, 19320. [CrossRef]
11. Iglesias, A.I.; Ong, J.S.; Khawaja, A.P.; Gharahkhani, P.; Tedja, M.S.; Verhoeven, V.J.M.; Bonnemaijer, P.W.M.; Wolfs, R.C.W.; Young, T.L.; Jansonius, N.M.; et al. International Glaucoma Genetics Consortium (IGGC) and Consortium for Refractive Error and Myopia (CREAM). Determining Possible Shared Genetic Architecture between Myopia and Primary Open-Angle Glaucoma. *Investig. Ophthalmol. Vis. Sci.* **2019**, *60*, 3142–3149. [CrossRef] [PubMed]
12. Shin, J.; Lee, J.W.; Kim, E.A.; Caprioli, J. The effect of corneal biomechanical properties on rebound tonometer in patients with normal-tension glaucoma. *Am. J. Ophthalmol.* **2015**, *159*, 144–154. [CrossRef] [PubMed]
13. Wu, W.; Dou, R.; Wang, Y. Comparison of Corneal Biomechanics between Low and High Myopic Eyes-A Meta-analysis. *Am. J. Ophthalmol.* **2019**, *207*, 419–425. [CrossRef]
14. Leung, C.K.; Mohamed, S.; Leung, K.S.; Cheung, C.Y.L.; Chan, S.L.W.; Cheng, D.K.Y.; Lee, A.K.C.; Leung, G.Y.O.; Rao, S.K.; Lam, D.S.C. Retinal nerve fiber layer measurements in myopia: An optical coherence tomography study. *Investig. Ophthalmol. Vis. Sci.* **2006**, *47*, 5171–5176. [CrossRef] [PubMed]
15. Kang, S.H.; Hong, S.W.; Im, S.K.; Lee, S.H.; Ahn, M.D. Effect of myopia on the thickness of the retinal nerve fiber layer measured by Cirrus HD optical coherence tomography. *Investig. Ophthalmol. Vis. Sci.* **2010**, *51*, 4075–4083. [CrossRef]
16. Leung, C.K.; Yu, M.; Weinreb, R.N.; Mak, H.K.; Lai, G.; Ye, C.; Lam, D.S.C. Retinal nerve fiber layer imaging with spectral-domain optical coherence tomography: Interpreting the RNFL maps in healthy myopic eyes. *Investig. Ophthalmol. Vis. Sci.* **2012**, *53*, 7194–7200. [CrossRef]
17. Chong, G.T.; Lee, R.K. Glaucoma versus red disease: Imaging and glaucoma diagnosis. *Curr. Opin. Ophthalmol.* **2012**, *23*, 79–88. [CrossRef]
18. Zemborain, Z.Z.; Jarukasetphon, R.; Tsamis, E.; De Moraes, C.G.; Ritch, R.; Hood, D.C. Optical Coherence Tomography Can Be Used to Assess Glaucomatous Optic Nerve Damage in Most Eyes With High Myopia. *J. Glaucoma* **2020**, *29*, 833–845. [CrossRef] [PubMed]
19. Zhang, Y.; Wen, W.; Sun, X. Comparison of Several Parameters in Two Optical Coherence Tomography Systems for Detecting Glaucomatous Defects in High Myopia. *Investig. Ophthalmol. Vis. Sci.* **2016**, *57*, 4910–4915. [CrossRef]
20. Seol, B.R.; Jeoung, J.W.; Park, K.H. Glaucoma Detection Ability of Macular Ganglion Cell-Inner Plexiform Layer Thickness in Myopic Preperimetric Glaucoma. *Investig. Ophthalmol. Vis. Sci.* **2015**, *56*, 8306–8313. [CrossRef]
21. Kim, N.R.; Lee, E.S.; Seong, G.J.; Kang, S.Y.; Kim, J.H.; Hong, S.; Kim, C.Y. Comparing the ganglion cell complex and retinal nerve fibre layer measurements by Fourier domain OCT to detect glaucoma in high myopia. *Br. J. Ophthalmol.* **2011**, *95*, 1115–1121. [CrossRef]
22. Akashi, A.; Kanamori, A.; Nakamura, M.; Fujihara, M.; Yamada, Y.; Negi, A. The ability of macular parameters and circumpapillary retinal nerve fiber layer by three SD-OCT instruments to diagnose highly myopic glaucoma. *Investig. Ophthalmol. Vis. Sci.* **2013**, *54*, 6025–6032. [CrossRef] [PubMed]
23. Hwang, Y.H.; Kim, M.K.; Kim, D.W. Segmentation Errors in Macular Ganglion Cell Analysis as Determined by Optical Coherence Tomography. *Ophthalmology* **2016**, *123*, 950–958. [CrossRef]
24. Biswas, S.; Lin, C.; Leung, C.K. Evaluation of a Myopic Normative Database for Analysis of Retinal Nerve Fiber Layer Thickness. *JAMA Ophthalmol.* **2016**, *134*, 1032–1039, Erratum in *JAMA Ophthalmol.* **2016**, *134*, 1336. [CrossRef] [PubMed]
25. Baek, S.U.; Kim, K.E.; Kim, Y.K.; Park, K.H.; Jeoung, J.W. Development of Topographic Scoring System for Identifying Glaucoma in Myopic Eyes: A Spectral-Domain OCT Study. *Ophthalmology* **2018**, *125*, 1710–1719. [CrossRef]
26. Kim, Y.W.; Lee, J.; Kim, J.S.; Park, K.H. Diagnostic Accuracy of Wide-Field Map from Swept-Source Optical Coherence Tomography for Primary Open-Angle Glaucoma in Myopic Eyes. *Am. J. Ophthalmol.* **2020**, *218*, 182–191. [CrossRef]
27. Vuori, M.L.; Mäntyjärvi, M. Tilted disc syndrome may mimic false visual field deterioration. *Acta Ophthalmol.* **2008**, *86*, 622–625. [CrossRef] [PubMed]
28. Kumar, R.S.; Baskaran, M.; Singh, K.; Aung, T. Clinical characterization of young chinese myopes with optic nerve and visual field changes resembling glaucoma. *J. Glaucoma* **2012**, *21*, 281–286. [CrossRef]
29. Doshi, A.; Kreidl, K.O.; Lombardi, L.; Sakamoto, D.K.; Singh, K. Nonprogressive glaucomatous cupping and visual field abnormalities in young Chinese males. *Ophthalmology* **2007**, *114*, 472–479. [CrossRef]
30. Han, J.C.; Lee, E.J.; Kim, S.H.; Kee, C. Visual Field Progression Pattern Associated with Optic Disc Tilt Morphology in Myopic Open-Angle Glaucoma. *Am. J. Ophthalmol.* **2016**, *169*, 33–45. [CrossRef]

31. Lin, F.; Chen, S.; Song, Y.; Li, F.; Wang, W.; Zhao, Z.; Gao, X.; Wang, P.; Jin, L.; Liu, Y.; et al. Glaucoma Suspects with High Myopia Study Group. Classification of Visual Field Abnormalities in Highly Myopic Eyes without Pathologic Change. *Ophthalmology* **2022**, *129*, 803–812. [CrossRef] [PubMed]
32. Fannin, L.A.; Schiffman, J.C.; Budenz, D.L. Risk factors for hypotony maculopathy. *Ophthalmology* **2003**, *110*, 1185–1191. [CrossRef] [PubMed]
33. Stamper, R.L.; McMenemy, M.G.; Lieberman, M.F. Hypotonous maculopathy after trabeculectomy with subconjunctival 5-fluorouracil. *Am. J. Ophthalmol.* **1992**, *114*, 544–553. [CrossRef] [PubMed]
34. Costa, V.P.; Arcieri, E.S. Hypotony maculopathy. *Acta Ophthalmol. Scand.* **2007**, *85*, 586–597. [CrossRef] [PubMed]
35. Kao, S.T.; Lee, S.H.; Chen, Y.C. Late-onset Hypotony Maculopathy after Trabeculectomy in a Highly Myopic Patient with Juvenile Open-angle Glaucoma. *J. Glaucoma* **2017**, *26*, e137–e141. [CrossRef]
36. Saheb, H.; Ahmed, I.I. Micro-invasive glaucoma surgery: Current perspectives and future directions. *Curr. Opin. Ophthalmol.* **2012**, *23*, 96–104. [CrossRef]
37. Huth, A.; Viestenz, A. High myopia in vitrectomized eyes: Contraindication for minimally invasive glaucoma surgery implant? *Ophthalmologe* **2020**, *117*, 461–466. [CrossRef]
38. Fea, A.; Sacchi, M.; Franco, F.; Laffi, G.L.; Oddone, F.; Costa, G.; Serino, F.; Giansanti, F. Effectiveness and Safety of XEN45 in Eyes with High Myopia and Open Angle Glaucoma. *J. Glaucoma* **2023**, *32*, 178–185. [CrossRef] [PubMed]
39. Agrawal, P.; Martin, K.R. Ciliary body position variability in glaucoma patients assessed by scleral transillumination. *Eye* **2008**, *22*, 1499–1503. [CrossRef]
40. Gavrilova, B.; Roters, S.; Engels, B.F.; Konen, W.; Krieglstein, G.K. Late hypotony as a complication of viscocanalostomy: A case report. *J. Glaucoma* **2004**, *13*, 263–267. [CrossRef]
41. Kora, Y.; Koike, M.; Suzuki, Y.; Inatomi, M.; Fukado, Y.; Ozawa, T. Errors in IOL power calculations for axial high myopia. *Ophthalmic Surg.* **1991**, *22*, 78–81. [CrossRef] [PubMed]
42. Tehrani, M.; Krummenauer, F.; Blom, E.; Dick, H.B. Evaluation of the practicality of optical biometry and applanation ultrasound in 253 eyes. *J. Cataract. Refract. Surg.* **2003**, *29*, 741–746. [CrossRef] [PubMed]
43. Elhusseiny, A.M.; Salim, S. Cataract surgery in myopic eyes. *Curr. Opin. Ophthalmol.* **2023**, *34*, 64–70. [CrossRef] [PubMed]
44. Omoto, M.; Sugawara, K.; Torii, H.; Yotsukura, E.; Masui, S.; Shigeno, Y.; Nishi, Y.; Negishi, K. Investigating the Prediction Accuracy of Recently Updated Intraocular Lens Power Formulas with Artificial Intelligence for High Myopia. *J. Clin. Med.* **2022**, *11*, 4848. [CrossRef] [PubMed]
45. Ripandelli, G.; Scassa, C.; Parisi, V.; Gazzaniga, D.; D'Amico, D.J.; Stirpe, M. Cataract surgery as a risk factor for retinal detachment in very highly myopic eyes. *Ophthalmology* **2003**, *110*, 2355–2361. [CrossRef]
46. Cionni, R.J.; Barros, M.G.; Osher, R.H. Management of lens-iris diaphragm retropulsion syndrome during phacoemulsification. *J. Cataract. Refract. Surg.* **2004**, *30*, 953–956. [CrossRef]
47. Wang, D.; Yu, X.; Li, Z.; Ding, X.; Lian, H.; Mao, J.; Zhao, Y.; Zhao, Y.E. The Effect of Anterior Capsule Polishing on Capsular Contraction and Lens Stability in Cataract Patients with High Myopia. *J. Ophthalmol.* **2018**, *2018*, 8676451. [CrossRef]
48. Wang, W.; Xu, D.; Liu, X.; Xu, W. Case series: "Double arch" changes caused by capsule contraction syndrome after cataract surgery in highly myopic eyes. *BMC Ophthalmol.* **2021**, *21*, 367. [CrossRef]
49. Vetter, J.M.; Holzer, M.P.; Teping, C.; Weingartner, W.E.; Gericke, A.; Stoffelns, B.; Pfeiffer, N.; Sekundo, W. Intraocular pressure during corneal flap preparation: Comparison among four femtosecond lasers in porcine eyes. *J. Refract. Surg.* **2011**, *27*, 427–433. [CrossRef]
50. Bamashmus, M.A.; Saleh, M.F. Post-LASIK interface fluid syndrome caused by steroid drops. *Saudi J. Ophthalmol.* **2013**, *27*, 125–128. [CrossRef]
51. Shoji, N.; Ishida, A.; Haruki, T.; Matsumura, K.; Kasahara, M.; Shimizu, K. Interface Fluid Syndrome Induced by Uncontrolled Intraocular Pressure without Triggering Factors after LASIK in a Glaucoma Patient: A Case Report. *Medicine* **2015**, *94*, e1609. [CrossRef]
52. Kim, C.Y.; Jung, Y.H.; Lee, E.J.; Hyon, J.Y.; Park, K.H.; Kim, T.W. Delayed-onset interface fluid syndrome after LASIK following phacotrabeculectomy. *BMC Ophthalmol.* **2019**, *19*, 74. [CrossRef]
53. Grzybowski, A.; Kanclerz, P.; Tsubota, K.; Lanca, C.; Saw, S.M. A review on the epidemiology of myopia in school children worldwide. *BMC Ophthalmol.* **2020**, *20*, 27. [CrossRef]
54. Rudnicka, A.R.; Kapetanakis, V.V.; Wathern, A.K.; Logan, N.S.; Gilmartin, B.; Whincup, P.H.; Cook, D.G.; Owen, C.G. Global variations and time trends in the prevalence of childhood myopia, a systematic review and quantitative meta-analysis: Implications for aetiology and early prevention. *Br. J. Ophthalmol.* **2016**, *100*, 882–890. [CrossRef]
55. Modjtahedi, B.S.; Abbott, R.L.; Fong, D.S.; Lum, F.; Tan, D. Task Force on Myopia. Reducing the Global Burden of Myopia by Delaying the Onset of Myopia and Reducing Myopic Progression in Children: The Academy's Task Force on Myopia. *Ophthalmology* **2021**, *128*, 816–826. [CrossRef] [PubMed]
56. VanderVeen, D.K.; Kraker, R.T.; Pineles, S.L.; Hutchinson, A.K.; Wilson, L.B.; Galvin, J.A.; Lambert, S.R. Use of Orthokeratology for the Prevention of Myopic Progression in Children: A Report by the American Academy of Ophthalmology. *Ophthalmology* **2019**, *126*, 623–636. [CrossRef]

57. Pineles, S.L.; Kraker, R.T.; VanderVeen, D.K.; Hutchinson, A.K.; Galvin, J.A.; Wilson, L.B.; Lambert, S.R. Atropine for the Prevention of Myopia Progression in Children: A Report by the American Academy of Ophthalmology. *Ophthalmology* **2017**, *124*, 1857–1866. [CrossRef] [PubMed]
58. Lim, W.S.; Ho, H.Y.; Ho, H.C.; Chen, Y.W.; Lee, C.K.; Chen, P.J.; Lai, F.; Roger Jang, J.S.; Ko, M.L. Use of multimodal dataset in AI for detecting glaucoma based on fundus photographs assessed with OCT: Focus group study on high prevalence of myopia. *BMC Med. Imaging* **2022**, *22*, 206. [CrossRef] [PubMed]

Disclaimer/Publisher's Note: The statements, opinions and data contained in all publications are solely those of the individual author(s) and contributor(s) and not of MDPI and/or the editor(s). MDPI and/or the editor(s) disclaim responsibility for any injury to people or property resulting from any ideas, methods, instructions or products referred to in the content.

Review

Corneal Biomechanical Measures for Glaucoma: A Clinical Approach

Abdelrahman M. Elhusseiny [1,2], Giuliano Scarcelli [3,4] and Osamah J. Saeedi [3,4,*]

1. Department of Ophthalmology, Harvey and Bernice Jones Eye Institute, University of Arkansas for Medical Sciences, Little Rock, AR 72205, USA; amelhusseiny@uams.edu
2. Department of Ophthalmology, Boston Children's Hospital, Harvard Medical School, Boston, MA 02114, USA
3. Fischell Department of Bioengineering, University of Maryland, College Park, MD 20742, USA; scarc@umd.edu
4. Department of Ophthalmology and Visual Sciences, University of Maryland School of Medicine, Baltimore, MD 21201, USA
* Correspondence: osaeedi@som.umaryland.edu

Abstract: Over the last two decades, there has been growing interest in assessing corneal biomechanics in different diseases, such as keratoconus, glaucoma, and corneal disorders. Given the interaction and structural continuity between the cornea and sclera, evaluating corneal biomechanics may give us further insights into the pathogenesis, diagnosis, progression, and management of glaucoma. Therefore, some authorities have recommended baseline evaluations of corneal biomechanics in all glaucoma and glaucoma suspects patients. Currently, two devices (Ocular Response Analyzer and Corneal Visualization Schiempflug Technology) are commercially available for evaluating corneal biomechanics; however, each device reports different parameters, and there is a weak to moderate agreement between the reported parameters. Studies are further limited by the inclusion of glaucoma subjects taking topical prostaglandin analogues, which may alter corneal biomechanics and contribute to contradicting results, lack of proper stratification of patients, and misinterpretation of the results based on factors that are confounded by intraocular pressure changes. This review aims to summarize the recent evidence on corneal biomechanics in glaucoma patients and insights for future studies to address the current limitations of the literature studying corneal biomechanics.

Keywords: glaucoma; corneal biomechanics; ocular biomechanics; corneal hysteresis; ocular response analyzer; Corvis ST; Brillouin

1. Introduction

Glaucoma is a heterogeneous group of disorders characterized by progressive optic neuropathy with subsequent visual field defects if left untreated. It is a leading cause of irreversible vision loss worldwide, with an estimated global prevalence of 2.4–3.5% in people above the age of 40 and a higher prevalence in African origin populations [1,2]. Although intraocular pressure (IOP) reduction is the only proven modifiable risk factor in reducing the glaucoma progression [3], 30–57.1% of primary open-angle glaucoma (POAG) patients have normal IOP (normal tension glaucoma, NTG) [4–7]. Furthermore, ocular hypertension (OHT) patients may never develop glaucomatous damage [8]. Other risk factors such as thinner corneas, high myopia, genetic susceptibility, vascular factors, and family history have been postulated to contribute to glaucoma development and progression [8–10].

From a biomechanical perspective, the stress (applied force) of IOP does not cause glaucomatous damage itself, but rather the resulting strain (deformation) to the ocular tissues, specifically to the peripapillary sclera and the optic nerve head, where retinal ganglion cell (RGC) axons are most vulnerable; thus, determination of the modulus of elasticity (i.e., resistance to tissue deformation under applied stress) is critical to understanding the impact of IOP on ocular tissues and ultimately patient-specific susceptibility to the development and progression of the disease [11–13]. While the direct determination of the peripapillary scleral modulus is impractical clinically, given the overlying conjunctiva and posterior location, the corneal modulus is routinely measured in clinical settings to aid in the measurement of IOP and, more recently, to assist in glaucoma diagnosis and management [11–13]. Furthermore, the corneal biomechanical properties indicate the corneal capacity to dissipate energy from routine changes in IOP that may result from eye movement, blinking, or head movement. Hence, altered or impaired corneal biomechanical properties could increase the exposure of the optic nerve to IOP fluctuation and ultimately result in greater susceptibility to glaucomatous damage [14,15].

Several devices have been developed for in vivo evaluation of corneal biomechanics. These devices include the Ocular Response Analyzer (ORA; Reichert, NY, USA), the Corneal Visualization Scheimpflug Technology (Corvis ST, Oculus, Wetzlar, Germany), and, more recently, Brillouin microscopy (BM). Each device provides distinct sets of parameters for assessing corneal biomechanics. The ORA primarily measures corneal hysteresis (CH) and corneal resistance factor (CRF). These values are calculated based on applanation pressure data obtained when an air jet is applied to the cornea [16]. The CH has shown promise as a predictive factor for glaucoma development in individuals at risk of the disease: "glaucoma suspects". Moreover, CH has been associated with the rates of glaucoma progression and visual field loss in diagnosed glaucomatous patients [16].

In contrast, Corvis ST offers a more detailed evaluation of corneal biomechanical properties. It achieves this by directly visualizing corneal deformation and geometrical changes caused by the air jet. The latest Corvis ST software (1.3r1538) provides information on approximately 37 parameters [17]. An emerging method, BM, offers a non-contact three-dimensional evaluation of corneal biomechanics. It relies on the light scattering and is independent of the IOP. However, its application in evaluating corneal biomechanics in glaucoma patients remains unexplored.

The purpose of this review is to provide an overview of the current modalities assessing corneal biomechanics, the evidence of their relevance to glaucoma, and challenges in assessing corneal biomechanics in glaucoma patients that may help direct future research in this field.

2. Foundational Concepts in Corneal Biomechanics

A foundation of biomechanical principles and definitions is essential for a better understanding of the parameters reported by each device. Stiffness is the measure of the resistance of a specific material to being deformed when a certain force is applied to it [18]. As more than 90% of the cornea is composed of stacked collagen fibrils lamellae within the corneal stroma, the corneal stiffness is affected by collagen fibers' thickness and density. Factors such as age, diabetes, cross-linking, and glaucoma affect collagen density and alter corneal stiffness [19]. Furthermore, corneal biomechanics is most frequently described in the context of the stress–strain relationship. Stress is the force applied to a specific area (stress = force/cross-sectional area) and describes the inner resistance of the material when deformed. Strain is the deformation resulting from the applied force (stress) (strain = elongation (difference in length upon deformation/original length). In the eye, the IOP exerts pressure on the inner structures, including the cornea and lamina cribrosa, creating stresses throughout the corneal thickness and in multiple directions [18].

In the case of linearly elastic materials, the relationship between stress and strain is linearly proportional. The stress–strain slope defines the elastic modulus (Young's modulus), so the higher the slope, the more force is needed to deform a stiffer material [20]. However, the elastic modulus of the cornea is non-linear, and the J-shaped stress–strain curve is different from that of linear elastic material [21].

Furthermore, IOP is a major confounding factor where the higher the IOP, the stiffer the cornea. For example, an originally softer cornea will exhibit a stiffer behavior in the presence of a high IOP than an originally stiffer cornea at a lower IOP. This highlights the importance of accounting for the role of IOP in the determination of corneal biomechanics [21,22].

The cornea also is characterized by being anisotropic and viscoelastic. Anisotropy is described when a substance has mechanical properties depending on the direction of the applied force. In other words, the biomechanical measures tested differ along different corneal meridians [18]. Material elasticity is described as the ability of the substance to return to its original form in a way similar to the deformation when the force was applied. However, viscosity means that part of the applied force is lost to internal friction as heat [18]. The cornea is a viscoelastic material, and part of the applied stress is lost. Therefore, the deformation during the loading phase differs from that during the unloading phase, and the difference is defined as mechanical CH. As a result of its viscoelastic properties, the corneal biomechanical response differs based on the rate of the applied stress, so the faster the rate of the applied force, the stiffer the corneal biomechanical response. In clinical settings, if the IOP is measured by air puff, the result will depend on the rate at which the air jet is applied [18].

3. In Vivo Clinical Assessment of Corneal Biomechanics

Evaluation of corneal biomechanics is challenging given the non-linearity of the corneal elastic modulus, anisotropy, and viscoelastic characteristics of the cornea. Several devices have been developed to evaluate corneal biomechanics; however, each has advantages and limitations.

3.1. Ocular Response Analyzer

In 2005, the ORA was introduced in clinical practice as the first device for evaluating corneal biomechanical behavior in vivo [23]. It is a non-contact tonometer that uses an air jet applied to the cornea's central 3–6 mm, causing corneal deformational changes. Those bi-directional changes are monitored by an advanced infrared electro-optical system that detects the corneal surface's infrared reflection during its deformation. Once the cornea is applanated (first applanation event) in response to the air jet, the piston releasing the air shuts down, allowing the cornea to return to its original form. The pressure at which the cornea applanates is defined as P1. However, due to the piston inertia, the air pressure continues to increase to its highest level (Pmax), causing further indentation of the cornea, which becomes slightly concave. As the piston produces the air shut-off, the air pressure subsequently decreases. Hence, the cornea returns to its original shape, gradually passing through a second applanation event before returning to its initial convex configuration. The air pressure at the second applanation event is defined as P2. Due to the corneal viscoelastic properties, the P2 (unloading pressure) value is always smaller than the P1 (loading pressure) value, and the difference between both is known as CH (measured in mmHg) (Figure 1).

The Goldmann-correlated IOP (IOPg) measured by the ORA was the average of P1 and P2 values. In contrast, the corneal-compensated IOP (IOPcc) was developed through empirical investigation to compensate for corneal factors in measuring the IOP, presumably producing more accurate IOP measurements, especially after refractive surgery [24]. Another parameter that the ORA reports is the CRF. The CRF is calculated based on this equation: CRF = a (P1 − bP2) + d, where a, b, and d are constants [20,25,26]. The CRF was developed to maximize correlation with the central corneal thickness (CCT). It should be noted that all four parameters reported by the ORA are calculated based on P1 and P2.

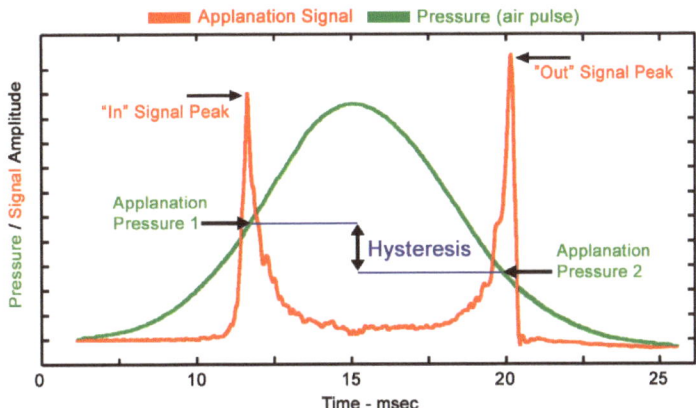

Figure 1. The applanation signal and air pulse pressure diagram was obtained over the course of 1 measurement. Applanation pressure 1 was the pressure at which the cornea reached a specific applanation state on inward movement, and applanation pressure 2 was the pressure at which the cornea passed through this applanation state on outward movement. The difference between these two pressures was the "corneal hysteresis" parameter, which was the main output by the machine. The image was obtained from http://www.reichert.com/ (accessed on 23 March 2023).

Previous studies have shown that CH is affected by age, CCT, diabetes status, keratoconus, and glaucoma [27–29]. An ex vivo study on rabbit eyes by Bao and colleagues found IOP to be highly correlated with CRF and weakly correlated with CH [30]. Another study by Touboul and colleagues concluded that CH was moderately dependent on IOP and CCT [27]. Although several studies have evaluated the corneal biomechanics using the ORA in different diseases, a lot of those studies did not account for the confounding nature of IOP and CCT on the reported ORA parameters, which makes it hard to interpret the results and draw a solid conclusion [20].

3.2. Corneal Visualization Scheimpflug Technology (Corvis ST)

In 2010, Oculus introduced Corvis ST as a non-contact method for in vivo assessment of corneal biomechanics. It is based on combining bidirectional dynamic applanation as in ORA and recording the deformational corneal changes through an ultra-high-speed Scheimpflug camera. A single concentric air jet is applied to the cornea, which is subsequently deformed, starting with an inward deformation phase, then the applanation phase, and then the highest concavity phase before returning to its original shape and passing through a second applanation phase. The ultra-high speed (4300 frames/second) Scheimpflug camera takes 140 images of the horizontal corneal meridian during the 32 milliseconds duration of the air jet. The images are further analyzed to report the dynamic corneal response parameters (DCR). Although both ORA and Corvis ST use air puffs, there is a difference between them. In the ORA, the air jet is variable depending on the P1, while the air pressure is constant in the Corvis ST [18,25,31,32].

The Corvis ST reports several parameters (Figure 2); however, it may also be confounded by the IOP [20,26,31].

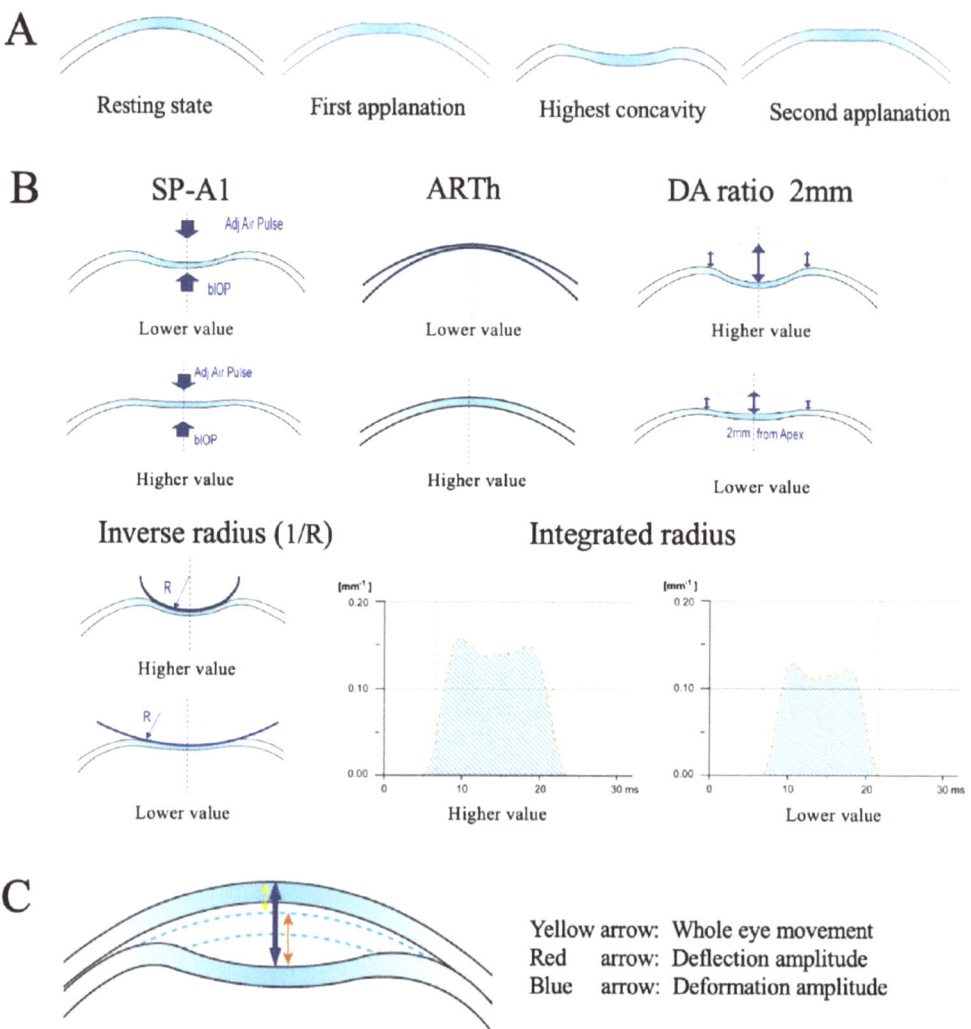

Figure 2. Schematic diagrams of the ocular biomechanical parameters provided by Corvis ST. (**A**) Cornea deformation during the Corvis ST measurement. From left to right: resting state before the measurement; first applanation; highest concavity; and second applanation. (**B**) Graphs illustrating SP-A1, ARTh, DA ratio 2 mm, inverse radius, and integrated radius. Lower values of SP-A1 and ARTh indicate a more deformable cornea, whereas higher values of DA ratio 2 mm, inverse radius, and integrated radius indicate a more deformable cornea. (**C**) Correlation between deformation amplitude and whole eye movement. Deformation amplitude is the sum of whole eye movement and pure corneal deformation named deflection amplitude. Yellow arrow: whole eye movement; Red arrow: deflection amplitude; and Blue arrow: deformation amplitude. Reprinted with permission from Wu N, Chen Y, and Sun X. Association Between Ocular Biomechanics Measured With Corvis ST and Glaucoma Severity in Patients With Untreated Primary Open Angle Glaucoma. Transl Vis Sci Technol. 2022;11(6):10. doi:10.1167/tvst.11.6.10 [33].

The applanation times (AT), lengths (AL), and velocities (AV) are recorded during the inward (first applanation, designated #1, e.g., A1T, A1L, A1V) and outward (second applanation, designated #2, e.g., A2T, A2L, A2V) phases of corneal applanation. The maximum corneal deformation during the air puff is known as maximum deformation (in millimeters). The curvature radius highest concavity (radius HC) is the corneal radius of curvature in millimeters at the highest corneal concavity, whereas the maximum inverse radius is 1/radius HC. It should be noted that the higher the radius HC, the more resistance to deformation, i.e., stiffer cornea, whereas the higher the inverse radius, the less resistance to deformation, i.e., softer cornea. The integrated inverse radius (IIR) is the integrated sum of the inverse concave radius between the first and second applanation [26,32]. During the air puff, a whole-eye movement (WEM) occurs. The parameters that compensate for WEM are known as "deflection" parameters, whereas those that do not compensate for WEM are "deformation" parameters. For example, the maximum deformation amplitude (DA Max) is the displacement of the corneal apex in the anterior–posterior plane.

In contrast, the maximum deflection amplitude (DeflAmpMax) is the DA Max minus WEM [26,32]. The DA ratios 1 or 2 mm are the DA Max divided by the DA at 1 or 2 mm away from the apex. The higher the DA ratio, the softer the cornea because the deformation occurs at the center but is limited at the periphery in a softer cornea. The peak distance (PD) is the distance between the two corneal peaks at the time of the highest corneal concavity. The Ambrosio Relational Thickness to the horizontal profile (ARTh) is the quotient corneal thickness at the thinnest point of the horizontal meridian, and the thickness changes [17,26,32,34]. The corneal biomechanical index (CBI) was developed using statistical methods to enhance keratoconus detection and screening [35]. Biomechanically corrected IOP (bIOP) compensates for the effects of CCT on IOP measurements. However, a recent study by Matsuura and colleagues demonstrated that bIOP was significantly associated with CH ($p < 0.001$). On the other hand, IOPcc measured by the ORA was not significantly associated with CH but was significantly associated with CCT [36].

Recently, more parameters have been developed for a more accurate evaluation of corneal stiffness, including the Stiffness Parameter at the first Applanation (SP-A1), Stiffness Parameter at Highest Concavity (SP-HC), and Stress–Strain Index (SSI). Those parameters have been proposed to be heavily affected by corneal stiffness rather than the IOP [31]. The SSI was developed in 2019 by Eliasy and colleagues for mapping the overall corneal stiffness [37]. Zhang and colleagues have reported that the SSI map values demonstrated small fluctuations with IOP and CCT [38]. Pillunat and colleagues developed a novel parameter called biomechanical glaucoma factor (BGF) based on several Corvis ST parameters, including DA ratio progression, HCT, Pachymetry slope, bIOP, and Pachymetry [39]. Although a study by Fujishiro and colleagues demonstrated a significant correlation between Corvis ST measurements (DA ratio, SP-A1, and inverse radius) and ORA-measured CH, the correlation was weak to moderate. The authors have suggested that the optimal model for calculating CH using Corvis ST parameters is: CH = $-76.3 + 4.6 \times A1T + 1.9 \times A2T + 3.1 \times DA + 0.016 \times CCT$ [17].

3.3. Brillouin Microscopy

The BM is a non-contact device that uses a different approach for evaluating corneal biomechanics based on light scattering (Brillouin scattering) arising from the interaction between light photons and acoustic phonons (thermodynamic fluctuations). Upon Brillouin interaction, the scattered light acquires a frequency shift related to the longitudinal elastic modulus of the sample without needing any mechanical perturbation [40]. Unlike previous methods, BM provides a non-contact, non-perturbative three-dimensional (3D) mapping of the corneal elastic modulus. Furthermore, its novel elasticity metrics enable distinguishing ectatic from normal corneas in vivo with previously unattainable mechanical sensitivity [41,42]. The measurement, as currently performed, is also independent of IOP [43] and thus may help solve the problem of the confounding effect of IOP on corneal biomechanical measures. This is particularly an issue in glaucoma, where glaucoma patients may have high IOP or lower treated IOP.

In vivo corneal biomechanics have never been evaluated in glaucoma patients using BM. However, several studies have used BM to evaluate corneal biomechanics in keratoconus, collagen cross-linking efficacy, and in vivo crystalline lens evaluation [44–47]. In an ex vivo study, Scarcelli and colleagues reported that the keratoconic corneas had a significantly lower Brillouin frequency shift in the cone area compared to normal corneas ($p < 0.001$) [44]. However, there were no significant differences in the mean Brillouin frequency in an area outside the cone compared to corresponding areas in normal healthy corneas. In the most recent in vivo studies by Zhang and colleagues, motion-tracking was introduced to enhance Brillouin measurement sensitivity [42]; they retrospectively compared the corneal biomechanics of early keratoconus patients to healthy controls. They reported a statistically significant reduction in the Brillouin frequency shift of keratoconic corneas compared to normal corneas ($p < 0.001$), demonstrating the great potential of mechanical metrics to identify the earliest stage of ectasia progression [41]. Further studies are needed to evaluate the utility of the BM in other disease conditions, including different types of glaucoma [46].

Other methods using ultrasound or optical coherence tomography (OCT) are being developed to evaluate corneal biomechanics, such as ultrasound elastography, OCT elastography, and electronic speckle pattern interferometry [25,48].

4. Clinical Studies Measuring Corneal Hysteresis and Corneal Resistance Factor in Glaucoma Patients

Several studies have evaluated the ORA parameters (CH and CRF) in different types of glaucoma and OHT (Table 1) [49–74].

Table 1. Studies evaluating corneal hysteresis in glaucoma/ocular hypertension (OHT) patients compared to healthy controls using the ocular response analyzer.

Study	Number of Patients		Prostaglandin Therapy in the Glaucoma Group	Parameters That Were Significantly Different between Study Groups
	Glaucoma/OHT	Healthy controls		
Kirwan and colleagues, 2006 [49]	8 (CG)	42	Not reported	- CG eyes had a statistically significantly lower mean CH (6.3 mmHg) compared to normal controls (12.5 mmHg)
Sullivan-Mee and colleagues, 2008 [74]	99 (primary glaucoma) 58 (OHT) 70 (GS)	71	- Glaucoma: 33% use PGA alone and 40% used PGA with adjunct - OH group: 31% used PGA alone and 7% used PGA with adjunct	- Glaucoma group had a significantly lower mean CH (8.1 mmHg) compared to OHT (8.9 mmHg), GS (8.9 mmHg), and normal (9.7 mmHg). - OHT group had a significantly higher CRF (10.2 mmHg) compared to glaucoma (8.3 mmHg), GS (8.5 mmHg), and normal (9.2 mmHg)
Mangouritsas and colleagues, 2009 [51]	108 (POAG)	74	- 42.6% were treated with one medication and 57.4% were treated with >1 medication. However, medication details were not reported	- POAG eyes had a statistically significantly lower mean CH (8.9 mmHg) compared to normal controls (10.9 mmHg)
Sun and colleagues, 2009 [52]	40 (unilateral CPACG)	40	- 8/40 used PGA	- CPACG eyes had a statistically significantly lower mean CH (6.8 mmHg) compared to the fellow eyes (10.5 mmHg) and normal controls (10.5 mmHg)

Table 1. Cont.

Study	Number of Patients		Prostaglandin Therapy in the Glaucoma Group	Parameters That Were Significantly Different between Study Groups
Abibtol and colleagues, 2010 [53]	58 (OAG-HTG)	75	- All patients were treated with glaucoma medication. However, no details were reported	- Glaucomatous eyes had a statistically significantly lower mean CH (8.7 mmHg) compared to normal controls (10.4 mmHg)
Ayala, 2011 [63]	30 (POAG) 30 (PXG)	30	- POAG and PXG patients were on glaucoma medications. However, no details were reported	- CH was significantly lower in PXG compared to normal subjects and POAG - No significant difference in CH between normal controls and POAG
Narayanaswamy and colleagues, 2011 [54]	162 (POAG-HTG and NTG) 131 (PACG)	150	- Patients using medications were not excluded. However, details were not reported	- After adjusting for age, sex, and IOP, CH was significantly lower in PACG (9.4 mmHg) compared to normal controls (10.1 mmHg) - No difference in CH between POAG and normal controls
Kaushik and colleagues, 2012 [55]	36 (POAG-HTG) 18 (POAG-NTG) 101 (GS) 38 (OHT) 59 (PACD)	71	- Patients on any topical ophthalmic treatment were excluded from the study	- CH was significantly lower in HTG (7.9 mmHg) and NTG (8.0 mmHg) compared to normal controls (9.5 mmHg) - CRF was lowest in NTG (7.8 mmHg) and highest in HTG (11.1 mmHg)
Grise-Dulac and colleagues, 2012 [56]	38 (POAG-HTG) 14 (NTG) 27 (OHT)	22	- Not reported	- NTG eyes had a statistically significantly lower mean CH (9.8 mmHg) compared to normal controls (11.05 mmHg) - NTG eyes had a statistically significantly lower mean CRF (9.5 mmHg) compared to normal controls (11.00 mmHg) and HTG (11.1 mmHg)
Derty-Morel and colleagues, 2012 [57]	59 (POAG)	55	- Not reported	- African healthy controls and POAG patients had a significantly lower CH compared to Caucasian normal and POAG patients
Morita and colleagues, 2012 [58]	83 (NTG)	83	- Not reported	- NTG eyes had a statistically significantly lower mean CH and CRF (9.2 mmHg and 8.9 mmHg, respectively) compared to normal controls (10.8 mmHg and 10.6 mmHg, respectively)

Table 1. Cont.

Study	Number of Patients		Prostaglandin Therapy in the Glaucoma Group	Parameters That Were Significantly Different between Study Groups
Cankaya and colleagues, 2012 [59]	64 (PEX) 78 (PXG)	102	- 12/78: PGA alone - 34/78: PGA and other medications	- PXG eyes had a statistically significantly lower mean CH (6.9 mmHg) compared to normal controls (9.4 mmHg) and PEX eyes (8.5 mmHg) - CRF was not significantly different in PXG eyes (9.5 mmHg) compared to the control group (9.8 mmHg) and PEX (9.3 mmHg)
Beyazyildiz and colleagues, 2014 [60]	66 (POAG) 46 (PXG)	50	- POAG: 54% - PXG: 63%	- PXG eyes had a statistically significantly lower mean CH (7.6 mmHg) compared to normal controls (9.6 mmHg) and POAG eyes (9.1 mmHg) - CRF was significantly lower in PXG eyes (9.0 mmHg) compared to the control group (9.8 mmHg) and POAG (10.1 mmHg)
Shin and colleagues, 2015 [61]	97 (POAG-NTG)	89	- 47/97	- NTG eyes had a statistically significantly lower mean CH and CRF (9.9 mmHg and 9.7 mmHg, respectively) compared to normal controls (10.5 mmHg and 10.5 mmHg, respectively)
Hussnain and colleagues, 2015 [62]	322 (POAG)	1418	- Not reported	- POAG eyes had a statistically significantly lower mean CH (9.5 mmHg) compared to normal controls (9.9 mmHg)
Yazgan and colleagues, 2015 [64]	43 eyes (PEX) 30 eyes (PXG)	45 eyes	- 17/30	- PXG eyes had a statistically significantly lower mean CH (6.8 mmHg) compared to normal controls (10.3 mmHg) and PEX eyes (8.2 mmHg) - CRF was significantly higher in the control group (10.3 mmHg) compared to PEX (7.9 mmHg) and PXG (7.9 mmHg)
Dana and colleagues, 2015 [65]	37 eyes (POAG)	21 eyes	- Not reported	- POAG eyes had a lower CH (9.8 mmHg) and CRF (10.3 mmHg) compared to control eyes (11.0 and 11.6, respectively)
Pillunat and colleagues, 2016 [66]	48 (POAG-HTG) 38 (POAG-NTG) 18 (OHT)	44	- Patients were on topical medications; however, details were not reported	- POAG eyes had a statistically significantly lower mean CH (8.9 mmHg) and CRF (9.07 mmHg) compared to OHT (CH: 10.2 mmHg, CRF: 10.7 mmHg) and normal controls (CH: 9.7 mmHg, CRF: 10.2 mmHg)

Table 1. Cont.

Study	Number of Patients		Prostaglandin Therapy in the Glaucoma Group	Parameters That Were Significantly Different between Study Groups
Perucho-Gonzalez and colleagues, 2016 [67]	78 (PCG)	53	- Not reported	- PCG eyes had a statistically significant lower CH (8.5 vs. 11.3 mmHg) and CRF (9.8 vs. 11.02 mmHg) compared to controls
Perucho-Gonzalez and colleagues, 2017 [68]	66 (PCG)	94	- Not reported	- PCG eyes had a statistically significant higher AME (9.0 vs. 3.2), PAE (3.1 vs. 0.9), and PME (30.8 vs. 7.5) compared to controls - PCG eyes had a statistically significant lower CH (8.5 vs. 11.1 mmHg) and CRF (9.9 vs. 10.7 mmHg) compared to controls
Park and colleagues, 2018 [69]	95 (POAG-NTG)	93	- Patients on glaucoma medications were excluded	- NTG eyes had significantly lower CH (10.5 mmHg) and CRF (10.1 mmHg) compared to normal controls (10.8 and 10.6 mmHg, respectively)
Potop and colleagues, 2020 [73]	79 eyes (POAG regardless of IOP) 68 eyes (OHT)	67 eyes	- Patients on glaucoma medications were not excluded	- POAG eyes had lower CH (8.5 mmHg) compared to OHT (9.6 mmHg) and normal controls (11.7 mmHg)
Aoki and colleagues, 2021 [70]	68 (POAG)	68	- Patients on glaucoma medications were not excluded	- CH was significantly lower in POAG (8.9 mmHg) compared to normal eyes (9.9 mmHg)
Rojananuangnit, 2021 [71]	272 (POAG) 143 (NTG) 48 (PACG) 30 (OHT)	465	- POAG: 414/434 eyes - NTG: 141/143 eyes - PACG: 46/74 eyes - OHT: 24/44 eyes	- CH in OHT (10.1 mmHg) was significantly higher than POAG (8.74) and PACG (9.09 mmHg) - No statistically significant difference in CH between OHT (10.1 mmHg) and NTG (9.5 mmHg) - The CH was significantly lower in the glaucoma groups compared to normal controls
Del Buey-Sayas and colleagues, 2021 [72]	491 (Glaucoma or GS)	574	- Not reported	- CH in glaucoma patients (9.6 mmHg) is lower than in the control group (10.7 mmHg) and all forms of GS

CG: congenital glaucoma, GS: glaucoma suspect, OAG: open-angle glaucoma, POAG: primary open-angle glaucoma, CPACG: chronic primary angle-closure glaucoma, HTG: high-tension glaucoma, NTG: normal-tension glaucoma, PCG: primary congenital glaucoma, COAG: chronic open-angle glaucoma, PEX: pseudoexfoliation syndrome, PXG: pseudoexfoliation glaucoma, PACD: primary angle-closure disease, PACG: primary angle-closure glaucoma, CH: corneal hysteresis, CRF: corneal resistance factor, AME: anterior maximum elevation, PAE: posterior apex elevation, and PME: posterior maximum elevation.

Most studies have reported that CH is lower in glaucoma/OHT patients compared to healthy controls. This may reflect corneal biomechanical differences in glaucoma but may be confounded by IOP. For example, in cases of increased IOP, the tension on the cornea increases, and its ability to dissipate energy decreases, resulting in smaller differences between the P1 and P2 and, accordingly, a lower CH [20].

Several factors must be considered in interpreting the results of these studies. A common misconception is to interpret CH/CRF values as parameters for corneal stiffness, although both are parameters for elasticity and viscosity rather than purely elasticity parameters. In other words, low CH by itself does not mean a soft or stiff cornea [15,20]. Second, some studies have evaluated CH/CRF in POAG without stratifying them into high-tension glaucoma (HTG) and NTG, which have different biomechanical profiles [50,51,54]. Age and diabetes status have further been reported to affect CH; therefore, any interpretation of the results should consider adjusting for those factors [75,76].

The CH and CRF have been reported to differ among different types and stages of glaucoma. In a study of 894 subjects, Rojananuangnit retrospectively compared CH in glaucoma patients (POAG-HTG, POAG-NTG, primary angle-closure glaucoma (PACG), and OHT) to normal controls. He reported that the mean CH was significantly lower in POAG-HTG compared to POAG-NTG and OHT. However, the difference was not statistically significant between POAG-HTG and PACG. In POAG-HTG and PACG, mean CH was significantly different between different stages of glaucoma, being lower in more severe stages of the disease. For example, the mean CH in the POAG-HTG severe stage was 7.92 mmHg compared to 9.22 mmHg in the early stage and 8.74 in the moderate stage ($p < 0.001$). In PACG, the mean CH was statistically significantly lower in the severe stage (8.45) compared to the early stage (9.85) ($p = 0.004$) but not when compared to the moderate stage (9.04, $p = 0.2$) [71]. In contrast, in a study of 49 patients, Yang and colleagues compared the CH in POAG-HTG versus POAG-NTG and reported no significant difference (10.11 mmHg versus 10.17 mmHg, respectively) ($p = 0.81$) [77]. A cross-sectional study of 162 subjects by Beyazyildiz and colleagues found that the mean CH was significantly lower in pseudoexfoliative glaucoma (PXG) (7.6 mmHg) compared to POAG patients (9.1 mmHg) and normal controls (9.6 mmHg) ($p < 0.001$). The CRF was also significantly lower in PXG patients (9.0 mmHg) compared to POAG patients (10.1 mmHg) and healthy controls (9.8 mmHg) [60].

Other studies have investigated the relationship between CH and glaucomatous structural changes. In a multicenter prospective study (EPIC-Norfolk Eye Study), Khawaja and colleagues evaluated the association between CH and Heidelberg retina tomograph (HRT) and Glaucoma Detection with Variable Corneal Compensation scanning laser polarimeter (GDxVCC). They reported that the CH was positively correlated with HRT rim area and GDxVCC-derived retinal nerve fiber layer (RNFL) thickness and modulation and negatively correlated with the HRT-derived linear cup-to-disc ratio [78]. Another prospective study by Wells and colleagues found that the CH was significantly correlated with the mean cup depth in glaucoma patients [79].

Further work has studied the potential association between CH and structural glaucoma progression. Wong and colleagues demonstrated that lower CH is significantly associated with anterior lamina cribrosa displacement, suggesting lower CH could be a risk factor for glaucoma progression [80]. Jammal and Medeiros measured the neuroretinal rim by the OCT of the Bruch's membrane opening minimum rim width (MRW) and correlated it with baseline CH in 118 glaucomatous eyes. They reported that lower baseline CH was associated with faster loss of neuroretinal rim and that for each one mmHg lower baseline CH, the MRW loss was faster by -0.38 µ/year [81]. Radcliffe and colleagues reported that eyes with optic disc hemorrhage—another potentially important sign of glaucoma progression—have significantly lower CH (8.7 mmHg) compared to those without disc hemorrhage (9.2 mmHg) ($p = 0.002$) [82].

Finally, lower CH has been associated with a higher chance of visual field progression. Medeiros and colleagues conducted a prospective longitudinal study, including 114 glaucomatous eyes, with a mean follow-up of 4 years to demonstrate the role of baseline CH on visual field progression. They reported that the visual field index declined at a 0.25% faster rate annually per one mmHg lower CH [83]. Another prospective study by Kamalipour and colleagues included 248 glaucomatous and glaucoma suspect eyes with a mean follow-up of 4.8 years. They reported that for each one mmHg lower baseline CH, there was a faster decline in the 10-2 visual field mean deviation (MD) (0.07 dB/year) and 1.35 increased odds of visual field progression. However, there was no statistically significant correlation between the CH and 24-2 visual field MD [84]. Chan and colleagues reported that for each one mmHg decline in the CRF over time, the visual field MD declined by a 0.14 dB faster rate annually ($p = 0.007$) [85]. A recent study reported that lower baseline CH was associated with more rapid rates of optic nerve microvasculature loss in POAG patients [86].

Lower CH has been proposed as a risk factor for the development of glaucoma in glaucoma suspects. In a prospective study by Susanna and colleagues following up 287 eyes identified as glaucoma suspects, 44 eyes developed visual field defects during the follow-up. They demonstrated that baseline CH was significantly lower in patients who developed glaucoma (9.5 mmHg) compared to those who did not develop glaucoma (10.2 mmHg) ($p = 0.01$). They further demonstrated a 21% increased glaucoma risk for each one mmHg lower baseline CH [87].

5. Clinical Studies Using Corneal Visualization Scheimpflug Technology (Corvis ST) in Evaluating Corneal Biomechanics in Glaucoma Patients

While several studies have reported promising results for corneal biomechanical biomarkers using Corvis ST in glaucoma patients, the results have shown differing and sometimes conflicting results. Several studies have reported that glaucoma patients have less deformable corneas than normal controls [88–90], whereas other studies reported the reverse—more deformable corneas in glaucoma patients compared to controls [91,92]. This discrepancy may be related to certain limitations in the development of the technology as well as in the study design. For example, some studies drew conclusions based on corneal stiffness parameters that may be more dependent and confounded by IOP. Other limitations in the current literature include the inclusion of patients on chronic prostaglandin analogue (PGA) therapy, which may alter corneal biomechanics. Topical PGA alters the expression of matrix and tissue metalloproteinases, causing structural changes that may affect corneal stiffness and biomechanical properties. Finally, glaucoma at high pressures (HTG) and glaucoma at normal pressures (NTG) may have different biomechanical properties, and some prior work has not stratified POAG patients in one classification or another [14,31]. A recent meta-analysis by Catania and colleagues included six prospective studies comparing the Corvis ST parameters between POAG-HTG versus normal controls. They concluded that POAG-HTG patients had stiffer corneas than normal controls based on significantly lower DA, PD, HCT, A1V, and A2T and significantly higher radius HC compared to healthy controls [14].

It should be noted that most of these factors were correlated strongly with IOP [31]. On the other hand, factors such as SP-A1, SP-HC, or SSI, which were less affected by IOP, were not included in their analysis [14]. Table 2 summarizes major studies evaluating corneal biomechanics in glaucoma/OHT patients using Corvis ST [33,39,70,88–107].

Table 2. Studies evaluating corneal biomechanics in glaucoma/ocular hypertension (OHT) patients compared to healthy control using Corvis ST.

Study	Number of Patients		Prostaglandin Therapy in the Glaucoma Group	Parameters Evaluated *	Parameters That Were Significantly Different between Study Groups	Conclusion
	Glaucoma/OHT	Healthy controls				
Leung and colleagues, 2013 [93]	101 glaucomatous eyes 39 glaucoma suspect eyes	40	PGAs were used, but the exact number of patients on PGA was not reported	5 parameters - A1L - A2L - A1V - A2V - DA	None of the five factors were statistically significantly different between both groups	
Salvetat and colleagues, 2015 [88]	85 (POAG)	79	33/87 patients	10 parameters - A1T - A1L - A1V - A2T - A2L - A2V - HCT - DA - PD - Radius HC	- A1T was higher in the POAG group ($p = 0.007$). - The A1V ($p = 0.04$), A2T ($p < 0.001$), A2V ($p = 0.014$), and DA HC ($p < 0.001$) were lower in the POAG group.	POAG eyes have less deformable corneas than controls
Wang and colleagues, 2015 [89]	37 (POAG-HTG)	36	Patients on glaucoma medications were not excluded from the study	10 parameters - A1T - A1L - A1V - A2T - A2L - A2V - HCT - DA - CCR - PD	- A1T, A2V, and PD were higher in the POAG ($p < 0.05$). - DA, A1V, and A2T were lower in the POAG ($p < 0.05$).	POAG eyes have less deformable cornea compared to controls
Coste and colleagues, 2015 [90]	37 (COAG)	19	Not reported	7 parameters - DA - A1T - A2T - HCT - A1L - A2L - Corneal velocity	- DA was significantly lower in the COAG group - HCT was significantly shorter in the COAG group	Corneal deformation is lower in glaucomatous patients compared to controls
Lee and colleagues, 2016 [94]	34 (POAG-HTG) 26 (POAG-NTG)	61	79.5% were on glaucoma medications	10 parameters - A1T - A1L - A1V - A2T - A2L - A2V - HCT - DA - PD - Radius HC	- A2V ($p = 0.001$) and PD ($p = 0.005$) were greater in the glaucoma group - HCT was shorter in the glaucoma group ($p = 0.002$)	

Table 2. Cont.

Study	Number of Patients	Prostaglandin Therapy in the Glaucoma Group	Parameters Evaluated *	Parameters That Were Significantly Different between Study Groups	Conclusion
Tian and colleagues, 2016 [95]	42 (POAG-HTG) 60	34/42 patients	10 parameters - A1T - A1L - A1V - A2T - A2L - A2V - HCT - DA - PD - Radius HC	- DA was significantly lower in the POAG ($p < 0.001$). - A1V, A2T, and PD were lower in the POAG group	- Corneal deformation is lower in glaucomatous patients compared to controls
Wu and colleagues, 2016 [108]	69 19	35/69 (treatment naïve) 34/69 (at least 2 years of PGA therapy)	10 parameters - A1T - A1L - A1V - A2T - A2L - A2V - DA - PD - Radius HC - HCT	- After adjusting for age, gender, IOP, CCT, axial length, and corneal curvature, DA was significantly lower in treatment-naïve POAG compared to POAG on PGA therapy and controls	- Treatment-naïve POAG eyes have less deformable corneas compared to patients on PGA therapy
Jung and colleagues, 2017 [96]	136 (OAG) 75	82/136 patients	9 parameters - A1L - A1V - A2L - A2V - DA - PD - Radius HC - WEM - DFA	- DA was smaller compared to controls ($p = 0.03$)	- Corneal deformation is lower in glaucomatous eyes compared to controls
Hong and colleagues, 2019 [91]	80 (POAG-NTG) 155	76% were on glaucoma medications but they did not specify the number	10 parameters - A1T - A1L - A1V - A2T - A2L - A2V - HCT - DA - PD - Radius HC	- A1V was significantly higher in the NTG group	- NTG has more deformable corneas compared to controls
Miki and colleagues, 2019 [92]	75 (POAG-medically controlled) 47	Mean number of topical medications was 1.8 ± 1.2. However, % of eyes that used prostaglandin analogues was not specified	8 parameters - A1T - A1V - A2T - A2V - HC DFA - PD - Radius HC - WEM	- Glaucoma was negatively correlated with A1T, A2T, radius HC, and WEM	- Eyes with medically controlled POAG are more deformable compared to normal controls

Table 2. Cont.

Study	Number of Patients	Prostaglandin Therapy in the Glaucoma Group	Parameters Evaluated *	Parameters That Were Significantly Different between Study Groups	Conclusion
Pillunat and colleagues, 2019 [39]	70 (POAG-NTG)	70	115/140 eyes	They used five parameters (DA ratio progression, HCT, Pachymetry slope, biomechanically corrected IOP, Pachymetry) to calculate Dresden BGF	- The BGF was statistically higher in the NTG (0.67) compared to normal controls (0.33) ($p < 0.001$). - DA ratio progression was higher and HCT was shorter in NTG compared to controls	- Using a cut-off of 0.5 BGF, NTG can be differentiated from normal controls and correctly classified in 76% of eyes - NTG eyes may have stiffer corneas with reduced viscoelastic capability
Vinciguerra and colleagues, 2020 [97]	41 (POAG-HTG) 33 (POAG-NTG) 45 (OHT)	37	37/41 patients (POAG-HTG) 23/33 patients (POAG-NTG) 31/45 patients (OHT)	4 parameters - SP-A1 - SP-HC - Inverse concave radius - DA ratio	- SP-A1 and SP-HC were significantly lower in NTG compared to the other three groups - Inverse concave radius and DA ratio were significantly higher in NTG compared to the other 3 groups	NTG eyes have a more deformable cornea compared to HTG, OHT, and controls
Miki and colleagues, 2020 [98]	35 (POAG-NTG)	35	0 (All patients were treatment-naïve)	10 parameters - A1T - A1V - A2T - A2V - HC DFA - PD - Radius HC - DA ratio 1 mm - Integrated radius - WEM Max	- A1T, A2T, and radius HC were significantly smaller in NTG compared to controls - PD, DA 1 mm, and integrated radius were significantly larger in NTG compared to controls	Corneas of untreated NTG eyes are more deformable compared to controls
Pradhan and colleagues, 2020 [100]	29 (POAG including NTG) 32 (PXG)	33	0 (All patients were treatment-naïve)	7 parameters - A1L - A1V - A2L - A2V - DA - PD - Radius HC	- After adjusting for IOP, there was no difference in any parameter between the three groups	No difference in corneal deformability between POAG, PXG, and controls

Table 2. Cont.

Study	Number of Patients	Prostaglandin Therapy in the Glaucoma Group	Parameters Evaluated *	Parameters That Were Significantly Different between Study Groups	Conclusion
Pradhan and colleagues, 2020 [101]	27 (PXG) 14 (PXF + OHT) 29 (PXF)	0 (All patients were treatment-naïve)	7 parameters - A1L - A1V - A2L - A2V - DA - PD - Radius HC	- DA and corneal velocities were significantly lower in PXG and PXF + OHT compared to PXF and normal controls - After adjusting for IOP and age, there was no difference in any parameter between the four groups	No difference in corneal deformability between PXG, PXF, PXF + OHT, and controls
Jung and colleagues, 2020 [99]	46 (POAG-HTG) 54 (POAG-NTG)	32/46 in HTG 38/54 in NTG	7 parameters - A1L - A1V - A2L - A2V - PD - DA - Radius HC	- A1V and DA were smaller in HTG compared to NTG and controls - Radius HC was larger in HTG compared to controls	Eyes with POAG-HTG have less deformable corneas compared to NTG and controls
Aoki and colleagues, 2021 [70]	68 (POAG)	56/68	BGF	- No statistical difference in BGF between POAG eyes (0.61) and normal controls (0.51)	- BGF is not useful in differentiating POAG eyes from normal controls
Wei and colleagues, 2021 [102]	45 (POAG-HTG) 49 (POAG-NTG)	Several glaucoma patients were on PGA, but they did not report a specific number	19 parameters - Max inverse concave radius - DAR 2mm - DAR 1mm - Integrated radius - SP-A1 - A1-DFL - HC-DFL - A2-DFL - A1-DFA - HC-DFA - A2-DFA - DFA Max - WEM - A1-DF Area - HC-DF Area - A2-DF Area - A1-dDFL - A2-dDFL - dDFL Max - HC-dDFL	- Maximum inverse concave radius and DAR (1 and 2 mm) were significantly higher in NTG eyes compared to controls - Integrated radius and DAR 2 mm were significantly higher in NTG compared to HTG - SPA-1 was significantly lower in NTG compared to HTG - No significant difference in any of the parameters between HTG and normal controls	- NTG eyes have more deformable corneas compared to HTG and controls - No difference in corneal deformability between HTG and controls

Table 2. Cont.

Study	Number of Patients	Prostaglandin Therapy in the Glaucoma Group	Parameters Evaluated *	Parameters That Were Significantly Different between Study Groups	Conclusion
Silva and colleagues, 2022 [103]	61 eyes (POAG) 32 eyes (Amyloidotic glaucoma) 37 eyes (OHT)	72% (POAG) 59% (Amyloidotic glaucoma) 59% (OHT) 53 eyes	14 parameters - A1T - A1V - A2T - A2V - A1-DFL - A2-DFL - PD - Radius HC - DA HC - HC-DFA - SSI - SP-A1 - DA ratio - Integrated radius	- Eyes with OHT had significantly higher SPA-1 compared to POAG, and SSI compared to amyloidotic glaucoma - Eyes with amyloidotic glaucoma had lower HC-DFA and higher integrated radius compared to controls	Eyes with OHT have less deformable corneas compared to POAG, Amyloidotic glaucoma, and controls
Zarei and colleagues, 2022 [104]	66 eyes (POAG-HTG) 21 eyes (POAG-NTG) 26 eyes (PXG) 46 eyes (PACG)	70 eyes	31 parameters - A1T - A1V - A2T - A2V - HCT - PD - Radius HC - A1-DA - HC-DA - A2-DA - A1-DFL - HC-DFL - A2-DFL - A1-DFA - HC-DFA - A2-DFA - DFA Max - WEM Max - A1-DF area - HC- DF area - A1-dArc length - HC-dArc length - A2-dArc length - dArc length Max - DA ratio Max - ARTh - Integrated radius - SP-A1 - SSI - CBI	- Radius indices were lower in HTG, NTG, and PXG compared to controls - Max inverse radius and integrated radius were higher in PACG compared to controls	Altered corneal biomechanics in different types of glaucoma

Table 2. *Cont.*

Study	Number of Patients	Prostaglandin Therapy in the Glaucoma Group	Parameters Evaluated *	Parameters That Were Significantly Different between Study Groups	Conclusion
Xu and colleagues, 2022 [107]	113 (POAG-HTG) 108 (POAG-NTG)	113 47/113 (POAG-HTG) 42/108 (POAG-NTG)	5 parameters ** - DA - DA ratio - Integrated radius - SPA-1 - SSI	- DA was higher in the NTG compared to HTG ($p = 0.03$) but not when compared to controls ($p = 0.93$) - No significant difference between three groups in DA ratio, integrated radius, SP-A1, and SSI measurements	Based on Corvis ST *** results, NTG eyes have more deformable corneas compared to HTG but not when compared to controls
Wu and colleagues, 2022 [33]	55 (POAG-HTG) 47 (POAG-NTG)	51	0 (All patients were treatment-naïve)	13 parameters - A1T - A1V - A2T - A2V - HCT - DA - PD - Radius HC - SP-A1 - Integrated radius - ARTh - DA ratio 2 mm - WEM	- DA was significantly higher and A1T and HC time were significantly lower in NTG, and HTG compared to normal controls - Comparing NTG and HTG, only A1V was significantly different being lower in HTG	NTG eyes have more deformable corneas compared to HTG and normal controls - HTG eyes have more deformable corneas compared to normal controls
Vieira and colleagues, 2022 [106]	70 (POAG-HTG) 16 (PXG) 23 (OHT)	37	All glaucoma patients and 92.9% of OHT were medically treated. However, details were not reported	8 parameters - A1L - A1V - A2L - A2V - PD - Radius HC - DA - CSI	- OHT has significantly higher A1L, A2V and lower A1V compared to POAG and PXG	Eyes with OHT have stiffer corneas compared to healthy controls, POAG, PXG

Table 2. Cont.

Study	Number of Patients	Prostaglandin Therapy in the Glaucoma Group	Parameters Evaluated *	Parameters That Were Significantly Different between Study Groups	Conclusion
Halkiadakis and colleagues, 2022 [105]	30 (POAG-HTG) 25 (OHT)	25	POAG and OHT were medically treated but details were not reported	15 parameters - A1T - A2T - A1L - A2L - A1V - A2V - HCT - Radius HC - HCC - DA - DA 2mm - SP-A1 - Inverse concave radius	- A2T was the only parameter that was statistically significantly different between groups being shorter in POAG ($p = 0.048$)	- Corneas of POAG may have altered viscoelasticity based on reduced A2T

POAG: primary open-angle glaucoma, HTG: high-tension glaucoma, NTG: normal-tension glaucoma, COAG: chronic open-angle glaucoma, OAG: open-angle glaucoma, PXF: pseudoexfoliation syndrome, PXG: pseudoexfoliation glaucoma, PACG: primary angle-closure glaucoma, PGA: prostaglandin analogues, A1T: first applanation time, A1L: length of the flattened cornea in A1, A1V: first applanation velocity, A2T: second applanation time, A2L: length of the flattened cornea in A2, A2V: second applanation velocity, DA: deformation amplitude, HCT: time from start until cornea reaching the highest concavity, PD: peak distance, HC: highest concavity, CCR: central curvature radius of the cornea at the highest concavity, WEM: whole eye movement, DFA: deflection amplitude, DF area: area displaced as a result of corneal deformation in the horizontal section analyzed, IOP: intraocular pressure, SP-A1: stiffness parameter at first applanation, SP-HC: stiffness parameter at highest concavity, HCC: central curvature radius of the cornea at the highest concavity, Max: maximum, mm: millimeter, dDFL: delta deflection arc length, ARTh: Ambrosio relational thickness to the horizontal profile, CBI: Corvis biomechanical index, SSI: stress–strain index, CSI: concavity shape index, and BGF: biomechanical glaucoma factor. *: Apart from intraocular pressure or central corneal thickness. **: There was no separate statistical analysis for HTG compared to NTG. ***: In addition to Corvis ST, a corneal indentation device was also used to evaluate corneal stiffness and showed that the corneal stiffness in NTG was lower than that of HTG ($p = 0.001$) and controls ($p = 0.023$).

Few studies compared the corneal biomechanics of OHT patients versus POAG patients using Corvis ST. Silva and colleagues demonstrated that OHT eyes had less deformable "stiffer" corneas based on significantly higher SP-A1 compared to POAG patients ($p = 0.04$), although subjects were not stratified into HTG and NTG [103]. On the other hand, a study by Vinciguerra and colleagues reported that NTG eyes had more deformable "softer" corneas compared to those with OHT and HTG based on significantly lower SP-A1, SP-HC, and higher DA ratio and inverse concave radius [97].

BGF is a summary metric developed by Pillunat and colleagues, composed of several Corvis ST parameters. They proposed that a cut-off value of 0.5 may help differentiate NTG eyes from normal healthy control, with an area under the curve (AUC) of 0.8 and a sensitivity of 76% [39]. However, a larger study by Aoki and colleagues reported that BGF was not a helpful parameter for differentiating POAG and normal controls with an AUC of 0.61 [70]. In their cohort, they reported a higher AUC (0.7) for ORA-measured CH [70]. However, they did not differentiate between HTG and NTG and included patients on topical glaucoma medications. BGF has also been combined with anterior chamber parameters in evaluating PACG patients in a retrospective study that showed that median BGF was significantly lower in the PACG patients (6.2) compared to controls (6.6) ($p < 0.001$). The anterior chamber volume and BGF combination had the highest AUC (0.93), potentially improving PACG detection [109].

Limited literature is currently available about the potential association of Corvis ST measurements and glaucoma severity with varying results. A study by Wu and colleagues investigated the relationship between corneal biomechanics as measured by Corvis ST and visual field changes. They reported that the shorter the WEM, the worse the MD ($p = 0.02$) and the higher the pattern's standard deviation (PSD) ($p = 0.03$) in NTG. However, these associations were not found in HTG [33]. Similarly, Bolivar and colleagues reported no significant association between Corvis ST parameters and visual field MD or PSD in POAG patients [110]. It should be noted that both studies were performed on treatment-naïve patients [33,110]. In contrast, Hirasawa and colleagues reported significant correlations between A1V and A2T and glaucomatous visual field defects [111]. Vinciguerra and colleagues reported that in POAG eyes (HTG and NTG), there was a significant negative correlation between DA ratio ($p = 0.01$) and inverse concave radius ($p = 0.02$) and MD and a significant positive correlation between SP-A1 ($p = 0.01$) and SP-HC ($p = 0.03$) and MD [97]. There was also a significant association between PSD and Corvis ST measurements. Notably, both Hirasawa and Vinciguerra included patients on glaucoma medications, including PGA, which may have confounded the results [97,111]. The contradictory results may be explained at least in part by the inclusion of treatment-naïve patients in the first two studies and treated patients in the last two. A prospective study by Qassim and colleagues investigated the correlation between SP-A1 and the risk of glaucoma progression in 228 glaucoma suspects. They demonstrated that the higher the SP-A1, the faster the rate of RNFL and RGC loss ($p < 0.001$). They reported that patients with higher SP-A1 and lower CCT had 2.9-folds increased risk of RNFL loss > 1 μm/year ($p = 0.006$) [112].

6. Prostaglandin Analogues-Induced Corneal Biomechanical Changes

There is growing evidence about the effect of PGA on ocular biomechanics related to the mechanism of increased uveoscleral outflow. Several studies have demonstrated that topical PGA therapy may result in increased matrix metalloproteinase expression and decreased expression of tissue inhibitors of metalloproteinases, which subsequently results in structural changes in the outer coat of the eye. These changes are hypothesized to reduce the ocular stiffness and increase permeability to aqueous outflow, reducing the IOP [31,113–115]. This same effect could result in corneal biomechanical changes that may confound the accuracy of serial IOP readings.

There is a controversy about the effect of topical PGA on CH and CRF. A study by Tsikripis and colleagues evaluated 108 POAG eyes on latanoprost with or without timolol. They reported that the mean CH significantly increased in both groups, whereas CRF did not change [116]. Another study by Meda and colleagues evaluated 70 eyes of 35 patients treated with long-term PGA. The PGA therapy was stopped for six weeks in one eye of each patient. They reported that cessation of PGA increased CH and CRF [117].

Using Corvis ST, Wu and colleagues compared the changes in corneal biomechanics of treatment-naïve POAG patients versus POAG under chronic PGA therapy (for at least two years) versus normal controls. Although they concluded that long-term PGA therapy induces deformational changes based on the significant increase in the DA, this may be related to the fact that DA inversely correlates with IOP. Hence, the measured increase in DA may have been because of decreased IOP with PGA therapy [98,108,118]. Similarly, Sanchez-Barahona and colleagues reported that three months of PGA therapy in treatment-naïve POAG patients resulted in significant changes in corneal biomechanics based on significant changes in A1T ($p = 0.001$), A2T ($p = 0.001$), and DA ($p = 0.0003$) [119]. However, all three parameters are strongly affected by IOP, and changes may also be explained by IOP reduction with PGA therapy. Therefore, future studies need to evaluate the PGA-induced biomechanical changes using more recent parameters that are more heavily affected by stiffness than IOP changes, such as SP-A1, SP-HC, SSI, and DA ratio [32,120].

On the other hand, Zheng and colleagues [121] studied the effect of travoprost on rabbit cornea biomechanics and showed that topical PGA resulted in softer corneas with decreased resistance to deformation. This raises the question of the actual IOP-lowering effect of the PGA therapy since, theoretically, part of it may be related to measurement artifacts with softer corneas that may underestimate IOP measurement [31]. However, further studies are needed to better assess the IOP-lowering effect of PGA independent of its effect on corneal biomechanics. Another ex vivo study investigated the effects of travoprost and tafluprost on rabbit corneas. The authors demonstrated a significant decrease in the tangent modulus by almost 30% and increased interfibril spacing after PGA therapy [122].

7. Effect of Other Topical Anti-Glaucoma Medications on Corneal Biomechanics

Limited literature is available about the effect of IOP-lowering medications other than PGA on corneal biomechanics. One study by Aydemir and colleagues [123] reported that there was a statistically significant difference in the CH between patients on benzalkonium chloride containing brimonidine (8.77 mmHg) compared to healthy controls (10.26 mmHg) ($p = 0.02$). However, there was no difference in the CH or CRF between purite-containing brimonidine and the control group. This study highlights the effect of preservative agents on corneal biomechanics.

8. Conclusions

The biomechanical characterization of the cornea has emerged as an exciting frontier in glaucoma diagnosis and management. However, the limitations of existing methods and the weak to moderate agreement between the reported parameters limited their widespread use in clinical practice. Additionally, studies of corneal biomechanics in glaucoma are further limited by their inclusion of glaucoma subjects taking topical PGA, which may alter corneal biomechanics, leading to contradicting results. Furthermore, some studies lack proper patient stratification and sometimes misinterpret results due to reported factors that are confounded by IOP changes. It is important to note that corneal biomechanical properties are dynamic metrics and can change over time with age, corneal trauma, or surgery.

There is a clear need for a more robust measure of corneal biomechanics that can more accurately determine the modulus of elasticity. New devices such as BM represent a promising, novel approach to evaluating corneal biomechanical properties in glaucoma patients independent of IOP.

Author Contributions: Conceptualization, A.M.E., G.S., and O.J.S.; methodology, A.M.E. and O.J.S.; formal analysis, A.M.E., G.S. and O.J.S.; data curation, A.M.E.; writing—original draft preparation, A.M.E.; writing: A.M.E.—review and editing, G.S. and O.J.S.; supervision, O.J.S. All authors have read and agreed to the published version of the manuscript.

Funding: This research received no external funding.

Institutional Review Board Statement: Not applicable.

Informed Consent Statement: Not applicable.

Data Availability Statement: Data used in this review article are publicly available.

Conflicts of Interest: GS has patents (UMD and MGH) related to Brillouin technology and Equity and is a consultant for Intelon Optics. The other authors declare no conflict of interest.

References

1. Zhang, N.; Wang, J.; Li, Y.; Jiang, B. Prevalence of primary open angle glaucoma in the last 20 years: A meta-analysis and systematic review. *Sci. Rep.* **2021**, *11*, 13762. [CrossRef] [PubMed]
2. Tham, Y.C.; Li, X.; Wong, T.Y.; Quigley, H.A.; Aung, T.; Cheng, C.Y. Global prevalence of glaucoma and projections of glaucoma burden through 2040: A systematic review and meta-analysis. *Ophthalmology* **2014**, *121*, 2081–2090. [CrossRef] [PubMed]
3. Heijl, A.; Leske, M.C.; Bengtsson, B.; Hyman, L.; Bengtsson, B.; Hussein, M. Reduction of intraocular pressure and glaucoma progression: Results from the Early Manifest Glaucoma Trial. *Arch. Ophthalmol.* **2002**, *120*, 1268–1279. [CrossRef] [PubMed]
4. Rotchford, A.P.; Johnson, G.J. Glaucoma in Zulus: A population-based cross-sectional survey in a rural district in South Africa. *Arch. Ophthalmol.* **2002**, *120*, 471–478. [CrossRef] [PubMed]
5. Klein, B.E.; Klein, R.; Sponsel, W.E.; Franke, T.; Cantor, L.B.; Martone, J.; Menage, M.J. Prevalence of glaucoma. The Beaver Dam Eye Study. *Ophthalmology* **1992**, *99*, 1499–1504. [CrossRef] [PubMed]
6. Dielemans, I.; Vingerling, J.R.; Wolfs, R.C.; Hofman, A.; Grobbee, D.E.; de Jong, P.T. The prevalence of primary open-angle glaucoma in a population-based study in The Netherlands. The Rotterdam Study. *Ophthalmology* **1994**, *101*, 1851–1855. [CrossRef]
7. Bonomi, L.; Marchini, G.; Marraffa, M.; Bernardi, P.; De Franco, I.; Perfetti, S.; Varotto, A.; Tenna, V. Prevalence of glaucoma and intraocular pressure distribution in a defined population. The Egna-Neumarkt Study. *Ophthalmology* **1998**, *105*, 209–215. [CrossRef]

8. Gordon, M.O.; Beiser, J.A.; Brandt, J.D.; Heuer, D.K.; Higginbotham, E.J.; Johnson, C.A.; Keltner, J.L.; Miller, J.P.; Parrish, R.K., II; Wilson, M.R.; et al. The Ocular Hypertension Treatment Study: Baseline Factors That Predict the Onset of Primary Open-Angle Glaucoma. *Arch. Ophthalmol.* **2002**, *120*, 714–720. [CrossRef]
9. Xu, L.; Wang, Y.; Wang, S.; Wang, Y.; Jonas, J.B. High myopia and glaucoma susceptibility the Beijing Eye Study. *Ophthalmology* **2007**, *114*, 216–220. [CrossRef]
10. Wareham, L.K.; Calkins, D.J. The Neurovascular Unit in Glaucomatous Neurodegeneration. *Front. Cell Dev. Biol.* **2020**, *8*, 452. [CrossRef]
11. Burgoyne, C.F.; Downs, J.C.; Bellezza, A.J.; Suh, J.K.; Hart, R.T. The optic nerve head as a biomechanical structure: A new paradigm for understanding the role of IOP-related stress and strain in the pathophysiology of glaucomatous optic nerve head damage. *Prog. Retin. Eye Res.* **2005**, *24*, 39–73. [CrossRef] [PubMed]
12. Liu, B.; McNally, S.; Kilpatrick, J.I.; Jarvis, S.P.; O'Brien, C.J. Aging and ocular tissue stiffness in glaucoma. *Surv. Ophthalmol.* **2018**, *63*, 56–74. [CrossRef] [PubMed]
13. Lee, K.M.; Kim, T.W.; Lee, E.J.; Girard, M.J.A.; Mari, J.M.; Weinreb, R.N. Association of Corneal Hysteresis With Lamina Cribrosa Curvature in Primary Open Angle Glaucoma. *Investig. Ophthalmol. Vis. Sci.* **2019**, *60*, 4171–4177. [CrossRef] [PubMed]
14. Catania, F.; Morenghi, E.; Rosetta, P.; Paolo, V.; Vinciguerra, R. Corneal Biomechanics Assessment with Ultra High Speed Scheimpflug Camera in Primary Open Angle Glaucoma Compared with Healthy Subjects: A meta-analysis of the Literature. *Curr. Eye Res.* **2022**, *48*, 161–171. [CrossRef] [PubMed]
15. Roberts, C.J. Corneal hysteresis and beyond: Does it involve the sclera? *J. Cataract Refract. Surg.* **2021**, *47*, 427–429. [CrossRef]
16. Jammal, A.A.; Medeiros, F.A. Corneal hysteresis: Ready for prime time? *Curr. Opin. Ophthalmol.* **2022**, *33*, 243–249. [CrossRef]
17. Fujishiro, T.; Matsuura, M.; Fujino, Y.; Murata, H.; Tokumo, K.; Nakakura, S.; Kiuchi, Y.; Asaoka, R. The Relationship Between Corvis ST Tonometry Parameters and Ocular Response Analyzer Corneal Hysteresis. *J. Glaucoma* **2020**, *29*, 479–484. [CrossRef]
18. Roberts, C.J.; Liu, J. *Corneal Biomechanics: From Theory to Practice*; Kugler Publications: Amsterdam, The Netherlands, 2017.
19. Blackburn, B.J.; Jenkins, M.W.; Rollins, A.M.; Dupps, W.J. A Review of Structural and Biomechanical Changes in the Cornea in Aging, Disease, and Photochemical Crosslinking. *Front. Bioeng. Biotechnol.* **2019**, *7*, 66. [CrossRef]
20. Roberts, C.J. Concepts and misconceptions in corneal biomechanics. *J. Cataract Refract. Surg.* **2014**, *40*, 862–869. [CrossRef]
21. Elsheikh, A.; Alhasso, D.; Rama, P. Biomechanical properties of human and porcine corneas. *Exp. Eye Res.* **2008**, *86*, 783–790. [CrossRef]
22. Elsheikh, A.; Wang, D.; Pye, D. Determination of the modulus of elasticity of the human cornea. *J. Refract. Surg.* **2007**, *23*, 808–818. [CrossRef] [PubMed]
23. Luce, D.A. Determining in vivo biomechanical properties of the cornea with an ocular response analyzer. *J. Cataract Refract. Surg.* **2005**, *31*, 156–162. [CrossRef]
24. Pepose, J.S.; Feigenbaum, S.K.; Qazi, M.A.; Sanderson, J.P.; Roberts, C.J. Changes in corneal biomechanics and intraocular pressure following LASIK using static, dynamic, and noncontact tonometry. *Am. J. Ophthalmol.* **2007**, *143*, 39–47. [CrossRef] [PubMed]
25. Yuan, A.; Pineda, R. Developments in Imaging of Corneal Biomechanics. *Int. Ophthalmol. Clin.* **2019**, *59*, 1–17. [CrossRef] [PubMed]
26. Esporcatte, L.P.G.; Salomão, M.Q.; Lopes, B.T.; Vinciguerra, P.; Vinciguerra, R.; Roberts, C.; Elsheikh, A.; Dawson, D.G.; Ambrósio, R. Biomechanical diagnostics of the cornea. *Eye Vis.* **2020**, *7*, 9. [CrossRef]
27. Touboul, D.; Roberts, C.; Kérautret, J.; Garra, C.; Maurice-Tison, S.; Saubusse, E.; Colin, J. Correlations between corneal hysteresis, intraocular pressure, and corneal central pachymetry. *J. Cataract Refract. Surg.* **2008**, *34*, 616–622. [CrossRef]
28. Vellara, H.R.; Patel, D.V. Biomechanical properties of the keratoconic cornea: A review. *Clin. Exp. Optom.* **2015**, *98*, 31–38. [CrossRef]
29. Şahin, A.; Bayer, A.; Özge, G.k.; Mumcuoglu, T. Corneal Biomechanical Changes in Diabetes Mellitus and Their Influence on Intraocular Pressure Measurements. *Investig. Ophthalmol. Vis. Sci.* **2009**, *50*, 4597–4604. [CrossRef]
30. Bao, F.; Deng, M.; Wang, Q.; Huang, J.; Yang, J.; Whitford, C.; Geraghty, B.; Yu, A.; Elsheikh, A. Evaluation of the relationship of corneal biomechanical metrics with physical intraocular pressure and central corneal thickness in ex vivo rabbit eye globes. *Exp. Eye Res.* **2015**, *137*, 11–17. [CrossRef]
31. Shen, S.R.; Fleming, G.P.; Jain, S.G.; Roberts, C.J. A Review of Corneal Biomechanics and Scleral Stiffness in Topical Prostaglandin Analog Therapy for Glaucoma. *Curr. Eye Res.* **2022**, *48*, 172–181. [CrossRef]
32. Lopes, B.T.; Bao, F.; Wang, J.; Liu, X.; Wang, L.; Abass, A.; Eliasy, A.; Elsheikh, A. Review of in-vivo characterisation of corneal biomechanics. *Med. Innovel Technol. Devices* **2021**, *11*, 100073. [CrossRef]
33. Wu, N.; Chen, Y.; Sun, X. Association Between Ocular Biomechanics Measured With Corvis ST and Glaucoma Severity in Patients With Untreated Primary Open Angle Glaucoma. *Transl. Vis. Sci. Technol.* **2022**, *11*, 10. [CrossRef] [PubMed]
34. Salomão, M.Q.; Hofling-Lima, A.L.; Faria-Correia, F.; Lopes, B.T.; Rodrigues-Barros, S.; Roberts, C.J.; Ambrósio, R. Dynamic corneal deformation response and integrated corneal tomography. *Indian J. Ophthalmol.* **2018**, *66*, 373–382. [CrossRef] [PubMed]
35. Vinciguerra, R.; Ambrósio, R., Jr.; Elsheikh, A.; Roberts, C.J.; Lopes, B.; Morenghi, E.; Azzolini, C.; Vinciguerra, P. Detection of Keratoconus With a New Biomechanical Index. *J. Refract. Surg.* **2016**, *32*, 803–810. [CrossRef]
36. Matsuura, M.; Murata, H.; Fujino, Y.; Yanagisawa, M.; Nakao, Y.; Tokumo, K.; Nakakura, S.; Kiuchi, Y.; Asaoka, R. Relationship between novel intraocular pressure measurement from Corvis ST and central corneal thickness and corneal hysteresis. *Br. J. Ophthalmol.* **2020**, *104*, 563–568. [CrossRef]

37. Eliasy, A.; Chen, K.J.; Vinciguerra, R.; Lopes, B.T.; Abass, A.; Vinciguerra, P.; Ambrósio, R., Jr.; Roberts, C.J.; Elsheikh, A. Determination of Corneal Biomechanical Behavior in-vivo for Healthy Eyes Using CorVis ST Tonometry: Stress-Strain Index. *Front. Bioeng. Biotechnol.* **2019**, *7*, 105. [CrossRef]
38. Zhang, H.; Eliasy, A.; Lopes, B.; Abass, A.; Vinciguerra, R.; Vinciguerra, P.; Ambrósio, R., Jr.; Roberts, C.J.; Elsheikh, A. Stress-Strain Index Map: A New Way to Represent Corneal Material Stiffness. *Front. Bioeng. Biotechnol.* **2021**, *9*, 640434. [CrossRef]
39. Pillunat, K.R.; Herber, R.; Spoerl, E.; Erb, C.; Pillunat, L.E. A new biomechanical glaucoma factor to discriminate normal eyes from normal pressure glaucoma eyes. *Acta. Ophthalmol.* **2019**, *97*, e962–e967. [CrossRef]
40. Prevedel, R.; Diz-Muñoz, A.; Ruocco, G.; Antonacci, G. Brillouin microscopy: An emerging tool for mechanobiology. *Nat. Methods* **2019**, *16*, 969–977. [CrossRef]
41. Zhang, H.; Asroui, L.; Tarib, I.; Dupps, W.J., Jr.; Scarcelli, G.; Randleman, J.B. Motion-Tracking Brillouin Microscopy Evaluation of Normal, Keratoconic, and Post-Laser Vision Correction Corneas. *Am. J. Ophthalmol.* **2023**, *254*, 128–140. [CrossRef]
42. Zhang, H.; Asroui, L.; Randleman, J.B.; Scarcelli, G. Motion-tracking Brillouin microscopy for in-vivo corneal biomechanics mapping. *Biomed. Opt. Express.* **2022**, *13*, 6196–6210. [CrossRef] [PubMed]
43. Nair, A.; Ambekar, Y.S.; Zevallos-Delgado, C.; Mekonnen, T.; Sun, M.; Zvietcovich, F.; Singh, M.; Aglyamov, S.; Koch, M.; Scarcelli, G.; et al. Multiple Optical Elastography Techniques Reveal the Regulation of Corneal Stiffness by Collagen XII. *Investig. Ophthalmol. Vis. Sci.* **2022**, *63*, 24. [CrossRef] [PubMed]
44. Scarcelli, G.; Besner, S.; Pineda, R.; Yun, S.H. Biomechanical characterization of keratoconus corneas ex vivo with Brillouin microscopy. *Investig. Ophthalmol. Vis. Sci.* **2014**, *55*, 4490–4495. [CrossRef] [PubMed]
45. Scarcelli, G.; Besner, S.; Pineda, R.; Kalout, P.; Yun, S.H. In vivo biomechanical mapping of normal and keratoconus corneas. *JAMA Ophthalmol.* **2015**, *133*, 480–482. [CrossRef] [PubMed]
46. Seiler, T.G.; Shao, P.; Eltony, A.; Seiler, T.; Yun, S.H. Brillouin Spectroscopy of Normal and Keratoconus Corneas. *Am. J. Ophthalmol.* **2019**, *202*, 118–125. [CrossRef]
47. Yun, S.H.; Chernyak, D. Brillouin microscopy: Assessing ocular tissue biomechanics. *Curr. Opin. Ophthalmol.* **2018**, *29*, 299–305. [CrossRef]
48. Ferguson, T.J.; Singuri, S.; Jalaj, S.; Ford, M.R.; De Stefano, V.S.; Seven, I.; Dupps, W.J., Jr. Depth-resolved Corneal Biomechanical Changes Measured Via Optical Coherence Elastography Following Corneal Crosslinking. *Transl. Vis. Sci. Technol.* **2021**, *10*, 7. [CrossRef]
49. Kirwan, C.; O'Keefe, M.; Lanigan, B. Corneal hysteresis and intraocular pressure measurement in children using the reichert ocular response analyzer. *Am. J. Ophthalmol.* **2006**, *142*, 990–992. [CrossRef]
50. Sullivan-Mee, M.; Katiyar, S.; Pensyl, D.; Halverson, K.D.; Qualls, C. Relative importance of factors affecting corneal hysteresis measurement. *Optom. Vis. Sci.* **2012**, *89*, E803–E811. [CrossRef]
51. Mangouritsas, G.; Morphis, G.; Mourtzoukos, S.; Feretis, E. Association between corneal hysteresis and central corneal thickness in glaucomatous and non-glaucomatous eyes. *Acta Ophthalmol.* **2009**, *87*, 901–905. [CrossRef]
52. Sun, L.; Shen, M.; Wang, J.; Fang, A.; Xu, A.; Fang, H.; Lu, F. Recovery of corneal hysteresis after reduction of intraocular pressure in chronic primary angle-closure glaucoma. *Am. J. Ophthalmol.* **2009**, *147*, 1061–1066.e2. [CrossRef] [PubMed]
53. Abitbol, O.; Bouden, J.; Doan, S.; Hoang-Xuan, T.; Gatinel, D. Corneal hysteresis measured with the Ocular Response Analyzer in normal and glaucomatous eyes. *Acta Ophthalmol.* **2010**, *88*, 116–119. [CrossRef] [PubMed]
54. Narayanaswamy, A.; Su, D.H.; Baskaran, M.; Tan, A.C.; Nongpiur, M.E.; Htoon, H.M.; Wong, T.Y.; Aung, T. Comparison of ocular response analyzer parameters in chinese subjects with primary angle-closure and primary open-angle glaucoma. *Arch. Ophthalmol.* **2011**, *129*, 429–434. [CrossRef]
55. Kaushik, S.; Pandav, S.S.; Banger, A.; Aggarwal, K.; Gupta, A. Relationship between corneal biomechanical properties, central corneal thickness, and intraocular pressure across the spectrum of glaucoma. *Am. J. Ophthalmol.* **2012**, *153*, 840–849.e2. [CrossRef]
56. Grise-Dulac, A.; Saad, A.; Abitbol, O.; Febbraro, J.L.; Azan, E.; Moulin-Tyrode, C.; Gatinel, D. Assessment of corneal biomechanical properties in normal tension glaucoma and comparison with open-angle glaucoma, ocular hypertension, and normal eyes. *J. Glaucoma* **2012**, *21*, 486–489. [CrossRef]
57. Detry-Morel, M.; Jamart, J.; Hautenauven, F.; Pourjavan, S. Comparison of the corneal biomechanical properties with the Ocular Response Analyzer® (ORA) in African and Caucasian normal subjects and patients with glaucoma. *Acta. Ophthalmol.* **2012**, *90*, e118–e124. [CrossRef] [PubMed]
58. Morita, T.; Shoji, N.; Kamiya, K.; Fujimura, F.; Shimizu, K. Corneal biomechanical properties in normal-tension glaucoma. *Acta. Ophthalmol.* **2012**, *90*, e48–e53. [CrossRef] [PubMed]
59. Cankaya, A.B.; Anayol, A.; Özcelik, D.; Demirdogen, E.; Yilmazbas, P. Ocular response analyzer to assess corneal biomechanical properties in exfoliation syndrome and exfoliative glaucoma. *Graefes. Arch. Clin. Exp. Ophthalmol.* **2012**, *250*, 255–260. [CrossRef]
60. Beyazyıldız, E.; Beyazyıldız, Ö.; Arifoğlu, H.B.; Altıntaş, A.K.; Köklü, S.G. Comparison of ocular response analyzer parameters in primary open angle glaucoma and exfoliation glaucoma patients. *Indian J. Ophthalmol.* **2014**, *62*, 782–787. [CrossRef]
61. Shin, J.; Lee, J.W.; Kim, E.A.; Caprioli, J. The effect of corneal biomechanical properties on rebound tonometer in patients with normal-tension glaucoma. *Am. J. Ophthalmol.* **2015**, *159*, 144–154. [CrossRef]
62. Hussnain, S.A.; Alsberge, J.B.; Ehrlich, J.R.; Shimmyo, M.; Radcliffe, N.M. Change in corneal hysteresis over time in normal, glaucomatous and diabetic eyes. *Acta. Ophthalmol.* **2015**, *93*, e627–e630. [CrossRef] [PubMed]

63. Ayala, M. Corneal hysteresis in normal subjects and in patients with primary open-angle glaucoma and pseudoexfoliation glaucoma. *Ophthalmic. Res.* **2011**, *46*, 187–191. [CrossRef] [PubMed]
64. Yazgan, S.; Celik, U.; Alagöz, N.; Taş, M. Corneal biomechanical comparison of pseudoexfoliation syndrome, pseudoexfoliative glaucoma and healthy subjects. *Curr. Eye Res.* **2015**, *40*, 470–475. [CrossRef] [PubMed]
65. Dana, D.; Mihaela, C.; Raluca, I.; Miruna, M.; Catalina, I.; Miruna, C.; Schmitzer, S.; Catalina, C. Corneal hysteresis and primary open angle glaucoma. *Rom. J. Ophthalmol.* **2015**, *59*, 252–254.
66. Pillunat, K.R.; Hermann, C.; Spoerl, E.; Pillunat, L.E. Analyzing biomechanical parameters of the cornea with glaucoma severity in open-angle glaucoma. *Graefes. Arch. Clin. Exp. Ophthalmol.* **2016**, *254*, 1345–1351. [CrossRef]
67. Perucho-González, L.; Martínez de la Casa, J.M.; Morales-Fernández, L.; Bañeros-Rojas, P.; Saenz-Francés, F.; García-Feijoó, J. Intraocular pressure and biomechanical corneal properties measure by ocular response analyser in patients with primary congenital glaucoma. *Acta. Ophthalmol.* **2016**, *94*, e293–e297. [CrossRef]
68. Perucho-González, L.; Sáenz-Francés, F.; Morales-Fernández, L.; Martínez-de-la-Casa, J.M.; Méndez-Hernández, C.D.; Santos-Bueso, E.; Brookes, J.L.; García-Feijoó, J. Structural and biomechanical corneal differences between patients suffering from primary congenital glaucoma and healthy volunteers. *Acta. Ophthalmol.* **2017**, *95*, e107–e112. [CrossRef]
69. Park, K.; Shin, J.; Lee, J. Relationship between corneal biomechanical properties and structural biomarkers in patients with normal-tension glaucoma: A retrospective study. *BMC Ophthalmol.* **2018**, *18*, 7. [CrossRef]
70. Aoki, S.; Miki, A.; Omoto, T.; Fujino, Y.; Matsuura, M.; Murata, H.; Asaoka, R. Biomechanical Glaucoma Factor and Corneal Hysteresis in Treated Primary Open-Angle Glaucoma and Their Associations with Visual Field Progression. *Investig. Ophthalmol. Vis. Sci.* **2021**, *62*, 4. [CrossRef]
71. Rojananuangnit, K. Corneal Hysteresis in Thais and Variation of Corneal Hysteresis in Glaucoma. *Clin. Optom.* **2021**, *13*, 287–299. [CrossRef]
72. Del Buey-Sayas, M.; Lanchares-Sancho, E.; Campins-Falcó, P.; Pinazo-Durán, M.D.; Peris-Martínez, C. Corneal Biomechanical Parameters and Central Corneal Thickness in Glaucoma Patients, Glaucoma Suspects, and a Healthy Population. *J. Clin. Med.* **2021**, *10*, 2637. [CrossRef] [PubMed]
73. Potop, V.; Coviltir, V.; Schmitzer, S.; Corbu, C.; Ionescu, I.C.; Burcel, M.; Dăscălescu, D. The Relationship Between Corneal Hysteresis and Retinal Ganglion Cells—A Step Forward in Early Glaucoma Diagnosis. *Med. Sci. Monit.* **2020**, *26*, e924672. [CrossRef] [PubMed]
74. Sullivan-Mee, M.; Billingsley, S.C.; Patel, A.D.; Halverson, K.D.; Alldredge, B.R.; Qualls, C. Ocular Response Analyzer in subjects with and without glaucoma. *Optom. Vis. Sci.* **2008**, *85*, 463–470. [CrossRef] [PubMed]
75. Liang, L.; Zhang, R.; He, L.Y. Corneal hysteresis and glaucoma. *Int. Ophthalmol.* **2019**, *39*, 1909–1916. [CrossRef] [PubMed]
76. Daxer, A.; Misof, K.; Grabner, B.; Ettl, A.; Fratzl, P. Collagen fibrils in the human corneal stroma: Structure and aging. *Investig. Ophthalmol. Vis. Sci.* **1998**, *39*, 644–648.
77. Yang, Y.; Ng, T.K.; Wang, L.; Wu, N.; Xiao, M.; Sun, X.; Chen, Y. Association of 24-Hour Intraocular Pressure Fluctuation with Corneal Hysteresis and Axial Length in Untreated Chinese Primary Open-Angle Glaucoma Patients. *Transl. Vis. Sci. Technol.* **2020**, *9*, 25. [CrossRef]
78. Khawaja, A.P.; Chan, M.P.Y.; Broadway, D.C.; Garway-Heath, D.F.; Luben, R.; Yip, J.L.Y.; Hayat, S.; Khaw, K.-T.; Foster, P.J. Corneal Biomechanical Properties and Glaucoma-Related Quantitative Traits in the EPIC-Norfolk Eye Study. *Investig. Ophthalmol. Vis. Sci.* **2014**, *55*, 117–124. [CrossRef]
79. Wells, A.P.; Garway-Heath, D.F.; Poostchi, A.; Wong, T.; Chan, K.C.; Sachdev, N. Corneal hysteresis but not corneal thickness correlates with optic nerve surface compliance in glaucoma patients. *Investig. Ophthalmol. Vis. Sci.* **2008**, *49*, 3262–3268. [CrossRef]
80. Wong, B.J.; Moghimi, S.; Zangwill, L.M.; Christopher, M.; Belghith, A.; Ekici, E.; Bowd, C.; Fazio, M.A.; Girkin, C.A.; Weinreb, R.N. Relationship of Corneal Hysteresis and Anterior Lamina Cribrosa Displacement in Glaucoma. *Am. J. Ophthalmol.* **2020**, *212*, 134–143. [CrossRef]
81. Jammal, A.A.; Medeiros, F.A. Corneal Hysteresis and Rates of Neuroretinal Rim Change in Glaucoma. *Ophthalmol. Glaucoma* **2022**, *5*, 483–489. [CrossRef]
82. Radcliffe, N.M.; Tracer, N.; De Moraes, C.G.V.; Tello, C.; Liebmann, J.M.; Ritch, R. Relationship between optic disc hemorrhage and corneal hysteresis. *Can. J. Ophthalmol.* **2020**, *55*, 239–244. [CrossRef] [PubMed]
83. Medeiros, F.A.; Meira-Freitas, D.; Lisboa, R.; Kuang, T.M.; Zangwill, L.M.; Weinreb, R.N. Corneal hysteresis as a risk factor for glaucoma progression: A prospective longitudinal study. *Ophthalmology* **2013**, *120*, 1533–1540. [CrossRef] [PubMed]
84. Kamalipour, A.; Moghimi, S.; Eslani, M.; Nishida, T.; Mohammadzadeh, V.; Micheletti, E.; Girkin, C.A.; Fazio, M.A.; Liebmann, J.M.; Zangwill, L.M.; et al. A Prospective Longitudinal Study to Investigate Corneal Hysteresis as a Risk Factor of Central Visual Field Progression in Glaucoma. *Am. J. Ophthalmol.* **2022**, *240*, 159–169. [CrossRef] [PubMed]
85. Chan, E.; Yeh, K.; Moghimi, S.; Proudfoot, J.; Liu, X.; Zangwill, L.; Weinreb, R.N. Changes in Corneal Biomechanics and Glaucomatous Visual Field Loss. *J. Glaucoma* **2021**, *30*, e246–e251. [CrossRef]
86. Mohammadzadeh, V.; Moghimi, S.; Nishida, T.; Mahmoudinezhad, G.; Kamalipour, A.; Micheletti, E.; Zangwill, L.; Weinreb, R.N. Effect of Corneal Hysteresis on the Rates of Microvasculature Loss in Glaucoma. *Ophthalmol. Glaucoma* **2022**, *6*, 177–186. [CrossRef]

87. Susanna, C.N.; Diniz-Filho, A.; Daga, F.B.; Susanna, B.N.; Zhu, F.; Ogata, N.G.; Medeiros, F.A. A Prospective Longitudinal Study to Investigate Corneal Hysteresis as a Risk Factor for Predicting Development of Glaucoma. *Am. J. Ophthalmol.* **2018**, *187*, 148–152. [CrossRef]
88. Salvetat, M.L.; Zeppieri, M.; Tosoni, C.; Felletti, M.; Grasso, L.; Brusini, P. Corneal Deformation Parameters Provided by the Corvis-ST Pachy-Tonometer in Healthy Subjects and Glaucoma Patients. *J. Glaucoma* **2015**, *24*, 568–574. [CrossRef]
89. Wang, W.; Du, S.; Zhang, X. Corneal Deformation Response in Patients With Primary Open-Angle Glaucoma and in Healthy Subjects Analyzed by Corvis ST. *Investig. Ophthalmol. Vis. Sci.* **2015**, *56*, 5557–5565. [CrossRef]
90. Coste, V.; Schweitzer, C.; Paya, C.; Touboul, D.; Korobelnik, J.F. Evaluation of corneal biomechanical properties in glaucoma and control patients by dynamic Scheimpflug corneal imaging technology. *J. Fr. Ophthalmol.* **2015**, *38*, 504–513. [CrossRef]
91. Hong, K.; Wong, I.Y.H.; Singh, K.; Chang, R.T. Corneal Biomechanics Using a Scheimpflug-Based Noncontact Device in Normal-Tension Glaucoma and Healthy Controls. *Asia Pac. J. Ophthalmol.* **2019**, *8*, 22–29. [CrossRef]
92. Miki, A.; Yasukura, Y.; Weinreb, R.N.; Yamada, T.; Koh, S.; Asai, T.; Ikuno, Y.; Maeda, N.; Nishida, K. Dynamic Scheimpflug Ocular Biomechanical Parameters in Healthy and Medically Controlled Glaucoma Eyes. *J. Glaucoma* **2019**, *28*, 588–592. [CrossRef] [PubMed]
93. Leung, C.K.; Ye, C.; Weinreb, R.N. An ultra-high-speed Scheimpflug camera for evaluation of corneal deformation response and its impact on IOP measurement. *Investig. Ophthalmol. Vis. Sci.* **2013**, *54*, 2885–2892. [CrossRef] [PubMed]
94. Lee, R.; Chang, R.T.; Wong, I.Y.; Lai, J.S.; Lee, J.W.; Singh, K. Novel Parameter of Corneal Biomechanics That Differentiate Normals From Glaucoma. *J. Glaucoma* **2016**, *25*, e603–e609. [CrossRef] [PubMed]
95. Tian, L.; Wang, D.; Wu, Y.; Meng, X.; Chen, B.; Ge, M.; Huang, Y. Corneal biomechanical characteristics measured by the CorVis Scheimpflug technology in eyes with primary open-angle glaucoma and normal eyes. *Acta Ophthalmol.* **2016**, *94*, e317–e324. [CrossRef]
96. Jung, Y.; Park, H.L.; Yang, H.J.; Park, C.K. Characteristics of corneal biomechanical responses detected by a non-contact scheimpflug-based tonometer in eyes with glaucoma. *Acta Ophthalmol.* **2017**, *95*, e556–e563. [CrossRef] [PubMed]
97. Vinciguerra, R.; Rehman, S.; Vallabh, N.A.; Batterbury, M.; Czanner, G.; Choudhary, A.; Cheeseman, R.; Elsheikh, A.; Willoughby, C.E. Corneal biomechanics and biomechanically corrected intraocular pressure in primary open-angle glaucoma, ocular hypertension and controls. *Br. J. Ophthalmol.* **2020**, *104*, 121–126. [CrossRef]
98. Miki, A.; Yasukura, Y.; Weinreb, R.N.; Maeda, N.; Yamada, T.; Koh, S.; Asai, T.; Ikuno, Y.; Nishida, K. Dynamic Scheimpflug Ocular Biomechanical Parameters in Untreated Primary Open Angle Glaucoma Eyes. *Investig. Ophthalmol. Vis. Sci.* **2020**, *61*, 19. [CrossRef]
99. Jung, Y.; Park, H.L.; Oh, S.; Park, C.K. Corneal biomechanical responses detected using corvis st in primary open angle glaucoma and normal tension glaucoma. *Medicine* **2020**, *99*, e19126. [CrossRef]
100. Pradhan, Z.S.; Deshmukh, S.; Dixit, S.; Sreenivasaiah, S.; Shroff, S.; Devi, S.; Webers, C.A.B.; Rao, H.L. A comparison of the corneal biomechanics in pseudoexfoliation glaucoma, primary open-angle glaucoma and healthy controls using Corvis ST. *PLoS ONE* **2020**, *15*, e0241296. [CrossRef]
101. Pradhan, Z.S.; Deshmukh, S.; Dixit, S.; Gudetti, P.; Devi, S.; Webers, C.A.B.; Rao, H.L. A comparison of the corneal biomechanics in pseudoexfoliation syndrome, pseudoexfoliation glaucoma, and healthy controls using Corvis® Scheimpflug Technology. *Indian J. Ophthalmol.* **2020**, *68*, 787–792. [CrossRef]
102. Wei, Y.H.; Cai, Y.; Choy, B.N.K.; Li, B.B.; Li, R.S.; Xing, C.; Wang, X.; Tian, T.; Fang, Y.; Li, M.; et al. Comparison of corneal biomechanics among primary open-angle glaucoma with normal tension or hypertension and controls. *Chin. Med. J.* **2021**, *134*, 1087–1092. [CrossRef] [PubMed]
103. Silva, N.; Ferreira, A.; Baptista, P.M.; Figueiredo, A.; Reis, R.; Sampaio, I.; Beirão, J.; Vinciguerra, R.; Menéres, P.; Menéres, M.J. Corneal Biomechanics for Ocular Hypertension, Primary Open-Angle Glaucoma, and Amyloidotic Glaucoma: A Comparative Study by Corvis ST. *Clin. Ophthalmol.* **2022**, *16*, 71–83. [CrossRef] [PubMed]
104. Zarei, R.; Zamani, M.H.; Eslami, Y.; Fakhraei, G.; Tabatabaei, M.; Esfandiari, A.R. Comparing corneal biomechanics and intraocular pressure between healthy individuals and glaucoma subtypes: A cross-sectional study. *Ann. Med. Surg.* **2022**, *82*, 104677. [CrossRef] [PubMed]
105. Halkiadakis, I.; Tzimis, V.; Gryparis, A.; Markopoulos, I.; Konstadinidou, V.; Zintzaras, E.; Tzakos, M. Evaluation of Corvis ST tonometer with the updated software in glaucoma practice. *Int. J. Ophthalmol.* **2022**, *15*, 438–445. [CrossRef]
106. Vieira, M.J.; Pereira, J.; Castro, M.; Arruda, H.; Martins, J.; Sousa, J.P. Efficacy of corneal shape index in the evaluation of ocular hypertension, primary open-angle glaucoma and exfoliative glaucoma. *Eur. J. Ophthalmol.* **2022**, *32*, 275–281. [CrossRef]
107. Xu, Y.; Ye, Y.; Chen, Z.; Xu, J.; Yang, Y.; Fan, Y.; Liu, P.; Chong, I.T.; Yu, K.; Lam, D.C.C.; et al. Corneal Stiffness and Modulus of Normal-Tension Glaucoma in Chinese. *Am. J. Ophthalmol.* **2022**, *242*, 131–138. [CrossRef]
108. Wu, N.; Chen, Y.; Yu, X.; Li, M.; Wen, W.; Sun, X. Changes in Corneal Biomechanical Properties after Long-Term Topical Prostaglandin Therapy. *PLoS ONE* **2016**, *11*, e0155527. [CrossRef]
109. Chou, C.C.; Shih, P.J.; Wang, C.Y.; Jou, T.S.; Chen, J.P.; Wang, I.J. Corvis Biomechanical Factor Facilitates the Detection of Primary Angle Closure Glaucoma. *Transl. Vis. Sci. Technol.* **2022**, *11*, 7. [CrossRef]
110. Bolivar, G.; Sanchez-Barahona, C.; Ketabi, S.; Kozobolis, V.; Teus, M.A. Corneal Factors Associated with the Amount of Visual Field Damage in Eyes with Newly Diagnosed, Untreated, Open-angle Glaucoma. *Ophthalmol. Ther.* **2021**, *10*, 669–676. [CrossRef]

111. Hirasawa, K.; Matsuura, M.; Murata, H.; Nakakura, S.; Nakao, Y.; Kiuchi, Y.; Asaoka, R. Association between Corneal Biomechanical Properties with Ocular Response Analyzer and Also CorvisST Tonometry, and Glaucomatous Visual Field Severity. *Transl. Vis. Sci. Technol.* **2017**, *6*, 18. [CrossRef]
112. Qassim, A.; Mullany, S.; Abedi, F.; Marshall, H.; Hassall, M.M.; Kolovos, A.; Knight, L.S.W.; Nguyen, T.; Awadalla, M.S.; Chappell, A.; et al. Corneal Stiffness Parameters Are Predictive of Structural and Functional Progression in Glaucoma Suspect Eyes. *Ophthalmology* **2021**, *128*, 993–1004. [CrossRef] [PubMed]
113. Honda, N.; Miyai, T.; Nejima, R.; Miyata, K.; Mimura, T.; Usui, T.; Aihara, M.; Araie, M.; Amano, S. Effect of latanoprost on the expression of matrix metalloproteinases and tissue inhibitor of metalloproteinase 1 on the ocular surface. *Arch. Ophthalmol.* **2010**, *128*, 466–471. [CrossRef] [PubMed]
114. Kim, J.W.; Lindsey, J.D.; Wang, N.; Weinreb, R.N. Increased human scleral permeability with prostaglandin exposure. *Investig. Ophthalmol. Vis. Sci.* **2001**, *42*, 1514–1521.
115. Lindsey, J.D.; Crowston, J.G.; Tran, A.; Morris, C.; Weinreb, R.N. Direct matrix metalloproteinase enhancement of transscleral permeability. *Investig. Ophthalmol. Vis. Sci.* **2007**, *48*, 752–755. [CrossRef] [PubMed]
116. Tsikripis, P.; Papaconstantinou, D.; Koutsandrea, C.; Apostolopoulos, M.; Georgalas, I. The effect of prostaglandin analogs on the biomechanical properties and central thickness of the cornea of patients with open-angle glaucoma: A 3-year study on 108 eyes. *Drug Des. Dev. Ther.* **2013**, *7*, 1149–1156. [CrossRef]
117. Meda, R.; Wang, Q.; Paoloni, D.; Harasymowycz, P.; Brunette, I. The impact of chronic use of prostaglandin analogues on the biomechanical properties of the cornea in patients with primary open-angle glaucoma. *Br. J. Ophthalmol.* **2017**, *101*, 120–125. [CrossRef]
118. Vinciguerra, R.; Elsheikh, A.; Roberts, C.J.; Ambrósio, R., Jr.; Kang, D.S.; Lopes, B.T.; Morenghi, E.; Azzolini, C.; Vinciguerra, P. Influence of Pachymetry and Intraocular Pressure on Dynamic Corneal Response Parameters in Healthy Patients. *J. Refract. Surg.* **2016**, *32*, 550–561. [CrossRef]
119. Sánchez-Barahona, C.; Bolívar, G.; Katsanos, A.; Teus, M.A. Latanoprost treatment differentially affects intraocular pressure readings obtained with three different tonometers. *Acta Ophthalmol.* **2019**, *97*, e1112–e1115. [CrossRef]
120. Chong, J.; Dupps, W.J., Jr. Corneal biomechanics: Measurement and structural correlations. *Exp. Eye Res.* **2021**, *205*, 108508. [CrossRef]
121. Zheng, X.; Wang, Y.; Zhao, Y.; Cao, S.; Zhu, R.; Huang, W.; Yu, A.; Huang, J.; Wang, Q.; Wang, J.; et al. Experimental Evaluation of Travoprost-Induced Changes in Biomechanical Behavior of Ex-Vivo Rabbit Corneas. *Curr. Eye Res.* **2019**, *44*, 19–24. [CrossRef]
122. Zhu, R.; Zheng, X.; Guo, L.; Zhao, Y.; Wang, Y.; Wu, J.; Yu, A.; Wang, J.; Bao, F.; Elsheikh, A. Biomechanical Effects of Two Forms of PGF2α on Ex-vivo Rabbit Cornea. *Curr. Eye Res.* **2021**, *46*, 452–460. [CrossRef] [PubMed]
123. Aydemir, G.A.; Demirok, G.; Eksioglu, U.; Yakin, M.; Ornek, F. The Effect of Long-Term Usage of Single-Agent Antiglaucomatous Drops with Different Preservatives on Cornea Biomechanics. *Beyoglu Eye J.* **2021**, *6*, 24–30. [CrossRef] [PubMed]

Disclaimer/Publisher's Note: The statements, opinions and data contained in all publications are solely those of the individual author(s) and contributor(s) and not of MDPI and/or the editor(s). MDPI and/or the editor(s) disclaim responsibility for any injury to people or property resulting from any ideas, methods, instructions or products referred to in the content.

Systematic Review

Premium Intraocular Lenses in Glaucoma—A Systematic Review

Ashley Shuen Ying Hong [1,†], Bryan Chin Hou Ang [2,3,*,†], Emily Dorairaj [4] and Syril Dorairaj [5]

1. Yong Loo Lin School of Medicine, National University of Singapore, Singapore 119228, Singapore; hongashley@gmail.com
2. Department of Ophthalmology, National Healthcare Group Eye Institute, Tan Tock Seng Hospital, Singapore 308433, Singapore
3. Department of Ophthalmology, National Healthcare Group Eye Institute, Woodlands Health Campus, Singapore 768024, Singapore
4. Charles E. Schmidt College of Medicine, Florida Atlantic University, Boca Raton, FL 33431, USA; emilyadorairaj@gmail.com
5. Department of Ophthalmology, Mayo Clinic, Jacksonville, FL 32224, USA; dorairaj.syril@mayo.edu
* Correspondence: drbryanang@gmail.com
† These authors contributed equally to this work.

Abstract: The incidence of both cataract and glaucoma is increasing globally. With increasing patient expectation and improved technology, premium intraocular lenses (IOLs), including presbyopia-correcting and toric IOLs, are being increasingly implanted today. However, concerns remain regarding the use of premium IOLs, particularly presbyopia-correcting IOLs, in eyes with glaucoma. This systematic review evaluates the use of premium IOLs in glaucoma. A comprehensive search of the MEDLINE database was performed from inception until 1 June 2023. Initial search yielded 1404 records, of which 12 were included in the final review of post-operative outcomes. Studies demonstrated high spectacle independence for distance and good patient satisfaction in glaucomatous eyes, with positive outcomes also in post-operative visual acuity, residual astigmatism, and contrast sensitivity. Considerations in patient selection include anatomical and functional factors, such as the type and severity of glaucomatous visual field defects, glaucoma subtype, presence of ocular surface disease, ocular changes after glaucoma surgery, and the reliability of disease monitoring, all of which may be affected by, or influence, the outcomes of premium IOL implantation in glaucoma patients. Regular reviews on this topic are needed in order to keep up with the rapid advancements in IOL technology and glaucoma surgical treatments.

Keywords: premium intraocular lens; glaucoma; multifocal intraocular lens; extended depth of focus intraocular lens; toric intraocular lens

1. Introduction

Premium intraocular lenses (IOLs) are broadly considered to include presbyopia-correcting IOLs (multifocal IOLs (MFIOLs), extended depth of focus (EDOF) IOLs, accommodative IOLs), and toric IOLs for astigmatism correction. Compared with traditional monofocal IOLs, premium IOLs offer the benefit of better unaided visual acuity, greater spectacle independence, and higher patient satisfaction. In recent years, significant technological advances have been made in cataract surgery and IOL technology, resulting in more precise and predictable refractive outcomes [1]. Due to increasing patient expectation and demand for spectacle independence, premium IOLs have been increasingly adopted in clinical practice in recent years [2]. Global revenue from premium IOL usage is expected to grow at a compound annual growth rate (CAGR) of 9.03% from USD 1.5 billion in 2021 to USD 2.5 billion in 2028, compared to a 6.2% CAGR revenue growth for traditional IOLs [3].

While cataract is the most prevalent cause of reversible loss of vision, glaucoma remains the leading cause of irreversible blindness, characterized by a progressive optic neuropathy with degeneration of retinal ganglion cells and visual field loss [4]. It is estimated that one in five people undergoing cataract surgery have glaucoma or ocular hypertension, with the incidence of both cataract and glaucoma increasing with age [5]. Despite their benefits, many surgeons traditionally exercise caution in implanting premium IOLs, particularly presbyopia-correcting lenses, in patients with glaucoma. Concerns arise regarding contrast sensitivity (CS) loss and subjective visual disturbances such as glares and haloes, which may be more debilitating in patients with glaucomatous visual loss. Pathological changes in glaucoma may also potentially interact with the optical effects of MFIOLs [6]. However, advancements in premium IOL technology and development are enabling improved visual and refractive outcomes, with reduced side effects and less compromise to contrast sensitivity. More studies have also begun to report outcomes of premium IOLs in eyes with glaucoma and associated glaucomatous visual field loss.

However, there has not been a recent systematic review of literature regarding the use of premium IOLs in glaucoma patients, with the last extensive review published more than a decade ago [6]. This systemic review aims to summarize the current available literature reporting the surgical outcomes of premium IOL implantation in eyes with glaucoma. Pre-operative considerations and patient selection factors will also be discussed.

1.1. Overview of IOL Types

1.1.1. Monofocal IOLs

Monofocal IOLs provide excellent outcomes for distant vision, with the benefit of generally low cost and low frequency of photic phenomena such as glares and haloes [7]. They are the safest IOL choice for patients with pre-existing ocular pathology as they do not split light. However, as they only provide one focus point, they fail to deliver spectacle independence for near and intermediate vision.

Monovision is a surgical option correcting distant vision in the dominant eye while the non-dominant eye is corrected for near or intermediate vision, relying on neural adaptation to achieve a broader range of functional vision [8]. Generally utilizing monofocal IOLs, this option has proven a cost-effective method to provide spectacle independence, while avoiding the photic adverse effect caused by MFIOLs. It is best suited for patients prioritising spectacle independence. However, it is also associated with loss of depth perception and suboptimal vision at intermediate distances [8].

1.1.2. Multifocal IOLs (MFIOLs)

MFIOLs, first approved by the United States Food and Drug Administration (US FDA) in 1997 [9], come in varying optical designs, such as diffractive, refractive, bifocal, trifocal, or hybrid IOLs, and provide multiple focal points. Refractive MFIOLs, such as the Mplus (Oculentis, GmbH, Berlin, Germany), achieves multifocality via light refraction based on Snell's law [10]. Their design is rotationally symmetrical, with two or more concentric rings of different curvature radii and optical power on the front surface of the lens [11]. However, the lens performance is influenced by IOL centration as this affects the light percentage passing through the various optical zones [11]. Additionally, near visual acuity (VA) also depends on pupil size due to the near focus zone of the MFIOL being concentrically allocated [12].

Diffractive MFIOLs, such as the Acrysof ReSTOR (Alcon Laboratories, Fort Worth, TX, USA) and the Tecnis (Abbott Medical Optics, Johnson & Johnson Vision, Santa Ana, CA, USA), achieve multifocality via light interference based on the Huygens–Fresnel principle [13]. They feature multiple concentric rings with diffractive micro-structures separated by steps 2 μm in height that work independently regardless of pupil size, creating diffractive wave patterns that can focalize light rays on two or more foci. However, as light rays pass through multiple diffractive surfaces, this causes energy loss and reduces contrast sensitivity and increases the frequency of glares and haloes in patients [14].

While both diffractive and refractive MFIOLs produce similar uncorrected visual acuity, diffractive MFIOLs have been observed to provide better unaided near visual acuity [15,16]. Furthermore, refractive MFIOLs tend to cause more halo or glare symptoms due to light scattering at the transitional zone between the distant and near focus of the MFIOL [17]. Various studies [14–18] have shown that compared to traditional monofocal IOLs, MFIOLs offer greater spectacle independence but have a higher risk of glare, halo, and lower contrast sensitivity [19,20]. In patients with MFIOLs, more than a third of patients may report increased glares and haloes [21], and multiple meta-analyses [20] have reported a decrease in CS. However, some studies [22] continue to produce conflicting results. There are also several contraindications to MFIOL implantation, such as corneal aberrations, asymmetric capsulorrhexis, haptics deformation, or lens subluxation, all of which can lead to IOL decentration, an increase in higher order aberrations, and diminished object contrast discrimination [10].

1.1.3. Extended Depth of Focus (EDOF)

The Extended Depth of Focus (EDOF) technology, applied in IOLs such as the Tecnis Symfony (Johnson & Johnson Vision, Santa Ana, CA, US), was first approved by the US FDA in 2016. This recent innovation creates a single elongated focal point to enhance depth of focus and range of vision [21], effectively providing satisfactory near and intermediate vision while addressing limitations of MFIOLs, including negative photic phenomena such as glares and haloes [23,24]. Higher order aspheric monofocal IOLs, which are designed to provide improved intermediate vision, achieve this by redistributing power from the periphery to the centre of the IOL, enhancing depth of focus [25,26]. The structure of EDOF IOLs is based on a diffractive echelette design forming a step structure to achieve constructive interference of light from different lens zones to produce a novel light diffraction pattern [27]. Image quality is further enhanced through proprietary achromatic technology and negative spherical aberration correction [28]. Since 2016, several types of EDOF IOLs have been made commercially available, including the Mini WELL (Sifi Medtech, Catania, Italy), IC-8 (AcuFocus Inc., Irvine, CA, USA) and Wichterle Intraocular Lens-Continuous Focus (Medicem, Kamenné Zehrovice, Czech Republic).

EDOF IOLs enhance correction of chromatic aberration and maintain good CS that may be comparable to that of monofocal IOLs [21,29,30]. In terms of visual outcomes, EDOF IOLs have demonstrated better near [27] and intermediate vision [31] compared to monofocal IOLs, but worse outcomes than trifocal IOLs. However, there have been conflicting results regarding CS outcomes following EDOF implantation. Certain studies [27] have demonstrated a decrease in CS in eyes with EDOF IOLs under scotopic conditions, compared to eyes with monofocal IOLs. However, Pedrotti et al. [29] reported no significant difference, while Mencucci et al. [32] reported that EDOF IOLs performed significantly better than trifocal IOLs under both photopic and scotopic conditions.

1.1.4. Accommodative IOLs

Accommodative IOLs aim to preserve the ocular dioptric system accommodation capacity that is lost after cataract extraction [14]. These are dynamic IOLs that act independently of pupil size, creating a pseudo-accommodative phenomenon via anterior displacement of the lens optic plate, increasing the dioptric power of the eye and improving spectacle independence [33]. The design structure includes single optic, dual optic, and curvature change IOLs. They provide good far and intermediate vision as well as CS but are limited at near visual acuity and have post-operative outcome variability, necessitating further correction for near vision, and they may also have a higher risk of capsular contraction and opacification [34].

A recent review by Ong et al. [35] showed that accommodative IOLs had better near visual acuity at 6 months compared to monofocal IOLs but had greater posterior capsular opacification affecting distance visual acuity. Compared to MFIOLs, the optical plate in accommodative IOLs maintains the same power in every point without any transition areas, resulting in decreased adverse photic phenomenon such as glares, halos, blurs, and glows [17,36,37].

Accommodative IOL are advantageous for glaucoma patients as these lenses do not decrease CS [34]. However, they have increased risk of capsular contraction, commonly seen in pseudo-exfoliation patients, a condition associated with weakened zonules, further decreasing functionality of IOL accommodative system [34].

1.1.5. Toric IOLs

Toric IOLs are astigmatism-correcting lenses that allow for a specific focal point. Zvornicanin et al. [2] conducted a review of recent trials utilizing premium IOL in eyes with cataract without any other ocular pathology and demonstrated a UDVA of 0.3 logMAR in 70–95% of patients, with a residual astigmatism of 1 D or less noted in 67–88% of patients, and spectacle independence reported in 60–85% of patients.

2. Materials and Methods

2.1. Search Strategy

A literature search was performed in MEDLINE bibliographic database from inception to 1 June 2023. The following key terms were utilized in combination: "multifocal*", "bifocal*", "trifocal*", "diffractive*", "EDOF", "extended depth of focus". The detailed search strategy is available in Supplementary Material (Table S1). References of sources and previous reviews were hand-searched to identify additional relevant articles. Articles were viewed through Rayyan (Qatar Computing Research Institute, Doha, Qatar) and duplicates were identified and removed.

2.2. Study Selection

Studies which reported surgical outcomes of premium IOL implantation in glaucoma eyes were shortlisted for data extraction. The article sieve was conducted by two independent reviewers (ASYH, ED), and each article was reviewed by both reviewers who were blinded to each other's decisions. Disputes were resolved through consensus discussion between the reviewers, followed by arbitration from a third reviewer if necessary. The inclusion and exclusion criteria are listed in Table 1.

Table 1. Inclusion and exclusion criteria.

Inclusion Criteria	Exclusion Criteria
Population: 1. Patients with visually significant cataract (unilateral or bilateral) and glaucoma. Studies: 1. Randomized controlled trials, case series, prospective and retrospective studies. Interventional Arm: 1. Phacoemulsification or femtosecond laser cataract surgery with premium IOL implantation, performed with or without concurrent glaucoma surgery.	Population 1. Primary surgeries other than phacoemulsification with intraocular lens implant and glaucoma surgery (e.g., corneal inlays); 2. Secondary surgeries; 3. Concomitant ocular pathology besides glaucoma: keratopathy, maculopathy, retinopathy, optic neuropathy, as well as any ocular condition that is deemed to confound visual acuity assessment; 4. Non-visually significant cataracts (e.g., clear lens). Studies 1. Non-published studies; 2. Studies not written in English. Interventional Arm 1. Clear lens extraction or refractive lens exchange; 2. Extracapsular cataract extraction; 3. Intracapsular cataract extraction.

2.3. Data Extraction

For each included trial, two reviewers (ASYH, ED) extracted data at the longest point of follow up, abstracted them into an Excel spreadsheet (Microsoft Corp, Albuquerque, NM, USA), and checked for conflicting data entries. Differences were discussed and resolved with a third reviewer where necessary.

Data extracted included study characteristics, baseline patient information (age, gender, type of pathology and surgery), baseline visual field parameters, Pre-operative intraocular pressure, and number of glaucoma medications. For continuous variables, mean and standard deviation were abstracted. For categorical variables, frequency and percentages were abstracted.

All outcomes pertained to surgical results following premium IOL implantation in eyes with glaucoma. Primary outcomes included uncorrected and corrected distance and near visual acuity. Secondary outcomes included spectacle independence, photic phenomena (glares and haloes), astigmatism, contrast sensitivity, and patient satisfaction.

2.4. Quality and Risk of Bias Assessment

Assessments on the risk of bias and certainty of evidence was performed on the final list of studies included in our review. Risk of bias was ascertained at the study level and assessed by 2 reviewers independently and in duplicate. Conflicts were resolved by consensus, with arbitration by a third reviewer if necessary.

An assessment of the methodology quality of the cohort studies was performed using the domains of the Newcastle–Ottawa Scale [38], considering (1) the selection of cohorts; (2) the comparability of cohorts; and (3) the assessment of outcomes. Studies with <5 stars are considered low quality, 5–7 stars moderate quality, and >7 stars high quality. Non-controlled trials (such as case reports and case series) included in this study were assessed using the modified Newcastle–Ottawa scale [39] based on four domains: selection; ascertainment; causality; and reporting.

3. Results

Our search yielded 1404 records in total. After screening based on title and abstract, 1391 references were excluded. Full-text assessment was performed on the remaining 13 records; 2 were not retrieved, and 1 additional study was included from citation searching (Figure 1). Twelve studies [22,40–50] fulfilled the inclusion criteria.

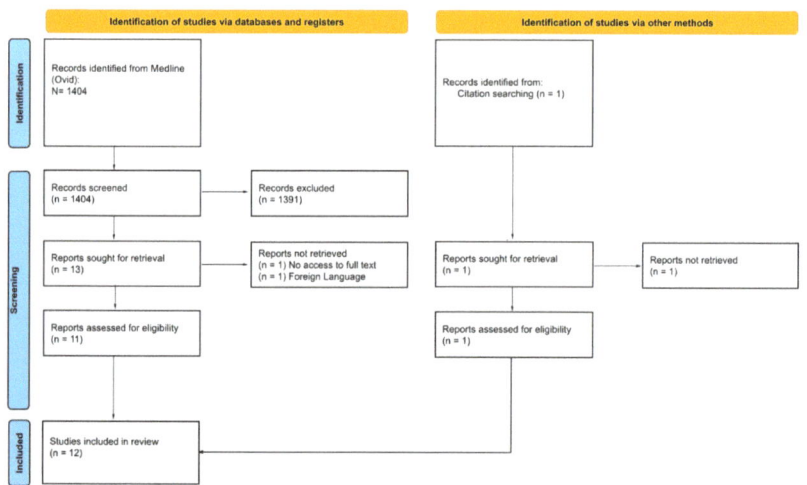

Figure 1. Preferred reporting items for systematic reviews and meta analyses (PRISMA) flow diagram.

3.1. Methodology Quality and Risk of Bias

The results of the risk of bias assessment conducted using the Newcastle–Ottawa Scale [1] for cohort studies and modified Newcastle–Ottawa scale [2] for case results are shown below (Tables 2 and 3).

Table 2. Quality assessment using the Newcastle–Ottawa scale of included cohort studies (included studies: [3–10]).

Author (Year)	Selection				Comparability	Outcome			Total Score
	Representativeness of Exposed Cohort (Maximum:*)	Selection of Non-Exposed Cohort (Maximum:*)	Ascertainment of Exposure (Maximum:*)	Demonstration that the Current Outcome of Interest Was Not Present at Start of Study (Maximum:*)	Comparability of Cohorts on the Basis of the Design or Analysis (Maximum:*)	Assessment of Outcome (Maximum:*)	Was Follow Up Long Enough for Outcomes to Occur (Maximum:*)	Adequacy of Follow Up of Cohorts (Maximum:*)	
Ichioka 2022 [4]	*		*	*	*	*	*	*	7
Sanchez-Sanchez 2021 [7]	*	*	*	*	*	*	*	*	8
Takai 2021 [10]	*	*	*	*	*	*	*	*	8
Lopez Caballero 2022 [6]	*		*	*	*	*	*	*	8
Ichioka 2021 [3]	*		*	*	*	*	*	*	7
Kamath 2000 [8]	*	*	*	*	*	*	*	*	8
Rementeria-Capelo 2022 [5]	*	*	*	*	*	*	*	*	8
Kerr 2023 [9]	*		*	*	*	*	*	*	7

Table 3. Quality assessment using the modified Newcastle–Ottawa scale (M-NOS) for other included studies (included studies: [11–14]).

Author (Year)	* Selection	Ascertainment		Causality				Reporting
	Q1	Q2	Q3	Q4	Q5	Q6	Q7	Q8
Bissen Miyajima 2023 [11]	No	Yes	Yes	No	No	No	Yes	Yes
Brown 2015 [12]	No	Yes	Yes	No	No	No	Yes	Yes
Ouchi 2015 [13]	No	Yes	Yes	No	No	No	Yes	Yes
Ferguson 2023 [14]	No	Yes	Yes	No	No	No	Yes	Yes

* M-NOS components.

All eight cohort studies [3–10] had high NOS ranks and the mean NOS score was 7.625. Three studies lost points because there was absence of a control group with non-glaucomatous eyes. The findings of both reviewers were similar, regardless of whether the material appeared biased.

1. *Selection*: (Question 1) Does/Do the patient(s) represent(s) the whole experience of the investigator (center), or is the selection method unclear to the extent that other patients with similar presentations may not have been reported?
2. *Ascertainment:* (Question 2) Was the exposure adequately ascertained? (Question 3) Was the outcome adequately ascertained?
3. *Causality:* (Question 4) Were other alternative causes that may explain the observation ruled out? (Question 5) Was there a challenge/rechallenge phenomenon? (Question 6) Was there a dose-response effect? (Question 7) Was follow-up long enough for outcomes to occur?
4. *Reporting:* (Question 8) Is/Are the case(s) described in sufficient detail to allow other investigators to replicate the research or to allow practitioners to make inferences related to their own practice?

3.2. Patient Characteristics

The qualitative analysis included a pooled total of 399 glaucomatous eyes from 12 studies. The mean age was 73.8 years old, with a male-to-female ratio of 19:20. Three studies originated from the USA, four studies from Japan, three studies from Spain, one study from Australia, and one study from the UK. Three studies [40,41,49,50] reported on EDOF IOL outcomes, six studies [22,42,44–47] reported on toric IOL outcomes, one study [48] reported on MFIOL outcomes, and one study [43] reported outcomes from EDOF, bifocal, and trifocal IOL implantation. Study characteristics (author, publication year, sample size, age range, IOL type, surgery type, and outcomes) were extracted and summarized in Tables S2–S4 (Supplementary Material).

3.3. Surgical Outcomes from Trials

3.3.1. Spectacle Independence

Five studies [22,40,43,49,50] reported on spectacle independence in glaucomatous eyes. Ferguson et al. [40] implanted 52 eyes with mild open angle glaucoma with EDOF non-toric or toric IOLs (AcrySof IQ Vivity/AcrySof IQ Vivity Toric, Alcon Laboratories, Fort Worth, TX, USA) and demonstrated a high rate of spectacle independence post-operatively in both toric and non-toric IOL groups (spectacle independence rates: 92% for distance tasks; 50% for intermediate tasks; and 38% for near tasks). Ouchi et al. [22] implanted 15 eyes (11 patients) with coexisting ocular pathologies (including 4 glaucoma eyes—1 with acute angle closure glaucoma; 3 with normal tension glaucoma) with MFIOLs (LENTIS MPlus LS-313MF30 and the LENTIS Mplus Toric LU-313MFT (Oculentis GmbH, Berlin, Germany)). All patients were completely spectacle independent for distance vision. For distance vision, 11 patients (100%) rated their quality of vision as 4 or higher (very good or good) among 5 items, and 7 patients (64%) rated it as 5 (very good). For near vision, the results of the glaucoma patients were not individually shared and thus not reported in this review. Of note, however, is that Ouchi et al. included only a small sample size of glaucoma eyes (4/15 eyes with coexisting ocular pathology). Sanchez-Sanchez et al. [43] implanted bifocal (AcrySof ReSTOR +3.00, Alcon Laboratories, Fort Worth, TX, USA) and trifocal (AcrySof Panoptix, Alcon Laboratories, Fort Worth, TX, USA, Fine Vision PhysIOL, Liege, Belgium) IOLs in nine patients with glaucoma (77.8% Bifocal Lens; 22.2% Trifocal Lens) and in nine patients with pre-perimetric glaucoma (77.8% Bifocal Lens; 22.2% Trifocal Lens). A total of 68% of patients achieved spectacle independence for distance tasks, 89% for intermediate tasks, and 56% for near tasks. Rementeria-Capelo et al. [49] evaluated visual outcomes in 25 control patients and 25 study patients with ocular pathology. Study patients included six patients with glaucoma and two with ocular hypertension undergoing bilateral combined iStent and cataract surgery with EDOF IOL (AcrySof IQ Vivity; Alcon Laboratories, Fort Worth, TX, USA). All study patients were spectacle independent for distance. For near vision, the results of the glaucoma patients were not individually shared and thus not reported in this review. Of note, however, is that Rementería-Capelo et al. [49] included only a small sample size of glaucoma eyes (8/25 study patients). Kerr et al. [50] implanted an EDOF IOL (AcrySof IQ Vivity; Alcon Laboratories, Fort Worth, TX, USA) in 32 glaucomatous eyes (29 with primary open angle glaucoma; 3 with secondary open-angle glaucoma) and a monofocal IOL (Clareon/SN6Atx/SN60WF; Alcon Laboratories, Fort Worth, TX, USA) in 26 glaucomatous eyes (23 with primary open angle glaucoma; 3 with secondary open angle glaucoma), with a trans-trabecular micro-bypass stent (iStent; Glaukos Corp., San Clemente, CA, USA)) or Schlemm canal microstent (Hydrus Microstent; Ivantis, USA) concurrently implanted in 14 eyes in the EDOF group. In the EDOF group, spectacle independence was high, with 13 patients never requiring spectacles, 3 patients rarely requiring spectacles for distance and intermediate activities, and 7 patients never requiring spectacles for near activities. Spectacle independence for intermediate and near activities was significantly better in the EDOF group compared to the monofocal group; the number of patients in the monofocal group always requiring spectacles for intermediate and near activities was 4 and 11, respectively.

In summary, five studies reported high spectacle independence rates for distance vision. These results were consistent across different IOL types (including bifocal, trifocal, and toric IOLs) and observed also in studies examining outcomes of premium IOL implantation in combination with glaucoma surgery. For near vision, specific outcomes for glaucomatous eyes were not reported in two out of the five studies. In the remaining three studies, the results varied for near vision.

3.3.2. Contrast Sensitivity

Five studies [22,40,41,43,49] reported CS outcomes following premium IOL implantation in glaucomatous eyes. Sanchez-Sanchez et al. [43] reported that patients with glaucoma implanted with MFIOLs had poorer monocular visual acuity than healthy controls and lower contrast sensitivity values at high spatial frequencies—at 12 cycles per degree, binocular CS values for healthy, glaucoma, and pre-perimetric glaucoma eyes were 2.11, 1.87, and 2.05, respectively. However, there was no clinically significant difference in CS between patients with pre-perimetric glaucoma and healthy controls. Ouchi et al. [49] demonstrated that even after MFIOL implantation, CS in all eyes with various ocular pathologies including glaucoma patients (a prior history of acute glaucoma, NTG) were still comparable to those of normal healthy subjects. Ferguson et al. [40] showed favorable CS results following EDOF IOL implantation: mean binocular mesopic CS achieved was 1.76 ± 0.16 at a spatial frequency of 1 cycle-per-degree (cpd). Bissen-Miyajima et al. [41] evaluated outcomes of diffractive EDOF IOLs (Symfony®, models ZXR00V and ZXV150-375, Johnson and Johnson Surgical Vision, Santa Ana, CA, USA) in 16 NTG eyes and demonstrated that their post-operative visual function was mostly comparable to those of normal eyes following implantation of the same IOLs, with post-operative CS within the normal range, except for four eyes at 18 cycles per degree. Rementeria-Capelo et al. [49] showed no difference in CS after EDOF IOL implantation in all patients in both control and study groups with glaucoma and ocular hypertensive patients.

In summary, five studies have demonstrated that post-operative CS values in glaucomatous eyes remained comparable to healthy subjects after MFIOL and EDOF implantation. However, one study [43] observed that glaucoma patients implanted with MFIOLs had poorer monocular visual acuity and lower CS at high spatial frequencies compared to healthy controls and patients with pre-perimetric glaucoma.

3.3.3. Visual Acuity

All 12 studies reported visual acuity outcomes. Kamath et al. [48] reported outcomes following AMO Array MFIOL (Allergan Medical Optics) implantation in 81 eyes (70 patients) with ocular pathology (study group), including 11 glaucomatous eyes and 6 eyes with ocular hypertension. The control group had implantation of monofocal IOL of similar design (AMO SI-40NB) and included 12 glaucomatous eyes. Within the study group, 29% achieved an Uncorrected Distance Visual Acuity (UDVA) of $\geq 6/12$, while 82% achieved $\geq$N8 near vision, and 24% achieved $\geq$both 6/12 and N8 vision. Within the study group, 94% achieved a Best Corrected Visual Acuity (BCVA) of $\geq 6/12$, while 88% achieved a BCVA of $\geq$N8 for near vision, and 88% achieved a BCVA of $\geq$both 6/12 and N8 near vision. Takai et al. [44] compared the post-operative refractive status in 20 eyes (20 patients) implanted with toric (10 eyes) and non-toric (10 eyes) IOLs during combined cataract surgery and micro-hook ab interno trabeculectomy. This study showed that the mean Uncorrected Visual Acuity (UCVA) of the Toric IOL group (logMAR 0.23 ± 0.25) was significantly better than that of the non-toric IOL group (logMAR 0.45 ± 0.26) at 3 months post-operatively ($p < 0.05$). Bissen-Miyajima et al. [41] showed that the post-operative visual outcomes (distance-corrected visual acuity, contrast sensitivity) of glaucoma patients following EDOF IOL implantation was almost comparable to those of normal eyes with the same IOLs implanted and were within normal ranges. Ichioka et al. (2021) [46] investigated the effect of toric IOL implantation on visual acuity and astigmatism in 20 POAG eyes with a pre-existing corneal astigmatism of -1.5 D, following combined cataract surgery with micro-hook ab interno trabeculotomy. Post-operatively, the logMAR UCVA was significantly better in the toric group (toric, 0.07 ± 0.07; non-toric, 0.33 ± 0.30; $p = 0.0020$). Ichioka et al. (2022) [42] investigated outcomes of toric IOL implantation on visual acuity and astigmatism in 18 POAG eyes, with a pre-existing corneal astigmatism of -1.5 D, following combined iStent implantation and cataract surgery. Pre-operatively, both groups had similar logarithm of the minimum angle of resolution (logMAR) UCVAs; post-operatively, the logMAR UCVA was significantly better in the toric group (non-toric, 0.45 ± 0.31; toric, 0.14 ± 0.15; $p = 0.021$). Ferguson et al. [40] implanted 52 POAG eyes with an EDOF IOL—(AcrySof IQ Vivity; Alcon Laboratories, Fort Worth, TX, USA) and showed favorable UDVA and Uncorrected Intermediate Visual Acuity (UIVA) at 4 months post-operatively. The mean binocular UDVA and CDVA were 0.03 ± 0.12 LogMAR and -0.06 ± 0.07 LogMAR, respectively. The mean UIVA and UNVA were 0.18 ± 0.12 LogMAR and 0.31 ± 0.18 LogMAR, respectively. A total of 85% of the subjects achieved $\geq 20/25$ UDVA, and 77% of the subjects achieved $\geq 20/32$ UIVA at 4 months post-operatively. Lopez-Caballero [45] compared 26 eyes undergoing iStent implantation and phacoemulsification with implantation with the AcrySof toric IOL (Alcon Laboratories, Fort Worth, TX, USA) in the study group and 41 eyes undergoing isolated phacoemulsification with toric IOL implantation in the control group. Toric IOLs were also implanted in patients with advanced visual field damage (control group: 13 mild glaucoma, 17 moderate glaucoma, 11 severe glaucoma; study group: 11 mild glaucoma, 7 moderate glaucoma, 8 severe glaucoma). There were 39 POAG, 1 closed angle glaucoma, and 2 pseudo-exfoliation glaucoma patients in the control group, while 18 POAG, 3 close angle glaucoma, 2 pseudo-exfoliation glaucoma, and 3 pigmentary glaucoma patients were included in the study group. Despite severe visual field loss, patients still achieved excellent uncorrected post-operative vision in eyes targeted for emmetropia (0.04 LogMar in the cataract group and 0.03 in the combined surgery group). Brown et al. [47] implanted AcrySof toric IOLs (Alcon Laboratories, Fort Worth, TX, USA) in 126 eyes of 87 patients with glaucoma and corneal astigmatism. The UDVA was 0.04 ± 0.08 logMAR for all eyes, and 98% of all eyes achieved an UDVA of Snellen's 20/40 or better, with 76% achieving 20/25 or better and 47% achieving 20/20. The CDVA for all eyes was 0.01 ± 0.03 logMAR post-operatively. Rementeria-Capelo et al. [49] reported excellent visual acuity results in all patients in both the control and study groups after EDOF implantation, with both groups achieving a mean binocular uncorrected visual acuity better than 0.0 logMAR. Statistically significant differences were only found for uncorrected monocular acuity and at the +2.5 D value of the defocus curve, although these differences were unlikely to be clinically relevant. Monocular UDVA was better in the control group (-0.01 ± 0.07) compared with the study group (0.03 ± 0.08), $p = 0.027$. There were no other statistically significant differences in DVA, with an uncorrected binocular acuity of -0.06 ± 0.06 for the control group and -0.05 ± 0.06 for the study group. Kerr et al. [50] implanted EDOF IOLs (AcrySof IQ Vivity; Alcon Laboratories, Fort Worth, TX, USA) in 32 glaucomatous eyes and monofocal IOLs (Clareon/SN6ATx/SN60WF; Alcon) in 26 glaucomatous eyes. UIVA (0.06 ± 0.16 versus 0.39 ± 0.10 LogMAR; $p < 0.001$) and UNVA outcomes (0.29 ± 0.10 versus 0.55 ± 0.18 LogMAR; $p < 0.001$) were significantly better in the EDOF group than in the monofocal group, respectively.

In summary, all studies showed excellent visual acuity results. In studies [40,44,46] comparing toric and non-toric IOLs, glaucomatous eyes implanted with toric IOLs showed better UCVA results. Studies on EDOF IOLs generally showed favorable UDVA and UIVA outcomes. However, one study [49] showed that UDVA was still better in the control group with normal eyes compared to the study group. Another study [43], examining bifocal and trifocal IOLs in 38 patients (9 glaucoma, 9 pre-perimetric glaucoma, 11 healthy), also showed that healthy patients had statistically better monocular UDVA, CDVA, and LCVA than patients with glaucoma for all values, except for binocular 10% contrast-corrected VA. Excellent uncorrected post-operative visual results were achieved even when toric IOLs were implanted in patients with advanced visual field damage.

3.3.4. Astigmatism

Five studies [42,44–47] reported astigmatism outcomes. Ichioka et al. 2021) [46] included 20 POAG eyes (20 patients) with pre-existing corneal astigmatism exceeding -1.5 D implanted with either non-toric (n = 10) or toric IOLs (n = 10). Post-operatively, residual astigmatism was significantly less in the toric IOL group compared to the non-toric IOL group (toric, -0.63 ± 0.56 D vs non-toric, -1.53 ± 0.74 D, $p = 0.0110$; toric, 70% of eyes vs non-toric, 10% of eyes had 1.0 D or less astigmatism). Vector analyses showed the post-operative centroid magnitude of astigmatism was less in the toric IOL group (0.23 D at 83 degrees) than the non-toric IOL group (1.03 D at 178 degrees). Takai et al. [44] showed that the mean absolute residual cylinder in the non-toric IOL group (2.25 ± 0.62 D) was significantly greater than that of the toric IOL group (1.30 ± 0.68 D) ($p < 0.05$). Post-operatively, 60% of eyes in the toric IOL group and 10% in the non-toric IOL group had an absolute astigmatism level of 1.5 D or less. Brown et al. [47] showed that toric IOLs can reliably reduce astigmatism and improve uncorrected vision in eyes with cataract and glaucoma. Astigmatism improved from 1.47 ± 1.10 D to 0.31 ± 0.37 D post-operatively. The residual cylinder was 1.00 D or less in 97% of eyes, 0.75 D or less in 90% of eyes, and 0.50 D or less in 83% of eyes. Ichioka et al. (2022) [42] included 18 POAG eyes with pre-existing corneal astigmatism exceeding -1.5 D implanted with non-toric (n = 10) or toric (n = 10) IOLs. Astigmatism decreased significantly in the toric group post-operatively compared to the non-toric group (non-toric, -2.03 ± 0.63 D; toric, -0.67 ± 0.53 D; $p = 0.0014$) Vector analyses showed the post-operative centroid magnitude and confidence eclipses of astigmatism was less in the toric group (0.47 D at 173° ± 0.73 D) than the non-toric group (1.10 D at 2° ± 1.91 D). Post-operatively, 78% of eyes in the toric group had 1.0 D or less refractive astigmatism compared with 11% in the non-toric group. Lopez-Caballero [45] compared 26 eyes with iStent and toric IOL implantation in the study group and 41 eyes undergoing isolated phacoemulsification with toric IOL implantation in the control group. The mean post-operative refractive cylinder was 0.26 D in the control and 0.11 D in the iStent group.

In summary, studies have demonstrated that toric IOLs provide predictable and good astigmatism outcomes in glaucomatous eyes undergoing standalone cataract surgery or in combination with selected glaucoma surgeries.

3.3.5. Patient Satisfaction

Five studies [22,40,43,49,50] reported on patient satisfaction via visual questionnaires. Ferguson et al. [40] implanted 52 eyes (26 patients) with POAG, with EDOF non-toric or toric IOLs (AcrySof IQ Vivity or AcrySof IQ Vivity Toric (Alcon Laboratories, Fort Worth, TX, USA)) and found that 85% of subjects reported they would choose the same lens again. Sanchez-Sanchez et al. [43] concluded that MFIOLs may be implanted in patients with pre-perimetric glaucoma with little fear of patient dissatisfaction. Interestingly, Rementeria-Capelo et al. [49] reported that all patients in the study group (including glaucoma patients) had a higher satisfaction with their visual performance than patients in the control group (average satisfaction in control group: 3.52 ± 0.51; study group: 0.84 ± 0.37, 52% ($p = 0.016$), and patients in the control group reported greater difficulty in reading newspapers ($p = 0.030$). All patients stated they would undergo surgery again with the same type of IOL. Kerr et al. [50] implanted EDOF IOLs (AcrySof IQ Vivity; Alcon Laboratories, Fort Worth, TX, USA) in 32 glaucomatous eyes and monofocal IOLs (Clareon/SN6Atx/SN60WF; Alcon) in 26 glaucomatous eyes. Patient satisfaction was significantly higher in the EDOF group for distance, intermediate, and near vision than in the monofocal group. All EDOF patients were "very satisfied" with their distance and intermediate vision compared to 9/13 (69.2%) and 6/13 (46.2%) in the monofocal group, respectively. For near vision, 12/16 (75.0%) in the EDOF group were "very satisfied" with their unaided near vision compared to 5/17 (38.5%) in the monofocal group ($p = 0.059$). Patients who received an EDOF lens were more likely to report that they would choose the same lens again (16/16 (100%) in the EDOF group compared to 10/13 (76.9%) in the monofocal group; $p = 0.085$).

Ouchi et al. [22] evaluated outcomes of 11 patients (15 eyes with coexisting ocular pathologies, 1 eye with past history of acute angle closure of Aulhorn classification stage 3 glaucoma, and 3 NTG eyes of Aulhorn classification stage 3) that underwent implantation of LENTIS Mplus (Oculentis GmbH). No patient reported poor or very poor vision quality. For distance vision, all 11 patients (100%) rated their quality of vision as 4 or higher (very good or good) among 5 items. For near vision, all 11 patients (100%) rated their quality of vision as 3 (acceptable) or higher.

In summary, studies have suggested a high level of patient satisfaction following the implantation of various premium IOLs in glaucoma patients, with two studies [49,50] reporting higher post-operative satisfaction in glaucomatous eyes compared to control eyes and a high percentage of patients stating they would choose the same IOL again.

3.3.6. Glares and Haloes

Three studies reported on glare outcomes. Ferguson et al. [40] implanted 52 POAG eyes with EDOF non-toric or toric IOLs and assessed glares and haloes reported on a scale of 1–5 (1 = not at all, 5 = extremely). Subjects reported a mean response of 2.6 ± 1.3 when asked if they noted glare/halos in dim light situations. However, 65% of the subjects reported not being bothered or only having very little dissatisfaction with glare/halo symptoms. Bissen-Miyajima et al. [41] evaluated outcomes of diffractive EDOF IOLs in 16 NTG eyes. Most cases reported the severity of glares, halos, and starbursts as none, mild, or moderate. Only one subject reported severe halos and another reported severe starburst. Kerr et al. [50] implanted EDOF IOLs in 32 glaucomatous eyes and monofocal IOLs in 26 glaucomatous eyes. Most participants did not experience any photic phenomena (glares/halos/starbursts), and there was no significant difference in the incidence of photic phenomena between both groups. Seven patients in the EDOF group reported glare, compared to six patients in the monofocal group.

Overall, across the three studies examining EDOF IOL implantation, despite subjects often reporting at least some level of glare or halo effects, many were not significantly bothered by these symptoms. There was no significant difference in photic phenomena between EDOF and monofocal IOLs in glaucomatous eyes. However, in one study, ocular pathology appeared to be associated with an increased incidence of halos.

4. Discussion: Considerations in Premium IOL Implantation in Glaucoma

Published literature has discussed a range of factors that may be considered when contemplating premium IOL implantation in glaucoma eyes.

4.1. Contrast Sensitivity

Contrast sensitivity (CS) is the measure of an individual's ability to detect a difference in luminance between two areas [51] and, in this context, may be affected by various factors including both the IOL and glaucoma. Decreased CS in glaucoma patients have been well documented [51–53]. Already in early disease, patients begin to lose retinal ganglion cells and retinal nerve fibre layer (RNFL) thickness, resulting in structural and functional changes causing a decrease in contrast sensitivity (CS). Glaucoma preferentially affects CS more than visual acuity (VA), and the decrease in CS correlates with the degree of structural and functional glaucomatous damage [51]. CS losses occur even in individuals with minimal or no field loss (<3 dB) and a relatively good VA (0.3 log MAR or better) [53]. CS has been found to correlate with visual field (VF) sensitivity and affects vision-related quality of life [53,54]. At mesopic levels, CS is correlated with visual field loss and affects glaucoma patients ability to perform daily activities and negatively impacts their quality of life [55].

MFIOLs, particularly refractive MFIOLs [6], have been shown to result in decrease in CS, where the mesopic CS is worse than photopic sensitivity, and where the loss is greater at higher compared to lower spatial frequencies after MFIOL implantation [56].

Farid et al. [19] stated that as the amount of light energy in focus at any given focal distance is reduced, out-of-focus light is superimposed, and approximately 18% of transmitted light in diffractive IOLs (which may vary depending on IOL design) is lost to higher orders of diffraction that are never focused on the retina. As a result, patients with multifocal IOLs may experience glare and halos, as well as reduced contrast sensitivity.

Hence, while MFIOLs may be considered in patients with early glaucoma or controlled ocular hypertension, they should be avoided in patients with uncontrolled and advanced glaucoma [18]. Cao et al. [20] found that both refractive and diffractive MFIOL subgroups had a lower CS, and Hawkins et al. [51] showed that reduced CS is significantly correlated with visual field losses in patients with glaucoma.

Nonetheless, studies implanting MFIOLs in glaucomatous eyes have shown promising results. Sanchez Sanchez et al. [43] found no clinically significant difference in CS between eyes with pre-perimetric glaucoma and healthy controls after MFIOL implantation. Ouchi et al. [22] showed that following MFIOL implantation, CS in all eyes with ocular pathologies were comparable to that of healthy eyes, and concluded that with careful case selection, sectorial refractive MFIOL may be effective in eyes with concurrent ocular pathology. Furthermore, ongoing developments in IOL technology such as IOL asphericity, used both in monofocal IOLs and premium IOLs today, have also improved CS outcomes after cataract surgery. Trueb et al. [57] demonstrated that eyes implanted with the aspheric AcrySof IQ IOL had better photopic and mesopic CS at medium and high spatial frequencies than in eyes implanted with the spherical AcrySof SN60AT IOL. Deshpande et al. [58] demonstrated that the optical design of aspheric IOLs reduced spherical aberrations and increased CS. Alternatively, accommodative IOLs may be considered advantageous as these lenses do not depend on pupil size and do not decrease CS [34]. However, they have increased risk of capsular contraction, commonly seen in pseudo-exfoliation patients, a condition associated with weakened zonules, further decreasing the functionality of IOL accommodative system [34].

4.2. Glaucomatous Visual Field (VF) Defects

The severity, extent, and location of glaucomatous VF defects are also considered when deciding if a glaucoma patient would benefit from premium IOLs.

Studies have utilized Octopus 101 autoperimetry, Goldmann manual perimetry, frequency doubling technology matrix perimetry, the automated Esterman binocular field test, and Humphrey Visual Field 30-2 perimetry testing in exploring visual field outcomes after MFIOL implantation.

Prior reviews have suggested that only glaucoma suspects and ocular hypertensive patients with no optic disc or visual field damage who have been stable for a longer period of time should be candidates for MFIOL implantation [6,34]. In addition, good control and stability of visual field damage, with no evidence of progression, has also been suggested to be a necessary prerequisite for premium IOL implantation in glaucomatous eyes [34]. With respect to lens choice, two recent studies [19,59] have shown that MFIOLs decrease the mean deviation (MD) of visual field tests with the Humphrey field analyzer (Carl Zeiss Meditec, Jena, Germany) compared to diffractive bifocal IOLs, monofocal IOLs or phakic eyes.

Farid et al. [19] demonstrated a significant depression of approximately 2 dB in HVF 10-2 testing in healthy eyes which underwent diffractive MFIOL implantation compared to monofocal IOL implantation. Aychoua et al. [59] demonstrated similar results on HVF 30-2 testing and concluded that it was likely due to reduction in differential light sensitivity [60]. Kang et al. [61] supported this finding, showing that patients with MFIOLs (3M diffractive bifocal IOL) have greater reduction of visual field on the Goldmann manual perimetry, compared to patients with monofocal IOLs, and this was reflected across different spot sizes and intensities.

However, Bi et al. [62] examined differences in Octopus 101 autoperimetry results between patients with MFIOLs (AcrySof ReSTOR SA60D3) and patients with monofocal IOLs (AcrySof SN60AT) and found no significant difference between the two groups.

4.3. Glaucoma Subtype

The varying characteristics of different subtypes of glaucoma may also influence the decision of premium IOL implantation in eyes with glaucoma. Within the spectrum of angle-closure disease, primary angle closure suspects (PACS) have no glaucomatous nerve damage nor functional visual defects, may experience angle opening after cataract extraction, and may benefit from premium IOLs. It should be noted however, that prior laser peripheral iridotomy is associated with a higher risk of zonulysis [63] and lens subluxation [64,65], which would pose challenges to premium IOL implantation. Eyes with established primary angle closure glaucoma (PACG) have functional VF defects and are also at a higher risk of disease progression compared to eyes with POAG [66,67], and this should be taken into consideration, even if there is only mild disease severity at baseline. In addition, angle-closure disease is associated with shorter axial length. This may increase the risk of refractive surprise and may be a consideration when choosing premium IOLs, although this may be much less of a concern today given the significant improvements in biometry and lens calculation methods.

Pseudo-exfoliation glaucoma (PXG) is a form of secondary open angle glaucoma that arises due to the deposition of extracellular material in the anterior chamber, trabecular meshwork, and other tissues in the eye [34], resulting in raised IOP and glaucomatous optic nerve damage. These eyes have poorly dilating pupils, a higher risk of zonular instability, a higher risk of uncontrolled IOP post-operatively [68], and may experience greater post-operative inflammation from vascular instability [69]. These factors are likely to pose challenges to premium IOL implantation and diminish their benefit with suboptimal post-operative refractive outcomes. First, intra-operatively, poorly dilating pupils limit the view of axis markings on toric IOLs and require additional pupil maneuvering during surgery. Second, zonular dialysis, posterior capsule rupture [70], and vitreous loss [71,72] which may occur during surgery often precludes the implantation of premium IOLs in these eyes. Third, the increased risk of posterior capsular opacification and capsular phimosis over time, with progressively weakening zonular support, has been shown to lead to a higher rate of IOL subluxation [73]. PXG is associated with significant IOL axis misalignment [74] and has been identified as the most common cause of IOL dislocation, often occurring less than a decade even after uncomplicated cataract surgery [75,76]. Late in-the-bag spontaneous intraocular lens dislocation [77] and progressive IOL decentration are not uncommon [78]. Hence, premium IOLs, which require precise centration and whose refractive outcomes are particularly affected by anterior capsule phimosis and zonulysis, have been mostly avoided in PXG eyes [34,79]. PXG eyes have an increased risk of refractive surprise after phacoemulsification: Manoharan et al. [80] analyzed refractive outcomes of phacoemulsification cataract surgery in glaucoma patients and showed that the odds of refractive surprise being greater than ±1.0 D were higher in patients with pseudo-exfoliation glaucoma (n = 23 eyes) compared with patients without glaucoma (OR = 7.27, p = 0.0138). Fourth, PXG has a greater risk of progression compared to other glaucoma subtypes following cataract surgery, the result of uncontrolled IOP and greater IOP fluctuation post-operatively [81]. Hence, implantation of premium IOLs even in PXG eyes with mild glaucoma should be cautioned against. Finally, the higher rate and greater severity of post-operative inflammation, iritis, and cellular precipitates [71] in these eyes may compromise visual acuity and reduce the benefit of premium IOL implantation in eyes with PXG.

4.4. Ocular Surface Disease (OSD)

OSD has been found to be present in 59% of glaucoma patients [82], with its incidence related to previous or ongoing conjunctival inflammation and scarring, tear film instability, advanced age, and the use of chronic anti-glaucoma medications. Glaucoma itself is associated with dry eye disease (DED), with approximately 11% of 5 million Americans over 50 years old found to have coexisting glaucoma and DED [83]. Untreated POAG patients also have a higher risk of ocular surface disease due to a 22% lower basal tear turnover rate, compared to patients without glaucoma [84]. The severity of OSD appears to increase with worsening glaucoma severity and results in decreasing quality of life [85]. Chronic anti-glaucoma medications themselves are a risk factor for OSD [83]. Prostaglandin analogues, for example, have been associated with meibomian gland dystrophy (MGD) [86], which leads to unstable tear film and fluctuating vision.

Severe OSD and DED are associated with inaccurate biometry and topography, which may result in inaccurate IOL selection [87,88]. Hence, the ocular surface would need to be carefully evaluated and optimized prior to biometry to ensure accurate ocular measurements and IOL power calculation, which is crucial in premium IOL selection. Post-operatively, OSD would still require continual management to ensure maximal visual benefit from premium IOL implantation. Fortunately, the IOP-lowering effect of cataract surgery itself may reduce the glaucoma medication burden post-operatively, thereby alleviating OSD and improving patients' quality of life after surgery [89].

4.5. Axial Length (AL) and Anterior Chamber Depth (ACD)

Successful pseudo-phakic rehabilitation after combined cataract and glaucoma surgery requires accurate IOL power calculation, which depends on precise measurements of the AL and ACD [90]. Eyes with glaucoma are associated with extremes in AL, with open angle glaucoma associated with high myopia [91] and increased AL [92], while angle-closure eyes have been shown to have shorter ALs [93]. Extreme ALs may, in the past, have been associated with increased risk of IOL calculation errors; however, modern IOL formulae have proven accurate even for extreme axial lengths, and they may lessen the influence of AL on the accuracy of IOL power calculations [94,95]. In a network meta-analysis, Lu et al. [95] investigated eight formulae: including Barrett Universal II, Kane, SRK/T, Hoffer Q, Haigas, Holladay I, Hill-Radial Basis Function (RBF 3.0) and Ladas Super Formula (LSF) used for IOL power calculations in eyes with PACG and found no significant differences in outcomes among all the above formulae. Studies have also shown that greater changes in ACD may occur post-operatively in eyes with a pre-operatively shallow ACD (<2.5 mm) and a shorter AL (<23 mm), and that this was associated with a higher risk of refractive error after cataract surgery [96].

4.6. Type of Glaucoma Surgery

4.6.1. Combined Phacoemulsification and Trabeculectomy Surgery

Combined phacoemulsification and trabeculectomy surgery ("phacotrabeculectomy") improves visual acuity and minimizes the post-operative IOP spikes and morbidity that may occur in a two-stage operation [97,98]. Previous studies have shown that phacotrabeculectomy can provide an IOP reduction that is not inferior to that of trabeculectomy alone [99,100], with visual outcomes following this procedure being comparable to those obtained with phacoemulsification alone [101,102]. However, they have a higher risk of complications such as hypotony, hyphema, and anterior chamber shallowing compared to phacoemulsification alone [103], and they are also associated with surgically induced astigmatism which may affect refractive outcomes [104]. There are also various aspects of phacotrabeculectomy surgery that may affect the post-operative ocular characteristics and refractive status of the eye, thus increasing the risk of refractive surprises that may therefore preclude the implantation of premium IOLs in glaucoma eyes.

First, multiple studies [105–107] have suggested a greater risk of myopic shift and myopic refractive surprise following phacotrabeculectomy compared to phacoemulsification alone. Chan et al. [107] demonstrated that phacotrabeculectomy was more likely to have an IOL prediction error of >1.00 D ($p = 0.02$) and a myopic shift of >0.50 D or 1.00 D ($p = 0.03$ or 0.02, respectively) post-operatively. This observed myopic shift may be attributed to the trabeculectomy surgery itself [108–110]. While Muallem et al. [108] and Zhang et al. [110] assessed the performance of ocular biometry via contact A-scan ultrasonography, and Tan et al. [109] and Zhang et al. [110] performed calculations using early generation IOL, Yeh et al. [111], using the Haigis formula, also found a higher frequency of myopic surprise (>1.0 D) in post-trabeculectomy eyes with pre-operative IOP <9 mm Hg despite the use of laser interferometry as well as modern biometry and IOL prediction formulae. Additionally, post-operative myopic surprises appeared to be associated with IOP spikes—37.5% of eyes with pre-operatively low IOPs and subsequent IOP spikes experienced myopic surprises of over 1 D. These refractive changes after trabeculectomy surgery are likely to negatively affect the accuracy and predictability of visual outcomes following premium lens implantation in phacotrabeculectomy surgery.

Second, despite its overall good safety profile, phacotrabeculectomy has its associated complications, including hypotony and hypotonous maculopathy, hyphema, anterior chamber shallowing, and visual field wipeout. The frequency of these complications are higher than that of phacoemulsification alone [103,112,113], with many potentially causing refractive errors or even persistent visual loss after surgery [114,115].

Third, changes in AL and ACD [116–118] have been reported to occur after trabeculectomy surgery, thus increasing the risk of IOL power calculation errors [107,119] despite accurate pre-operative ocular biometry measurements [120]. A decrease in ACD may cause the eye to be more myopic post-operatively, while a decrease in AL would render the eye more hyperopic [106]. Studies have demonstrated a decrease in AL after phacotrabeculectomy; Poon et al. [121] showed that phacotrabeculectomy resulted in a mean decrease in AL of 0.16 ± 0.15 mm in PACG and 0.16 ± 0.11 mm in POAG eyes ($p = 0.96$), as well as an increase in ACD of 1.41 ± 0.91 mm in PACG, and 0.87 ± 0.86 mm in POAG eyes ($p = 0.04$). Law et al. [122] demonstrated that AL reduction following phacotrabeculectomy (117 (57) microm) was significantly larger than that following cataract surgery alone (75 (38) microm, $p < 0.02$), and this correlated significantly with post-operative IOP ($p < 0.002$). This decrease in AL is likely due to the effect of the trabeculectomy surgery itself, as demonstrated in a number of previous studies [116,117,123]. Factors that may affect AL include post-operative IOP changes and methods used to measure AL. Studies [116,123,124] have shown that the magnitude of AL change may depend on the magnitude of IOP change. Studies that use applanation ultrasound have reported larger reductions in AL after trabeculectomy [116,117,125], and different methods used to measure AL may explain disparity between studies, with any difference possibly being due to avoidance of globe indentation with the non-contact method, in contrast to contact ultrasonic biometry that may underestimate AL in soft, post-trabeculectomy eyes [116–118,123,126]. Francis et al. [127] utilized non-contact optical biometry and found a small but statistically significant decrease in AL following trabeculectomy, but with a difference less than that seen in the above-mentioned studies [116]. Of note, however, Zhang et al. [110] obtained AL measurements of patients undergoing phacoemulsification after prior trabeculectomy using both non-contact and contact methods of AL measurement and found no significant difference in either mean AL or mean ACD measured by both methods.

Lastly, keratometry readings may be affected by trabeculectomy surgery [125,128,129]. Various studies have demonstrated astigmatism, superior steepening, or superior flattening, with effects lasting up to 12 months post-operatively [104,117,122,129–135]. Trabeculectomies performed in the superior quadrant may induce the axis of corneal astigmatism to fall at the meridian of the scleral flap, causing a with-the-rule (WTR) corneal astigmatism [104,130,136]. Hong et al. [137] reported a shift of astigmatism at the vertical meridian from +2.17 D to a −1.72 D over 12 months after a combined triple procedure of a trabeculectomy, extracapsular cataract extraction, and IOL implantation. A study that evaluated keratometric changes after trabeculectomy revealed induced WTR astigmatism with a mean of 0.81 ± 1.08 D at the post-operative month three which tends to resolve within one year [134].

In comparing post-trabeculectomy patients undergoing standalone cataract surgery and patients undergoing phacotrabeculectomy, Bae et al. [106] found that IOL power prediction was less accurate in OAG patients undergoing cataract surgery post-trabeculectomy (CAT group) or undergoing combined phacotrabeculectomy surgery (CCT group) compared to OAG patients undergoing standalone cataract surgery (OC group). A prior trabeculectomy resulted in larger refractive prediction error, while combined trabeculectomy resulted in a slight myopic shift post-operatively. In the CAT group, the large mean absolute error (MAE) could have been due to contact A-scan ultrasonography used to measure AL as these softer, post-trabeculectomy eyes were more susceptible to deformation, thereby resulting in a falsely shorter AL measurement and greater myopic refractive surprise [111]. In the CCT group, the mean error was approximately −0.40 D, which was statistically greater than that in the OC group. Phacotrabeculectomy causes a greater IOP change compared to staged cataract surgery following prior trabeculectomy (6.61 versus 0.59 mmHg) [128], thus potentially causing greater AL, ACD, and keratometry changes [113,124].

In summary, biometric changes may occur after trabeculectomy surgery and should be a consideration in patients undergoing phacotrabeculectomy or standalone cataract surgery after prior trabeculectomy. AL, ACD, and keratometry are primary determinants in IOL power calculations; hence, post-operative changes in these biometric measurements may lead to increased risk of post-operative refractive surprise and IOL power calculation errors [127,138], although some studies have demonstrated that post-phacotrabeculectomy AL and keratometric changes may effectively negate each other's effect [139]. Standalone cataract surgeries performed much later (at least 3–6 months [118]) after initial trabeculectomy surgery may allow refractive outcomes and ocular measurements to stabilize, thereby not precluding the use of premium IOLs (particularly toric IOLs) in these patients. The use of non-contact optical biometry [38], such as laser interferometry and modern biometry formulae [140], may provide more accurate IOL power calculation for eyes undergoing combined phacotrabeculectomy.

4.6.2. MIGS (Minimally Invasive Glaucoma Surgery)

MIGS are a newer spectrum of surgical techniques and devices which have emerged rapidly in recent years. Most share common characteristics, including having a microinvasive approach, minimal tissue trauma, at least modest efficacy, as well as a more rapid post-operative recovery and higher safety profile compared to traditional glaucoma surgeries [141]. MIGS can be classified based on the site of anatomical intervention and augmentation, including (1) angle-based MIGS, where trabecular outflow is increased by bypassing the trabecular meshwork and directing aqueous humor into Schlemm's canal; (2) subconjunctival MIGS, where a drainage pathway is created into the sub-Tenon's space; and (3) suprachoroidal MIGS, where uveoscleral outflow is increased via implantation of suprachoroidal shunts [142]. Premium IOL implantation, as explained in the aforementioned sections, requires refractive neutrality and ocular biometric stability of surgical procedures. A number of studies have examined refractive outcomes following cataract surgery and monofocal IOL implantation with MIGS. However, few have reported results of premium IOL implantation in this context.

Angle-based MIGS bypass the resistance to aqueous outflow at the level of the trabecular meshwork, through microstenting, micro-incisions, and viscodilation [143]. Microstenting options include the iStent series. The first-generation iStent (Glaukos Corp., San Clemente, CA, USA) was introduced in 2012, with clinical trials demonstrating its efficacy when implanted in combination with cataract extraction. Samuelson et al. [144] showed significant IOP reduction after iStent and cataract surgery compared to cataract surgery alone, with similar safety profiles. Scott et al. [145] demonstrated that combined iStent and cataract surgery is likely to be a refractively neutral procedure—80% and 95% of eyes achieved target refraction within ± 0.5 D and ± 1.00 D, respectively. Manoharan et al. [80] demonstrated no difference in refractive outcomes in glaucoma patients who underwent combined phacoemulsification with iStent compared to phacoemulsification alone. The later iteration of the iStent, the iStent Inject (Glaukos Corp., San Clemente, CA, USA), achieved US FDA approval in 2018. Ang et al. [146] demonstrated minimal influence of iStent inject implantation on the MAE (-0.13 ± 0.08 D), with 82.8% of eyes achieving a post-operative refraction within 0.5 D of target in combined iStent inject implantation and phacoemulsification in Asian eyes with NTG. Ioannidis et al. [147] concluded too that the iStent inject device is refractively neutral—73.9% and 98.9% of eyes were within 0.5 D and 1.0 D of the target refraction, respectively. Trabecular bypass stents would not be expected to impact refractive outcomes, given that multiple studies [40,42,45,144] have demonstrated the refractive neutrality of this device. Other angle-based MIGS include micro-incision options, such as the Trabectome and Kahook Dual Blade. The Trabectome (NeoMedix, Inc., Tustin, CA, USA) was FDA approved in April 2004 [148]. Luebke et al. [149] demonstrated no difference in refractive outcomes between patients undergoing combined trabectome–cataract surgery compared to cataract surgery alone. Refractive outcomes following Kahook Dual Blade (New World Medical, Rancho Cucamonga, CA, USA) surgery, a goniotomy procedure, was also studied by Sieck et al. [150], who demonstrated no difference in refractive outcomes of glaucomatous patients undergoing phacoemulsification with or without KDB goniotomy.

Subconjunctival MIGS options include the XEN45 Gel Stent (Allergan, Dublin, CA, USA) and Preserflo Microshunt (Santen Co., Japan). The XEN45 Gel Stent was FDA approved in 2016 [148], and Grover et al. [151] demonstrated a significant reduction in IOP and medication use with a good safety profile, thus offering XEN45 as a MIGS option for patients with refractory open-angle glaucoma. Bormann et al. [152] compared refractive changes after surgery between trabeculectomy and the XEN45, showing that the SIA was nearly similar in both groups (0.75 ± 0.60 diopters after TE and 0.81 ± 0.56 diopters after XEN; $p = 0.57$).

Suprachoroidal MIGS procedures aim to take advantage of the uveoscleral pathway to reduce IOP, not being subject to an episcleral venous pressure floor [142]. As they have the most potential to alter the AL of the eye, they may result in post-operative refractive surprises. Although there are no US FDA-approved suprachoroidal MIGS at present, there are multiple devices in the investigational pipeline, such as the iStent Supra (Glaukos Corporation, San Clemente, CA, USA) and the MINIject (iStar Medical, Wavre, Belgium) [142]. Previously, the Cypass (Alcon, Ft. Worth, TX, USA) showed significant IOP and medication reduction when combined with cataract extraction [153] but was withdrawn from the market in 2018 when 5-year data from the COMPASS-XT study suggested a significantly increased rate of endothelial cell loss [154]. While there are no formal published studies to support these claims, there have been numerous independent accounts on various platforms reporting myopic surprises after combined Cypass and cataract extraction, with anecdotal accounts reporting myopic shifts of between 1.00 D to 3.00 D [155].

As largely refractively neutral surgeries with a high safety profile, angle-based MIGS appears less subject to confounding factors that may influence post-operative refractive outcomes; hence, it may potentially be favorable to the implantation of premium IOLs. Toric IOL implantation appears suitable in angle-based MIGS, with studies demonstrating good refractive results and spectacle-free outcomes following cataract extraction and toric IOL implantation with Tanito microhook trabeculotomy (TMH) [44,46] and iStent implantation [45–47]. Subconjunctival MIGS, performed ab internally without requiring conjunctival closure, is less likely to induce significant SIA, resulting in more predictable and consistent refractive outcomes, which contrasts with trabeculectomy which increases WTR astigmatism post-operatively [127].

4.7. Functional and Structural Monitoring of Glaucoma

MFIOLs may affect the sensitivity of investigative tests for glaucomasu progression, including visual field assessment and optic nerve imaging. Inoue et al. [156] reported that diffractive MFIOLs may cause wavy artifacts on OCT imaging. Aychoua et al. [59] reported that patients with diffractive MFIOLs (Tecnis® Multifocal ZM900; AMO, AT LISA® 809M; CarlZeiss Meditec, Jena, Germany) demonstrated a clinically relevant reduction in visual sensitivity (with a lower mean deviation of 2 dB, compared to monofocal IOLs) as assessed with standard automated perimetry, and the study concluded that the reduction was likely related to the MFIOL design, as opposed to patients' pseudo-phakic status. Furthermore, in patients with posterior eye segment changes, the visualization of macula and the optic nerve may be impaired after both toric and MFIOL implantation, and this could pose difficulties in later diagnostic or therapeutic procedures [157].

5. Conclusions

Overall, few studies have explored the use of premium IOLs in glaucoma patients, with some studies including only a small number of glaucomatous eyes. However, results thus far are promising. Studies have demonstrated high spectacle independence (especially for distance vision), contrast sensitivity comparable to that of healthy subjects, and excellent visual acuity results. Toric IOLs have been shown to provide good visual and refractive outcomes in eyes undergoing selected glaucoma surgeries and in eyes with ocular hypertension and incipient glaucoma, with high levels of patient satisfaction post-operatively. However, there may currently be insufficient evidence to support the safety of premium IOL implantation in patients with advanced glaucoma. Caution should be applied when cataract extraction is combined with certain glaucoma surgeries as there may be additional intra- and post-operative factors that influence the quality of vision after surgery.

Premium IOLs may confer greater benefit with appropriately chosen patient profiles [158] and may be recommended for certain patients with glaucoma, depending on the disease severity and type of visual field deficit. Specifically, premium IOLs tend to have better outcomes in glaucoma suspects, patients with ocular hypertension, and glaucoma patients with early, well-controlled disease. IOL selection should be individualized according to the patient's desired refractive outcome, visual expectations, subtype of glaucoma, type and severity of glaucomatous visual field defect, presence of ocular surface disease, and type of glaucoma surgery. Regular reviews on this topic are necessary given the rapid advancements in both IOL technology and glaucoma surgical treatment [159].

Supplementary Materials: The following supporting information can be downloaded at https://www.mdpi.com/article/10.3390/bioengineering10090993/s1: Table S1: Detailed Search Strategy; Table S2: Patient and Study Characteristics; Table S3: Visual Results Table; Table S4: Refractive Outcomes Table.

Author Contributions: Conceptualization, B.C.H.A. and A.S.Y.H.; writing—original draft preparation, A.S.Y.H. and B.C.H.A.; writing—review and editing, A.S.Y.H., B.C.H.A., E.D. and S.D.; results interpretation, A.S.Y.H., B.C.H.A. and E.D. All authors have read and agreed to the published version of the manuscript.

Funding: This research received no external funding.

Conflicts of Interest: The authors declare no conflict of interest.

References

1. Grzybowski, A. Recent developments in cataract surgery. *Ann. Transl. Med.* **2020**, *8*, 1540. [CrossRef]
2. Zvorničanin, J.; Zvorničanin, E. Premium intraocular lenses: The past, present and future. *J. Curr. Ophthalmol.* **2018**, *30*, 287–296. [CrossRef] [PubMed]
3. Research, G.V. Global Intraocular Lens Market Outlook Report 2023–2028: Adoption & Penetration of Premium IOLs Gaining Momentum—ResearchAndMarkets.com. Available online: https://www.businesswire.com/news/home/20230421005192/en/Global-Intraocular-Lens-Market-Outlook-Report-2023-2028-Adoption-Penetration-of-Premium-IOLs-Gaining-Momentum---ResearchAndMarkets.com (accessed on 6 June 2023).
4. Ling, J.D.; Bell, N.P. Role of Cataract Surgery in the Management of Glaucoma. *Int. Ophthalmol. Clin.* **2018**, *58*, 87–100. [CrossRef] [PubMed]
5. Leibowitz, H.M.; Krueger, D.E.; Maunder, L.R.; Milton, R.C.; Kini, M.M.; Kahn, H.A.; Nickerson, R.J.; Pool, J.; Colton, T.L.; Ganley, J.P.; et al. The Framingham Eye Study monograph: An ophthalmological and epidemiological study of cataract, glaucoma, diabetic retinopathy, macular degeneration, and visual acuity in a general population of 2631 adults, 1973–1975. *Surv. Ophthalmol.* **1980**, *24*, 335–610. [PubMed]
6. Ichhpujani, P.; Bhartiya, S.; Sharma, A. Premium IOLs in Glaucoma. *J. Curr. Glaucoma Pract.* **2013**, *7*, 54–57. [CrossRef]
7. Mencucci, R.; Cennamo, M.; Venturi, D.; Vignapiano, R.; Favuzza, E. Visual outcome, optical quality, and patient satisfaction with a new monofocal IOL, enhanced for intermediate vision: Preliminary results. *J. Cataract Refract. Surg.* **2020**, *46*, 378–387. [CrossRef] [PubMed]
8. Evans, B.J. Monovision: A review. *Ophthalmic Physiol. Opt.* **2007**, *27*, 417–439. [CrossRef] [PubMed]
9. Kumar, B.V.; Phillips, R.P.; Prasad, S. Multifocal intraocular lenses in the setting of glaucoma. *Curr. Opin. Ophthalmol.* **2007**, *18*, 62–66. [CrossRef]
10. Grzybowski, A.; Kanclerz, P.; Tuuminen, R. Multifocal intraocular lenses and retinal diseases. *Graefes Arch. Clin. Exp. Ophthalmol.* **2020**, *258*, 805–813. [CrossRef]
11. Nuzzi, R.; Tripoli, F.; Ghilardi, A. Evaluation of the Effects of Multifocal Intraocular Lens Oculentis LENTIS Mplus LS-313 MF30 on Visual Performance in Patients Affected by Bilateral Cataract and Treated with Phacoemulsification. *J. Ophthalmol.* **2022**, *2022*, 1315480. [CrossRef] [PubMed]
12. Artigas, J.M.; Menezo, J.L.; Peris, C.; Felipe, A.; Díaz-Llopis, M. Image quality with multifocal intraocular lenses and the effect of pupil size: Comparison of refractive and hybrid refractive-diffractive designs. *J. Cataract Refract. Surg.* **2007**, *33*, 2111–2117. [CrossRef]
13. Palomino Bautista, C.; Carmona González, D.; Castillo Gómez, A.; Bescos, J.A. Evolution of visual performance in 250 eyes implanted with the Tecnis ZM900 multifocal IOL. *Eur. J. Ophthalmol.* **2009**, *19*, 762–768. [CrossRef] [PubMed]
14. Nuzzi, R.; Tridico, F. Comparison of visual outcomes, spectacles dependence and patient satisfaction of multifocal and accommodative intraocular lenses: Innovative perspectives for maximal refractive-oriented cataract surgery. *BMC Ophthalmol.* **2017**, *17*, 12. [CrossRef]
15. Javitt, J.C.; Steinert, R.F. Cataract extraction with multifocal intraocular lens implantation: A multinational clinical trial evaluating clinical, functional, and quality-of-life outcomes. *Ophthalmology* **2000**, *107*, 2040–2048. [CrossRef] [PubMed]
16. Martínez Palmer, A.; Gómez Faiña, P.; España Albelda, A.; Comas Serrano, M.; Nahra Saad, D.; Castilla Céspedes, M. Visual function with bilateral implantation of monofocal and multifocal intraocular lenses: A prospective, randomized, controlled clinical trial. *J. Refract. Surg.* **2008**, *24*, 257–264. [CrossRef] [PubMed]
17. Pieh, S.; Lackner, B.; Hanselmayer, G.; Zöhrer, R.; Sticker, M.; Weghaupt, H.; Fercher, A.; Skorpik, C. Halo size under distance and near conditions in refractive multifocal intraocular lenses. *Br. J. Ophthalmol.* **2001**, *85*, 816–821. [CrossRef] [PubMed]
18. Balgos, M.; Vargas, V.; Alió, J.L. Correction of presbyopia: An integrated update for the practical surgeon. *Taiwan J. Ophthalmol.* **2018**, *8*, 121–140. [CrossRef] [PubMed]

19. Farid, M.; Chak, G.; Garg, S.; Steinert, R.F. Reduction in mean deviation values in automated perimetry in eyes with multifocal compared to monofocal intraocular lens implants. *Am. J. Ophthalmol.* **2014**, *158*, 227–231.e221. [CrossRef] [PubMed]
20. Cao, K.; Friedman, D.S.; Jin, S.; Yusufu, M.; Zhang, J.; Wang, J.; Hou, S.; Zhu, G.; Wang, B.; Xiong, Y.; et al. Multifocal versus monofocal intraocular lenses for age-related cataract patients: A system review and meta-analysis based on randomized controlled trials. *Surv. Ophthalmol.* **2019**, *64*, 647–658. [CrossRef]
21. Akella, S.S.; Juthani, V.V. Extended depth of focus intraocular lenses for presbyopia. *Curr. Opin. Ophthalmol.* **2018**, *29*, 318–322. [CrossRef]
22. Ouchi, M.; Kinoshita, S. Implantation of refractive multifocal intraocular lens with a surface-embedded near section for cataract eyes complicated with a coexisting ocular pathology. *Eye* **2015**, *29*, 649–655. [CrossRef]
23. de Vries, N.E.; Webers, C.A.; Verbakel, F.; de Brabander, J.; Berendschot, T.T.; Cheng, Y.Y.; Doors, M.; Nuijts, R.M. Visual outcome and patient satisfaction after multifocal intraocular lens implantation: Aspheric versus spherical design. *J. Cataract Refract. Surg.* **2010**, *36*, 1897–1904. [CrossRef] [PubMed]
24. Sachdev, G.S.; Ramamurthy, S.; Sharma, U.; Dandapani, R. Visual outcomes of patients bilaterally implanted with the extended range of vision intraocular lens: A prospective study. *Indian J. Ophthalmol.* **2018**, *66*, 407–410. [CrossRef] [PubMed]
25. Auffarth, G.U.; Gerl, M.; Tsai, L.; Janakiraman, D.P.; Jackson, B.; Alarcon, A.; Dick, H.B. Clinical evaluation of a new monofocal IOL with enhanced intermediate function in patients with cataract. *J. Cataract Refract. Surg.* **2021**, *47*, 184–191. [CrossRef] [PubMed]
26. Unsal, U.; Sabur, H. Comparison of new monofocal innovative and standard monofocal intraocular lens after phacoemulsification. *Int. Ophthalmol.* **2021**, *41*, 273–282. [CrossRef]
27. Liu, J.; Dong, Y.; Wang, Y. Efficacy and safety of extended depth of focus intraocular lenses in cataract surgery: A systematic review and meta-analysis. *BMC Ophthalmol.* **2019**, *19*, 198. [CrossRef]
28. Gatinel, D.; Loicq, J. Clinically Relevant Optical Properties of Bifocal, Trifocal, and Extended Depth of Focus Intraocular Lenses. *J. Refract. Surg.* **2016**, *32*, 273–280. [CrossRef]
29. Pedrotti, E.; Bruni, E.; Bonacci, E.; Badalamenti, R.; Mastropasqua, R.; Marchini, G. Comparative Analysis of the Clinical Outcomes with a Monofocal and an Extended Range of Vision Intraocular Lens. *J. Refract. Surg.* **2016**, *32*, 436–442. [CrossRef]
30. Pedrotti, E.; Carones, F.; Aiello, F.; Mastropasqua, R.; Bruni, E.; Bonacci, E.; Talli, P.; Nucci, C.; Mariotti, C.; Marchini, G. Comparative analysis of visual outcomes with 4 intraocular lenses: Monofocal, multifocal, and extended range of vision. *J. Cataract Refract. Surg.* **2018**, *44*, 156–167. [CrossRef]
31. Guo, Y.; Wang, Y.; Hao, R.; Jiang, X.; Liu, Z.; Li, X. Comparison of Patient Outcomes following Implantation of Trifocal and Extended Depth of Focus Intraocular Lenses: A Systematic Review and Meta-Analysis. *J. Ophthalmol.* **2021**, *2021*, 1115076. [CrossRef] [PubMed]
32. Mencucci, R.; Favuzza, E.; Caporossi, O.; Savastano, A.; Rizzo, S. Comparative analysis of visual outcomes, reading skills, contrast sensitivity, and patient satisfaction with two models of trifocal diffractive intraocular lenses and an extended range of vision intraocular lens. *Graefes Arch. Clin. Exp. Ophthalmol.* **2018**, *256*, 1913–1922. [CrossRef] [PubMed]
33. Montés-Micó, R.; Ferrer-Blasco, T.; Charman, W.N.; Cerviño, A.; Alfonso, J.F.; Fernández-Vega, L. Optical quality of the eye after lens replacement with a pseudoaccommodating intraocular lens. *J. Cataract Refract. Surg.* **2008**, *34*, 763–768. [CrossRef] [PubMed]
34. Iancu, R.; Corbu, C. Premium intraocular lenses use in patients with cataract and concurrent glaucoma: A review. *Maedica* **2013**, *8*, 290–296. [PubMed]
35. Ong, H.S.; Evans, J.R.; Allan, B.D. Accommodative intraocular lens versus standard monofocal intraocular lens implantation in cataract surgery. *Cochrane Database Syst. Rev.* **2014**, Cd009667. [CrossRef] [PubMed]
36. Petermeier, K.; Messias, A.; Gekeler, F.; Szurman, P. Effect of +3.00 diopter and +4.00 diopter additions in multifocal intraocular lenses on defocus profiles, patient satisfaction, and contrast sensitivity. *J. Cataract Refract. Surg.* **2011**, *37*, 720–726. [CrossRef] [PubMed]
37. Woodward, M.A.; Randleman, J.B.; Stulting, R.D. Dissatisfaction after multifocal intraocular lens implantation. *J. Cataract Refract. Surg.* **2009**, *35*, 992–997. [CrossRef]
38. Wells, G.A.; Shea, B.; O'Connell, D.; Peterson, J.; Welch, V.; Losos, M.; Tugwell, P. The Newcastle–Ottawa Scale (NOS) for Assessing the Quality of Nonrandomised Studies in Meta-Analyses. Available online: https://www.ohri.ca/programs/clinical_epidemiology/oxford.asp (accessed on 5 June 2023).
39. Murad, M.H.; Sultan, S.; Haffar, S.; Bazerbachi, F. Methodological quality and synthesis of case series and case reports. *BMJ Evid. Based. Med.* **2018**, *23*, 60–63. [CrossRef]
40. Ferguson, T.J.; Wilson, C.W.; Shafer, B.M.; Berdahl, J.P.; Terveen, D.C. Clinical Outcomes of a Non-Diffractive Extended Depth-of-Focus IOL in Eyes with Mild Glaucoma. *Clin. Ophthalmol.* **2023**, *17*, 861–868. [CrossRef]
41. Bissen-Miyajima, H.; Ota, Y.; Yuki, K.; Minami, K. Implantation of diffractive extended depth-of-focus intraocular lenses in normal tension glaucoma eyes: A case series. *Am. J. Ophthalmol. Case Rep.* **2023**, *29*, 101792. [CrossRef]
42. Ichioka, S.; Ishida, A.; Takayanagi, Y.; Manabe, K.; Matsuo, M.; Tanito, M.; Tanito, M. Roles of Toric intraocular Lens implantation on visual acuity and astigmatism in glaucomatous eyes treated with iStent and cataract surgery. *BMC Ophthalmol.* **2022**, *22*, 487. [CrossRef] [PubMed]

43. Sánchez-Sánchez, C.; Rementería-Capelo, L.A.; Puerto, B.; López-Caballero, C.; Morán, A.; Sánchez-Pina, J.M.; Contreras, I. Visual Function and Patient Satisfaction with Multifocal Intraocular Lenses in Patients with Glaucoma and Dry Age-Related Macular Degeneration. *J. Ophthalmol.* **2021**, *2021*, 9935983. [CrossRef]
44. Takai, Y.; Sugihara, K.; Mochiji, M.; Manabe, K.; Tsutsui, A.; Tanito, M. Refractive Status in Eyes Implanted with Toric and Nontoric Intraocular Lenses during Combined Cataract Surgery and Microhook Ab Interno Trabeculotomy. *J. Ophthalmol.* **2021**, *2021*, 5545007. [CrossRef]
45. López-Caballero, C.; Sánchez-Sánchez, C.; Puerto, B.; Blázquez, V.; Sánchez-Pina, J.M.; Contreras, I. Refractive outcomes of toric intraocular lens in combined trabecular micro bypass stent implantation and cataract surgery in glaucomatous eyes. *Int. Ophthalmol.* **2022**, *42*, 2711–2718. [CrossRef] [PubMed]
46. Ichioka, S.; Manabe, K.; Tsutsui, A.; Takai, Y.; Tanito, M. Effect of Toric Intraocular Lens Implantation on Visual Acuity and Astigmatism Status in Eyes Treated With Microhook Ab Interno Trabeculotomy. *J. Glaucoma* **2021**, *30*, 94–100. [CrossRef]
47. Brown, R.H.; Zhong, L.; Bozeman, C.W.; Lynch, M.G. Toric Intraocular Lens Outcomes in Patients With Glaucoma. *J. Refract. Surg.* **2015**, *31*, 366–372. [CrossRef]
48. Kamath, G.G.; Prasad, S.; Danson, A.; Phillips, R.P. Visual outcome with the array multifocal intraocular lens in patients with concurrent eye disease. *J. Cataract Refract. Surg.* **2000**, *26*, 576–581. [CrossRef] [PubMed]
49. Rementería-Capelo, L.A.; Lorente, P.; Carrillo, V.; Sánchez-Pina, J.M.; Ruiz-Alcocer, J.; Contreras, I. Patient Satisfaction and Visual Performance in Patients with Ocular Pathology after Bilateral Implantation of a New Extended Depth of Focus Intraocular Lens. *J. Ophthalmol.* **2022**, *2022*, 4659309. [CrossRef] [PubMed]
50. Kerr, N.M.; Moshegov, S.; Lim, S.; Simos, M. Visual Outcomes, Spectacle Independence, and Patient-Reported Satisfaction of the Vivity Extended Range of Vision Intraocular Lens in Patients with Early Glaucoma: An Observational Comparative Study. *Clin. Ophthalmol.* **2023**, *17*, 1515–1523. [CrossRef] [PubMed]
51. Hawkins, A.S.; Szlyk, J.P.; Ardickas, Z.; Alexander, K.R.; Wilensky, J.T. Comparison of contrast sensitivity, visual acuity, and Humphrey visual field testing in patients with glaucoma. *J. Glaucoma* **2003**, *12*, 134–138. [CrossRef] [PubMed]
52. Fatehi, N.; Nowroozizadeh, S.; Henry, S.; Coleman, A.L.; Caprioli, J.; Nouri-Mahdavi, K. Association of Structural and Functional Measures With Contrast Sensitivity in Glaucoma. *Am. J. Ophthalmol.* **2017**, *178*, 129–139. [CrossRef] [PubMed]
53. Richman, J.; Lorenzana, L.L.; Lankaranian, D.; Dugar, J.; Mayer, J.; Wizov, S.S.; Spaeth, G.L. Importance of visual acuity and contrast sensitivity in patients with glaucoma. *Arch. Ophthalmol.* **2010**, *128*, 1576–1582. [CrossRef]
54. Burton, R.; Crabb, D.P.; Smith, N.D.; Glen, F.C.; Garway-Heath, D.F. Glaucoma and reading: Exploring the effects of contrast lowering of text. *Optom. Vis. Sci.* **2012**, *89*, 1282–1287. [CrossRef] [PubMed]
55. Korth, M.J.; Jünemann, A.M.; Horn, F.K.; Bergua, A.; Cursiefen, C.; Velten, I.; Budde, W.M.; Wisse, M.; Martus, P. Synopsis of various electrophysiological tests in early glaucoma diagnosis--temporal and spatiotemporal contrast sensitivity, light- and color-contrast pattern-reversal electroretinogram, blue-yellow VEP. *Klin. Monbl. Augenheilkd.* **2000**, *216*, 360–368. [CrossRef] [PubMed]
56. Yeu, E.; Cuozzo, S. Matching the Patient to the Intraocular Lens: Preoperative Considerations to Optimize Surgical Outcomes. *Ophthalmology* **2021**, *128*, e132–e141. [CrossRef] [PubMed]
57. Trueb, P.R.; Albach, C.; Montés-Micó, R.; Ferrer-Blasco, T. Visual acuity and contrast sensitivity in eyes implanted with aspheric and spherical intraocular lenses. *Ophthalmology* **2009**, *116*, 890–895. [CrossRef]
58. Deshpande, R.; Satijia, A.; Dole, K.; Mangiraj, V.; Deshpande, M. Effects on ocular aberration and contrast sensitivity after implantation of spherical and aspherical monofocal intraocular lens—A comparative study. *Indian J. Ophthalmol.* **2022**, *70*, 2862–2865. [CrossRef] [PubMed]
59. Aychoua, N.; Junoy Montolio, F.G.; Jansonius, N.M. Influence of multifocal intraocular lenses on standard automated perimetry test results. *JAMA Ophthalmol.* **2013**, *131*, 481–485. [CrossRef]
60. Wilensky, J.T.; Hawkins, A. Comparison of contrast sensitivity, visual acuity, and Humphrey visual field testing in patients with glaucoma. *Trans. Am. Ophthalmol. Soc.* **2001**, *99*, 213–217, discussion 217–218.
61. Kang, S.G.; Lee, J.H. The change of visual acuity and visual field by diminished illumination in eyes with multifocal intraocular lens. *Korean J. Ophthalmol.* **1994**, *8*, 72–76. [CrossRef]
62. Bi, H.; Cui, Y.; Ma, X.; Cai, W.; Wang, G.; Ji, P.; Xie, X. Early clinical evaluation of AcrySof ReSTOR multifocal intraocular lens for treatment of cataract. *Ophthalmologica* **2008**, *222*, 11–16. [CrossRef]
63. Athanasiadis, Y.; de Wit, D.W.; Nithyanandrajah, G.A.; Patel, A.; Sharma, A. Neodymium:YAG laser peripheral iridotomy as a possible cause of zonular dehiscence during phacoemulsification cataract surgery. *Eye* **2010**, *24*, 1424–1425. [CrossRef]
64. Mutoh, T.; Barrette, K.F.; Matsumoto, Y.; Chikuda, M. Lens dislocation has a possible relationship with laser iridotomy. *Clin. Ophthalmol.* **2012**, *6*, 2019–2022. [CrossRef]
65. Hu, R.; Wang, X.; Wang, Y.; Sun, Y. Occult lens subluxation related to laser peripheral iridotomy: A case report and literature review. *Medicine* **2017**, *96*, e6255. [CrossRef] [PubMed]
66. Pan, Y.; Varma, R. Natural history of glaucoma. *Indian J. Ophthalmol.* **2011**, *59*, S19–S23. [CrossRef] [PubMed]
67. Sun, X.; Dai, Y.; Chen, Y.; Yu, D.Y.; Cringle, S.J.; Chen, J.; Kong, X.; Wang, X.; Jiang, C. Primary angle closure glaucoma: What we know and what we don't know. *Prog. Retin. Eye Res.* **2017**, *57*, 26–45. [CrossRef] [PubMed]

68. Levkovitch-Verbin, H.; Habot-Wilner, Z.; Burla, N.; Melamed, S.; Goldenfeld, M.; Bar-Sela, S.M.; Sachs, D. Intraocular pressure elevation within the first 24 hours after cataract surgery in patients with glaucoma or exfoliation syndrome. *Ophthalmology* **2008**, *115*, 104–108. [CrossRef]
69. Law, S.K.; Riddle, J. Management of cataracts in patients with glaucoma. *Int. Ophthalmol. Clin.* **2011**, *51*, 1–18. [CrossRef]
70. Drolsum, L.; Haaskjold, E.; Davanger, M. Pseudoexfoliation syndrome and extracapsular cataract extraction. *Acta Ophthalmol.* **1993**, *71*, 765–770. [CrossRef] [PubMed]
71. Drolsum, L.; Ringvold, A.; Nicolaissen, B. Cataract and glaucoma surgery in pseudoexfoliation syndrome: A review. *Acta Ophthalmol. Scand.* **2007**, *85*, 810–821. [CrossRef] [PubMed]
72. Scorolli, L.; Scorolli, L.; Campos, E.C.; Bassein, L.; Meduri, R.A. Pseudoexfoliation syndrome: A cohort study on intraoperative complications in cataract surgery. *Ophthalmologica* **1998**, *212*, 278–280. [CrossRef]
73. Ritch, R. Exfoliation syndrome. *Curr. Opin. Ophthalmol.* **2001**, *12*, 124–130. [CrossRef] [PubMed]
74. Riedl, J.C.; Rings, S.; Schuster, A.K.; Vossmerbaeumer, U. Intraocular lens dislocation: Manifestation, ocular and systemic risk factors. *Int. Ophthalmol.* **2023**, *43*, 1317–1324. [CrossRef] [PubMed]
75. Shingleton, B.J.; Neo, Y.N.; Cvintal, V.; Shaikh, A.M.; Liberman, P.; O'Donoghue, M.W. Outcome of phacoemulsification and intraocular lens implantion in eyes with pseudoexfoliation and weak zonules. *Acta Ophthalmol.* **2017**, *95*, 182–187. [CrossRef] [PubMed]
76. Jehan, F.S.; Mamalis, N.; Crandall, A.S. Spontaneous late dislocation of intraocular lens within the capsular bag in pseudoexfoliation patients. *Ophthalmology* **2001**, *108*, 1727–1731. [CrossRef]
77. Davis, D.; Brubaker, J.; Espandar, L.; Stringham, J.; Crandall, A.; Werner, L.; Mamalis, N. Late in-the-bag spontaneous intraocular lens dislocation: Evaluation of 86 consecutive cases. *Ophthalmology* **2009**, *116*, 664–670. [CrossRef] [PubMed]
78. Mönestam, E.I. Incidence of dislocation of intraocular lenses and pseudophakodonesis 10 years after cataract surgery. *Ophthalmology* **2009**, *116*, 2315–2320. [CrossRef]
79. Shingleton, B.J.; Marvin, A.C.; Heier, J.S.; O'Donoghue, M.W.; Laul, A.; Wolff, B.; Rowland, A. Pseudoexfoliation: High risk factors for zonule weakness and concurrent vitrectomy during phacoemulsification. *J. Cataract Refract. Surg.* **2010**, *36*, 1261–1269. [CrossRef]
80. Manoharan, N.; Patnaik, J.L.; Bonnell, L.N.; SooHoo, J.R.; Pantcheva, M.B.; Kahook, M.Y.; Wagner, B.D.; Lynch, A.M.; Seibold, L.K. Refractive outcomes of phacoemulsification cataract surgery in glaucoma patients. *J. Cataract Refract. Surg.* **2018**, *44*, 348–354. [CrossRef]
81. Moon, Y.; Sung, K.R.; Kim, J.M.; Shim, S.H.; Yoo, C.; Park, J.H. Risk Factors Associated With Glaucomatous Progression in Pseudoexfoliation Patients. *J. Glaucoma* **2017**, *26*, 1107–1113. [CrossRef]
82. Stewart, W.C.; Stewart, J.A.; Nelson, L.A. Ocular surface disease in patients with ocular hypertension and glaucoma. *Curr. Eye Res.* **2011**, *36*, 391–398. [CrossRef]
83. Zhang, X.; Vadoothker, S.; Munir, W.M.; Saeedi, O. Ocular Surface Disease and Glaucoma Medications: A Clinical Approach. *Eye Contact Lens.* **2019**, *45*, 11–18. [CrossRef] [PubMed]
84. Kuppens, E.V.; van Best, J.A.; Sterk, C.C.; de Keizer, R.J. Decreased basal tear turnover in patients with untreated primary open-angle glaucoma. *Am. J. Ophthalmol.* **1995**, *120*, 41–46. [CrossRef] [PubMed]
85. Skalicky, S.E.; Goldberg, I.; McCluskey, P. Ocular surface disease and quality of life in patients with glaucoma. *Am. J. Ophthalmol.* **2012**, *153*, 1–9.e2. [CrossRef] [PubMed]
86. Mocan, M.C.; Uzunosmanoglu, E.; Kocabeyoglu, S.; Karakaya, J.; Irkec, M. The Association of Chronic Topical Prostaglandin Analog Use With Meibomian Gland Dysfunction. *J. Glaucoma* **2016**, *25*, 770–774. [CrossRef] [PubMed]
87. Donaldson, K.; Parkhurst, G.; Saenz, B.; Whitley, W.; Williamson, B.; Hovanesian, J. Call to action: Treating dry eye disease and setting the foundation for successful surgery. *J. Cataract Refract. Surg.* **2022**, *48*, 623–629. [CrossRef] [PubMed]
88. Starr, C.E.; Gupta, P.K.; Farid, M.; Beckman, K.A.; Chan, C.C.; Yeu, E.; Gomes, J.A.P.; Ayers, B.D.; Berdahl, J.P.; Holland, E.J.; et al. An algorithm for the preoperative diagnosis and treatment of ocular surface disorders. *J. Cataract Refract. Surg.* **2019**, *45*, 669–684. [CrossRef]
89. Fu, L.; Chan, Y.K.; Li, J.; Nie, L.; Li, N.; Pan, W. Long term outcomes of cataract surgery in severe and end stage primary angle closure glaucoma with controlled IOP: A retrospective study. *BMC Ophthalmol.* **2020**, *20*, 160. [CrossRef]
90. Mehta, R.; Tomatzu, S.; Cao, D.; Pleet, A.; Mokhur, A.; Aref, A.A.; Vajaranant, T.S. Refractive Outcomes for Combined Phacoemulsification and Glaucoma Drainage Procedure. *Ophthalmol. Ther.* **2022**, *11*, 311–320. [CrossRef]
91. Chen, S.J.; Lu, P.; Zhang, W.F.; Lu, J.H. High myopia as a risk factor in primary open angle glaucoma. *Int. J. Ophthalmol.* **2012**, *5*, 750–753. [CrossRef]
92. Kuzin, A.A.; Varma, R.; Reddy, H.S.; Torres, M.; Azen, S.P. Ocular biometry and open-angle glaucoma: The Los Angeles Latino Eye Study. *Ophthalmology* **2010**, *117*, 1713–1719. [CrossRef]
93. Sherpa, D.; Badhu, B.P. Association between axial length of the eye and primary angle closure glaucoma. *Kathmandu Univ. Med. J. KUMJ* **2008**, *6*, 361–363. [CrossRef] [PubMed]
94. Wang, Q.; Jiang, W.; Lin, T.; Wu, X.; Lin, H.; Chen, W. Meta-analysis of accuracy of intraocular lens power calculation formulas in short eyes. *Clin. Exp. Ophthalmol.* **2018**, *46*, 356–363. [CrossRef] [PubMed]
95. Lu, W.; Hou, Y.; Yang, H.; Sun, X. A systemic review and network meta-analysis of accuracy of intraocular lens power calculation formulas in primary angle-closure conditions. *PLoS ONE* **2022**, *17*, e0276286. [CrossRef] [PubMed]

96. Ning, X.; Yang, Y.; Yan, H.; Zhang, J. Anterior chamber depth—A predictor of refractive outcomes after age-related cataract surgery. *BMC Ophthalmol.* **2019**, *19*, 134. [CrossRef] [PubMed]
97. Kang, Y.S.; Sung, M.S.; Heo, H.; Ji, Y.S.; Park, S.W. Long-term outcomes of prediction error after combined phacoemulsification and trabeculectomy in glaucoma patients. *BMC Ophthalmol.* **2021**, *21*, 60. [CrossRef]
98. Krupin, T.; Feitl, M.E.; Bishop, K.I. Postoperative intraocular pressure rise in open-angle glaucoma patients after cataract or combined cataract-filtration surgery. *Ophthalmology* **1989**, *96*, 579–584. [CrossRef]
99. Murthy, S.K.; Damji, K.F.; Pan, Y.; Hodge, W.G. Trabeculectomy and phacotrabeculectomy, with mitomycin-C, show similar two-year target IOP outcomes. *Can. J. Ophthalmol.* **2006**, *41*, 51–59. [CrossRef]
100. Tsai, H.Y.; Liu, C.J.; Cheng, C.Y. Combined trabeculectomy and cataract extraction versus trabeculectomy alone in primary angle-closure glaucoma. *Br. J. Ophthalmol.* **2009**, *93*, 943–948. [CrossRef] [PubMed]
101. Tzu, J.H.; Shah, C.T.; Galor, A.; Junk, A.K.; Sastry, A.; Wellik, S.R. Refractive outcomes of combined cataract and glaucoma surgery. *J. Glaucoma* **2015**, *24*, 161–164. [CrossRef]
102. Vaideanu, D.; Mandal, K.; Hildreth, A.; Fraser, S.G.; Phelan, P.S. Visual and refractive outcome of one-site phacotrabeculectomy compared with temporal approach phacoemulsification. *Clin. Ophthalmol.* **2008**, *2*, 569–574. [CrossRef]
103. El Sayed, Y.M.; Elhusseiny, A.M.; Albalkini, A.S.; El Sheikh, R.H.; Osman, M.A. Mitomycin C-augmented Phacotrabeculectomy Versus Phacoemulsification in Primary Angle-closure Glaucoma: A Randomized Controlled Study. *J. Glaucoma* **2019**, *28*, 911–915. [CrossRef] [PubMed]
104. Hugkulstone, C.E. Changes in keratometry following trabeculectomy. *Br. J. Ophthalmol.* **1991**, *75*, 217–218. [CrossRef] [PubMed]
105. Ong, C.; Nongpiur, M.; Peter, L.; Perera, S.A. Combined Approach to Phacoemulsification and Trabeculectomy Results in Less Ideal Refractive Outcomes Compared With the Sequential Approach. *J. Glaucoma* **2016**, *25*, e873–e878. [CrossRef] [PubMed]
106. Bae, H.W.; Lee, Y.H.; Kim, D.W.; Lee, T.; Hong, S.; Seong, G.J.; Kim, C.Y. Effect of trabeculectomy on the accuracy of intraocular lens calculations in patients with open-angle glaucoma. *Clin. Exp. Ophthalmol.* **2016**, *44*, 465–471. [CrossRef]
107. Chan, J.C.; Lai, J.S.; Tham, C.C. Comparison of postoperative refractive outcome in phacotrabeculectomy and phacoemulsification with posterior chamber intraocular lens implantation. *J. Glaucoma* **2006**, *15*, 26–29. [CrossRef] [PubMed]
108. Muallem, M.S.; Nelson, G.A.; Osmanovic, S.; Quinones, R.; Viana, M.; Edward, D.P. Predicted refraction versus refraction outcome in cataract surgery after trabeculectomy. *J. Glaucoma* **2009**, *18*, 284–287. [CrossRef] [PubMed]
109. Tan, H.Y.; Wu, S.C. Refractive error with optimum intraocular lens power calculation after glaucoma filtering surgery. *J. Cataract Refract. Surg.* **2004**, *30*, 2595–2597. [CrossRef]
110. Zhang, N.; Tsai, P.L.; Catoira-Boyle, Y.P.; Morgan, L.S.; Hoop, J.S.; Cantor, L.B.; WuDunn, D. The effect of prior trabeculectomy on refractive outcomes of cataract surgery. *Am. J. Ophthalmol.* **2013**, *155*, 858–863. [CrossRef]
111. Yeh, O.L.; Bojikian, K.D.; Slabaugh, M.A.; Chen, P.P. Refractive Outcome of Cataract Surgery in Eyes With Prior Trabeculectomy: Risk Factors for Postoperative Myopia. *J. Glaucoma* **2017**, *26*, 65–70. [CrossRef]
112. Li, S.W.; Chen, Y.; Wu, Q.; Lu, B.; Wang, W.Q.; Fang, J. Angle parameter changes of phacoemulsification and combined phacotrabeculectomy for acute primary angle closure. *Int. J. Ophthalmol.* **2015**, *8*, 742–747. [CrossRef] [PubMed]
113. Iwasaki, K.; Takamura, Y.; Arimura, S.; Tsuji, T.; Matsumura, T.; Gozawa, M.; Inatani, M. Prospective Cohort Study on Refractive Changes after Trabeculectomy. *J. Ophthalmol.* **2019**, *2019*, 4731653. [CrossRef] [PubMed]
114. Hong, S.; Park, K.; Ha, S.J.; Yeom, H.Y.; Seong, G.J.; Hong, Y.J. Long-term intraocular pressure control of trabeculectomy and triple procedure in primary open angle glaucoma and chronic primary angle closure glaucoma. *Ophthalmologica* **2007**, *221*, 395–401. [CrossRef] [PubMed]
115. Chen, D.Z.; Koh, V.; Sng, C.; Aquino, M.C.; Chew, P. Complications and outcomes of primary phacotrabeculectomy with mitomycin C in a multi-ethnic asian population. *PLoS ONE* **2015**, *10*, e0118852. [CrossRef] [PubMed]
116. Cashwell, L.F.; Martin, C.A. Axial length decrease accompanying successful glaucoma filtration surgery. *Ophthalmology* **1999**, *106*, 2307–2311. [CrossRef] [PubMed]
117. Kook, M.S.; Kim, H.B.; Lee, S.U. Short-term effect of mitomycin-C augmented trabeculectomy on axial length and corneal astigmatism. *J. Cataract Refract. Surg.* **2001**, *27*, 518–523. [CrossRef]
118. Pakravan, M.; Alvani, A.; Esfandiari, H.; Ghahari, E.; Yaseri, M. Post-trabeculectomy ocular biometric changes. *Clin. Exp. Optom.* **2017**, *100*, 128–132. [CrossRef]
119. Olsen, T. Sources of error in intraocular lens power calculation. *J. Cataract Refract Surg.* **1992**, *18*, 125–129. [CrossRef] [PubMed]
120. Drexler, W.; Findl, O.; Menapace, R.; Rainer, G.; Vass, C.; Hitzenberger, C.K.; Fercher, A.F. Partial coherence interferometry: A novel approach to biometry in cataract surgery. *Am. J. Ophthalmol.* **1998**, *126*, 524–534. [CrossRef]
121. Poon, L.Y.; Lai, I.C.; Lee, J.J.; Tsai, J.C.; Lin, P.W.; Teng, M.C. Comparison of surgical outcomes after phacotrabeculectomy in primary angle-closure glaucoma versus primary open-angle glaucoma. *Taiwan J. Ophthalmol.* **2015**, *5*, 28–32. [CrossRef]
122. Law, S.K.; Mansury, A.M.; Vasudev, D.; Caprioli, J. Effects of combined cataract surgery and trabeculectomy with mitomycin C on ocular dimensions. *Br. J. Ophthalmol.* **2005**, *89*, 1021–1025. [CrossRef]
123. Németh, J.; Horóczi, Z. Changes in the ocular dimensions after trabeculectomy. *Int. Ophthalmol.* **1992**, *16*, 355–357. [CrossRef] [PubMed]
124. Volzhanin, A.V.; Petrov, S.Y.; Safonova, D.M.; Averich, V.V. On refraction shift after trabeculectomy. *Vestn. Oftalmol.* **2022**, *138*, 147–155. [CrossRef] [PubMed]

125. Husain, R.; Li, W.; Gazzard, G.; Foster, P.J.; Chew, P.T.; Oen, F.T.; Phillips, R.; Khaw, P.T.; Seah, S.K.; Aung, T. Longitudinal changes in anterior chamber depth and axial length in Asian subjects after trabeculectomy surgery. *Br. J. Ophthalmol.* **2013**, *97*, 852–856. [CrossRef]
126. Dickens, M.A.; Cashwell, L.F. Long-term effect of cataract extraction on the function of an established filtering bleb. *Ophthalmic Surg. Lasers* **1996**, *27*, 9–14. [CrossRef] [PubMed]
127. Francis, B.A.; Wang, M.; Lei, H.; Du, L.T.; Minckler, D.S.; Green, R.L.; Roland, C. Changes in axial length following trabeculectomy and glaucoma drainage device surgery. *Br. J. Ophthalmol.* **2005**, *89*, 17–20. [CrossRef]
128. Rosen, W.J.; Mannis, M.J.; Brandt, J.D. The effect of trabeculectomy on corneal topography. *Ophthalmic Surg.* **1992**, *23*, 395–398. [CrossRef]
129. Claridge, K.G.; Galbraith, J.K.; Karmel, V.; Bates, A.K. The effect of trabeculectomy on refraction, keratometry and corneal topography. *Eye* **1995**, *9 Pt 3*, 292–298. [CrossRef]
130. Investigators, A.A.G.I.S. The Advanced Glaucoma Intervention Study: 8. Risk of cataract formation after trabeculectomy. *Arch. Ophthalmol.* **2001**, *119*, 1771–1779. [CrossRef]
131. Cunliffe, I.A.; Dapling, R.B.; West, J.; Longstaff, S. A prospective study examining the changes in factors that affect visual acuity following trabeculectomy. *Eye* **1992**, *6 Pt 6*, 618–622. [CrossRef]
132. Dietze, P.J.; Oram, O.; Kohnen, T.; Feldman, R.M.; Koch, D.D.; Gross, R.L. Visual function following trabeculectomy: Effect on corneal topography and contrast sensitivity. *J. Glaucoma* **1997**, *6*, 99–103. [CrossRef]
133. Vernon, S.A.; Zambarakji, H.J.; Potgieter, F.; Evans, J.; Chell, P.B. Topographic and keratometric astigmatism up to 1 year following small flap trabeculectomy (microtrabeculectomy). *Br. J. Ophthalmol.* **1999**, *83*, 779–782. [CrossRef] [PubMed]
134. Egrilmez, S.; Ates, H.; Nalcaci, S.; Andac, K.; Yagci, A. Surgically induced corneal refractive change following glaucoma surgery: Nonpenetrating trabecular surgeries versus trabeculectomy. *J. Cataract Refract. Surg.* **2004**, *30*, 1232–1239. [CrossRef] [PubMed]
135. Willekens, K.; Pinto, L.A.; Delbeke, H.; Vandewalle, E.; Stalmans, I. Trabeculectomy With Moorfields Conjunctival Closure Technique Offers Safety Without Astigmatism Induction. *J. Glaucoma* **2016**, *25*, e531–e535. [CrossRef]
136. Kumari, R.; Saha, B.C.; Puri, L.R. Keratometric astigmatism evaluation after trabeculectomy. *Nepal. J. Ophthalmol.* **2013**, *5*, 215–219. [CrossRef]
137. Hong, Y.J.; Choe, C.M.; Lee, Y.G.; Chung, H.S.; Kim, H.K. The effect of mitomycin-C on postoperative corneal astigmatism in trabeculectomy and a triple procedure. *Ophthalmic Surg. Lasers* **1998**, *29*, 484–489. [CrossRef]
138. Belov, D.F.; Nikolaenko, V.P. Calculation of intraocular lens power after trabeculectomy. *Vestn. Oftalmol.* **2021**, *137*, 61–66. [CrossRef] [PubMed]
139. Pakravan, M.; Alvani, A.; Yazdani, S.; Esfandiari, H.; Yaseri, M. Intraocular lens power changes after mitomycin trabeculectomy. *Eur. J. Ophthalmol.* **2015**, *25*, 478–482. [CrossRef] [PubMed]
140. Iijima, K.; Kamiya, K.; Iida, Y.; Kasahara, M.; Shoji, N. Predictability of combined cataract surgery and trabeculectomy using Barrett Universal II formula. *PLoS ONE* **2022**, *17*, e0270363. [CrossRef]
141. Saheb, H.; Ahmed, I.I. Micro-invasive glaucoma surgery: Current perspectives and future directions. *Curr. Opin. Ophthalmol.* **2012**, *23*, 96–104. [CrossRef]
142. Pereira, I.C.F.; van de Wijdeven, R.; Wyss, H.M.; Beckers, H.J.M.; den Toonder, J.M.J. Conventional glaucoma implants and the new MIGS devices: A comprehensive review of current options and future directions. *Eye* **2021**, *35*, 3202–3221. [CrossRef]
143. Shah, M. Micro-invasive glaucoma surgery—An interventional glaucoma revolution. *Eye Vis.* **2019**, *6*, 29. [CrossRef] [PubMed]
144. Samuelson, T.W.; Katz, L.J.; Wells, J.M.; Duh, Y.J.; Giamporcaro, J.E. Randomized evaluation of the trabecular micro-bypass stent with phacoemulsification in patients with glaucoma and cataract. *Ophthalmology* **2011**, *118*, 459–467. [CrossRef] [PubMed]
145. Scott, R.A.; Ferguson, T.J.; Stephens, J.D.; Berdahl, J.P. Refractive outcomes after trabecular microbypass stent with cataract extraction in open-angle glaucoma. *Clin. Ophthalmol.* **2019**, *13*, 1331–1340. [CrossRef] [PubMed]
146. Ang, B.C.H.; Chiew, W.; Yip, V.C.H.; Chua, C.H.; Han, W.S.; Tecson, I.O.C.; Ogle, J.J.; Lim, B.A.; Hee, O.K.; Tay, E.L.Y.; et al. Prospective 12-month outcomes of combined iStent inject implantation and phacoemulsification in Asian eyes with normal tension glaucoma. *Eye Vis.* **2022**, *9*, 27. [CrossRef]
147. Ioannidis, A.S.; Töteberg-Harms, M.; Hamann, T.; Hodge, C. Refractive Outcomes After Trabecular Micro-Bypass Stents (iStent Inject) with Cataract Extraction in Open-Angle Glaucoma. *Clin. Ophthalmol.* **2020**, *14*, 517–524. [CrossRef]
148. Lim, R. The surgical management of glaucoma: A review. *Clin. Exp. Ophthalmol.* **2022**, *50*, 213–231. [CrossRef] [PubMed]
149. Luebke, J.; Boehringer, D.; Neuburger, M.; Anton, A.; Wecker, T.; Cakir, B.; Reinhard, T.; Jordan, J.F. Refractive and visual outcomes after combined cataract and trabectome surgery: A report on the possible influences of combining cataract and trabectome surgery on refractive and visual outcomes. *Graefes Arch. Clin. Exp. Ophthalmol.* **2015**, *253*, 419–423. [CrossRef]
150. Sieck, E.G.; Capitena Young, C.E.; Epstein, R.S.; SooHoo, J.R.; Pantcheva, M.B.; Patnaik, J.L.; Lynch, A.M.; Kahook, M.Y.; Seibold, L.K. Refractive outcomes among glaucoma patients undergoing phacoemulsification cataract extraction with and without Kahook Dual Blade goniotomy. *Eye Vis.* **2019**, *6*, 28. [CrossRef]
151. Grover, D.S.; Flynn, W.J.; Bashford, K.P.; Lewis, R.A.; Duh, Y.J.; Nangia, R.S.; Niksch, B. Performance and Safety of a New Ab Interno Gelatin Stent in Refractory Glaucoma at 12 Months. *Am. J. Ophthalmol.* **2017**, *183*, 25–36. [CrossRef]
152. Bormann, C.; Busch, C.; Rehak, M.; Schmidt, M.; Scharenberg, C.; Ziemssen, F.; Unterlauft, J.D. Refractive Changes after Glaucoma Surgery-A Comparison between Trabeculectomy and XEN Microstent Implantation. *Life* **2022**, *12*, 1889. [CrossRef] [PubMed]

153. Hoeh, H.; Ahmed, I.I.; Grisanti, S.; Grisanti, S.; Grabner, G.; Nguyen, Q.H.; Rau, M.; Yoo, S.; Ianchulev, T. Early postoperative safety and surgical outcomes after implantation of a suprachoroidal micro-stent for the treatment of open-angle glaucoma concomitant with cataract surgery. *J. Cataract Refract. Surg.* **2013**, *39*, 431–437. [CrossRef]
154. Lass, J.H.; Benetz, B.A.; He, J.; Hamilton, C.; Von Tress, M.; Dickerson, J.; Lane, S. Corneal Endothelial Cell Loss and Morphometric Changes 5 Years after Phacoemulsification with or without CyPass Micro-Stent. *Am. J. Ophthalmol.* **2019**, *208*, 211–218. [CrossRef] [PubMed]
155. Sarkisian, S.R., Jr.; Radcliffe, N.; Harasymowycz, P.; Vold, S.; Patrianakos, T.; Zhang, A.; Herndon, L.; Brubaker, J.; Moster, R.; Francis, B. Visual outcomes of combined cataract surgery and minimally invasive glaucoma surgery. *J. Cataract Refract. Surg.* **2020**, *46*, 1422–1432. [CrossRef] [PubMed]
156. Inoue, M.; Bissen-Miyajima, H.; Yoshino, M.; Suzuki, T. Wavy horizontal artifacts on optical coherence tomography line-scanning images caused by diffractive multifocal intraocular lenses. *J. Cataract Refract. Surg.* **2009**, *35*, 1239–1243. [CrossRef] [PubMed]
157. Braga-Mele, R.; Chang, D.; Dewey, S.; Foster, G.; Henderson, B.A.; Hill, W.; Hoffman, R.; Little, B.; Mamalis, N.; Oetting, T.; et al. Multifocal intraocular lenses: Relative indications and contraindications for implantation. *J. Cataract Refract. Surg.* **2014**, *40*, 313–322. [CrossRef]
158. Visser, N.; Bauer, N.J.; Nuijts, R.M. Toric intraocular lenses: Historical overview, patient selection, IOL calculation, surgical techniques, clinical outcomes, and complications. *J. Cataract Refract. Surg.* **2013**, *39*, 624–637. [CrossRef] [PubMed]
159. Hu, R.; Racette, L.; Chen, K.S.; Johnson, C.A. Functional assessment of glaucoma: Uncovering progression. *Surv. Ophthalmol.* **2020**, *65*, 639–661. [CrossRef]

Disclaimer/Publisher's Note: The statements, opinions and data contained in all publications are solely those of the individual author(s) and contributor(s) and not of MDPI and/or the editor(s). MDPI and/or the editor(s) disclaim responsibility for any injury to people or property resulting from any ideas, methods, instructions or products referred to in the content.

MDPI AG
Grosspeteranlage 5
4052 Basel
Switzerland
Tel.: +41 61 683 77 34

Bioengineering Editorial Office
E-mail: bioengineering@mdpi.com
www.mdpi.com/journal/bioengineering

Disclaimer/Publisher's Note: The title and front matter of this reprint are at the discretion of the Guest Editors. The publisher is not responsible for their content or any associated concerns. The statements, opinions and data contained in all individual articles are solely those of the individual Editors and contributors and not of MDPI. MDPI disclaims responsibility for any injury to people or property resulting from any ideas, methods, instructions or products referred to in the content.

www.ingramcontent.com/pod-product-compliance
Lightning Source LLC
LaVergne TN
LVHW072322090526
838202LV00019B/2331